Current Research in Cell Signaling

Current Research in Cell Signaling

Editor: Peter Madison

MURPHY & MOORE

www.murphy-moorepublishing.com

www.murphy-moorepublishing.com

ⓂⓂ MURPHY & MOORE

Cataloging-in-Publication Data

Current research in cell signaling / edited by Peter Madison.
   p. cm.
Includes bibliographical references and index.
ISBN 978-1-63987-143-8
1. Cellular signal transduction. 2. Cell interaction. 3. Cytology. I. Madison, Peter.
QP517.C45 C87 2022
571.74--dc23

Murphy & Moore Publishing,
1 Rockefeller Plaza,
New York City, NY 10020, USA

ISBN 978-1-63987-143-8 (Hardback)

Contents

Preface

This book has been a concerted effort by a group of academicians, researchers and scientists, who have contributed their research works for the realization of the book. This book has materialized in the wake of emerging advancements and innovations in this field. Therefore, the need of the hour was to compile all the required researches and disseminate the knowledge to a broad spectrum of people comprising of students, researchers and specialists of the field.

The capacity of the cells to send, process and accept signals with its environment and with itself is termed as cell signaling. Cells of every living organism exhibit this property. Some of the major types of cell signals are autocrine, juxtacrine, endocrine, intracrine and paracrine. The chemical signals or physical stimuli are detected by proteins which are present within the cell or on its surface. These proteins are known as receptors. The cells are programmed to respond to particular extracellular signal molecules. Anomalies in cell signaling can lead to diseases such as cancer, diabetes and autoimmunity. This book unfolds the innovative aspects of cell signaling which will be crucial for the progress of this field in the future. Such selected concepts that redefine the research within this field have been presented herein. This book will serve as a valuable source of reference for graduate and post graduate students.

At the end of the preface, I would like to thank the authors for their brilliant chapters and the publisher for guiding us all-through the making of the book till its final stage. Also, I would like to thank my family for providing the support and encouragement throughout my academic career and research projects.

Editor

Subverting Toll-Like Receptor Signaling by Bacterial Pathogens

Victoria A. McGuire* and J. Simon C. Arthur

Division of Cell Signalling and Immunology, School of Life Sciences, University of Dundee, Dundee, UK

Edited by:
Olivier Dussurget,
University Paris Diderot, France

Reviewed by:
Cammie Lesser,
Harvard Medical School, USA
Elizabeth Hong-Geller,
Los Alamos National Laboratory, USA
Sinead Corr,
Trinity College Dublin, Ireland

*Correspondence:
Victoria A. McGuire
v.a.mcguire@dundee.ac.uk

Pathogenic bacteria are detected by pattern-recognition receptors (PRRs) expressed on innate immune cells, which activate intracellular signal transduction pathways to elicit an immune response. Toll-like receptors are, perhaps, the most studied of the PRRs and can activate the mitogen-activated protein kinase (MAPK) and Nuclear Factor-κB (NF-κB) pathways. These pathways are critical for mounting an effective immune response. In order to evade detection and promote virulence, many pathogens subvert the host immune response by targeting components of these signal transduction pathways. This mini-review highlights the diverse mechanisms that bacterial pathogens have evolved to manipulate the innate immune response, with a particular focus on those that target MAPK and NF-κB signaling pathways. Understanding the elaborate strategies that pathogens employ to subvert the immune response not only highlights the importance of these proteins in mounting effective immune responses, but may also identify novel approaches for treatment or prevention of infection.

Keywords: MAPK, NF-κB, bacterial effector, signaling, TLR, bacterial pathogen, virulence

INTRODUCTION

Innate immunity provides the first line of defense against invading pathogens. Recognition of microbial ligands, or pathogen-associated molecular patterns (PAMPs) by pattern-recognition receptors (PRRs), stimulates innate immune cells to upregulate the expression of cytokines, chemokines, and proteins that directly target microbes. Toll-like receptors (TLRs) have been well studied amongst the PRRs, with 10 described in human and 12 in mouse (1). TLRs on the cell surface recognize ligands from extracellular microbes, such as peptidoglycan by TLR1/TLR2, lipoprotein by TLR2/6, lipopolysaccharide (LPS) by TLR4, and flagellin by TLR5. TLR3, TLR7, TLR8, and TLR9 are located in intracellular vesicles where they recognize microbial nucleic acids.

Stimulation of all TLRs activates the mitogen-activated protein kinase (MAPK) and Nuclear Factor-κB (NF-κB) signaling pathways, both of which are critical for an effective immune response. The current understanding of the signaling events that trigger MAPK and NF-κB activation in response to TLR stimulation have been reviewed recently (1–4), but is summarized below and in **Figure 1**.

Following detection of PAMPs by a TLR, signaling is initiated by the recruitment of adaptor proteins to the cytoplasmic Toll and IL-1 Receptor (TIR) domain of the receptor. Two main pathways of TLR signaling exist, defined on their use of either the MyD88 (myeloid differentiation primary-response protein 88) or TRIF (TIR domain-containing adaptor protein inducing interferon α/β) adaptor, with all TLRs except TLR3 able to utilize the MyD88 pathway. MyD88 recruits IL-1 receptor-associated kinase (IRAK) 4, IRAK1 and IRAK2 to form a complex known as the Myddosome, which subsequently recruits the E3 ubiquitin ligase TNF receptor-associated factor 6 (TRAF6).

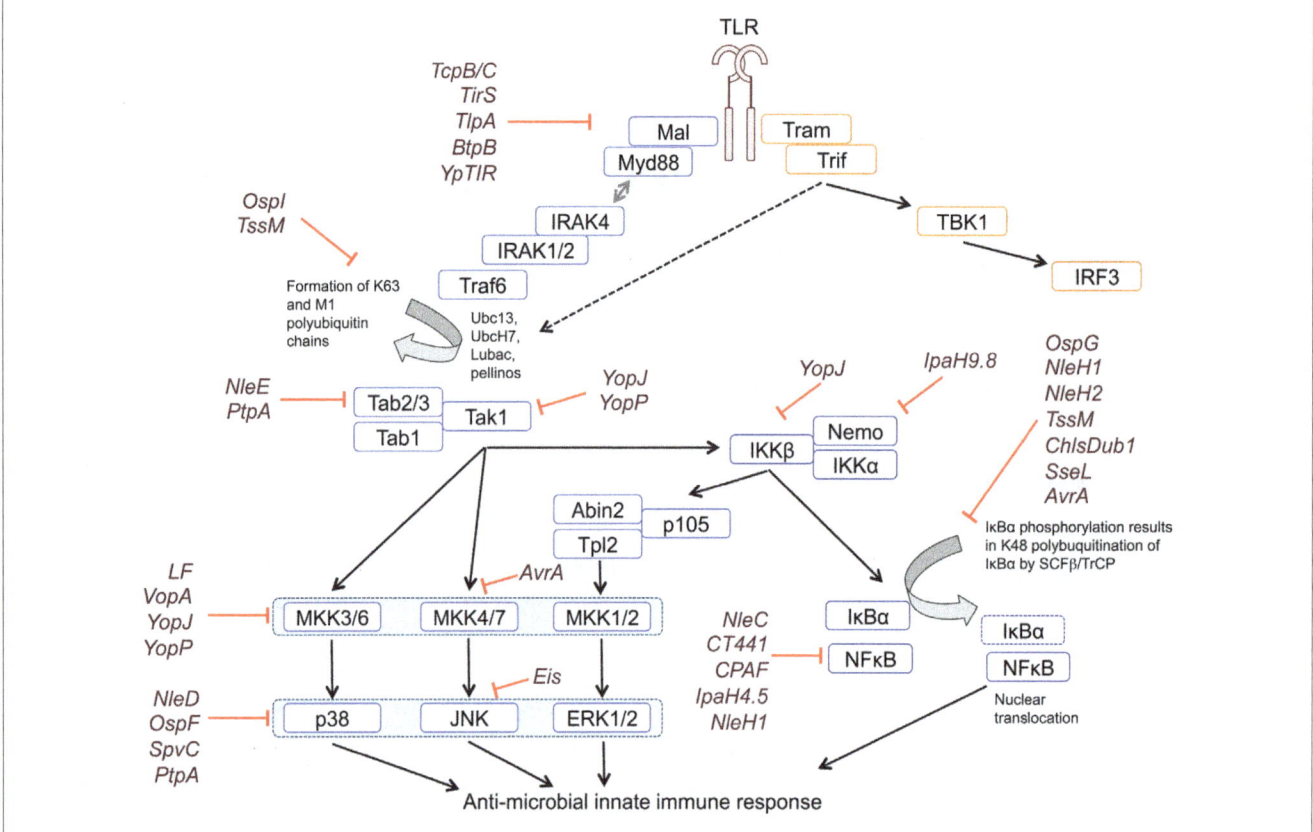

FIGURE 1 | Blockade of MAPK and NFκB signaling by bacterial effectors. TLR signaling is initiated by the recruitment of adaptor proteins to the TIR domain of the receptor. Recruitment of MyD88 facilitates Myddosome formation through binding of IRAK4, IRAK1, and IRAK2. IRAKs bind to and recruit the E3 ubiquitin ligase TRAF6, which – perhaps with input from other E3s – generates lysine-63 (K63) linked polyubiquitin chains. K63 linked polyubiquitin chains are used as a substrate by LUBAC to form M1-K63 hybrid polyubiquitin chains. K63 and M1-K63 polyubiquitin chains are necessary for downstream signaling mediated by TAK1. TAK1 phosphorylates and activates IKKα/β, which form the IKK complex together with NEMO/IKKγ. The IKK complex phosphorylates IκBα, resulting in its K48-linked polyubiquitination and proteasomal degradation, which releases the p65 NFκB subunit from inhibition. The IKK complex also phosphorylates p105, generating the p50 NFκB subunit, and enabling the active p50-p65 NFκB dimer to translocate to the nucleus. TAK1 also controls activation of the ERK1/2, p38, and JNK MAPK pathways by acting as a MAP3K for the p38 and JNK pathways and controlling the activation of ERK1/2 via Tpl2. Phosphorylation of p105 by the IKK complex releases Tpl2 from inhibition, allowing Tpl2 to activate ERK1/2 signaling. MAPKs phosphorylate their own downstream targets including other kinases and transcription factors that regulate transcription. Activation of TLR3 and TLR4 can also recruit the TRIF adaptor, which activates NFκB and MAPK signaling via both Receptor Interacting Protein 1 (RIP1) and TRAF6 upstream of TAK1, and activates IRF3 via IKKε and Tank-binding kinase 1 (TBK1). Bacterial effectors block signaling by interfering with different components of the signaling cascades, as indicated in the figure.

TRAF6 and/or other E3 ubiquitin ligases generate lysine-63 (K63) linked polyubiquitin chains that are used by the linear ubiquitin chain assembly complex (LUBAC) to generate linear (M1)-K63 hybrid polyubiquitin chains. The formation of both K63 and M1-K63 hybrid polyubiquitin chains is required to assemble the signaling complexes that activate downstream pathways. TAK1 plays a central role in activating downstream signaling pathways. First, it phosphorylates and activates IκB kinases (IKKα/β), which form the IKK complex together with NEMO/IKKγ. The IKK complex phosphorylates IκBα, enabling its recognition by the E3 ligase complex SCF-βTrCP (SKP1–cullin-1–F-box complex containing βTrCP), resulting in its K48-linked polyubiquitination and proteasomal degradation. Loss of IκBα releases the p65 NFκB subunit allowing it to translocate to the nucleus.

TAK1 also controls activation of the ERK1/2, p38, and JNK MAPK pathways. MAPK activation requires a cascade of at least three kinases. MAPKs are activated by a MAPK Kinase (MAP2K), which itself is activated by phosphorylation by an upstream MAPK Kinase Kinase (MAP3K). TAK1 acts as a MAP3K for the p38 and JNK pathways and via IKK it controls the activation of Tpl2, the MAP3K that activates ERK1/2 downstream of TLRs. Tpl2 activity is controlled by p105, which tethers it in an inactive complex with Abin2. Phosphorylation of p105 by the IKK complex, releases this complex, allowing Tpl2 to activate its substrates. MAPKs phosphorylate their own downstream targets including other kinases and transcription factors that regulate transcription (1–4).

Pathogenic bacteria have evolved elaborate strategies to perturb intracellular signaling pathways that activate the host immune response. This review describes the mechanisms bacteria use to inhibit TLR-dependent signaling, focusing on strategies that block MAPK and NF-κB signaling and

TABLE 1 | Mechanisms used by bacteria to inhibit TLR-dependent signaling by blocking MAPKs or NFκB.

Protein function	Protein	Bacterial species	Disease	Mechanism	Reference
TIR mimic	TlpA	*Salmonella* Enteritidis	Gastrointestinal disease	Postulated to compete with endogenous TIR domains to prevent signaling	Newman et al. (8)
	TirS	*Staphylococcus aureus*	Skin, respiratory tract, and GI tract infections	Blocks TLR2 signaling	Askarian et al. (9)
	TcpC	*Escherichia coli* CFT073 (UPEC)	Urinary tract infection	Binds MyD88 to prevent downstream signaling	Cirl et al. (10)
	TcpB/BtpA	*Brucella melitensis*	Brucellosis	Mimics Mal (TIRAP) to block TLR2/TLR4 signaling; targets Mal for proteasomal degradation	Cirl et al. (10), Radhakrishnan et al. (13), Sengupta et al. (14)
	BtpB	*Brucella melitensis*	Brucellosis	Interacts with MyD88 to block TLR signaling	Salcedo et al. (12)
	YpTIR	*Yersinia pestis*	Plague	Interacts with MyD88 to block TLR signaling	Rana et al. (11), Spear et al. (66)
Protease	LF	*Bacillus anthracis*	Anthrax	Cleaves MKKs within MAPK-docking domain	Duesbery et al. (15), Vitale et al. (16)
	NleD	*Escherichia coli* (EPEC/EHEC)	Gastrointestinal disease	Cleaves JNK and p38 within TxY dual phosphorylation motif	Baruch et al. (17)
	NleC	*Escherichia coli* (EPEC/EHEC)	Gastrointestinal disease	Cleaves amino-terminus of p65 NF-κB targeting it for proteasomal degradation	Yen and Ooka (18), Mühlen et al. (19), Baruch et al. (17), Pearson et al. (20)
	CT441	*Chlamydia* spp.	Urogenital tract infection, trachoma eye disease	Cleaves p65 NF-κB	Lad and Yang (21)
	CPAF	*Chlamydia* spp.	Urogenital tract infection, trachoma eye disease	Cleaves p65 NF-κB	Christian et al. (23)
Acetyltransferase	VopA	*Vibrio parahaemolyticus*	Gastrointestinal disease	*O*-acetylates MKKs in the activation loop to compete with phosphorylation; *N*-acetylates MKKs in the catalytic loop to disrupt ATP binding	Trosky et al. (24, 25)
	AvrA	*Salmonella* Typhimurium	Gastrointestinal disease	*O*-acetylates MKKs in the activation loop to compete with phosphorylation	Jones et al. (26)
	YopJ/YopP	*Yersinia* spp.	Plague/Yersiniosis	*O*-acetylates MKKs, TAK1 and IKKα and IKKβ in the activation loop to compete with phosphorylation	Orth et al. (28), Mittal et al. (30), Mukherjee et al. (29), Haase and Richter (32), Thiefes et al. (33), Paquette and Conlon (31), Meinzer et al. (34)
	Eis	*Mycobacterium tuberculosis*	Tuberculosis	*N*-acetylates DUSP16/MKP7 to block JNK activation	Kim et al. (35)
Phosphothreonine lyase	OspF	*Shigella* spp.	Dysentery	Removes phosphothreonine in the TxY activation loop of MAPKs	Li et al. (37)
	SpvC	*Salmonella* Typhimurium	Gastrointestinal disease	Removes phosphothreonine in the TxY activation loop of MAPKs	Mazurkiewicz et al. (38)
Kinase/phosphatase	OspG	*Shigella*	Dysentery	Binds to ubiquitin and E2-ubiquitin conjugates; prevents IκBα degradation	Kim et al. (43), Zhou et al. (44)
	NleH1	*Escherichia coli* (EPEC/EHEC)	Gastrointestinal disease	Inhibits IκBα degradation; binds to RPS3 to antagonize NF-κB activity	Gao and Wan (47), Royan et al. (46)
	NleH2	*Escherichia coli* (EPEC/EHEC)	Gastrointestinal disease	Inhibits IκBα degradation	Royan et al. (46)
	PtpA	*Mycobacterium tuberculosis*	Tuberculosis	Dephosphorylates p38 and JNK; competes with ubiquitin for TAB3 binding	Wang et al. (61)

(Continued)

TABLE 1 | Continued

Protein function	Protein	Bacterial species	Disease	Mechanism	Reference
E3 ligase	IpaH9.8	*Shigella*	Dysentery	Targets NEMO and MAPKK (Ste7) for degradation	Rohde et al. (48), Ashida et al. (50)
	IpaH4.5	*Shigella*	Dysentery	Targets NF-κB p65 for ubiquitination, preventing transcription	Wang et al. (51)
	IpaH0722	*Shigella*	Dysentery	Targets TRAF2 for ubiquitination, preventing PKC-induced NF-κB activity	Ashida et al. (52)
Deubiquitylase	SseL	*Salmonella* Typhimurium	Gastrointestinal disease	Prevents Lys48-linked ubiquitination and degradation of IκBα	Le Negrate et al. (57)
	ChlsDub1	*Chlamydia trachomatis*	Trachoma eye disease	Prevents Lys48-linked ubiquitination and degradation of IκBα	Le Negrate et al. (57)
	TssM	*Burkholderia pseudomallei*	Melioidosis	Prevents Lys63-linked ubiquitination of TRAF6/TRAF3 and Lys48-linked ubiquitination and degradation of IκBα	Shanks et al. (58) Tan et al. (60)
Glutamine deamidase	OspI	*Shigella flexneri*	Dysentery	Deamidates glutamine residue in Ubc13 to prevent TRAF6 binding	Sanada et al. (62)
Cysteine methyltransferase	NleE	*Escherichia coli* (EPEC)	Gastrointestinal disease	Targets Npl4 zinc finger domains of TAB2/3 to prevent binding to Lys63-linked polyubiquitin and TAK1 activity	Zhang et al. (63)

is summarised in **Table 1**. Interestingly, some intracellular bacteria can also activate MAPK or NF-κB pathways to their advantage at different stages of infection. For example, while within cellular vacuoles, *Salmonella* Typhimurium expresses the kinase SteC which phosphorylates MKK1/2 on Ser200 in the kinase domain (5). Phosphorylation of Ser200 causes MKK1/2 to autophosphorylate on Ser218 and Ser220, leading to activation of ERK1/2 and resulting in reorganization of the actin cytoskeleton, which restrains bacterial growth to control bacterial virulence (5). *S.* Typhimurium also uses SopE, SopE2, and SopB, which act redundantly to activate MAPK and NF-κB via Rho-family GTPases and stimulate inflammation (6). Infection of alveolar macrophages with *Legionalla pneumophila* causes Legionnaire's disease. *L. pneumophila* translocates the kinase LegK1 into macrophages where it activates NF-κB signaling to inhibit apoptosis and promote intracellular bacterial replication (7). LegK1 phosphorylates a number of proteins in both the canonical and non-canonical NF-κB pathway, including IκBα, IκBβ, IκBε, p100 (NFBK2), and p105 (NFKB1) (7). Phosphorylation of IκBα on serines 32 and 36 stimulate its degradation and promote translocation of NF-κB to the nucleus, while phosphorylation of p100 on serines 866 and 870 causes its cleavage to generate the p52 subunit and induce formation of the p52/RelB non-canonical NF-κB complex.

BLOCKING SIGNALING BY MIMICKING TIR:TIR INTERACTIONS

A number of bacteria target the initial stage of TLR activation by expressing TIR-containing proteins (Tcps) that interfere with TIR–TIR interactions. A bioinformatics screen for bacterial proteins with homology to human TIRs identified the first TIR-containing protein as TIR-like protein A (TlpA) from *Salmonella enterica* serovar Enteritidis (*Salmonella* Enteritidis), which

causes food-borne gastroenteritis (8). TlpA dose-dependently suppresses TLR/IL1 induced NF-κB activity and is thought to achieve this by competing with endogenous TIR domains to block downstream signaling (8). A similar mechanism is proposed for the *Staphylococcus aureus* TIR domain protein TirS which blocks TLR2-induced MAPK and NF-κB signaling (9).

Other Tcps, including TcpC from the uropathogenic *Escherichia coli* strain CFT073, TcpB/BtpA, and BtpB from *Brucella melitensis* which causes the chronic and debilitating zoonotic disease Brucellosis and ypTIR from the plague-causing *Yersinia pestis*, are all able to bind to MyD88 and prevent downstream signaling from TLRs (10–12). Additionally, TcpB was proposed to compete with the TIR-containing adaptor protein Mal/TIRAP to prevent TLR2- and TLR4-dependent signaling (13). TcpB downregulates Mal expression by targeting phosphorylated Mal for proteasomal degradation by a mechanism similar to the cellular SOCS1-mediated degradation of Mal (14).

BACTERIAL PROTEASES

Several bacterial proteins can inhibit signaling by selectively cleaving signaling enzymes. *Bacillus anthracis* lethal factor (LF) is a protease that forms part of the anthrax toxin. LF specifically targets MAPK kinases (MKKs) by cleaving within the MAPK-docking domain (D-domain), which is required for binding to downstream substrates. LF-induced proteolysis disrupts or removes the D-domain to generate kinases that are unable to interact with downstream MAPKs, thereby blocking their phosphorylation and activation. Although originally described to block MKK1/2 (15), LF is capable of cleaving all MKKs except MKK5 (16), resulting in reduced kinase activity for ERK, p38, and JNK MAPK pathways.

Enteropathogenic and enterohemorrhagic *E. coli* (EPEC/EHEC) are closely related bacteria that cause severe food-borne gastroenteritis. Both use type III secretion systems (T3SS) to inject effector proteins into the host cell. One of these, NleD, is a zinc metalloprotease that inactivates JNK and p38 by cleaving between the dual phosphorylation sites within the kinase activation loop (17). Proteolysis as a strategy to dampen the immune response is not restricted to MAPKs as EPEC/EHEC proteases also target the NF-κB signaling pathway. NleC, another zinc protease, cleaves the p65 subunit of NF-κB at its amino-terminus to promote its proteasomal degradation (17–20) and has also been shown to target other NF-κB components including IκBα, p50, and c-Rel (19, 20). The NF-κB p65 subunit is also a target of proteolysis by the *Chlamydia* proteases CT441 (21, 22) and Chlamydial protease-like activity factor (CPAF) (23). CT441 inhibits NFκB activation by cleaving p65 at residue 351/2, which lies between the Rel-homologous domain and the transactivation domain (21).

BACTERIAL ACETYLTRANSFERASES

Some bacterial effectors modify host signaling proteins to inhibit their activity. Vibrio outer protein A (VopA) is an acetyltransferase expressed by *Vibrio parahaemolyticus* that inhibits signaling by all MAPKs through both O-acetylation (serine and threonine acetylation) and N-acetylation (lysine acetylation) of MAP2Ks (24, 25). VopA acetylates MKK6 on three residues (Ser207, Lys210, and Thr211) in the activation loop and on Lys172 in the catalytic loop. Phosphorylation of Ser207 and Thr211 by MAP3Ks is critical for MKK6 activation, and acetylation of these resides by VopA blocks MKK6 activity. Lys172 coordinates the γ-phosphate of ATP, and its N-acetylation disrupts ATP binding to prevent phosphorylation of downstream substrates. This dual approach of preventing kinase activation and locking the kinase in an inactive state makes VopA an extremely potent inhibitor of MAPK signaling.

Salmonella Typhimurium expresses the O-acetyltranferase AvrA that modifies the threonine residue in the activation loop of MKK4 to prevent JNK activation (26). An interaction between MKK7 and AvrA was observed in a yeast-two-hybrid screen (27) suggesting that it can act on both MKKs that activate JNK. Although overexpressed AvrA inhibits both p38 and JNK phosphorylation, only JNK phosphorylation is inhibited during *S.* Typhimurium infection and JNK target genes are upregulated in cells infected with ΔAvrA, lending support for AvrA being targeted to the JNK signaling pathway (27).

Yersinia species deliver *Yersinia* outer proteins (Yops) into the host cell via a Type III secretion system. The *Y. pestis*/*Yersinia pseudotuberculosis* effector YopJ (YopP in *Yersinia enterocolitica*) inhibits MAPK signaling by blocking the phosphorylation and activation of MAP2Ks (28). YopJ O-acetylates critical residues in the MAP2K activation loop, as described for MKK6 and MKK2 (29–31). In addition to targeting MAP2Ks, YopJ/YopP also inhibits the MAP3K TAK1 (31–34). YopJ O-acetylates Thr184 and Thr187 in the activation loop of TAK1, preventing Thr187 autophosphorylation and thereby blocking kinase activation. Conflicting reports exist regarding the effect of YopP on the formation of the TAK1-TAB2/3 complex, with one showing that YopP interferes with TAK1-TAB2 binding (32), while a second

report demonstrated that it did not affect TAK1-TAB2/3 complex formation (33). YopP may also affect ubiquitination since overexpressed YopP blocks TRAF6-dependent polyubiquitination reactions, although the authors note that they were unable to reliably detect this effect on ubiquitination in *Yersinia*-infected cells (32). By acetylating both TAK1 and MKKs to prevent their activation, YopJ/P targets both the MAPK and NFκB signaling pathways. This dual targeting strategy is reinforced by the demonstration that YopJ also O-acetylates IKKα and IKKβ in the activation loop to inhibit their kinase activity and prevent activation of the NFκB pathway (28, 30).

Mitogen-activated protein kinases are inactivated by a number of different phosphatases of which the dual specificity phosphatase (DUSP) family members are key regulators of MAPK dephosphorylation in immunity. The Enhanced intracellular survival (Eis) protein of *Mycobacterium tuberculosis*, the causative agent of tuberculosis, targets the JNK pathway. Eis N-acetylates lysine 55 of the JNK-specific DUSP16, which is also known as MAPK phosphatase 7 or MKP7 (35). Lys55 lies within the substrate-docking domain of DUSP16, and its acetylation by Eis results in reduced JNK activity in cells. Similarly, DUSP1/MKP1 that has been acetylated on Lys57 by p300 reduces p38 activity (36). Acetylated DUSP1 binds more readily to p38, resulting in higher phosphatase activity and reduced p38 activity (36). Eis acetylation of DUSP16 is thought to act in a similar manner to reduce JNK activity.

BACTERIAL PHOSPHOTHREONINE LYASES

The OspF and SpvC proteins of *Shigella* and *S.* Typhimurium, respectively, target MAPK activation by specifically removing the phosphate group from phosphothreonine in the TxY activation loop (37, 38). Rather than acting as threonine-specific phosphatases, OspF and SpvC function as phosphothreonine lyases to irreversibly inactivate MAPKs via an eliminylation reaction whereby the threonine phosphate group is dephosphorylated by β-elimination to generate the unsaturated amino acid dehydrobutyrine (37, 39). The effect is irreversible as dehydrobutyrine lacks a hydroxyl group and cannot be phosphorylated. Although OspF and SpvC have activity against ERK, p38, and JNK (37, 38), OspF shows selectivity for ERK and p38 during *Shigella* infection (40, 41) and has actually been shown to potentiate JNK activity due to its phosphothreonine lyase activity on p38 disrupting a negative feedback loop between p38 and TAK1 (42).

BACTERIAL KINASES

Some bacteria express their own kinases. For example *Shigella* OspG is a serine/threonine kinase that binds to ubiquitin and E2-ubiquitin conjugates in the SCF-βTrCP complex, dampening the host immune response by reducing IκBα degradation (43, 44). The interaction between OspG and ubiquitin activates its kinase activity, which is required for it to inhibit NFκB signaling. Binding of OspG to E2-ubiquitin conjugates also represses ubiquitin transfer to E3 ligases, as it has been shown to stabilize a UbcH5b-ubiquitin complex (45).

OspG shares significant sequence homology with the NleH family of proteins in *E. coli* and like OspG, NleH1, and NleH2 can inhibit IκBα ubiquitination to prevent its degradation (46). However, NleH1/2 is regulated differently to OspG, since their kinase activity is not induced by ubiquitin (44). Instead, NleH1 binds to a novel subunit of NFκB, ribosomal protein S3 (RPS3), antagonizing its function of guiding p65 to specific promoters and thereby reducing its transcriptional activity (47).

BACTERIAL E3 LIGASES

In addition to expressing kinases and phosphatases that can interfere with ubiquitin-dependent signaling, bacteria also use their own E3 ligases and deubiquitinase. IpaH proteins belong to the Novel E3 Ligase (NEL) family of ubiquitin E3 ligases of which *Shigella* IpaH9.8 and *Salmonella* SspH1 were the first described members (48). Although a large number of bacterial E3 ligases have been identified (49) many of their ubiquitination targets are unknown. Using the yeast Saccharomyces cerevisiae as a model system, it was demonstrated that IpaH9.8 acts as an E3 ligase for the MAPKK Ste7 (48), and that in human cells IpaH9.8 mediates lysine-27 polyubiquitination of NEMO/IKKγ resulting in the degradation of both proteins (50). Other IpaH family members also possess E3 ligase activity, with *Shigella* IpaH4.5 ubiquitinating p65 to block NF-κB transcription (51) and IpaH0722 targeting TRAF2 for ubiquitin-dependent degradation to inhibit PKC-induced NF-κB activity (52). Interestingly reports are now emerging of bacterial E3 ligases targeting other aspects of immune signaling. For example, *Shigella* IpaH7.8 activates the inflammasome, resulting in cell death and enhanced bacterial replication (53).

BACTERIAL DEUBIQUITINASES (DUBs)

Deubiquitinases (DUBs) are proteases that remove ubiquitin from proteins. Both the AvrA and SseL proteins of *S.* Typhimurium and the ChlsDub1 protein of *Chlamydia trachomatis* possess DUB activity that inhibits K48-linked ubiquitination and degradation of IκBα, thus blocking NFκB activation (54–57). In addition to its DUB activity, ChlsDub1 has also been shown to have deNED-Dylating activity (59) that may contribute to suppressing IκBα degradation by antagonizing conjugation of the ubiquitin-like NEDD8 protein to the SCF-βTrCP complex, although this has not been formally demonstrated.

Burkholderia (Pseudomonas) pseudomallei, which causes melioidosis, expresses the effector protein TssM, which possesses DUB activity against both K63 and K48-linked polyubiquitin (58, 60). TssM overexpression causes reduced K48-linked

polyubiquitination of IκBα and reduced K63-linked polyubiquitination of TRAF6 to block NFκB-induced transcription (60).

OTHER BACTERIAL EFFECTORS

In addition to the above proteins, further bacterial effectors that block MAPK and NFκB activity via different mechanisms have started to emerge. *M. tuberculosis* tyrosine phosphatase PtpA dephosphorylates JNK and p38 to dampen cytokine expression (61). PtpA uses a novel mechanism whereby its phosphatase activity is stimulated by binding to ubiquitin via a novel ubiquitin-interacting motif-like (UIML) region (61). Mtb PtpA also suppresses NFκB activation by competitively binding to the Npl4 zinc finger domain (NZF) of TAB3, blocking its ability to bind to ubiquitin chains and thereby reducing Tak1 activity.

The *Shigella flexneri* type III effector OspI blocks TRAF6 mediated signaling by selectively deamidating a glutamine residue to glutamic acid in the E2 enzyme Ubc13 (62). The deamidation of Glu100 prevents Ubc13 from binding to TRAF6, inhibiting its E3 ligase activity and thereby blocking downstream signaling.

The EPEC NleE protein also uses a unique mechanism to inhibit bacterial-induced signaling. NleE is an S-adenosyl-L-methionine (SAM)-dependent cysteine methyltransferase that targets the Npl4 zinc finger (NZF) domains in TAB2/3 (63). Modification of cysteine residues in the zinc finger domains of TAB2/3 abolishes their ability to bind to Lys63-linked polyubiquitin chains and therefore blocks downstream TAK1 activation, consistent with the observed inhibition of IκBα phosphorylation and NFκB signaling in the presence of NleE (64, 65). NleE proteins from other pathogens, such as *S. flexneri* protein OspZ, were shown to be functionally interchangeable with NleE in blocking NFκB signaling and may also act as a cysteine methyltransferases (65).

SUMMARY AND FUTURE PERSPECTIVES

Bacterial pathogens have evolved diverse and elegant ways to block MAPK and NF-κB signaling downstream of TLR activation, enabling them to evade detection by the immune system and promote infection. Many bacteria employ strategies to simultaneously target a number of proteins within these, as well as other host pathways, to increase their chances of overcoming the immune response. Future discoveries in understanding how and why pathogens target particular proteins will not only demonstrate their importance in immunity, but will also help our understanding of how bacteria activate intracellular signaling pathways, and have the potential to identify new targets for the treatment or prevention of infection.

AUTHOR CONTRIBUTIONS

Both VM and SA wrote the article.

REFERENCES

1. Kawai T, Akira S. The role of pattern-recognition receptors in innate immunity: update on toll-like receptors. *Nat Immunol* (2010) **11**(5):373–84. doi:10.1038/ni.1863

2. Coll RC, O'Neill LA. New insights into the regulation of signalling by toll-like receptors and nod-like receptors. *J Innate Immun* (2010) **2**(5):406–21. doi:10.1159/000315469

3. Arthur JSC, Ley SC. Mitogen-activated protein kinases in innate immunity. *Nat Rev Immunol* (2013) **13**(9):679–92. doi:10.1038/nri3495

4. Cohen P. The TLR and IL-1 signalling network at a glance. *J Cell Sci* (2014) **127**(Pt 11):2383–90. doi:10.1242/jcs.149831

5. Odendall C, Rolhion N, Förster A, Poh J, Lamont DJ, Liu M, et al. The *Salmonella* kinase SteC targets the MAP kinase MEK to regulate the host actin cytoskeleton. *Cell Host Microbe* (2012) **12**(5):657–68. doi:10.1016/j.chom.2012.09.011

6. Bruno VM, Hannemann S, Lara-Tejero M, Flavell RA, Kleinstein SH, Galán JE. *Salmonella* Typhimurium type III secretion effectors stimulate innate immune responses in cultured epithelial cells. *PLoS Pathog* (2009) **5**(8):e1000538. doi:10.1371/journal.ppat.1000538

7. Ge J, Xu H, Li T, Zhou Y, Zhang Z, Li S, et al. A *Legionella* type IV effector activates the NF-kappaB pathway by phosphorylating the IkappaB family of inhibitors. *Proc Natl Acad Sci U S A* (2009) **106**(33):13725–30. doi:10.1073/pnas.0907200106

8. Newman RM, Salunkhe P, Godzik A, Reed JC. Identification and characterization of a novel bacterial virulence factor that shares homology with mammalian toll/interleukin-1 receptor family proteins. *Infect Immun* (2006) **74**(1):594–601. doi:10.1128/IAI.74.1.594-601.2006

9. Askarian F, van Sorge NM, Sangvik M, Beasley FC, Henriksen JR, Sollid JU, et al. A *Staphylococcus aureus* TIR domain protein virulence factor blocks TLR2-mediated NF-κB signaling. *J Innate Immun* (2014) **6**(4):485–98. doi:10.1159/000357618

10. Cirl C, Wieser A, Yadav M, Duerr S, Schubert S, Fischer H, et al. Subversion of toll-like receptor signaling by a unique family of bacterial toll/interleukin-1 receptor domain-containing proteins. *Nat Med* (2008) **14**(4):399–406. doi:10.1038/nm1734

11. Rana RR, Simpson P, Zhang M, Jennions M, Ukegbu C, Spear AM, et al. *Yersinia pestis* TIR-domain protein forms dimers that interact with the human adaptor protein MyD88. *Microb Pathog* (2011) **51**(3):89–95. doi:10.1016/j.micpath.2011.05.004

12. Salcedo SP, Marchesini MI, Degos C, Terwagne M, Von Bargen K, Lepidi H, et al. BtpB, a novel Brucella TIR-containing effector protein with immune modulatory functions. *Front Cell Infect Microbiol* (2013) **3**(July):28. doi:10.3389/fcimb.2013.00028

13. Radhakrishnan GK, Yu Q, Harms JS, Splitter GA. *Brucella* TIR domain-containing protein mimics properties of the toll-like receptor adaptor protein TIRAP. *J Biol Chem* (2009) **284**(15):9892–8. doi:10.1074/jbc.M805458200

14. Sengupta D, Koblansky A, Gaines J, Brown T, West AP, Zhang D, et al. Subversion of innate immune responses by *Brucella* through the targeted degradation of the TLR signaling adapter. *J Immunol* (2010) **184**(2):956–64. doi:10.4049/jimmunol.0902008

15. Duesbery NS, Webb CP, Leppla SH, Gordon VM, Klimpel KR, Copeland TD, et al. Proteolytic inactivation of MAP-kinase-kinase by anthrax lethal factor. *Science* (1998) **280**(5364):734–7. doi:10.1126/science.280.5364.734

16. Vitale G, Bernardi L, Napolitani G, Mock M, Montecucco C. Susceptibility of mitogen-activated protein kinase kinase family members to proteolysis by anthrax lethal factor. *J Biochem* (2000) **745**:739–45. doi:10.1042/bj3520739

17. Baruch K, Gur-Arie L, Nadler C, Koby S, Yerushalmi G, Ben-Neriah Y, et al. Metalloprotease type III effectors that specifically cleave JNK and NF-κB. *EMBO J* (2011) **30**(1):221–31. doi:10.1038/emboj.2010.297

18. Yen H, Ooka T, Iguchi A, Hayashi T, Sugimoto N, Tobe T. NleC, a type III secretion protease, compromises NF-kB activation by targeting p65/rela. *PLoS Pathog* (2010) **6**(12):e1001231. doi:10.1371/journal.ppat.1001231

19. Mühlen S, Ruchaud-Sparagano MH, Kenny B. Proteasome-independent degradation of canonical NFκB complex components by the NleC protein of pathogenic *Escherichia coli. J Biol Chem* (2011) **286**(7):5100–7. doi:10.1074/jbc.M110.172254

20. Pearson JS, Riedmaier P, Marchès O, Frankel G, Hartland EL. A type III effector protease NleC from enteropathogenic *Escherichia coli* targets NF-κB for degradation. *Mol Microbiol* (2011) **80**(1):219–30. doi:10.1111/j.1365-2958.2011.07568.x

21. Lad SP, Yang G, Scott DA, Wang G, Nair P, Mathison J, et al. Chlamydial CT441 Is a PDZ domain-containing tail-specific protease that interferes with the NF-κB pathway of immune response. *J Bacteriol* (2007) **189**(18):6619–25. doi:10.1128/JB.00429-07

22. Lad SP, Li J, da Silva Correia J, Pan Q, Gadwal S et al. Cleavage of p65/RelA of the NF-B pathway by Chlamydia. *Proc Natl Acad Sci U S A* (2007) **104**(8):2933–8. doi:10.1073/pnas.0608393104

23. Christian J, Vier J, Paschen SA, Häcker G. Cleavage of the NF-κB family protein p65/RelA by the chlamydial protease-like activity factor (CPAF) impairs proinflammatory signaling in cells infected with chlamydiae. *J Biol Chem* (2010) **285**(53):41320–7. doi:10.1074/jbc.M110.152280

24. Trosky JE, Mukherjee S, Burdette DL, Roberts M, McCarter L, Siegel RM, et al. Inhibition of MAPK signaling pathways by VopA from *Vibrio parahaemolyticus. J Biol Chem* (2004) **279**(50):51953–7. doi:10.1074/jbc.M407001200

25. Trosky JE, Li Y, Mukherjee S, Keitany G, Ball H, Orth K. VopA inhibits ATP binding by acetylating the catalytic loop of MAPK kinases. *J Biol Chem* (2007) **282**(47):34299–305. doi:10.1074/jbc.M706970200

26. Jones RM, Wu H, Wentworth C, Luo L, Collier-Hyams L, Neish AS. *Salmonella* AvrA coordinates suppression of host immune and apoptotic defenses via JNK pathway blockade. *Cell Host Microbe* (2008) **3**(4):233–44. doi:10.1016/j.chom.2008.02.016

27. Du F, Galán JE. Selective inhibition of type III secretion activated signaling by the *Salmonella* effector AvrA. *PLoS Pathog* (2009) **5**(9):1–12. doi:10.1371/journal.ppat.1000595

28. Orth K, Palmer LE, Bao ZQ, Stewart S, Rudolph AE, Bliska JB, et al. Inhibition of the mitogen-activated protein kinase kinase superfamily by a *Yersinia* effector. *Science* (1999) **285**(5435):1920–3. doi:10.1126/science.285.5435.1920

29. Mukherjee S, Keitany G, Li Y, Wang Y, Ball HL, Goldsmith EJ, et al. *Yersinia* YopJ acetylates and inhibits kinase activation by blocking phosphorylation. *Science* (2006) **312**:1211–4. doi:10.1126/science.1126867

30. Mittal R, Peak-Chew S-Y, McMahon HT. Acetylation of MEK2 and I kappa B kinase (IKK) activation loop residues by YopJ inhibits signaling. *Proc Natl Acad Sci U S A* (2006) **103**(49):18574–9. doi:10.1073/pnas.0608995103

31. Paquette N, Conlon J, Sweet C, Rus F, Wilson L, Pereira A, et al. Serine/threonine acetylation of TGF-β-activated kinase (TAK1) by *Yersinia pestis* YopJ inhibits innate immune signaling. *Proc Natl Acad Sci U S A* (2012) **109**(31):12710–5. doi:10.1073/pnas.1008203109

32. Haase R, Richter K, Pfaffinger G, Courtois G, Ruckdeschel K. *Yersinia* outer protein P suppresses TGF-β-activated kinase-1 activity to impair innate immune signaling in *Yersinia enterocolitica*-infected cells. *J Immunol* (2005) **175**(12):8209–17. doi:10.4049/jimmunol.175.12.8209

33. Thiefes A, Wolf A, Doerrie A, Grassl GA, Matsumoto K, Autenrieth I, et al. The *Yersinia enterocolitica* effector YopP inhibits host cell signalling by inactivating the protein kinase TAK1 in the IL-1 signalling pathway. *EMBO Rep* (2006) **7**(8):838–44. doi:10.1038/sj.embor.7400754

34. Meinzer U, Barreau F, Esmiol-Welterlin S, Jung C, Villard C, Léger T, et al. *Yersinia pseudotuberculosis* effector YopJ subverts the Nod2/RICK/TAK1 pathway and activates caspase-1 to induce intestinal barrier dysfunction. *Cell Host Microbe* (2012) **11**(4):337–51. doi:10.1016/j.chom.2012.02.009

35. Kim KH, An DR, Song J, Yoon JY, Kim HS, Yoon HJ, et al. *Mycobacterium tuberculosis* Eis protein initiates suppression of host immune responses by acetylation of DUSP16/MKP-7. *Proc Natl Acad Sci U S A* (2012) **109**(20):7729–34. doi:10.1073/pnas.1120251109

36. Cao W, Bao C, Padalko E, Lowenstein CJ. Acetylation of mitogen-activated protein kinase phosphatase-1 inhibits Toll-like receptor signaling. *J Exp Med* (2008) **205**(6):1491–503. doi:10.1084/jem.20071728

37. Li H, Xu H, Zhou Y, Zhang J, Long C, Li S, et al. The phosphothreonine lyase activity of a bacterial type III effector family. *Science* (2007) **315**(5814):1000–3. doi:10.1126/science.1138960

38. Mazurkiewicz P, Thomas J, Thompson JA, Liu M, Arbibe L, Sansonetti P, et al. SpvC is a Salmonella effector with phosphothreonine lyase activity on host mitogen-activated protein kinases. *Mol Microbiol* (2008) **67**(6):1371–83. doi:10.1111/j.1365-2958.2008.06134.x

39. Brennan DF, Barford D. Eliminylation: a post-translational modification catalyzed by phosphothreonine lyases. *Trends Biochem Sci* (2009) **34**(3):108–14. doi:10.1016/j.tibs.2008.11.005

40. Arbibe L, Kim DW, Batsche E, Pedron T, Mateescu B, Muchardt C, et al. An injected bacterial effector targets chromatin access for transcription factor NF-kappaB to alter transcription of host genes involved in immune responses. *Nat Immunol* (2007) **8**(1):47–56. doi:10.1038/ni1423

41. Kramer RW, Slagowski NL, Eze NA, Giddings KS, Morrison MF, Siggers KA, et al. Yeast functional genomic screens lead to identification of a role for a bacterial effector in innate immunity regulation. *PLoS Pathog* (2007) **3**(2):179–90. doi:10.1371/journal.ppat.0030021

42. Reiterer V, Grossniklaus L, Tschon T, Kasper CA, Sorg I, Arrieumerlou C. *Shigella flexneri* type III secreted effector OspF reveals new crosstalks of proinflammatory signaling pathways during bacterial infection. *Cell Signal* (2011) **23**(7):1188–96. doi:10.1016/j.cellsig.2011.03.006

43. Kim DW, Lenzen G, Page AL, Legrain P, Sansonetti PJ, Parsot C. The *Shigella flexneri* effector OspG interferes with innate immune responses by targeting ubiquitin-conjugating enzymes. *Proc Natl Acad Sci U S A* (2005) **102**(39):14046–51. doi:10.1073/pnas.0504466102

44. Zhou Y, Dong N, Hu L, Shao F. The *Shigella* type three secretion system effector OspG directly and specifically binds to host ubiquitin for activation. *PLoS One* (2013) **8**(2). doi:10.1371/journal.pone.0057558

45. Pruneda JN, Smith FD, Daurie A, Swaney DL, Villén J, Scott JD, et al. E2~Ub conjugates regulate the kinase activity of *Shigella* effector OspG during pathogenesis. *EMBO J* (2014) **33**(5):437–49. doi:10.1002/embj.201386386

46. Royan SV, Jones RM, Koutsouris A, Roxas JL, Falzari K, Weflen AW, Kim A, et al. Enteropathogenic *E. coli* non-LEE encoded effectors NleH1 and NleH2 attenuate NF-κB activation. *Mol Microbiol* (2010) **78**(5):1232–45. doi:10.1111/j.1365-2958.2010.07400.x

47. Gao X, Wan F, Mateo K, Callegari E, Wang D, Deng W, et al. Bacterial effector binding to ribosomal protein S3 subverts NF-κB function. *PLoS Pathog* (2009) **5**(12):e1000708. doi:10.1371/journal.ppat.1000708

48. Rohde JR, Breitkreutz A, Chenal A, Sansonetti PJ, Parsot C. Type III secretion effectors of the IpaH family are E3 ubiquitin ligases. *Cell Host Microbe* (2007) **1**(1):77–83. doi:10.1016/j.chom.2007.02.002

49. Hicks SW, Galán JE. Hijacking the host ubiquitin pathway: structural strategies of bacterial E3 ubiquitin ligases. *Curr Opin Microbiol* (2010) **13**(1):41–6. doi:10.1016/j.mib.2009.11.008

50. Ashida H, Kim M, Schmidt-Supprian M, Ma A, Ogawa M, Sasakawa C. A bacterial E3 ubiquitin ligase IpaH9.8 targets NEMO/IKKgamma to dampen the host NF-kappaB-mediated inflammatory response. *Nat Cell Biol* (2010) **12**(1):66–73. doi:10.1038/ncb2006

51. Wang F, Jiang Z, Li Y, He X, Zhao J, Yang X, et al. *Shigella flexneri* T3SS effector IpaH4.5 modulates the host inflammatory response via interaction with NF-κB p65 protein. *Cell Microbiol* (2013) **15**(3):474–85. doi:10.1111/cmi.12052

52. Ashida H, Nakano H, Sasakawa C. *Shigella* IpaH0722 E3 ubiquitin Ligase effector targets TRAF2 to Inhibit PKC-NF-κB activity in invaded epithelial cells. *PLoS Pathog* (2013) **9**(6):1–15. doi:10.1371/journal.ppat.1003409

53. Suzuki S, Mimuro H, Kim M, Ogawa M, Ashida H, Toyotome T, et al. *Shigella* IpaH7.8 E3 ubiquitin ligase targets glomulin and activates inflammasomes to demolish macrophages. *Proc Natl Acad Sci U S A* (2014) **111**(40):E4254–63. doi:10.1073/pnas.1324021111

54. Collier-Hyams LS, Zeng H, Sun J, Tomlinson AD, Bao ZQ, Chen H, et al. Cutting edge: *Salmonella* AvrA effector inhibits the key proinflammatory, anti-apoptotic NF-B pathway. *J Immunol* (2002) **169**(6):2846–50. doi:10.4049/jimmunol.169.6.2846

55. Ye Z, Petrof EO, Boone D, Claud EC, Sun J. *Salmonella* effector AvrA regulation of colonic epithelial cell inflammation by deubiquitination. *Am J Pathol* (2007) **171**(3):882–92. doi:10.2353/ajpath.2007.070220

56. Le Negrate G, Krieg A, Faustin B, Loeffler M, Godzik A, Krajewski S, et al. ChlaDub1 of chlamydia trachomatis suppresses NF-kappaB activation and inhibits IkappaBalpha ubiquitination and degradation. *Cell Microbiol* (2008) **10**(9):1879–92. doi:10.1111/j.1462-5822.2008.01178.x

57. Le Negrate G, Faustin B, Welsh K, Loeffler M, Krajewska M, Hasegawa P, et al. *Salmonella* secreted factor L deubiquitinase of *Salmonella* typhimurium Inhibits NF-κB, suppresses IκBα ubiquitination. *The J Immunol* (2008) **180**:5045–56. doi:10.4049/jimmunol.180.7.5045

58. Shanks J. Burtnick MN, Brett PJ, Waag DM, Spurgers KB, Ribot WJ, et al., *Burkholderia mallei* tssM encodes a putative deubiquitinase that is secreted and expressed inside infected RAW 264.7 murine macrophagest. *Infect Immun* (2009) **77**(4):1636–48. doi:10.1128/IAI.01339-08

59. Misaghi S, Balsara ZR, Catic A, Spooner E, Ploegh HL, Starnbach MN. Chlamydia trachomatis-derived deubiquitinating enzymes in mammalian cells during infection. *Mol Microbiol* (2006) **61**(1):142–50. doi:10.1111/j.1365-2958.2006.05199.x

60. Tan KS, Chen Y, Lim YC, Tan GG, Liu Y, Lim YT, et al. Suppression of host innate immune response by *Burkholderia pseudomallei* through the virulence factor TssM. *J Immunol* (2010) **184**(9):5160–71. doi:10.4049/jimmunol.0902663

61. Wang J, Li BX, Ge PP, Li J, Wang Q, Gao GF, et al. *Mycobacterium tuberculosis* suppresses innate immunity by coopting the host ubiquitin system. *Nat Immunol* (2015) **16**(3). doi:10.1038/ni.3096

62. Sanada T, Kim M, Mimuro H, Suzuki M, Ogawa M, Oyama A, et al. The *Shigella flexneri* effector OspI deamidates UBC13 to dampen the inflammatory response. *Nature* (2012) **483**(7391):623–6. doi:10.1038/nature10894

63. Zhang L, Ding X, Cui J, Xu H, Chen J, Gong YN, et al. Cysteine methylation disrupts ubiquitin-chain sensing in NF-κB activation. *Nature* (2011) **481**(7380):204–8. doi:10.1038/nature10690

64. Nadler C, Baruch K, Kobi S, Mills E, Haviv G, Farago M, et al. The type III secretion effector NleE inhibits NF-κB activation. *PLoS Pathog* (2010) **6**(1):403–4. doi:10.1371/journal.ppat.1000743

65. Newton HJ, Pearson JS, Badea L, Kelly M, Lucas M, Holloway G, et al. The type III effectors NieE and NleB from enteropathogenic E. coli and Ospz from *shigella* block nuclear translocation of NF- κB p65. *PLoS Pathog* (2010) **6**(5):1–16. doi:10.1371/journal.ppat.1000898

66. Spear AM, Rana RR, Jenner DC, Flick-Smith HC, Oyston PC, Simpson P, et al. A Toll/interleukin (IL)-1 receptor domain protein from *Yersinia pestis* interacts with mammalian IL-1/Toll-like receptor pathways but does not play a central role in the virulence of *Y. pestis* in a mouse model of bubonic plague. *Microbiology* (2012) **158**(6):1593–606. doi:10.1099/mic.0.055012-0

2

Modulation of Host Autophagy during Bacterial Infection: Sabotaging Host Munitions for Pathogen Nutrition

Pedro Escoll[1,2], Monica Rolando[1,2] and Carmen Buchrieser[1,2]*

[1] Institut Pasteur, Biologie des Bactéries Intracellulaires, Paris, France, [2] CNRS UMR 3525, Paris, France

Keywords: autophagosome, xenophagy, autophagy modulators, intracellular bacteria, bacterial nutrition

Edited by:
Abhay Satoskar,
The Ohio State University, USA

Reviewed by:
Eric Ghigo,
Centre national de la recherche
scientifique, France
Robert Heinzen,
National Institutes of Health, USA
Soubeyran Philippe,
Institut national de la santé et de la
recherche médicale, France

*Correspondence:
Carmen Buchrieser
cbuch@pasteur.fr

AUTOPHAGY IS A DEFENSE MECHANISM AGAINST INVADING PATHOGENS

Cellular homeostasis requires the balanced regulation of anabolic and catabolic processes. While anabolic metabolism consumes energy to build up cellular components, catabolic processes break down organic matters in order to provide energy for the cell and its anabolic processes. Autophagy is a highly conserved and regulated catabolic process by which the eukaryotic cell degrades unnecessary, undesirable, or dysfunctional cellular components, including organelles (1–3). Autophagy is induced by a variety of extra- and intracellular stress stimuli, such as nutrient starvation, oxidative stress, or accumulation of damaged organelles or toxic protein aggregates. Initiation of autophagy first leads to the formation of cup-shaped structures known as phagophores that engulf the undesirable or damaged cellular components. Subsequent elongation of phagophores form double-membrane vesicles called autophagosomes, which deliver their cargo to lysosomes where the content is degraded and recycled (1–3). Autophagy plays a central role in quality control of organelles and proteins, and additionally is a key mechanism to maintain cellular energy levels and nutrient homeostasis during starvation, promoting the recycling and salvage of cellular nutrients. Furthermore, the cellular autophagic machinery is also used to remove invading intracellular pathogens, a process called xenophagy (1, 2). In this case, phagophores engulf invading microbes forming autophagosomes and steering them toward lysosomal degradation. Thus, xenophagy is an innate immune mechanism against bacterial infection that has been shown to be essential to restrict intracellular growth of many bacteria such as *Salmonella enterica* serovar Typhimurium (4), *Mycobacterium tuberculosis* (5, 6), *Listeria monocytogenes* (7), or Group A *Streptococcus* (8).

Detection of bacterial components in the cytoplasm of mammalian cells induces autophagy via the activation of toll-like receptor 4 (TLR4) by bacterial lipopolysaccharide (LPS) and recognition of bacterial peptidoglycan by NOD1 and NOD2 (9, 10). TLR- and NOD-like receptor (NLR)-induced autophagy can be initiated during entry, uptake, or phagocytosis of bacteria by the host cell (10, 11), but bacteria can also be sensed by the Sequestosome-1-like receptors (SLRs) when they are already in the cytosol (1) (**Figure 1A**). In both cases, recruitment of autophagy proteins to the phagosome, such as the ULK1 complex, Beclin1, and ATG16L1, initiates membrane nucleation of the phagophore that will engulf the intracellular bacteria (10–12) (**Figure 1B**). ATG5–ATG12 associates with ATG16L1 and the ATG5–ATG12–ATG16L1 complex facilitates the addition of a phosphatidylethanolamine (PE) group to the carboxyl terminus of LC3, which function together with other factors to assemble, elongate, and allow the closure of nascent autophagosomes (1) (**Figure 1C**). In addition to this canonical mechanism of autophagy, phagosomes containing bacteria can recruit directly LC3, a process called LC3-associated phagocytosis (LAP). Upon delivery to phagosomes, LC3 promotes phagosome

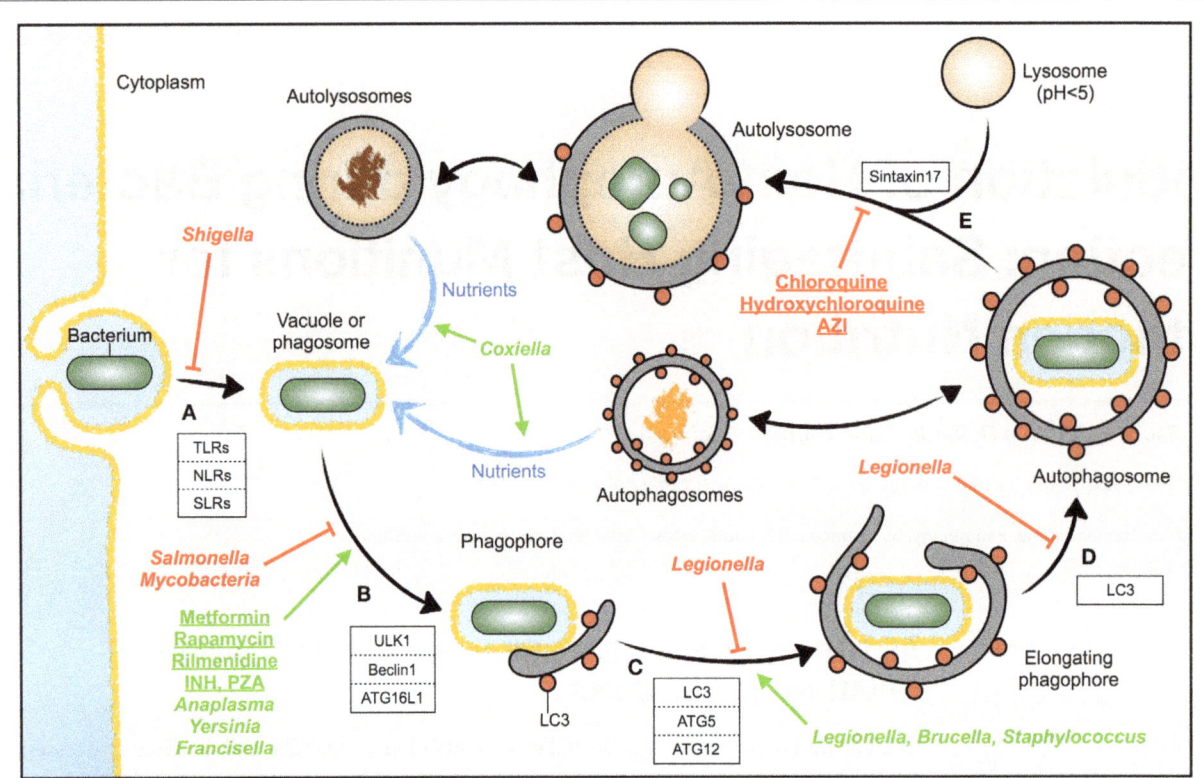

FIGURE 1 | Modulation of autophagy by drugs or intracellular bacteria. The different steps of the autophagic response during bacterial invasion are shown. The host factors known to participate in each step are depicted in white boxes. **(A)** Invading bacteria are sensed by immune receptors; **(B)** vesicle nucleation induced by specialized autophagy proteins; **(C)** phagophore elongation; **(D)** autophagophore completion; **(E)** autophagosome maturation by fusion with lysosomes, forming autolysosomes. Drugs (underlined) or intracellular bacteria (*cursiva*) inducing autophagy are displayed in green, while those inhibiting autophagy are displayed in red. The different steps where bacteria or drugs act are pointed with green arrows (activation) or red T-bars (inhibition). Blue arrows indicate nutrient flow, while doubled-headed arrows indicate the possibility that the content of cellular autophagosomes and autolysosomes can be diverted to the phagosome and used by pathogenic bacteria as a source of nutrients.

maturation and degradation of the content. Therefore, both LAP and canonical autophagy involve the enclosure of bacteria in an LC3-decorated compartment that is targeted for degradation by fusion with the lysosome (2). Membranes from the ER, the Golgi apparatus, the ER–mitochondria contact sites, or the plasma membrane contribute to the elongation of the double membrane of the phagophore in order to form the autophagosome (1) (**Figure 1D**). The attachment of syntaxin 17 to the autophagosomal membrane enables the fusion with lysosomes and represents the final maturation step of autophagosomes into autolysosomes (13) (**Figure 1E**), which normally leads to bacterial degradation in case of infection-induced autophagy (2).

PATHOGENIC INTRACELLULAR BACTERIA SUBVERT AND EXPLOIT THE AUTOPHAGY MACHINERY OF THE HOST

Xenophagy is a defense mechanism of the infected cell against invading bacteria, but intracellular pathogens have evolved mechanisms to inhibit or modulate the autophagy response of the host. For example, *M. tuberculosis* and *Salmonella* Typhimurium

inhibit autophagy initiation signaling upstream autophagosome formation (14, 15), whereas *Shigella flexneri* evades autophagy recognition by masking the bacterial surface (16) (**Figure 1**, *cursiva*, red).

In contrast to inhibition of autophagy, certain pathogenic intracellular bacteria induce autophagy and take advantage of it (17) (**Figure 1**, *cursiva*, green). These bacteria show defective replication in autophagy-deficient cells, and treatment of host cells with autophagy activators promotes bacterial replication. This observation raises the question, why a pathogen would increase a host defense mechanism like autophagy? In uninfected cells, augmentation of the autophagy rate is used to increase the intracellular pool of basic nutrients, to build new cellular structures. During infection, some intracellular bacteria have developed mechanisms to hijack the autophagosomes and redirect the by-products of the autophagic degradation toward microbial replication rather than for the use by the host cell (18). In most cases, these bacteria actively induce autophagy but, at the same time, block autophagosome maturation and fusion with the lysosome. In this case, augmentation of autophagy, rather than promoting bacterial clearance, promotes the acquisition of nutrients by the invading bacteria (18). Thus, certain bacteria may sabotage the host defense mechanism elicited by autophagosomes to use the autophagic vesicles as nutrient source for microbial growth.

An example is *Anaplasma phagocytophilum* that uses a secreted effector, Ats-1, to promote autophagosome nucleation and stimulates its own growth by using the nutrients contained in the autophagosomes (19). Indeed, autophagy induction using rapamycin favors bacterial infection, while autophagy inhibition decreases *A. phagocytophilum* replication (20). Another example is *Yersinia pseudotuberculosis*, a Gram-negative bacterium that replicates intracellularly by establishing a specialized compartment, the *Yersinia*-containing vacuole (YCVs), which accumulates autophagy markers (21). The stimulation of autophagy with rapamycin increases the size of the YCVs and the numbers of replicative bacteria in the YCVs, whereas autophagy inhibition restricts bacterial survival, suggesting that autophagy promotes *Y. pseudotuberculosis* replication (21). *Yersinia pestis* also replicates within YCVs decorated with autophagosome markers (22). The authors suggested that autophagosomes may provide a source of membrane, along with late endosomes, for the expansion of the YCV into a spacious compartment (22). The same mechanism was described for *Coxiella burnetii*, the causative agent of Q fever. *Coxiella*-replicative vacuoles (CRVs) are decorated with the autophagy proteins LC3, Beclin1, and Rab24, and overexpression of LC3 or Beclin1 increases the number and size of the CRVs (23, 24). Similar to *A. phagocytophilum* and *Y. pseudotuberculosis*, autophagy induction increases *C. burnetii* replication, while inhibition of autophagy blocks *Coxiella* vacuole formation (23, 25). Also, *Francisella tularensis*, a highly virulent Gram-negative bacterium responsible for tularemia, avoids xenophagy while inducing autophagy (26). It was shown that autophagy-derived radiolabeled amino acids are transferred from host proteins to *F. tularensis*, a process that was reduced when host cells were treated with autophagy inhibitors (26).

Other bacteria also co-opt the autophagic machinery for their benefit, although a direct relationship of host autophagy and pathogen nutrition has not been shown. *Brucella abortus* that causes brucellosis in humans replicates in ER-derived *Brucella*-containing vacuoles (BCVs). BCVs hijack autophagosome initiation factors, such as ULK1 or Beclin1, and become autophagosome-like compartments (27). Depletion of ULK1 and Beclin1, as well as pharmacological inhibition of autophagy, readily reduced BCV formation, suggesting that autophagy promotes *B. abortus* infection (27). *Staphylococcus aureus* was also reported to be sequestered in LC3-positive autophagosomes that evade the fusion with lysosomes (28). *S. aureus* uses α-toxin to induce autophagy by an ATG5-dependent mechanism that also involves reduction of cellular cAMP levels (29). Infection of cells depleted of ATG5 show decreased bacterial replication, showing that autophagy is necessary for *S. aureus* replication *in vitro* (28).

Thus, different pathogenic bacteria seem to employ a common strategy to subvert the autophagy machinery as they not only target autophagy proteins to block xenophagy set up by the cell to resist infection but also exploit autophagy to promote their own replication. One well studied example for this dual strategy is the Gram-negative intracellular bacterium *Legionella pneumophila*. After phagocytosis, the causative agent of Legionnaires' disease, forms a *Legionella*-containing vacuole (LCV) that recruits vesicles emerging from the endoplasmic reticulum (ER) and acquires autophagy markers like LC3, showing that LCVs rapidly become autophagosomes (30, 31). This process seems to be dependent on the T4SS bacterial effector LegA9, which promotes the recognition of the LCV by autophagy (32). Interestingly, inhibition of autophagy in permissive A/J mouse macrophages reduces *Legionella* survival at 2 h postinfection (30, 33), suggesting that routing the LCV to the autophagy pathway is beneficial for the bacteria. However, later, it has been shown that *L. pneumophila* is also restraining autophagy by secreting the specialized effectors, *Lp*SPL and RavZ, that inhibit, autophagosome formation and maturation, respectively (34, 35). The paradoxical existence in *Legionella* of bacterial effectors having opposite roles, on one hand, targeting the LCV to autophagy and, on the other hand, inhibiting autophagy may reflect the necessity for the bacteria to fine-tune host autophagy in a very balanced way. *Legionella* may need to target the LCV to autophagosomes, avoiding immediate killing (33), and at the same time, it needs to delay the maturation of the LCV-containing autophagosome into autolysosomes, gaining precious time for pathogen replication (30, 36).

AUTOPHAGY MODULATORS IN INFECTIOUS AND NON-INFECTIOUS DISEASES: SOME CONSIDERATIONS

Autophagy modulators are of great interest for medical purposes (37), as it was suggested that metabolic, neurodegenerative, infectious, and oncology diseases can benefit from autophagy modulation (3).

One can hypothesize that drugs inducing autophagy could increase bacterial clearance in infected cells. This hypothesis is supported by the fact that antibiotics widely and extensively used against the intracellular bacterium *M. tuberculosis*, isoniazid (INH), and pyrazinamide (PZA), although able to kill the bacteria directly *in vitro* (38, 39), have been recently shown to induce autophagy in the host cell promoting mycobacterial clearance (40). Moreover, autophagy is required for effective antimycobacterial drug action *in vivo*, suggesting that pharmacological modulation of autophagy could be a successful strategy against infections by intracellular bacteria (40, 41). This point of view was corroborated by another recent report showing that treatment of cystic fibrosis patients with the antibiotic azithromycin (AZI) was associated with opportunistic mycobacterial infections. AZI was shown to prevent lysosomal acidification and thereby impaired autophagic degradation of mycobacteria (42), suggesting that chronic use of the drug may predispose to mycobacterial disease. Thus, these reports suggest that induction of autophagy with drugs, such as INH or PZA, could successfully treat mycobacterial infections, while inhibition of autophagy with drugs, such as AZI, may in turn facilitate mycobacterial infections.

Similar to mycobacteria, several molecules inducing autophagy have been recently shown to reduce *Salmonella* Typhimurium replication in HeLa cells (43, 44). This direct relationship between drugs, modulating autophagy and the outcome of bacterial infection, emphasizes the essential role of autophagy in the host response to intracellular bacteria and seems to support pharmacological modulation of host autophagy during infection. Unfortunately, as shown above, the situation seems more complex than the conclusion "increase of cellular autophagy favors bacterial clearance."

The fact that autophagy inducers seem to be helpful in the treatment of *Mycobacteria* or *Salmonella* infections, but in turn might facilitate infections by *Anaplasma*, *Coxiella*, *Yersinia*, or *Francisella*, requires not only to be highly cautious in the use of autophagy modulators to treat infectious diseases but also to monitor the infectious risk during the use of autophagy modulators. Some autophagy modulators are already in use (**Figure 1**, underlined). Rapamycin, metformin, and rilmenidine, all autophagy inducers, are drugs approved and prescribed to prevent rejection of kidney transplants, to treat type 2 diabetes, and to treat hypertension, respectively (3, 37). In contrast, chloroquine and hydroxychloroquine, which are now under clinical trials as autophagy inhibitors for the treatment of certain resistant cancers, are drugs prescribed to treat malaria (3, 37). Moreover, hydroxychloroquine combined with doxycycline is currently used to treat *Coxiella*-induced chronic Q fever endocarditis (45). Some of these approved drugs might thus show a therapeutical benefit in case of infection.

In summary, the study of autophagy regulation during bacterial infection certainly shows the existence of a critical balance between a host-protective "immune-related" induction of autophagy (xenophagy) and a host-deleterious "metabolic-related" induction of autophagy by invading bacteria for nutritional theft of host energy resources. Results of clinical trials using autophagy modulators and a more profound understanding of the role of autophagy during infection are thus needed to correctly use autophagy modulators in the fight against infectious diseases.

AUTHOR CONTRIBUTIONS

All authors listed, have made substantial, direct and intellectual contribution to the work, and approved it for publication.

FUNDING

Work in CB's laboratory is financed by the Institut Pasteur, the Institut Carnot-Pasteur MI, the French Region Ile de France (DIM Malinf), and the Infect-ERA project EUGENPATH (ANR-13-IFEC-0003-02) and the ANR-10-LABX-62-IBEID. PE is financed by the Fondation pour la Recherche Médicale (FRM) grant No. DEQ20120323697.

REFERENCES

1. Deretic V, Saitoh T, Akira S. Autophagy in infection, inflammation and immunity. *Nat Rev Immunol* (2013) **13**:722–37. doi:10.1038/nri3532
2. Huang J, Brumell JH. Bacteria-autophagy interplay: a battle for survival. *Nat Rev Microbiol* (2014) **12**:101–14. doi:10.1038/nrmicro3160
3. Rubinsztein DC, Codogno P, Levine B. Autophagy modulation as a potential therapeutic target for diverse diseases. *Nat Rev Drug Discov* (2012) **11**:709–30. doi:10.1038/nrd3802
4. Birmingham CL, Smith AC, Bakowski MA, Yoshimori T, Brumell JH. Autophagy controls *Salmonella* infection in response to damage to the *Salmonella*-containing vacuole. *J Biol Chem* (2006) **281**:11374–83. doi:10.1074/jbc.M509157200
5. Gutierrez MG, Master SS, Singh SB, Taylor GA, Colombo MI, Deretic V. Autophagy is a defense mechanism inhibiting BCG and *Mycobacterium tuberculosis* survival in infected macrophages. *Cell* (2004) **119**:753–66. doi:10.1016/j.cell.2004.11.038
6. Zhao Z, Fux B, Goodwin M, Dunay IR, Strong D, Miller BC, et al. Autophagosome-independent essential function for the autophagy protein Atg5 in cellular immunity to intracellular pathogens. *Cell Host Microbe* (2008) **4**:458–69. doi:10.1016/j.chom.2008.10.003
7. Py BF, Lipinski MM, Yuan J. Autophagy limits *Listeria monocytogenes* intracellular growth in the early phase of primary infection. *Autophagy* (2007) **3**:117–25. doi:10.4161/auto.3618
8. Nakagawa I, Amano A, Mizushima N, Yamamoto A, Yamaguchi H, Kamimoto T, et al. Autophagy defends cells against invading Group A *Streptococcus*. *Science* (2004) **306**:1037–40. doi:10.1126/science.1103966
9. Xu Y, Jagannath C, Liu X-D, Sharafkhaneh A, Kolodziejska KE, Eissa NT. Toll-like receptor 4 is a sensor for autophagy associated with innate immunity. *Immunity* (2007) **27**:135–44. doi:10.1016/j.immuni.2007.05.022
10. Travassos LH, Carneiro LAM, Ramjeet M, Hussey S, Kim YG, Magalhães JG, et al. Nod1 and Nod2 direct autophagy by recruiting ATG16L1 to the plasma membrane at the site of bacterial entry. *Nat Immunol* (2010) **11**:55–62. doi:10.1038/ni.1823
11. Sanjuan MA, Dillon CP, Tait SWG, Moshiach S, Dorsey F, Connell S, et al. Toll-like receptor signalling in macrophages links the autophagy pathway to phagocytosis. *Nature* (2007) **450**:1253–7. doi:10.1038/nature06421
12. Kageyama S, Omori H, Saitoh T, Sone T, Guan J-L, Akira S, et al. The LC3 recruitment mechanism is separate from Atg9L1-dependent membrane

13. Itakura E, Kishi-Itakura C, Mizushima N. The hairpin-type tail-anchored SNARE syntaxin 17 targets to autophagosomes for fusion with endosomes/lysosomes. *Cell* (2012) **151**:1256–69. doi:10.1016/j.cell.2012.11.001
14. Shin DM, Jeon BY, Lee HM, Jin HS, Yuk JM, Song CH, et al. *Mycobacterium tuberculosis* eis regulates autophagy, inflammation, and cell death through redox-dependent signaling. *PLoS Pathog* (2010) **6**:e1001230. doi:10.1371/journal.ppat.1001230
15. Tattoli I, Sorbara MT, Vuckovic D, Ling A, Soares F, Carneiro LAM, et al. Amino acid starvation induced by invasive bacterial pathogens triggers an innate host defense program. *Cell Host Microbe* (2012) **11**:563–75. doi:10.1016/j.chom.2012.04.012
16. Ogawa M, Yoshimori T, Suzuki T, Sagara H, Mizushima N, Sasakawa C. Escape of intracellular *Shigella* from autophagy. *Science* (2005) **307**:727–31. doi:10.1126/science.1106036
17. Mostowy S, Cossart P. Bacterial autophagy: restriction or promotion of bacterial replication? *Trends Cell Biol* (2012) **22**:283–91. doi:10.1016/j.tcb.2012.03.006
18. Steele S, Brunton J, Kawula T. The role of autophagy in intracellular pathogen nutrient acquisition. *Front Cell Infect Microbiol* (2015) **5**:51. doi:10.3389/fcimb.2015.00051
19. Niu H, Xiong Q, Yamamoto A, Hayashi-Nishino M, Rikihisa Y. Autophagosomes induced by a bacterial Beclin 1 binding protein facilitate obligatory intracellular infection. *Proc Natl Acad Sci U S A* (2012) **109**:20800–7. doi:10.1073/pnas.1218674109
20. Niu H, Yamaguchi M, Rikihisa Y. Subversion of cellular autophagy by *Anaplasma phagocytophilum*. *Cell Microbiol* (2008) **10**:593–605. doi:10.1111/j.1462-5822.2007.01068.x
21. Moreau K, Lacas-Gervais S, Fujita N, Sebbane F, Yoshimori T, Simonet M, et al. Autophagosomes can support *Yersinia pseudotuberculosis* replication in macrophages. *Cell Microbiol* (2010) **12**:1108–23. doi:10.1111/j.1462-5822.2010.01456.x
22. Pujol C, Klein KA, Romanov GA, Palmer LE, Cirota C, Zhao Z, et al. *Yersinia pestis* can reside in autophagosomes and avoid xenophagy in murine macrophages by preventing vacuole acidification. *Infect Immun* (2009) **77**:2251–61. doi:10.1128/IAI.00068-09
23. Gutierrez MG, Vázquez CL, Munafó DB, Zoppino FCM, Berón W, Rabinovitch M, et al. Autophagy induction favours the generation and

formation in the autophagic response against *Salmonella*. *Mol Biol Cell* (2011) **22**:2290–300. doi:10.1091/mbc.E10-11-0893 maturation of the *Coxiella*-replicative vacuoles. *Cell Microbiol* (2005) 7:981–93. doi:10.1111/j.1462-5822.2005.00527.x

24. Vázquez CL, Colombo MI. *Coxiella burnetii* modulates Beclin 1 and Bcl-2, preventing host cell apoptosis to generate a persistent bacterial infection. *Cell Death Differ* (2010) 17:421–38. doi:10.1038/cdd.2009.129

25. Berón W, Gutierrez MG, Rabinovitch M, Colombo MI. *Coxiella burnetii* localizes in a Rab7-labeled compartment with autophagic characteristics. *Infect Immun* (2002) 70:5816–21. doi:10.1128/IAI.70.10.5816-5821.2002

26. Steele S, Brunton J, Ziehr B, Taft-Benz S, Moorman N, Kawula T. *Francisella tularensis* harvests nutrients derived via ATG5-independent autophagy to support intracellular growth. *PLoS Pathog* (2013) 9:e1003562. doi:10.1371/journal.ppat.1003562

27. Starr T, Child R, Wehrly TD, Hansen B, Hwang S, López-Otin C, et al. Selective subversion of autophagy complexes facilitates completion of the *Brucella* intracellular cycle. *Cell Host Microbe* (2012) 11:33–45. doi:10.1016/j.chom.2011.12.002

28. Mestre MB, Fader CM, Sola C, Colombo MI. Alpha-hemolysin is required for the activation of the autophagic pathway in *Staphylococcus aureus*-infected cells. *Autophagy* (2010) 6:110–25. doi:10.4161/auto.6.1.10698

29. Mestre MB, Colombo MI. cAMP and EPAC are key players in the regulation of the signal transduction pathway involved in the α-hemolysin autophagic response. *PLoS Pathog* (2012) 8:e1002664. doi:10.1371/journal.ppat.1002664

30. Amer AO, Swanson MS. Autophagy is an immediate macrophage response to *Legionella pneumophila*. *Cell Microbiol* (2005) 7:765–78. doi:10.1111/j.1462-5822.2005.00509.x

31. Kagan JC, Roy CR. *Legionella* phagosomes intercept vesicular traffic from endoplasmic reticulum exit sites. *Nat Cell Biol* (2002) 4:945–54. doi:10.1038/ncb883

32. Khweek AA, Caution K, Akhter A, Abdulrahman BA, Tazi M, Hassan H, et al. A bacterial protein promotes the recognition of the *Legionella pneumophila* vacuole by autophagy. *Eur J Immunol* (2013) 43:1333–44. doi:10.1002/eji.201242835

33. Amer AO, Byrne BG, Swanson MS. Macrophages rapidly transfer pathogens from lipid raft vacuoles to autophagosomes. *Autophagy* (2005) 1:53–8. doi:10.4161/auto.1.1.1589

34. Choy A, Dancourt J, Mugo B, O'Connor TJ, Isberg RR, Melia TJ, et al. The *Legionella* effector RavZ inhibits host autophagy through irreversible Atg8 deconjugation. *Science* (2012) 338:1072–6. doi:10.1126/science.1227026

35. Rolando M, Escoll P, Nora T, Botti J, Boitez V, Bedia C, et al. *Legionella pneumophila* S1P-lyase targets host sphingolipid metabolism and restrains autophagy. *Proc Natl Acad Sci U S A* (2016) 113(7):1901–6. doi:10.1073/pnas.1522067113

suggested by human genetics. *Proc Natl Acad Sci U S A* (2015) 112:E4281–7. doi:10.1073/pnas.1512289112

45. Raoult D, Houpikian P, Tissot Dupont H, Riss JM, Arditi-Djiane J, Brouqui P. Treatment of Q fever endocarditis: comparison of 2 regimens containing doxycycline and ofloxacin or hydroxychloroquine. *Arch Intern Med* (1999) 159:167–73. doi:10.1001/archinte.159.2.167

36. Joshi AD, Swanson MS. Secrets of a successful pathogen: *Legionella* resistance to progression along the autophagic pathway. *Front Microbiol* (2011) 2:138. doi:10.3389/fmicb.2011.00138

37. Vakifahmetoglu-Norberg H, Xia H-G, Yuan J. Pharmacologic agents targeting autophagy. *J Clin Invest* (2015) 125:5–13. doi:10.1172/JCI73937

38. Siddiqi SH, Hawkins JE, Laszlo A. Interlaboratory drug susceptibility testing of *Mycobacterium tuberculosis* by a radiometric procedure and two conventional methods. *J Clin Microbiol* (1985) 22:919–23.

39. Stottmeier KD, Beam RE, Kubica GP. Determination of drug susceptibility of mycobacteria to pyrazinamide in 7H10 agar. *Am Rev Respir Dis* (1967) 96:1072–5.

40. Kim JJ, Lee HM, Shin DM, Kim W, Yuk JM, Jin HS, et al. Host cell autophagy activated by antibiotics is required for their effective antimycobacterial drug action. *Cell Host Microbe* (2012) 11:457–68. doi:10.1016/j.chom.2012.03.008

41. Zullo AJ, Lee S. Old antibiotics target TB with a new trick. *Cell Host Microbe* (2012) 11:419–20. doi:10.1016/j.chom.2012.05.002

42. Renna M, Schaffner C, Brown K, Shang S, Tamayo MH, Hegyi K, et al. Azithromycin blocks autophagy and may predispose cystic fibrosis patients to mycobacterial infection. *J Clin Invest* (2011) 121:3554–63. doi:10.1172/JCI46095

43. Conway KL, Kuballa P, Song J-H, Patel KK, Castoreno AB, Yilmaz OH, et al. Atg16l1 is required for autophagy in intestinal epithelial cells and protection of mice from *Salmonella* infection. *Gastroenterology* (2013) 145:1347–57. doi:10.1053/j.gastro.2013.08.035

44. Kuo S-Y, Castoreno AB, Aldrich LN, Lassen KG, Goel G, Dančík V, et al. Small-molecule enhancers of autophagy modulate cellular disease phenotypes

3

Calcium Oscillations in Pancreatic α-cells Rely on Noise and ATP-Driven Changes in Membrane Electrical Activity

Virginia González-Vélez[1], Anthony Piron[2,3] and Geneviève Dupont[3,4]*

[1] *Department Basic Sciences, Universidad Autónoma Metropolitana-Azcapotzalco, CDMX, Mèxico, Mexico,* [2] *ULB Center for Diabetes Research, Faculté de Médecine, Université libre de Bruxelles (ULB), Brussels, Belgium,* [3] *Interuniversity Institute of Bioinformatics (IB2), Brussels, Belgium,* [4] *Unit of Theoretical Chronobiology, Faculté des Sciences, Université libre de Bruxelles (ULB), Brussels, Belgium*

Edited by:
Jianhua Xing,
University of Pittsburgh, United States

Reviewed by:
Linford Briant,
University of Oxford, United Kingdom
Pei-Chun Chen,
National Cheng Kung University,
Taiwan

*Correspondence:
Geneviève Dupont
gdupont@ulb.ac.be

In pancreatic α-cells, intracellular Ca^{2+} ($[Ca^{2+}]_i$) acts as a trigger for secretion of glucagon, a hormone that plays a key role in blood glucose homeostasis. Intracellular Ca^{2+} dynamics in these cells are governed by the electrical activity of voltage-gated ion channels, among which ATP-sensitive K^+ (K_{ATP}) channels play a crucial role. In the majority of α-cells, the global Ca^{2+} response to lowering external glucose occurs in the form of oscillations that are much slower than electrical activity. These Ca^{2+} oscillations are highly variable as far as inter-spike intervals, shapes and amplitudes are concerned. Such observations suggest that Ca^{2+} dynamics in α-cells are much influenced by noise. Actually, each Ca^{2+} increase corresponds to multiple cycles of opening/closing of voltage gated Ca^{2+} channels that abruptly become silent, before the occurrence of another burst of activity a few tens of seconds later. The mechanism responsible for this intermittent activity is currently unknown. In this work, we used computational modeling to investigate the mechanism of cytosolic Ca^{2+} oscillations in α-cells. Given the limited population of K_{ATP} channels in this cell type, we hypothesized that the stochastic activity of these channels could play a key role in the sporadic character of the action potentials. To test this assumption, we extended a previously proposed model of the α-cells electrical activity (Diderichsen and Göpel, 2006) to take Ca^{2+} dynamics into account. Including molecular noise on the basis of a Langevin type description as well as realistic dynamics of opening and closing of K_{ATP} channels, we found that stochasticity at the level of the activity of this channel is on its own not able to produce Ca^{2+} oscillations with a time scale of a few tens of seconds. However, when taking into account the intimate relation between Ca^{2+} and ATP changes together with the intrinsic noise at the level of the K_{ATP} channels, simulations displayed Ca^{2+} oscillations that are compatible with experimental observations. We analyzed the detailed mechanism and used computational simulations to identify the factors that can affect Ca^{2+} oscillations in α-cells.

Keywords: computational model, ATP-sensitive potassium channels, action potential, Langevin equation, stochastic channel, plasma-membrane Ca^{2+} ATPase

Calcium Oscillations in Pancreatic α-cells Rely on Noise and ATP-Driven Changes in Membrane...

15

INTRODUCTION

Pancreatic islets respond to changes in blood glucose levels so that β-cells secrete insulin when blood glucose is elevated and α-cells secrete glucagon when it is low. Glucagon mobilizes glucose from the liver and when normoglycemia is reestablished, glucagon release from α-cells is suppressed. Extrinsic and intrinsic factors are involved in glucagon secretion (Briant et al., 2016; Wendt and Eliasson, 2020). Individuals with diabetes often show an impaired glucagon secretion that contributes to their hyperglycaemia (D'Alessio, 2011; Gilon, 2020). However, the detailed mechanism by which α-cells regulate glucagon secretion is not fully understood (Yu et al., 2019).

Pancreatic α-cells are electrically excitable and stimulation of glucagon secretion is secondary to repetitive action potential (AP) firing. In a low glucose medium, AP's occur with a frequency of ~1–3 Hz. Depolarization of the α-cell plasma membrane allows Ca^{2+} to enter through voltage-gated Ca^{2+} channels, which leads to the exocytosis of secretory granules of glucagon. In agreement with this mechanism, electrical stimulation of α-cells leads to an increase in cell membrane capacitance due to granule fusion, a well-known Ca^{2+} dependent process that precedes glucagon release (Barg et al., 2000; Voets, 2000; Göpel et al., 2004; González-Vélez et al., 2012).

Electrical activity in α-cells is thus accompanied by an increase in intracellular Ca^{2+} concentration ($[Ca^{2+}]_i$), which results from the activation of voltage-gated Ca^{2+} channels. Interestingly, this rise in $[Ca^{2+}]_i$ occurs in the form of oscillations with an average frequency of ~0.5 min^{-1}, which is much lower than that of the AP's. Parallel measurements of electrical activity and $[Ca^{2+}]_i$ revealed that each oscillation corresponds to a burst of AP's and that the amplitude of the Ca^{2+} increase correlates with the frequency of AP's (MacDonald et al., 2007; Quoix et al., 2009; Le Marchand and Piston, 2010, 2012; Zhang et al., 2013; Kellard et al., 2020). These bursts of electrical activity are separated by quiescent periods during which $[Ca^{2+}]_i$ is close to basal level. Ca^{2+} oscillations are observed in most α-cells, in low or high glucose medium, although they are much reduced in both amplitude and frequency in high glucose. A key characteristic of these oscillations is their irregularity. Their shape, frequency and amplitude are extremely variable, not only from one cell to another but also in the course of time for one individual cell (Kellard et al., 2020). In this study, we investigated the mechanism responsible for the intermittency of electrical activity and thus for the existence of slow, irregular Ca^{2+} oscillations.

Plasma membrane ATP-sensitive K^+ channels (K_{ATP} channels) play a key role in controlling α-cells electrical activity, although the details of this control are still actively debated (Gilon, 2020; Zhang et al., 2020). When the amplitude of this current is relatively limited, voltage-gated Na^+ and Ca^{2+} channels can indeed initiate an AP. The number of K_{ATP} channels simultaneously active is however surprisingly low, of the order of 10 (Rorsman et al., 2014). With such a low number of channels, it is expected that fluctuations of molecular origin would play a key role in the dynamical evolution of the K_{ATP} current and thus of the whole voltage and Ca^{2+} dynamics (Gonze et al., 2018). The low K_{ATP} channel activity in α-cells also results in a very high input resistance, meaning that small currents as those associated with openings of a few ion channels may have a drastic effect on membrane voltage and electrical activity (Rorsman et al., 2014). In agreement with this reasoning, noise-induced APs have been observed in α-cells and theoretically simulated (Diderichsen and Göpel, 2006). Here, we pushed this observation forward and investigated the possibility that fluctuations related to the small K_{ATP} current might be responsible for the intermittent character of electrical activity and hence for the noisy Ca^{2+} oscillations. This hypothesis, which is ideally investigated by mathematical modeling, holds with the observations that α-cells activity is highly variable even at a given external glucose concentration and that there is no "typical" α-cell signature (Kellard et al., 2020). Importantly, we here focus on the mechanism responsible for the existence of slow, irregular Ca^{2+} oscillations and not on the actively debated mechanism of regulation of glucagon secretion. Regulation of glucagon secretion indeed involves both intrinsic mechanisms and paracrine signals. Here, we only take into account intrinsic processes. Since intermittent electrical activity and slow $[Ca^{2+}]_i$ oscillations have also been observed in *isolated* α-cells (Salehi et al., 2006; Quoix et al., 2009; Tuduri et al., 2009; Le Marchand and Piston, 2010), we indeed reasoned that paracrine signaling is not essential for their existence although it affects their characteristics (Briant et al., 2018b; Kellard et al., 2020).

Mathematical models have been developed to study α-cells electrical activity and glucagon secretion. Diderichsen and Göpel (2006) used experimental data from patch clamp experiments on pancreatic α-cells located on the surface of intact mouse islets to develop an accurate model of plasma membrane electrical activity. This model was extended to include first Ca^{2+} dynamics and secretion (Watts and Sherman, 2014), and later glucagon-like peptide 1 (GLP-1) and adrenaline effects (Montefusco and Pedersen, 2015) as well as the α-cell heterogeneity by introducing realistic cell-to-cell variations in the values of the parameters (Montefusco et al., 2020). A functional identification of the islet cell types based on their electrophysiological characteristics allowed to improve the agreement between experiments and simulations of these models (Briant et al., 2017). Diderichsen and Göpel's model was also re-used by Grubelnik et al. (2020) to study the link between the deformities in mitochondrial ultrastructure observed in α-cells of type 2 diabetes mellitus mice and glucagon secretion. On the other hand, Fridlyand and Philipson (2012) adapted a model initially developed to describe β-cells dynamics to propose a detailed description of α-cells electrical activity, Ca^{2+} changes, metabolism as well as paracrine and endocrine regulations. The effect of paracrine signaling (Watts et al., 2016; Briant et al., 2018b) on α-cells electrical activity was also investigated in models of pancreatic islets including β- and δ-cells. None of these studies have addressed the possible impact of the low number of K_{ATP} channels, nor the question of the mechanistic origin of cytosolic Ca^{2+} oscillations in α-cells.

The present study is based on the original Diderichsen and Göpel's model of α-cells electrical activity. This core model is sequentially extended to take into account Ca^{2+} dynamics and random fluctuations of the K_{ATP} current via the Langevin formalism. We found that stochasticity in this current can indeed induce intermittent electrical activity and Ca^{2+} oscillations, but only for unrealistically small values of the opening and closing rates of this K^+ channel. This theoretical prediction motivated us to further extend the model to take into account the variations of ATP concentrations that result from the activity of the plasma membrane Ca^{2+} ATPases. The resulting changes in ATP, which have been observed experimentally (Li et al., 2015), indeed slow down the dynamics of the K_{ATP} current and allow for intermittent electrical activity and Ca^{2+} oscillations resembling those observed experimentally. Finally, we used the model to investigate the sensitivity of calcium and electrical activities to key factors such as the number of K_{ATP} channels or the rate of Ca^{2+} pumping.

MODEL DESCRIPTION

We base our model (**Figure 1**) on the mathematical description of the electrical activity of pancreatic α-cells proposed by Diderichsen and Göpel (2006), which was carefully calibrated on experimental data (see also Briant et al., 2017). The original model incorporates an ATP-sensitive K^+ current (I_{KATP}) that couples the level of external glucose to the electrical properties of the α-cell plasma membrane, through the sensitivity of the intracellular ATP/ADP ratio to external glucose concentration. It also describes a voltage-gated Na^+ current (I_{Na}), a delayed rectifying (I_{KDR}) and a A-type K^+ current (I_{KA}), an unspecific leak current (I_{leak}) and a L-type (I_{CaL}) and a T-type Ca^{2+} current (I_{CaT}). We also consider an additional type of Ca^{2+} current, because N-type (Quesada et al., 2008; González-Vélez et al., 2010) or P/Q type (Rorsman et al., 2014) Ca^{2+} currents have been shown to play a key role in glucagon secretion in rodents.

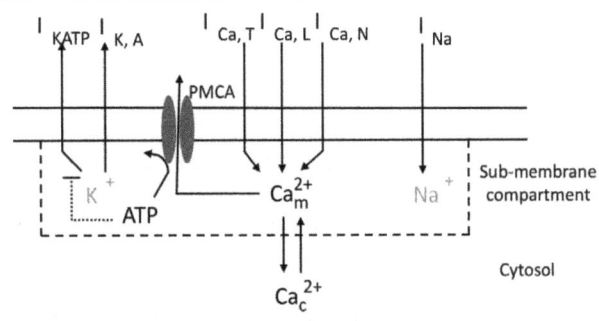

FIGURE 1 | Schematic representation of ionic currents, Ca^{2+} fluxes and ATP consumption in pancreatic α-cells. Transmembrane currents are: I_{KATP}, the ATP-sensitive K^+ current; I_{KA}, the high-voltage activated K^+ current; I_{CaL}, the high-voltage activated L-type Ca^{2+} current; I_{CaT}, the low-voltage activated T-type Ca^{2+} current; I_{CaN}, the high-voltage activated N-type Ca^{2+} current; I_{Na}, the voltage-gated Na^+ current. The model describes the evolution of Ca^{2+} concentration in a fictitious sub-membrane compartment (Ca_m^{2+}) and in the cytosol (Ca_c^{2+}). Exchanges between these 2 compartments occur by diffusion. Ca^{2+} is transported from the sub-membrane compartment into the extracellular medium by the ATP-consuming PMCA. ATP increases in the cytosol at a rate that depends on the concentration of extracellular glucose. ATP inhibits the ATP-sensitive K^+ current (I_{KATP}). K^+ and Na^+ concentrations (in gray) are not explicitly considered in the model.

Here, we arbitrarily chose to incorporate a N-type Ca^{2+} current (I_{CaN}). Ca^{2+} concentrations are described differently just below the plasma membrane and in the cytoplasm. Ca^{2+} entry via the L-, T- and N-type Ca^{2+} channels increases Ca^{2+} concentration in a hypothetical sub-plasma membrane compartment. From this compartment, Ca^{2+} diffuses into the cytoplasm or leaves the cell via the plasma membrane Ca^{2+} ATPase (PMCA). Ca^{2+} efflux from the cell is thus accompanied by the hydrolysis of ATP into ADP. The resulting decrease in ATP concentration provokes an increase in the K_{ATP} conductance, thereby providing a feedback from Ca^{2+} changes on the electrical properties of the membrane. The main features of the model are described here below. Additional information about the equations and the values of the parameters can be found in the **Supplementary Data** and in the original study of Diderichsen and Göpel (2006).

Model of Plasma Membrane Electrical Activity

Electrical activity is described by the following differential equation:

$$\frac{dV}{dt} = -(I_{CaT} + I_{CaL} + I_{CaN} + I_{Na} + I_{KA} + I_{KDR}$$
$$+ I_{KATP} + I_{leak})/C_m \qquad (1)$$

where V is the membrane voltage and C_m is the membrane capacitance set to 5 pF (Diderichsen and Göpel, 2006). For all currents, except for I_{KATP} (see section " ATP-Sensitive K^+ Current and ATP Evolution" below), we use the model proposed by Diderichsen and Göpel (2006) that is based on a Hodgkin-Huxley type description of ion channels and where the activation and inactivation functions have been fitted to experimental data. The high-voltage gated N-type Ca^{2+} current is described as in other studies (Csercsik et al., 2010; González-Vélez et al., 2010).

Ca^{2+} Dynamics

The evolution of sub-membrane Ca^{2+} concentration, noted Ca_m, is modeled as

$$\frac{dCa_m}{dt} = f_r \cdot \frac{-I_{CaT} - I_{CaL} - I_{CaN}}{2 \cdot F \cdot Vol/20} \cdot 10^6 +$$
$$V_b - V_p \frac{Ca_m^2}{Ca_m^2 + K_m^2} \cdot f_1(ATP) - \gamma(Ca_m - Ca_c) \qquad (2)$$

where f_r is the fraction of unbuffered Ca^{2+}, F the Faraday constant and Vol, the volume of an α-cell. It is assumed that the sub-membrane shell represents 1/20 of the total volume of the cytoplasm. The factor 10^6 allows to get concentrations expressed in μM. V_b stands for a basal rate of Ca^{2+} entry, which ensures a ~0.1 μM Ca^{2+} concentration when voltage-gated Ca^{2+} channels are closed. The third term represents Ca^{2+} pumping out of the cell through PMCA, which can be modeled by a Hill function with $n_H = 2$ (Dupont et al., 2016). This active transport is accompanied by ATP hydrolysis. Thus,

$$f_1(ATP) = \frac{ATP}{ATP + K_e} \qquad (3)$$

with K_e being the Michaelis-Menten constant of ATP hydrolysis by the PMCA.

Finally, the last term of Eq. 2 represents Ca^{2+} diffusion into the bulk of the cytoplasm. Given a Ca^{2+} diffusion coefficient of 13 $\mu m^2/s$ (Allbritton et al., 1992) and a diameter of 8 μm for an α-cell (Zimny and Blackard, 1975) in which the nucleus occupies ~70% of the cytoplasm, γ can be estimated to ~0.01 ms^{-1} (see **Supplementary Material**).

The evolution of intracellular ATP follows

$$\frac{dATP}{dt} = V_{gly}\frac{Glu}{Glu + K_{glu}} - \frac{V_p}{2}\frac{Ca_m^2}{Ca_m^2 + K_m^2} \cdot f_1(ATP) - k \cdot ATP \tag{4}$$

The first term of Eq. 4 is a phenomenological description of glycolysis in the form of Michaelis-Menten rate of ATP production from glucose. The second term represents ATP consumption by plasma membrane PMCA and is thus the same as in Eq. 2 except for the factor 2 that takes into account that one mole of ATP is hydrolyzed for two moles of Ca^{2+} transported. Rate constant k describes the consumption of ATP by other intracellular processes.

In agreement with Eq. 1, the evolution of cytoplasmic Ca^{2+} concentration is given by

$$\frac{dCa_c}{dt} = \frac{\gamma}{19}(Ca_m - Ca_c) \tag{5}$$

ATP-Sensitive K$^+$ Current and ATP Evolution

In the original model (Diderichsen and Göpel, 2006), the ATP-sensitive K$^+$ current is modeled as an ohmic ionic current with a constant conductance, i.e.,

$$I_{KATP} = g_{KATP}(V - V_K) \tag{6}$$

where V_K is the reversal potential for currents carried by potassium. The channel conductance, g_{KATP}, was considered as a constant parameter and an increase in the extracellular glucose concentration was simulated by a decrease in the value of this parameter.

The main goal of the present study is to assess the effect of noise on the K$_{ATP}$ current that arises in α-cells due to the small number of such channels. g_{KATP} is thus not considered as a constant parameter anymore. Instead, we stochastically simulate opening and closing of these channels on the basis of the Langevin formalism. These channels can flicker between an open and a closed state, with opening and closing rate constants denoted α and β, respectively. Considering a noise term, the deterministic evolution equation for the fraction of open channels (s) becomes:

$$\frac{ds}{dt} = \alpha(1 - s) \cdot f_2(ATP) - \beta s + \sqrt{\sigma}\xi(t) \tag{7}$$

where $\xi(t)$ is a random function of time. The last term is a noise function with 0 mean and

$$\sigma = \frac{\alpha(1 - s) \cdot f_2(ATP) + \beta s}{N_{KATP}} \tag{8}$$

with N_{KATP} the number of potentially openable K$_{ATP}$ channels (Fall et al., 2005).

$f_2(ATP)$ takes into account that K$_{ATP}$ channels are reversibly inhibited by ATP (Enkvetchakul et al., 2001). We chose the simple expression:

$$f_2(ATP) = \frac{K_{inh}}{ATP + K_{inh}} \tag{9}$$

where K_{inh} is a constant representing the concentration of ATP leading to half-maximal inhibition.

When taking into account the stochasticity of the K$_{ATP}$ channels, the g_{KATP} conductance appearing in Eq. 6 is now computed as

$$g_{KATP} = g_{KATP}^s \cdot s \cdot N_{KATP} \tag{10}$$

with g_{KATP}^s standing for the unitary conductance of a single ATP-sensitive K$^+$ channel.

Together with the equations of the original model (Diderichsen and Göpel, 2006) listed in the **Supplementary Data**, Eqs 1–10 constitute our computational model that has been integrated in Matlab, using an Euler integration scheme with $\Delta t = 0.012$ ms. Bifurcation diagrams have been established using the AUTO package of XPPAUT (Ermentrout, 2002).

Outline of the Modeling Approach

As explained in detail in the section "Results," we consider models of increasing complexity to investigate the possible impact of the fluctuations in the K$_{ATP}$ conductance on the existence of Ca^{2+} oscillations in α-cells.

Model 1 is the model proposed by Diderichsen and Göpel (2006), including N-type Ca^{2+} channels and two additional variables: subplasmalemmal Ca^{2+} (Ca_m, Eq. 2 with $f_1 = 1$) and cytosolic Ca^{2+} (Ca_c, Eq. 5). It is thus defined by Eqs 1, 2, and 5.

Model 2 moreover includes stochasticity in the K$_{ATP}$ conductance through Eqs 7, 8 and 10 that are considered in addition to the equations defining Model 1. f_1 and f_2 are taken equal to 1.

Model 3 allows to investigate the effect of the Ca^{2+}-induced variations in ATP concentration. It is similar to Model 2, except that it includes Eq. 4 to describe the evolution of [ATP]. Changes in the concentration of this nucleotide impact on the evolutions of subplasmalemmal Ca^{2+} via f_1 that is now given by Eq. 3 and of the K$_{ATP}$ channel via f_2 that is now given by Eq. 9.

RESULTS

Calcium Changes Induced by Electrical Activity

When α-cells are electrically active, [Ca^{2+}]$_i$ first rises just beneath the plasma membrane since Ca^{2+} is entering through voltage-gated channels located in this membrane. Sub-plasmalemmal Ca^{2+} is immediately buffered and diffuses in the cytosol where Ca^{2+} concentration is low at rest (~100 nM). This attenuates Ca^{2+} increases below the membrane and transmits signaling to the rest of the cell. Additionally, sub-membrane Ca^{2+} is pumped out of the cell by Ca^{2+} ATPases. *Model 1* takes these fluxes into account and combines a description of α-cell electrical activity with equations for the evolution of subplasmalemmal (Ca_m, Eq. 2) and cytosolic Ca^{2+} (Ca_c, Eq. 5). Simulations of *Model 1* show that each AP generates one spike of Ca^{2+} in the subplasmalemmal space (**Figures 2A,B**). Considering that this fictitious compartment occupies 1/20 of the α-cell volume, corresponding to a shell thickness of $\sim$100 nm, Ca^{2+} increases up to a level between 0.5 and 1 μM depending on the value of g_{KATP}. A supra-threshold increase in g_{KATP} induces repetitive changes in Ca_m, whose frequency and amplitude are fixed by the value of this conductance. Because of the relative slowness of diffusion, cytosolic Ca^{2+} can only increase during repetitive AP's (**Figure 2C**). When AP's are sustained long enough, cytosolic Ca^{2+} remains nearly constant at a steady level (not shown). Ca^{2+} concentration in the cytosol does not exceed 0.4 μM. In this model, repetitive electrical activity always leads to a sustained increase in cytosolic Ca^{2+}, which does not correspond to the slow Ca^{2+} oscillations observed experimentally in most α-cells.

The main characteristics of Ca^{2+} dynamics for different values of the K_{ATP} conductance are visible in the bifurcation diagram shown in **Figure 2D**. The shape of this diagram is most easily understood by looking at the companion bifurcation diagram showing how electrical activity depends on g_{KATP} (**Figure 2E**). Since Ca^{2+} concentration does not feedback on the cell electrical activity in *Model 1*, Ca^{2+} changes can be seen as a simple output of the latter activity. Starting from large values of the K_{ATP} conductance at which cells are hyperpolarized (right part of the diagram), a decrease in this conductance leads to depolarization and electrical firing, as observed in experiments upon the addition of the K_{ATP} channel blocker tolbutamide. A further decline in the K_{ATP} conductance abolishes electrical activity and the cell remains in a constantly depolarized state. This decline in activity allows the model to reproduce the observation that α-cells are electrically inactive in the presence of large concentrations of external glucose without considering paracrine signaling.

The bifurcation analysis allows to uncover the existence of a range of values of g_{KATP} for which multiple stable solutions can co-exist. This implies that different steady state solutions can be reached depending on the initial conditions. From 0.2265 to 0.2919 nS (blue region in **Figures 2D,E**), the system can potentially be in three states: a stable hyperpolarized state, an intermediate stable state ($\sim$ -35 mV) or an electrically active state with large amplitude AP's. However, the intermediate state, although stable from a physical point of view, is most probably of little physiological significance, as its basin of attraction is very limited. This implies that this state will rarely be reached, and that if reached, fluctuations of internal or external origin would push the system from this intermediate state to the hyperpolarized one or to the limit cycle corresponding to repetitive AP's. For larger values of g_{KATP} (from 0.2919 to 0.3715 nS, green region in **Figures 2D,E**), the labile, slightly polarized state coexist with the stable hyperpolarized one.

Figure 2C shows that this simple model predicts that when cells are in a stationary regime of repetitive AP's, cytosolic Ca^{2+} tends to reach a steady state, although very small changes of concentration occur at each AP ($\sim$20 nM). Such rapid and small changes are not expected to be visible experimentally.

In conclusion, slow Ca^{2+} oscillations resulting from intermittent electrical activity as observed in α-cells cannot result from repetitive AP's controlled by a constant K_{ATP} conductance.

Randomness in the K_{ATP} Conductance

During electrical activity, the number of simultaneously active K_{ATP} channels is very small, of the order of 10 (Rorsman et al., 2014). Thus, the fluctuations related to internal noise cannot be neglected and the deterministic description of their contribution to membrane current (Eq. 6) must be replaced by a stochastic description. We thus considered the transitions of the channel between an open and a closed state, with a noise on this process as described above (*Model 2* in section "Model of Plasma Membrane Electrical Activity"). Rate constants of channel opening and closing (α and β in Eq. 7) estimated from dwell time distributions are of the order of tenths of ms (Enkvetchakul et al., 2001). Assuming a unitary conductance (g^s_{KATP}) of 41 pS (Bokvist et al., 1999; Khan et al., 2001), the total number of K_{ATP} channels in the simulated cell needs to be around 60 to get an average of 5–10 simultaneously open channels during electrical activity, which corresponds to a global K_{ATP} conductance in the range 0.2–0.25 nS. This number fits in the large range of values estimated experimentally (Huang et al., 2011).

Simulations including noise at the level of K_{ATP} current display AP's that are variable in amplitude and frequency. When the average value of g_{KATP} is in the oscillatory range determined by the bifurcation diagram (**Figures 2D,E**), the simulated cell is electrically active (average cellular g_{KATP} = 0.14 nS in **Figures 3E–H**), while the cell is in a fluctuating hyperpolarized state when the average value of g_{KATP} corresponds to a stable steady state of the bifurcation diagram (average cellular g_{KATP} = 0.47 nS in **Figures 3A–D**). Model simulations never result in Ca^{2+} oscillations whatever the values taken for the unitary conductance of the K_{ATP} channel (g^s_{KATP}) and the total number of such channels (N_{KATP}) considered in the simulations. However, we observed that Ca^{2+} oscillations resembling experimental observations could be obtained in the simulations when ascribing to the rate constants of opening and closing of the K_{ATP} channels (parameters α and β) values at least 500 smaller than those reported from experiments (**Figures 3I–L**). In this case, an intermittent electrical activity generates bursts of AP's, with each AP leading to a sharp Ca^{2+} increase in sub-membrane Ca^{2+}. Sub-membrane Ca^{2+} diffuses in the cytoplasm, and because diffusion is slow, one burst of electrical activity involving multiple AP's leads to one Ca^{2+} peak in the cytoplasm. Simulations shown in panels I–K display strong resemblance with experimental observations (MacDonald et al., 2007; Quoix et al., 2009; Le Marchand and Piston, 2010, 2012; Zhang et al., 2013; Kellard et al., 2020). The correlation between the cytosolic Ca^{2+} spike and the period of electrical activity has been shown in isolated mouse α-cells (see for example Quoix et al., 2009) and in α-cells from intact islets (Kellard et al., 2020).

From a mechanistic point of view, when the rate constants of opening and closing of the K_{ATP} channels are small, K_{ATP} current variations last long enough to induce a change in the electrical properties of the membrane: when the cell is hyperpolarized,

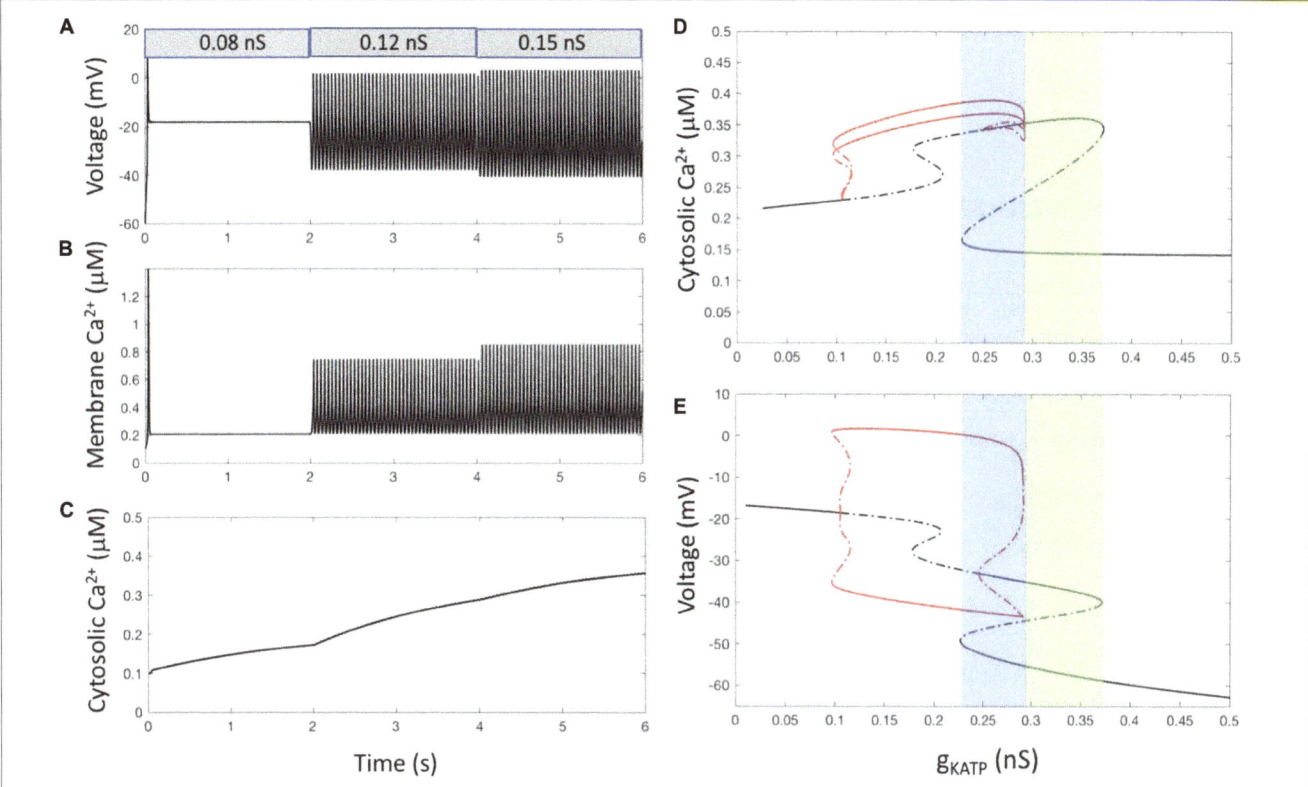

FIGURE 2 | α-cell Ca^{2+} dynamics resulting from electrical activity. **(A)** Membrane voltage for 3 values of g_{KATP}, simulating a decrease in external glucose. From $t = 0$ to 2 s, g_{KATP} = 0.08 nS and g_{leak} = 0.2 nS. From $t = 2$ to 4 s, g_{KATP} = 0.12 nS and g_{leak} = 0.2 nS. For $t > 4$ s, g_{KATP} = 0.15 nS and g_{leak} = 0.13 nS. The decrease in g_{leak} that accompanies the increase in g_{KATP} allows the model to reproduce the experimentally observed lower maximum voltage reached during repetitive AP's at high glucose (Montefusco and Pedersen, 2015). **(B)** Corresponding changes in [Ca^{2+}] in the sub-plasmalemmal compartment (Ca_m). **(C)** Corresponding changes in [Ca^{2+}] in the cytosol (Ca_c). Because diffusion is much slower than electrical activity, cytosolic concentration does not follow the Ca^{2+} spikes occurring just below the membrane. **(D,E)** Bifurcation diagrams showing the evolution of Ca_c **(D)** and voltage **(E)** when changing g_{KATP}. In both panels, black lines indicate steady states, while red lines show the minimum and maximum values reached during limit cycle oscillations. Plain lines indicate stable solutions and dashed lines instable ones. Bifurcation diagrams have been established using the AUTO package of xppaut (Ermentrout, 2002). For all panels, equations are those corresponding to *Model 1* (see section "Outline of the Modeling Approach") with the default values of parameters listed in **Supplementary Tables 1, 2** of the **Supplementary Material**, except for the indicated values of g_{KATP} and g_{leak}. In panels **(D,E)**, g_{leak} = 0.2 nS.

a random decrease in g_{KATP} of sufficient duration leads to a cell membrane depolarization that triggers electrical activity and Ca^{2+} entry. Similarly, when the cell membrane is electrically active, a random increase in g_{KATP} of sufficient duration leads to a decrease in V that brings the cell in a resting state. With the experimentally reported values of channel opening and closing rates, the random changes in g_{KATP} are too fast to induce AP's. When simulating a stepwise decrease in g_{KATP} from 0.47 to 0.2 nS in the absence of noise, the change in g_{KATP} must last at least 110 ms to induce electrical activity (not shown). This agrees with the numerical observation that if the fluctuation-driven changes of g_{KATP} have a characteristic time of a few ms as measured experimentally (Enkvetchakul et al., 2001), they will not induce AP's (**Figures 2A–C**).

Thus, simulations predict that Ca^{2+} oscillations may in principle arise from noise-induced changes in the cell K$_{ATP}$ current. However, the time scale of these changes must be much slower than that of the intrinsic dynamics of the voltage-gated ionic channels generating the AP's. This does not correspond to the reported rates of opening and closing of the K$_{ATP}$ channels

and raises the possibility that some physiological process drives slow changes in the opening of the K$_{ATP}$ channels.

Ca^{2+}-Driven [ATP] Variations Are Responsible for Slow Changes in Electrical Activity

K$_{ATP}$ channels are regulated by variations of the intracellular ATP concentration (Bokvist et al., 1999). On the other hand, α-cells show oscillations in Ca^{2+} and ATP submembrane concentrations when observed in constant hypoglycemic conditions (Li et al., 2015). These oscillations are in opposite phase, most probably because Ca^{2+} transport out of the cell is an ATP-consuming process. Here, we investigate the hypothesis that even at constant external glucose, ATP/ADP changes may trigger changes in membrane electrical activity and Ca^{2+} entry that would thus be responsible for the observed, slow cytosolic Ca^{2+} oscillations. To this end, we consider [ATP] as a variable in *Model 3*, as well as its relationship with Ca^{2+} dynamics and its inhibitory effect on K$_{ATP}$ channels conductance. For sake of simplicity,

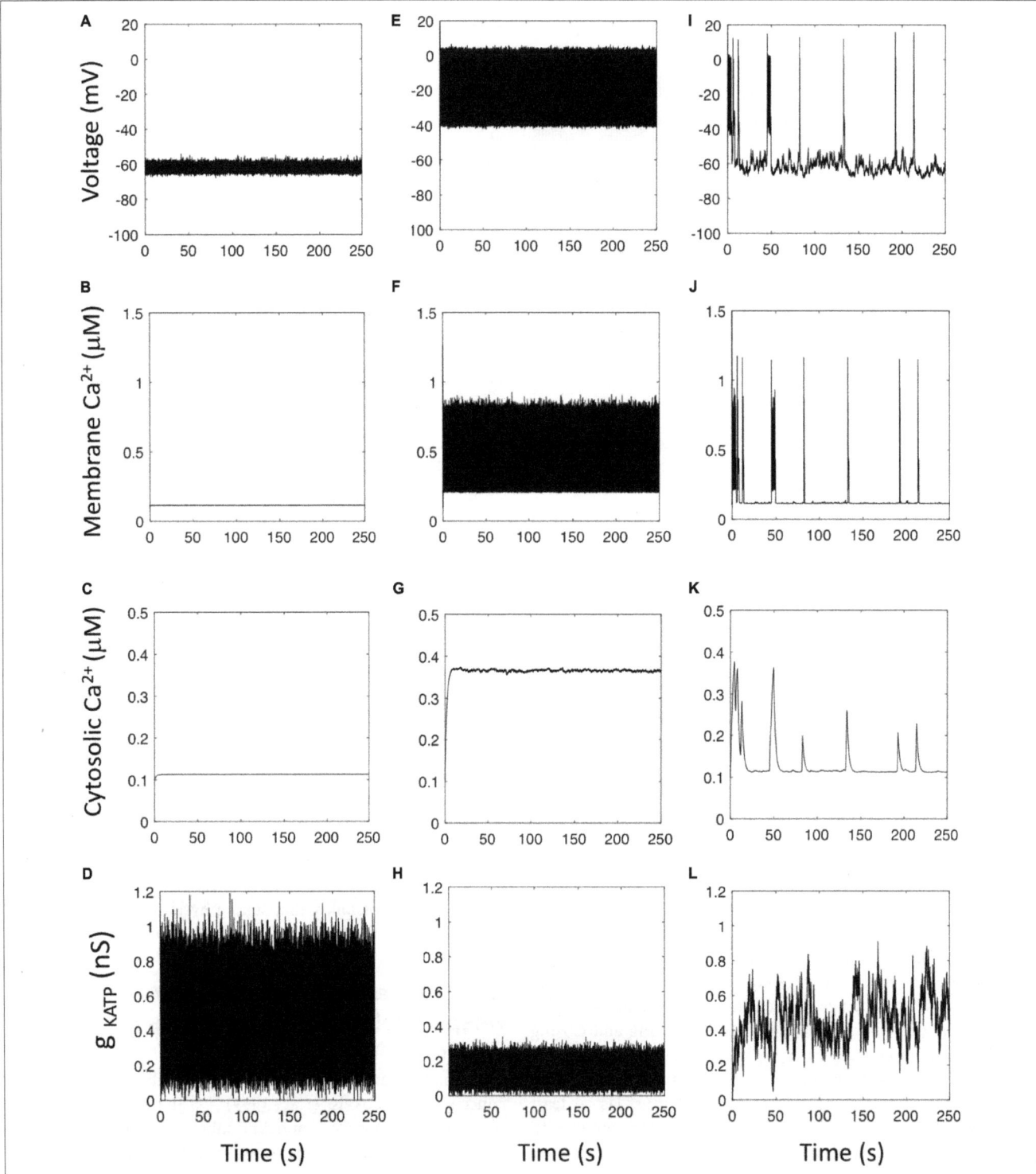

FIGURE 3 | Effect of stochasticity in the K_{ATP} current. Panels show the evolution of the main variables of the model when the conductance of the K_{ATP} channel is described by a Langevin equation (Eqs 7 and 8 with $f_2 = 1$) to take into account the randomness due to the small number of such channels in α-cells. **(A–D)** Rate constant of opening of the K_{ATP} channel, $\alpha = 0.1$ ms^{-1}, rate constant of closing of the K_{ATP} channel, $\beta = 0.4$ ms^{-1}; unitary conductance of the K_{ATP} channel$g^s_{KATP} = 0.041$ nS. The cell is in a quiet, hyperpolarized state despite the large variations of the total cell g_{KATP} conductance. **(E–H)** When the unitary conductance of the K_{ATP} channel equals 0.012 nS, the cell is in a depolarized state resulting in a constantly high Ca^{2+} concentration. **(I–L)** When considering slow rates of opening and closing of the K_{ATP} channel, $\alpha = 1\ 10^{-4}$ ms^{-1} and $\beta = 4\ 10^{-4}$ ms^{-1}, simulations predict the occurrence of intermittent electrical activity leading to irregular Ca^{2+} spikes in the cytosol. Equations are those corresponding to Model 3 (see section "Outline of the Modeling Approach") with the default values of parameters listed in the **Supplementary Material**, except for the indicated values of g^s_{KATP} and $g_{leak} = 0.2$ nS.

we consider that the conductance of these channels depends on [ATP] via an inhibitory function of the Michaelis-Menten type (Eq. 9). Keeping realistic values for the rates of opening and closing of the K_{ATP} channels (parameters α and β), simulations show highly variable Ca^{2+} oscillations that result from intermittent electrical activity (**Figure 4**). Each burst of electrical activity triggers a massive entry of Ca^{2+} in the subplasmalemmal compartment (Ca_m, **Figure 4B**), which then invades the cytoplasm (**Figure 4C**). Because PMCA are fully active, ATP is consumed, and the evolution of its concentration is the mirror image of that of cytosolic Ca^{2+} (**Figure 4D**). This reduced level of ATP favors a large K_{ATP} current, which in turn reduces electrical activity and Ca^{2+} entry, thus allowing ATP to rise again.

Although fluctuations of the K_{ATP} conductance are very rapid (**Figure 4E**), g_{KATP} is in average larger when [ATP] is low (periods of activity) than when [ATP] is high (quiescent periods) because of the inhibition of the K_{ATP} current by ATP. This is visible when computing the moving average of g_{KATP} (**Figure 4F**). The trend in g_{KATP} evolution indeed correlates with that of ATP. Such a trend in the moving average of g_{KATP} does not appear when ATP is considered as a constant (*Model 2*, not shown). As a consequence

of these trends, large, unlikely perturbations are necessary to switch from an inactive to an active period and *vice-versa*. Thus, the mutual interaction between ATP dynamics and electrical activity that occurs through the Ca^{2+} changes creates trends in the changes in electrical activity allowing for intermittent activity, despite the rapid fluctuations in g_{KATP}.

The cytosolic Ca^{2+} oscillations shown in **Figure 4C** have widely different durations, from ~15 s to ~2 min. Because of stochasticity, these values are different from one simulation to the other. Given that glucagon secretion is triggered by Ca^{2+} increases above a certain threshold (González-Vélez et al., 2012), it is interesting to investigate what controls the ratio of active *versus* inactive periods. As randomness is required to initiate changes from an active to a quiescent period and *vice-versa*, it can be expected that the number of K_{ATP} channels will play a key role in controlling the number of transitions and thus the activity ratio. To investigate the effect of this factor, we performed simulations with different values of N_{KATP} and computed the fraction of time during which cells exhibit electrical activity (**Figure 5C**) as well as the average number of Ca^{2+} spikes during a 500 s simulation (**Figure 5E**). In the simulations, changes in the

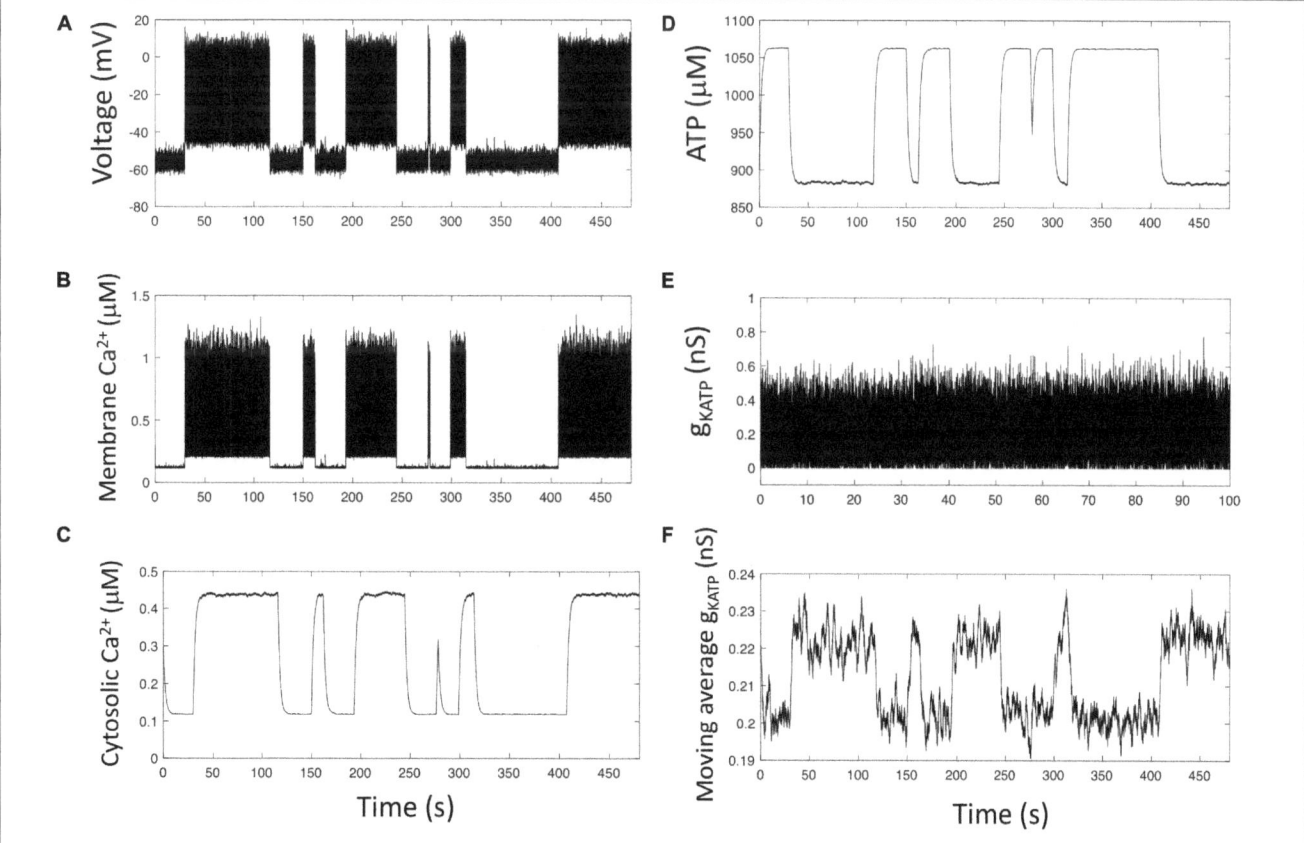

FIGURE 4 | Simulations predict that changes in membrane ATP and stochasticity in the K_{ATP} current generate slow Ca^{2+} oscillations. **(A–E)** Numerical simulations of *Model 3* (see section "Outline of the Modeling Approach") that describes the opening and closing of the K_{ATP} channel by a Langevin stochastic equation and takes into account the consumption of ATP by PMCA as well as the inhibition of K_{ATP} channels opening by ATP. **(F)** The moving average of g_{KATP} allows to visualize that, although highly random, g_{KATP} tends to increase when ATP concentration is low and to decrease when ATP concentration is large. There are thus long-lasting trends in g_{KATP} changes that can induce robust changes in α-cells electrical activity. Values of g_{KATP} are averaged on a 200 s period.

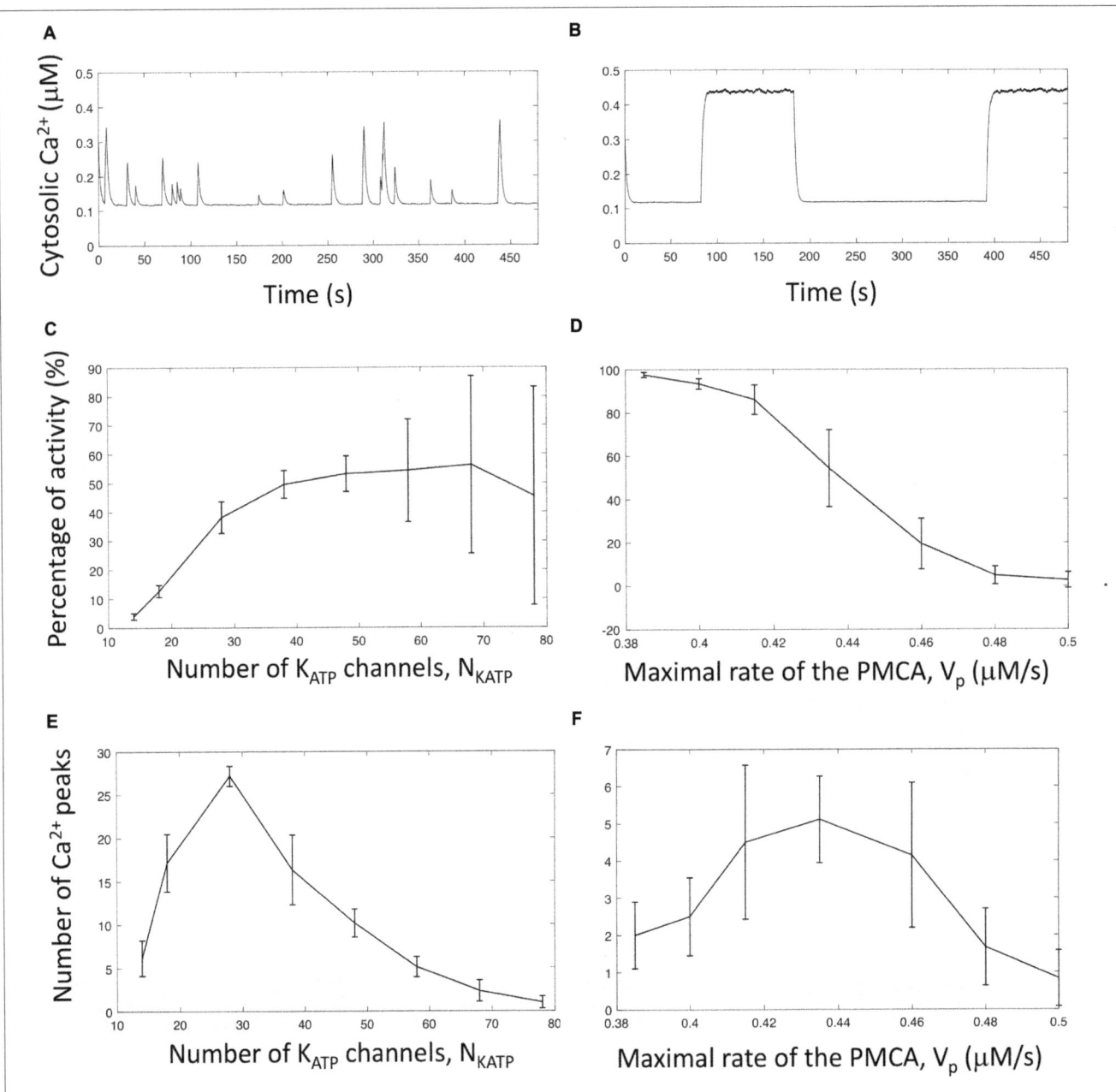

FIGURE 5 | Effect of the number of K_{ATP} channels and rate of PMCA on the pattern of Ca^{2+} oscillations in α-cells. **(A)** Time series of cytosolic Ca^{2+} for $N_{KATP} = 14$ and $g_{KATP}^s = 0.170$ nS. The maximal global cellular K_{ATP} conductance ($g_{KATP}^s \cdot N_{KATP}$) is the same as in **Figure 4**, but the number of participating channels is smaller. **(B)** Time series of cytosolic Ca^{2+} for $N_{KATP} = 68$ and $g_{KATP}^s = 0.03497$ nS. The maximal global cellular K_{ATP} conductance ($g_{KATP}^s \cdot N_{KATP}$) is the same as in **Figure 4**, but the number of participating channels is larger. **(C)** Normalized duration of activity as a function of N_{KATP}. **(D)** Normalized duration of activity as a function of the maximal velocity of the PMCA (V_p). **(E)** Number of cytosolic Ca^{2+} peaks in a 500 s period as a function of N_{KATP}. **(F)** Number of cytosolic Ca^{2+} peaks in a 500 s period as a function of the maximal velocity of the PMCA (V_p). Panels **(C,D)** show the times during which cytosolic Ca^{2+} exceeds 0.3 μM, divided by the total simulation time (500 s). These ratios are expressed in percentages. Six independent simulations have been run for $N_{KATP} \leq 58$, and 9 for $N_{KATP} > 58$. For panels **(E,F)**, a peak is counted when Ca_c crosses 0.3 μM with a positive slope. Six independent simulations have been run for all values of V_p.

values of N_{KATP} were accompanied by changes in the single channel unitary conductance (g_{KATP}^s) in such a way that the product $N_{KATP} \cdot g_{KATP}^s$ remains constant. Given this constraint, the global cell K_{ATP} conductance remains the same for all simulations, which allowed us to only investigate the effect of changes in randomness.

As visible in **Figure 5A**, when the number of channels is very small ($N_{KATP} = 14$), the cell membrane is most of the time hyperpolarized and infrequent, short-duration and low-amplitudes Ca^{2+} spikes occur. If N_{KATP} is smaller than 10, the cell is always in a resting state, although the average value of the cell K_{ATP} conductance (0.21 nS) is in the oscillatory domain (see **Figure 2D**). Fluctuations in g_{KATP} are indeed so important that random depolarizations of the cell membrane are never long enough to initiate an AP. From N_{KATP} = 10, infrequent Ca^{2+} spikes start to occur. They are very short in duration and amplitude because, once initiated, they are rapidly aborted by a fluctuation that repolarizes the cell. As a consequence, average Ca^{2+} and ATP concentrations are near their resting levels. These concentrations respectively increase and decrease when considering more channels in the simulations, i.e., when randomness is less pronounced. This is due to an increase in both the number of Ca^{2+} peaks (i.e., in the number of bursts of electrical activity, **Figure 5E**) and the duration of the peaks. Both changes are due to a possibly longer effect of random changes in g_{KATP} allowing to initiate changes in electrical activity. When N_{KATP} gets still larger, simulations exhibit a small number of Ca^{2+} spikes of very long duration (**Figure 5B**). Fluctuations indeed decrease and once in a state, active or resting, the cell tends to remain in this state. In consequence, the number of spikes on a 500 s period becomes smaller (**Figure 5E**), and the average time of activity larger (**Figure 5C**). It should be emphasized that for all simulations presented in **Figures 5C** and **E**, the average value of g_{KATP} remains approximatively the same (~0.21 nS). In the deterministic analysis, this value corresponds to repetitive AP's (**Figure 2D**). The different behaviors observed for different values of N_{KATP} thus exclusively rely on the characteristics of the noise. This computational observation provides an explanation for the widely different profiles of Ca^{2+} oscillations observed in single α-cells that probably express different number of K_{ATP} channels.

In a given cell characterized by a fixed number of K_{ATP} channels, bursts of electrical activity and Ca^{2+} oscillations are sensitive to factors affecting Ca^{2+} dynamics and particularly to the maximal rate of the PMCA, as shown in **Figures 5D** and **F**. This velocity indeed affects both Ca^{2+} pumping and ATP hydrolysis. When pumping is slower that in the control situation (**Figure 4**), Ca^{2+} and ATP concentrations at the membrane remain large, which tends to decrease the number of bursts of activity and increase their duration. Upon an increase in V_p, bursts become shorter. Thus, the number of spikes in a given period of time reaches a maximum value. When Vp further increases, the decrease in ATP concentration prevents this nucleotide from inducing trends in g_{KATP} changes and Ca^{2+} spikes finally disappear.

Simulations thus predict that the interplay between Ca^{2+} and ATP dynamics that occur through PMCA activity can induce a trend in the noise-initiated changes in K_{ATP} conductance (**Figure 6**). Rapid fluctuations in the cell K_{ATP} conductance occur because of the small number of such channels involved in the electrical activity of α-cells. The resulting changes in membrane Ca^{2+} induce changes in ATP concentration, because Ca^{2+} and ATP levels in the submembrane space are coupled via the activity of the PMCA. ATP changes in turn feedback on the conductance of the K_{ATP} channels. As the concentration of this nucleotide evolves slowly, these changes are in average maintained on a period of time that is sufficiently long to induce or refrain electrical activity. Thus, the combination of randomness at the level of the K_{ATP} current and of ATP changes can account for the observed long-duration changes in

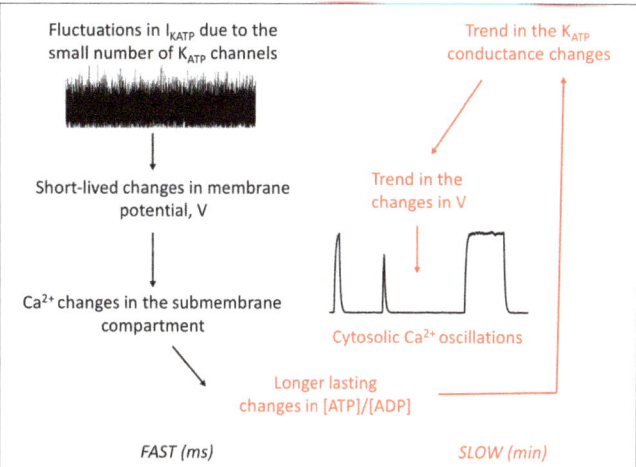

FIGURE 6 | Schematic representation of the interplay between randomness of the K_{ATP} current and ATP changes to generate irregular, slow Ca^{2+} oscillations in α-cells. Rapid fluctuations in the K_{ATP} current due to the small number of channels in the α-cell plasma membrane create fluctuations in membrane voltage that can induce a short-lived change in membrane voltage, which in turn modifies submembrane Ca^{2+} concentration. If the dynamics of ATP is not considered, this noise-induced change in Ca^{2+} would last some milliseconds. If ATP is considered, large fluctuations create a change in ATP concentration of larger duration, because ATP kinetics is much slower than electrical activity. Despite the rapid noise on V that is still present, this will create a trend in the changes in V because of the feedback exerted by ATP on the K_{ATP} current. Thus, periods of quiescence and activity will alternate, which is responsible for cytosolic Ca^{2+} oscillations by diffusion of the membrane Ca^{2+}.

electrical activity and thus for the Ca^{2+} oscillations observed in α-cells.

DISCUSSION

In low glucose conditions, more than half of α-cells display Ca^{2+} oscillations and these oscillations can persist in high glucose, although much reduced in frequency and amplitude. These Ca^{2+} oscillations are highly variable in shape, duration and interspike interval, not only among different cells but also in the course of time in a given individual cell. Such a randomness strongly suggests the presence of a high level of molecular noise, which generally results from a low number of copies of one or several biochemical species (Gonze et al., 2018). In direct agreement with this theoretical concept, α-cells electrical activity involves a limited number of K_{ATP} channels that play a crucial role in their electrical activity (Rorsman et al., 2014). The aim of this study was to investigate the link between the small numbers of K_{ATP} channels and the existence of irregular, slow cytosolic Ca^{2+} oscillations in α-cells in the rigorous framework of a mathematical model closely based on experimental observations. As Ca^{2+} oscillations are observed not only in α-cells of the intact pancreas but also in isolated cells (Salehi et al., 2006; Quoix et al., 2009; Tuduri et al., 2009; Le Marchand and Piston, 2010), they most probably result from an intrinsic mechanism, modulated by paracrine signaling.

We thus built a model of increasing complexity, on the basis of a previously proposed model of the α-cells electrical activity (Diderichsen and Göpel, 2006). This sequential approach showed that the randomness of the K_{ATP} current does not *per se* allow for Ca^{2+} oscillations. Random changes in this current are indeed too fast to induce any long-term change that could induce intermittent electrical activity. However, when taking into account the triangular relationship between electrical current, Ca^{2+} changes and ATP consumption, the model can account for the electrical and calcium dynamics experimentally observed in α-cells. Moreover, it also accounts for the fact that these dynamics are not stereotypic as quite moderate changes in the number of K_{ATP} channels can induce significant changes in the pattern of Ca^{2+} oscillations. The model predicts that factors that interfere with cell Ca^{2+} changes – such as pumping rates, diffusion coefficients or buffering capacities – may modify the durations during which repetitive AP's occur. This prediction is relevant for the impaired glucagon response to hypoglycemia and hyperglucagonemia observed in type 2 diabetes.

The simulations predict that the number of simultaneously active K_{ATP} channels during electrical activity is between ~ 5 and 10, in agreement with experimental observations (Zhang et al., 2020). This number is much smaller than the total number of channels considered in the simulations ($N_{KATP} = 58$), which are all potentially openable. Their intrinsically low open probability and their inhibition by ATP explain why only a small fraction are open simultaneously. Moreover, we expect our value of N_{KATP} to be underestimated since we did not take into account effective fatty acid metabolism in low glucose conditions (Briant et al., 2018a) nor the spatio-temporal dynamics of ATP. In any case, that intermittency in electrical activity can be obtained with the model over a relatively large range of values of N_{KATP}, as shown in **Figure 5**, indicates that the proposed mechanism for Ca^{2+} oscillations in α-cells is robust.

A similar role for stochasticity at the level of the K_{ATP} current has been put forward to account for the irregularity of the neuronal firing pattern in hippocampal CA3 neurons (Huang et al., 2007). However, in this case, the changes in K_{ATP} current are due to rapid, random fluctuations in the concentration of ATP, which results in an irregular frequency of AP's and not in intermittent electrical activity. In the more general context of networks dynamics, it is known that fluctuations can propagate along the different nodes of the network with a rate of decay or enhancement that depends on the network's structure (Maslov et al., 2007; Zhang et al., 2012).

Within this study, we did not investigate the behavior of the model when changing external glucose or considering paracrine signaling, which are left for further study. These limitations are due to the fact that the model used for electrical activity (Diderichsen and Göpel, 2006) does not reproduce one of the key characteristics of α-cells when raising external glucose, i.e., the reduction in the amplitude of the AP's (see Ashcroft and Rorsman, 2013 for example). This reduction leads to less activation of the Ca^{2+} channels linked to glucagon secretion. In **Figure 2**, this reduction was simulated by an artificial change in the leak conductance when changing the conductance of the K_{ATP} channels to simulate the changes in external glucose. However, given the reliability of the Diderichsen and Göpel's model and the many unknowns in the field of α-cells dynamics, we considered this unchanged model as a safe starting point to explore the impact of a stochastic K_{ATP} current and of changes in ATP concentration. However, we acknowledge that values of parameters for the ionic currents obtained with better methods of α-cell identification could improve the agreement between the simulations and the experiments (Briant et al., 2017; Montefusco et al., 2020).

The hydrolysis of ATP that parallels Ca^{2+} extrusion out of the cell plays a key role in the behavior predicted by the model. However, ATP synthesis is also sensitive to intracellular Ca^{2+} changes because mitochondrial metabolism is Ca^{2+}-sensitive (Wacquier et al., 2019). The ATP/ADP ratio is also controlled by glycolysis and fatty acid metabolism (Briant et al., 2018a). Further studies are required to address these interrelationships in a more accurate way (Olivos-Santes et al., 2019), although the qualitative results obtained with the relatively simple model used in this study should be robust towards these numerous possible model refinements.

As another limitation, we only considered noise at the level of the K_{ATP} channel, while all molecular processes, and especially Ca^{2+} dynamics, are subjected to noise. This simplification was based on the experimental observation of an unusually small K_{ATP} conductance in α-cells, in particular in comparison with pancreatic β-cells that are known to be equipped with the same channels but still display very regular voltage and calcium patterns. However, the details of our conclusions may be affected by adopting a full stochastic description of the cell dynamics. In particular, it may lead to less stereotypic cytosolic Ca^{2+} changes, as reported in the experiments.

An interesting perspective would be to couple the model presented in this study to our previously proposed model that relates glucagon secretion to cytosolic Ca^{2+} changes (González-Vélez et al., 2012). The latter model is able to predict glucagon secretion using as an input experimentally obtained Ca^{2+} traces. It was found that glucagon secretion does not correlate with the frequency of Ca^{2+} oscillations. This investigation could be pursued by using Ca^{2+} time series obtained by simulations of the present model instead of experimental data. This could lead to identify the key elements of electrical and/or calcium activities controlling glucagon secretion.

AUTHOR CONTRIBUTIONS

VG-V, AP, and GD contributed to the conceptualization, design of the study, development, simulation, and analysis of the models. GD and VG-V contributed to the preparation of the manuscript. All authors contributed to the article and approved the final version.

FUNDING

GD is research director at the Belgian FRS-FNRS. VG-V and GD acknowledge previous financial support of FNRS and CONACYT México through a bilateral cooperation project.

ACKNOWLEDGMENTS

We thank Vincent Verjans for scientific advice.

REFERENCES

Allbritton, N., Meyer, T., and Stryer, L. (1992). Range of messenger action of calcium ion and inositol 1,4,5-trisphosphate. *Science* 258, 1812–1815. doi: 10.1126/science.1465619

Ashcroft, F., and Rorsman, P. (2013). KATP channels and islet hormone secretion: new insights and controversies. *Nat. Rev. Endocrinol.* 9, 660–669. doi: 10.1038/nrendo.2013.166

Barg, S., Galvanovskis, J., Göpel, S., Rorsman, P., and Eliasson, L. (2000). Tight coupling between electrical activity and exocytosis in mouse glucagon-secreting α-cells. *Diabetes Metab. Res. Rev.* 49, 1500–1510. doi: 10.2337/diabetes.49.9.1500

Bokvist, K., Olsen, H., Hoy, M., Gotfredsen, C., Holmes, W., Buschard, K., et al. (1999). Characterisation of sulphonylurea and ATP-regulated Ca2+ channels in rat pancreatic α-cells. *Eur. J. Physiol.* 438, 428–436. doi: 10.1007/s004240051058

Briant, L., Dodd, M., Chibalina, M., Rorsman, N., Johnson, P., Carmeliet, P., et al. (2018a). CPT1a-dependent long-chain fatty acid oxidation contributes to maintaining glucagon secretion from pancreatic islets. *Cell Rep.* 23, 3300–3311. doi: 10.1016/j.celrep.2018.05.035

Briant, L., Reinbothe, T., Spiliotis, I., Miranda, C., Rodriguez, B., and Rorsman, P. (2018b). δ-cells and β-cells are electrically coupled and regulate α-cell activity via somatostatin. *J. Physiol.* 596, 197–215.

Briant, L., Salehi, A., Vergari, E., Zhang, Q., and Rorsman, P. (2016). Glucagon secretion from pancreatic α-cells. *Upsala J. Med. Sci.* 121, 113–119. doi: 10.3109/03009734.2016.1156789

Briant, L., Zhang, Q., Vergari, E., Kellard, J., Rodriguez, B., Ashcroft, F., et al. (2017). Functional identification of islet cell types by electrophysiological fingerprinting. *J. R. Soc. Interface* 14:20160999. doi: 10.1098/rsif.2016.0999

Csercsik, D., Farkas, I., Szederkenyi, G., Hrabovszky, E., Liposits, Z., and Hangos, K. (2010). Hodgkin-Huxley type modelling and parameter estimation of GnRH neurons. *Biosystems* 100, 198–207. doi: 10.1016/j.biosystems.2010.03.004

D'Alessio, D. (2011). The role of dysregulated glucagon secretion in type 2 diabetes. *Diabetes. Obes. Metab.* 13(Suppl. 1), 126–132. doi: 10.1111/j.1463-1326.2011.01449.x

Diderichsen, P., and Göpel, S. (2006). Modelling the electrical activity of pancreatic α-cells based on experimental data from intact mouse islets. *J. Biol. Phys.* 32, 209–229. doi: 10.1007/s10867-006-9013-0

Dupont, G., Falcke, M., Kirk, V., and Sneyd, J. (2016). *Models of Calcium Signalling.* Berlin: Springer International Publishing., doi: 10.1007/978-3-319-29647-0

Enkvetchakul, D., Loussouam, G., Makhina, E., and Nichols, C. (2001). ATP interaction with the open state of the KATP channel. *Biophys. J.* 80, 719–728. doi: 10.1016/s0006-3495(01)76051-1

Ermentrout, B. (2002). *Simulating, Analyzing, and Animating Dynamical Systems: A Guide to XPPAUT for Researchers and Students.* Philadelphia, PA: SIAM.

Fall, C., Marland, E., Wagner, J., and Tyson, J. (2005). *Computational Cell Biology.* New York, NY: Springer.

Fridlyand, L., and Philipson, L. (2012). A computational systems analysis of factors regulating a cell glucagon secretion. *Islets* 4, 262–283. doi: 10.4161/isl.22193

Gilon, P. (2020). The role of α-cells in islet function and glucose homeostasis in health and type 2 diabetes. *J. Mol. Biol.* 432, 1367–1394. doi: 10.1016/j.jmb.2020.01.004

González-Vélez, V., Dupont, G., Gil, A., Gonzalez, A., and Quesada, I. (2012). Model for glucagon secretion by pancreatic α-cells. *PLoS One* 7:e32282. doi: 10.1371/journal.pone.0032282

González-Vélez, V., Gil, A., and Quesada, I. (2010). Minimal state models for ionic channels involved in glucagon secretion. *Math. Biosci. Eng.* 7, 793–807. doi: 10.3934/mbe.2010.7.793

Gonze, D., Gérard, C., Wacquier, B., Woller, A., Tosenberg, A., and Dupont, G. (2018). Modeling-based investigation of the effect of noise in cellular systems. *Front. Mol. Biosc.* 5:34. doi: 10.3389/fmolb.2018.00034

Göpel, S., Zhang, Q., Eliasson, L., Ma, X. S., Galvanovskis, J., and Kanno, T. (2004). Capacitance measurements of exocytosis in mouse pancreatic α-, β- and δ-cells within intact islets of Langerhans. *J. Physiol.* 556, 711–726. doi: 10.113/jphysiol.2003.059675

Grubelnik, V., Markovic, R., Lipovsek, S., Leitinger, G., Gosak, M., Dolensek, J., et al. (2020). Modelling of dysregulated glucagon secretion in type 2 diabetes by considering mitochondrial alterations in pancreatic α-cells. *R. Soc. Open Sci.* 7:191171. doi: 10.1098/rsos.191171

Huang, C. W., Huang, C. W., Chen, J., Tsai, J., and Wu, S. (2007). Glucose and hippocampal neuronal excitability: role of ATP-sensitive potassium channels. *J. Neurosci. Res.* 85, 1468–1477. doi: 10.1002/jnr.21284

Huang, Y.-C., Gaisano, H., and Leung, Y.-M. (2011). Electrophysiological identification of mouse islet -cells. From isolated single α-cells to in situ assessment within pancreas slices. *Islets* 3, 139–143. doi: 10.4161/isl.3.4.16166

Kellard, A., Rorsman, N., Hill, T., Armour, S., van de Bunt, M., Rorsman, P., et al. (2020). Reduced somatostatin signalling leads to hypersecretion of glucagon in mice fed a high-fat diet. *Mol. Metab.* 40:101021. doi: 10.1016/j.molmet.2020.101021

Khan, F., Goforth, P., Zhang, M., and Satin, L. (2001). Insulin activates ATP-sensitive K+ channels in pancreatic β-cells through a phosphatidyl-inositol 3-kinase-dependent pathway. *Diabetes Metab. Res. Rev.* 50, 2192–2198. doi: 10.2337/diabetes.50.10.2192

Le Marchand, S., and Piston, D. (2010). Glucose suppression of glucagon secretion. *J. Biol. Chem.* 285, 14389–14398.

Le Marchand, S., and Piston, D. (2012). Glucose decouples intracellular Ca2+ activity from glucagon secretion in mouse pancreatic islet alpha-cells. *PLoS One* 7:e47084. doi: 10.1371/journal.pone.0047084

Li, J., Yu, Q., Ahooghalandari, P., Gribble, F., Reimann, F., Tengholm, A., et al. (2015). Submembrane ATP and Ca2+ kinetics in α-cells: unexpected signaling for glucagon secretion. *FASEB J.* 29, 3379–3388. doi: 10.1096/fj.14-265918

MacDonald, P., De Marinis, Y. Z., Ramracheya, R., Salehi, A., Ma, X., Johnson, P., et al. (2007). A KATP channel-dependent pathway within α cells regulates glucagon release from both rodent and human islets of Langerhans. *PLoS Biol.* 5:e143. doi: 10.1371/journal.pbio.0050143

Maslov, S., Sneppen, K., and Ispolatov, I. (2007). Spreading out of perturbations in reversible reaction networks. *New J. Phys.* 9:273. doi: 10.1088/1367-2630/9/8/273

Montefusco, F., Cortese, G., and Pedersen, M. (2020). Heterogeneous alpha-cell population modeling of glucose-induced inhibition of electrical activity. *J. Theor. Biol.* 485:110036. doi: 10.1016/j.jtbi.2019.110036

Montefusco, F., and Pedersen, M. G. (2015). Mathematical modelling of local calcium and regulated exocytosis during inhibition and stimulation of glucagon secretion from pancreatic alpha-cells. *J. Physiol.* 593, 4519–4530. doi: 10.1113/jp270777

Olivos-Santes, E., Romero-Campos, H., Dupont, G., and González-Vélez, V. (2019). A modelling study of glycolytic oscillations and electrical activity in pancreatic alpha cells. *Rev. Mex. De Ing. Biom.* 40, 1–10.

Quesada, I., Tuduri, E., Ripoll, C., and Nadal, A. (2008). Physiology of the pancreatic α-cell and glucagon secretion: role in glucose homeostasis and diabetes. *J. Endocrin.* 199, 5–19. doi: 10.1677/joe-08-0290

Quoix, N., Cheng-Xue, R., Mattart, L., Zeinoun, Z., Guiot, Y., Beauvois, M., et al. (2009). Glucose and pharmacological modulators of ATP-sensitive K+ channels control [Ca2+]c by different mechanisms in isolated mouse α-cells. *Diabetes Metab. Res. Rev.* 58, 412–421. doi: 10.2337/db07-1298

Rorsman, P., Ramracheya, R., Rorsam, N., and Zhang, Q. (2014). ATP-regulated potassium channels in pancreatic alpha and beta cells: similar functions but reciprocal effects on secretion. *Diabetologia* 57, 1749–1761. doi: 10.1007/s00125-014-3279-8

Salehi, A., Vieir, E., and Gylfe, E. (2006). Paradoxical stimulation of glucagon secretion by high glucose concentrations. *Diabetes Metab. Res. Rev.* 55, 2318–2323. doi: 10.2337/db06-0080

Tuduri, E., Marroqui, L., Soriano, S., Ropero, A., Batista, T., Piquer, S., et al. (2009). Inhibitory effects of leptin on pancreatic α-cell function. *Diabetes Metab. Res. Rev.* 58, 1616–1624. doi: 10.2337/db08-1787

Voets, T. (2000). Dissection of three Ca2+-dependent steps leading to secretion in chromaffin cells from mouse adrenal slice. *Neuron* 28, 537–545. doi: 10.1016/s0896-6273(00)00131-8

Wacquier, B., Combettes, L., and Dupont, G. (2019). Cytoplasmic and mitochondrial calcium signaling: a two-way relationship. *Cold Spring Harb. Perspect. Biol.* 11:a035139. doi: 10.1101/cshperspect.a035139

Watts, M., Ha, J., Kimchi, O., and Sherman, A. (2016). Paracrine regulation of glucagon secretion: the β/α/δ model. *Am. J. Physiol. Endocrinol. Metab.* 310, E597–E611.

Watts, M., and Sherman, A. (2014). Modeling the pancreatic α-cell: dual mechanisms of glucose suppression of glucagon secretion. *Biophys. J.* 106, 741–751. doi: 10.1016/j.bpj.2013.11.4504

Wendt, A., and Eliasson, L. (2020). Pancreatic -cells – The unsung heroes in islet function. *Sem. Cell and Dev. Biol.* 103, 41–50. doi: 10.1016/j.semcdb.2020.01.006

Yu, A., Shuai, H., Ahooghalandari, P., Gylfe, E., and Tengholm, A. (2019). Glucose controls glucagon secretion by directly modulating cAMP in alpha cells. *Diabetologia* 61, 1212–1224. doi: 10.1007/s00125-019-4857-6

Zhang, H., Chen, Y., and Chen, Y. (2012). Noise propagation in gene regulation networks involving interlinked positive and negative feedback loops. *PLoS One* 7:e51840. doi: 10.1371/journal.pone.0051840

Zhang, Q., Dou, H., and Rorsman, P. (2020). 'Resistance is futile?'– paradoxical inhibitory effects of KATP channel closure in glucagon-secreting α-cells. *J. Physiol.* 598, 4765–4780. doi: 10.1113/JP279775

Zhang, Q., Ramracheya, R., Lahmann, C., Tarasov, A., Bengtsson, M., Braha, O., et al. (2013). Role of KATP channels in glucose-regulated glucagon secretion and impaired counterregulation in Type 2 Diabetes. *Cell Metab.* 18, 871–882. doi: 10.1016/j.cmet.2013.10.014

Zimny, M., and Blackard, W. (1975). The surface structure of isolated pancreatic islet cells. *Cell Tissue Res.* 164, 467–471.

4

Unveiling the Role of the Integrated Endoplasmic Reticulum Stress Response in *Leishmania* Infection

K. L. Dias-Teixeira[1], R. M. Pereira[2], J. S. Silva[3], N. Fasel[4], B. H. Aktas[5] and U. G. Lopes[1]*

[1] *Institute of Biophysics Carlos Chagas Filho, Federal University of Rio de Janeiro, Rio de Janeiro, Brazil,* [2] *Institute of Microbiology Paulo de Goes, Federal University of Rio de Janeiro, Rio de Janeiro, Brazil,* [3] *Department of Biochemistry and Immunology, University of São Paulo, Ribeirão Preto, Brazil,* [4] *Department of Biochemistry, Faculty of Biology and Medicine, Center for Immunity and Infection Lausanne, University of Lausanne, Lausanne, Switzerland,* [5] *Laboratory of Translation, Department of Hematology, Brigham and Women's Hospital, Harvard Medical School, Boston, MA, USA*

Edited by:
Abhay Satoskar,
Ohio State University, USA

Reviewed by:
Patricia Talamás-Rohana,
Cinvestav, Mexico
Ranadhir Dey,
Food and Drug Administration, USA

*Correspondence:
U. G. Lopes
lopesu@biof.ufrj.br

The integrated endoplasmic reticulum stress response (IERSR) is an evolutionarily conserved adaptive mechanism that ensures endoplasmic reticulum (ER) homeostasis and cellular survival in the presence of stress including nutrient deprivation, hypoxia, and imbalance of Ca^+ homeostasis, toxins, and microbial infection. Three transmembrane proteins regulate integrated signaling pathways that comprise the IERSR, namely, IRE-1 that activates XBP-1, the pancreatic ER kinase (PERK) that phosphorylates the eukaryotic translation initiation factor 2 and transcription factor 6 (ATF6). The roles of IRE-1, PERK, and ATF4 in viral and some bacterial infections are well characterized. The role of IERSR in infections by intracellular parasites is still poorly understood, although one could anticipate that IERSR may play an important role on the host's cell response. Recently, our group reported the important aspects of XBP-1 activation in *Leishmania amazonensis* infection. It is, however, necessary to address the relevance of the other IERSR branches, together with the possible role of IERSR in infections by other *Leishmania* species, and furthermore, to pursue the possible implications in the pathogenesis and control of parasite replication in macrophages.

Keywords: *Leishmania*, ER stress, XBP-1, IFN-1, PERK, ATF4

INTRODUCTION

The endoplasmic reticulum (ER) is a dynamic tubular network involved in different processes such as protein folding, lipid synthesis, and the biogenesis of autophagosomes and peroxisomes (1). When the process of protein synthesis and/or folding is disturbed, the ER induces a transcriptional program, the integrated endoplasmic reticulum stress response (IERSR), leading to the increase of ER-chaperone expression, lipid synthesis, and the induction of other sets of gene products involved in the retrograde transport and degradation of unfolded proteins (ERAD) (2). These conserved adaptive responses reduce demand on the folding capacity or ER, increase ER's folding capacity, and clear this organelle off of unfolded proteins. However, during this process, a set of genes that also regulate the expression of cytokines and promote the resistance to oxidative stress are upregulated. The three branches that regulate the ER response are comprised by the activating transcription factor 6 (ATF6), inositol-requiring kinase 1 (IRE-1), and the protein kinase R (PKR)-like endoplasmic reticulum kinase (PERK). IRE-1 activates the X-box binding protein 1 (XBP-1), a transcriptional factor that plays a critical role in cellular homeostasis and regulates the expression of important cytokines related to the antiviral immunity response, such as IFN1-β. PERK phosphorylates eIF2α, which reduces overall protein synthesis while upregulating the expression of activating transcription factor 4 (ATF4), which drives the expression of genes that play a critical role in restoring cellular homeostasis, resistance to oxidative stress together with genes related to the autophagic pathway and the innate immunity response. Interestingly, both XBP-1 and ATF4 can be activated by toll-like receptors (TLRs). For instance, the engagement of TLR2 and TLR4 can specifically activate XBP-1 leading to the production of pro-inflammatory cytokines that restrain bacterial burden in infected macrophages (3). ATF4 can be directly activated by the TLR4-MyD88 pathway following stimulation of human monocytes with lipopolysaccharide (LPS) (4).

Viruses can selectively induce specific branches of the IERSR. For instances, human cytomegalovirus and hepatitis C activate the IERSR response, while some viruses, such as dengue virus and hepatitis C virus induce the IERSR trough the exploitation of the ER membranes during the replication process (5). Additionally, some viruses induce the IERSR and the inhibition of the translational process due to the phosphorylation of eIF2-α, reducing the production of cytokines and interfering with the host immune response. This process is highly induced by enteroviruses (6). Some viruses adapted the IERSR pathways to favor their infection directly. The phosphorylation of eIF2-α induces the translation of a specific set of proteins including ATF4. ATF4 can, for example, enhance human immunodeficiency virus (HIV) replication through a synergistic interaction with the HIV regulatory protein Tat (7).

The role of IERSR pathways in parasite infection is poorly investigated. Recently, it was reported that *Plasmodium berghei* induces the ER stress response and XBP-1 mRNA splicing and translation of the transcriptionally active XBP-1 spliced form (XBP-1s) in hepatocytes. This activation was demonstrated to be crucial for parasite replication inside hepatocytes and to the progression of the infection (8). XBP-1s can modulate the synthesis of phospholipids, such as phosphatidylcholine (PC), in hepatocytes. PC is a major component of membranes, and it has been demonstrated that malaria parasites uptake host-derived PC and, most probably, PC is also employed for enlarging the parasitophorous vacuole membrane (9). Most recently, we showed that induction of ER stress favors *Leishmania amazonensis* infection in a TLR2-dependent manner, culminating in the formation of XBP-1s. XBP-1 induces IFN-β expression and modulates the oxidative response of infected macrophages, thereby promoting parasite proliferation (10).

However, it will be important to test these observations in other *Leishmania* species and to address the relevance of the PERK/ATF4 and ATF6 branches of the IERSR during *Leishmania* infection.

THE ROLE OF XBP-1 IN *LEISHMANIA* INFECTION

We recently observed that *L. amazonensis* induces the activation of XBP-1 in macrophages. RAW 264.7 cells knocked down for XBP-1 exhibited reduced parasite load, likely due to impaired

translocation of the IRF3 transcription factor resulting in reduced IFN-1 expression (10). We also observed that infected XBP-1 knocked down macrophages produce higher nitric oxide levels and reduced Hemeoxygenase (HO)-1 expression compared to control macrophages. However, how XBP-1 controls oxidative stress in *L. amazonensis* infection requires further investigation. One mechanism that could induce this effect is the activation or repression of the NF-κB transcription factor. *L amazonensis* activates an NF-κB p50/p50 repressor homodimer, which promotes reduction in iNOS expression and favors parasite growth (11). The production of ROS can activate the ER stress response, which can suppress NF-κB activation in the later phase of IERSR (12). The protein A20, an ubiquitin-editing NF-κB inhibitor protein, may play an important role in this process, as this protein can negatively regulate NF-κB during oxidative stress (13). Additionally, it is important to understand if other *Leishmania* species induce the IERSR branches, and the role, if any, in pathogenesis. Experiments carried out by our group observed an induction of the XBP-1 spliced form in clinical samples from patients infected with *Leishmania braziliensis*, another *Leishmania* species widely found in Brazil and the main causative agent of cutaneous leishmaniasis. These data indicate that other *Leishmania* species can activate this pathway, and that IERSR may play a role in *Leishmania*-associated pathogenesis.

THE INDUCTION OF ER STRESS: THE ROLE OF TLRs in XBP-1 ACTIVATION IN *LEISHMANIA* INFECTION

The mechanism through which *L. amazonensis* induces ER stress is not understood. *Leishmania* parasitophorous vacuoles interact continuously with the ER compartment and may recruit components that are important for parasite intracellular survival (14). The inhibition of such membrane compartment fusion with the parasitophorous vacuole results in the reduction of infection (15). It is conceivable that such compartment fusions may favor the activation of IERSR branches in infection.

The contribution of TLR receptors in IERSR remains to be elucidated. There is evidence to suggest that TLRs play a role for the success of *L. amazonensis* infection that is linked with IERSR activation. For instance, when TLR2 KO macrophages were treated with the ER stress inductor thapsigargin, there was a reduction of the *L. amazonensis* proliferation compared to wild-type cells (10). Additional results obtained by our group showed that TLR2 was partially required for XBP-1 activation (splicing) due to *L. amazonensis* infection. However, the mechanism by which *L. amazonensis* induces XBP-1 activation and ER stress remains unclear.

THE PERK/ATF4 BRANCH OF IERSR and *LEISHMANIA* INFECTION: IS A FUNCTIONAL ROLE?

The PERK/ATF4 branch of IERSR plays an important role in certain cellular processes that are also exploited to establish *Leishmania* infection. For instance, *L. amazonensis* induces the

PI3K/AKT signaling pathway (16), and it has been reported that the PERK-eIF2α pathway and PI3K signaling increases ATF4 expression, nuclear localization, and transcriptional activity (17–19). Additionally, PERK can directly regulate the activation of the nuclear factor (erythroid-derived 2)-like 2 (NRF2), an important antioxidant transcription factor that regulates the expression of a number of antioxidative response genes (20). Additionally, ATF4 has an important role in the autophagy. PERK/eIF2α/ATF4 signaling can induce upregulation of cytoprotective autophagy genes, such as ATG5 and ATG7, which promote cellular survival (21). In addition, ATF4 controls the microtubule associated protein 1A/1B-light chain 3 (LC3) expression. LC3B is important to generate the autophagosome formation, a hallmark of the autophagic process (22, 23). In 2012, Cyrino et al. showed that *Leishmania* parasites induce LC3B conversion and suggested that autophagy favors *L. amazonensis* infection (24). ATF4 is upregulated by HIV-1 infection and enhances HIV replication, likely due to synergistic interactions with the HIV Tat protein. Importantly, the expression of ATF4 induces HIV reactivation in chronically infected cell lines (7). Recently, our group showed that the Tat viral protein also increases *L. amazonensis* infection, in a PKR-dependent manner (25). *L. amazonensis* is able to induce PKR, a pathway activated in viral infections (26). *L. amazonensis* can also modulate IFN-1 expression in a TLR2/PKR-dependent fashion to promote the infection by the parasite, another pathway that is shared in viral infections (27). Taken together, due to classical function of IERSR in viral infections, it is relevant to test the role of PERK/ATF4 in viral co-infection and *Leishmania*.

CONCLUSION REMARKS

It is well known that the IERSR can modulate viral and bacterial infection, promoting the induction of cytokines, including IFN-1, which can be determinant to the outcome of several infections. Recent work suggests that the IERSR is required for the development of intracellular parasites. For instance, the activation of XBP-1 in hepatocytes infected by *P. berghei* favor the infection by the parasite through the modulation of lipid synthesis. Corroborating this notion, it has been demonstrated that *L. amazonensis* activates XBP-1 leading to IFN-1 expression and the expression of antioxidative responsive genes, such as HO-1. Unveiling the mechanisms by which IERSR promote intracellular parasitic infection requires further investigation. These investigations would include determining the role of XBP-1 in resistance to oxidative stress due to *Leishmania* infection and examining other components of the ER stress signaling pathway,

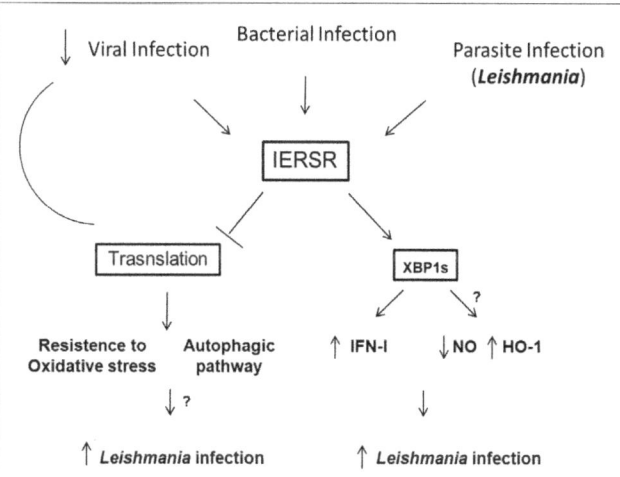

FIGURE 1 | Schematic model of the IERSR activation during *Leishmania* **infection.** Viral, bacterial, and parasite infections can activate the IERSR. Viral infections can induce the activation of IERSR and lead to the inhibition of the translational process that suppress the viral replication. In *Leishmania* infection, the activation of IERSR can induce the IFN-1 production, which favors the intracellular parasite growth. Additionally, XBP1s activation is important to promote the induction of HO-1 expression, promoting the parasite infection. On the other hand, *Leishmania* infection can be favored by the control of the antioxidative response and induction of autophagic process. Both processes can be induced by the activation of IERSR, and we may predict that *Leishmania* can trigger these pathways.

such as ATF6, in the context of parasitic infection. We can predict that the investigation of IERSR in intracellular parasitic infections may reveal novel drug targets. **Figure 1** shows the a schematic model of IERSR activation in *Leishmania* infection.

AUTHOR CONTRIBUTIONS

UL – project supervisor and wrote the paper, KD-T – performed experiments and wrote the paper, RP – revised the paper and discussed the results, JS – provided samples and supervised experiments, NF – revised the paper and contributed with the discussion, BA – supervised experiments, revised the paper, and discussed results.

FUNDING

This work was supported by the Conselho Nacional de Desenvolvimento Científico e Tecnológico, Brazil, to UL.

REFERENCES

1. Hetz C. The unfolded protein response: controlling cell fate decisions under ER stress and beyond. *Nat Rev Mol Cell Biol* (2012) 3(2):89–102. doi:10.1038/nrm3270
2. Martinon F, Glimcher LH. Regulation of innate immunity by signaling pathways emerging from the endoplasmic reticulum. *Curr Opin Immunol* (2011) 23(1):35–40. doi:10.1016/j.coi.2010.10.016
3. Martinon F, Chen X, Lee AH, Glimcher LH. TLR activation of the transcription factor XBP1 regulates innate immune responses in macrophages. *Nat Immunol* (2010) 11(5):411–8. doi:10.1038/ni.1857
4. Zhang C, Bai N, Chang A, Zhang Z, Yin J, Shen W, et al. ATF4 is directly recruited by TLR4 signaling and positively regulates TLR4-trigged cytokine production in human monocytes. *Cell Mol Immunol* (2013) 10(1):84–94. doi:10.1038/cmi.2012.57
5. El-Hage N, Luo G. Replication of hepatitis C virus RNA occurs in a membrane-bound replication complex containing nonstructural viral proteins and RNA. *J Gen Virol* (2003) 84:2761–9. doi:10.1099/vir.0.19305-0
6. Lloyd RE. Translational control by viral proteinases. *Virus Res* (2006) 119:76–88. doi:10.1016/j.virusres.2005.10.016

7. Caselli E, Benedetti S, Gentili V, Grigolato J, Di Luca D. Short communication: activating transcription factor 4 (ATF4) promotes HIV type 1 activation. *AIDS Res Hum Retroviruses* (2012) 28(8):907–12. doi:10.1089/AID.2011.0252

8. Inácio P, Zuzarte-Luís V, Ruivo MT, Falkard B, Nagaraj N, Rooijers K, et al. Parasite-induced ER stress response in hepatocytes facilitates Plasmodium liver stage infection. *EMBO Rep* (2015) 16(8):955–64. doi:10.15252/embr.201439979

9. Itoe MA, Sampaio JL, Cabal GG, Real E, Zuzarte-Luis V, March S, et al. Host cell phosphatidylcholine is a key mediator of malaria parasite survival during liver stage infection. *Cell Host Microbe* (2014) 16(6):778–86. doi:10.1016/j.chom.2014.11.006

10. Dias-Teixeira KL, Calegari-Silva TC, Dos Santos GR, Vitorino Dos Santos J, Lima C, Medina JM, et al. The integrated endoplasmic reticulum stress response in *Leishmania amazonensis* macrophage infection: the role of X-box binding protein 1 transcription factor. *FASEB J* (2016) 30(4):1557–65. doi:10.1096/fj.15-281550

11. Calegari-Silva TC, Pereira RM, De-Melo LD, Saraiva EM, Soares DC, Bellio M, et al. NF-kappaB-mediated repression of iNOS expression in *Leishmania amazonensis* macrophage infection. *Immunol Lett* (2009) 127(1):19–26. doi:10.1016/j.imlet.2009.08.009

12. Kitamura M, Hiramatsu N. The oxidative stress: endoplasmic reticulum stress axis in cadmium toxicity. *Biometals* (2010) 23(5):941–50. doi:10.1007/s10534-010-9296-2

13. Nakajima S, Kitamura M. Bidirectional regulation of NF-κB by reactive oxygen species: a role of unfolded protein response. *Free Radic Biol Med* (2013) 65:162–74. doi:10.1016/j.freeradbiomed.2013.06.020

14. Ndjamen B, Kang BH, Hatsuzawa K, Kima PE. *Leishmania* parasitophorous vacuoles interact continuously with the host cell's endoplasmic reticulum; parasitophorous vacuoles are hybrid compartments. *Cell Microbiol* (2010) 12(10):1480–94. doi:10.1111/j.1462-5822.2010.01483.x

15. Canton J, Ndjamen B, Hatsuzawa K, Kima PE. Disruption of the fusion of *Leishmania* parasitophorous vacuoles with ER vesicles results in the control of the infection. *Cell Microbiol* (2012) 14(6):937–48. doi:10.1111/j.1462-5822.2012.01767.x

16. Calegari-Silva TC, Vivarini ÁC, Miqueline M, Dos Santos GR, Teixeira KL, Saliba AM, et al. The human parasite *Leishmania amazonensis* downregulates iNOS expression via NF-κB p50/p50 homodimer: role of the PI3K/Akt pathway. *Open Biol* (2015) 5(9):150118. doi:10.1098/rsob.150118

17. Cao H, Yu S, Yao Z, Galson DL, Jiang Y, Zhang X, et al. Activating transcription factor 4 regulates osteoclast differentiation in mice. *J Clin Invest* (2010) 120:2755–66. doi:10.1172/JCI42106

18. Inageda K. Insulin modulates induction of glucose-regulated protein 78 during endoplasmic reticulum stress via augmentation of ATF4 expression in human neuroblastoma cells. *FEBS Lett* (2010) 584:3649–54. doi:10.1016/j.febslet.2010.07.040

26. Pereira RM, Teixeira KL, Barreto-de-Souza V, Calegari-Silva TC, De-Melo LD, Soares DC, et al. Novel role for the double-stranded RNA-activated protein kinase PKR: modulation of macrophage infection by the protozoan parasite *Leishmania*. *FASEB J* (2010) 24(2):617–26. doi:10.1096/fj.09-140053

27. Vivarini Ade C, Pereira RM, Teixeira KL, Calegari-Silva TC, Bellio M, Laurenti MD, et al. Human cutaneous leishmaniasis: interferon-dependent expression of double-stranded RNA – dependent protein kinase (PKR) via TLR2. *FASEB J* (2011) 25(12):4162–73. doi:10.1096/fj.11-185165

19. Lian N, Lin T, Liu W, Wang W, Li L, Sun S, et al. Transforming growth factor β suppresses osteoblast differentiation via the vimentin activating transcription factor 4 (ATF4) axis. *J Biol Chem* (2012) 287:35975–84. doi:10.1074/jbc.M112.372458

20. Dey S, Sayers CM, Verginadis II, Lehman SL, Cheng Y, Cerniglia GJ, et al. ATF4-dependent induction of hemeoxygenase 1 prevents anoikis and promotes metastasis. *J Clin Invest* (2015) 125(7):2592–608. doi:10.1172/JCI78031

21. Luo J-Q, Chen D-W, Yu B. Upregulation of amino acid transporter expression induced by L-leucine availability in L6 myotubes is associated with ATF4 signaling through mTORC1-dependent mechanism. *Nutrition* (2013) 29:284–90. doi:10.1016/j.nut.2012.05.008

22. Kabeya Y, Mizushima N, Ueno T, Yamamoto A, Kirisako T, Noda T, et al. LC3, a mammalian homologue of yeast Apg8p, is localized in autophagosome membranes after processing. *EMBO J* (2000) 19(21):5720–8. doi:10.1093/emboj/19.21.5720

23. Baehrecke EH. Autophagy: dual roles in life and death? *Nat Rev Mol Cell Biol* (2005) 6(6):505–10. doi:10.1038/nrm1666

24. Cyrino LT, Araújo AP, Joazeiro PP, Vicente CP, Giorgio S. In vivo and in vitro *Leishmania amazonensis* infection induces autophagy in macrophages. *Tissue Cell* (2012) 44(6):401–8. doi:10.1016/j.tice.2012.08.003

25. Vivarini Áde C, Pereira Rde M, Barreto-de-Souza V, Temerozo JR, Soares DC, Saraiva EM, et al. HIV-1 Tat protein enhances the intracellular growth of *Leishmania amazonensis* via the ds-RNA induced protein PKR. *Sci Rep* (2015) 5:16777. doi:10.1038/srep16777

5

Leishmania Infection Engages Non-Receptor Protein Kinases Differentially to Persist in Infected Hosts

Naixin Zhang and Peter E. Kima*

Department of Microbiology and Cell Science, University of Florida, Gainesville, FL, USA

Edited by:
Abhay Satoskar,
The Ohio State University, USA

Reviewed by:
Charles C. Caldwell,
University of Cincinnati, USA
Patricia Talamás-Rohana,
CINVESTAV, Mexico

*Correspondence:
Peter E. Kima
pkima@ufl.edu

Protein kinases play important roles in the regulation of cellular activities. In cells infected by pathogens, there is an increasing appreciation that dysregulated expression of protein kinases promotes the success of intracellular infections. In *Leishmania*-infected cells, expression and activation of protein kinases, such as the mitogen-activated protein kinases, kinases in the PI3-kinase signaling pathway, and kinases in the NF-κB-signaling pathway, are modulated in some manner. Several recent reviews have discussed our current understanding of the roles of these kinases in *Leishmania* infections. Apart from the kinases in the pathways enumerated above, there are other host cell protein kinases that are activated during the *Leishmania* infection of mammalian cells whose roles also appear to be significant. This review discusses recent observations on the Abl family of protein kinases and the protein kinase regulated by RNA in *Leishmania* infections.

Keywords: *Leishmania*, protein kinases, Abl kinase, PKR, phagocytosis

INTRODUCTION

Natural *Leishmania* infections are initiated by the deposition of promastigotes forms of *Leishmania* by sand flies at the site of their blood meal. Current understanding is that phagocytes, particularly neutrophils that are recruited earliest to the bite site become a sanctuary for the promastigotes (1). Once within mammalian cells, *Leishmania* commence to modify their gene expression profile, which culminates with their transformation into amastigote forms. By 24 h post-infection, the parasites are fully transformed into amastigote forms, which are the replicative form of *Leishmania* within mammalian cells and hosts. Some parasite species (*Leishmania tropica, Leishmania major, Leishmania mexicana,* and *Leishmania braziliensis*) replicate within inflammatory cells that are recruited to the bite site, which results in cutaneous lesions; other parasite species (*Leishmania donovani* and *Leishmania infantum*) traffic by still poorly understood mechanisms to the visceral organs where they take up residence and participate in the development of infected cell foci. Cells that are recruited to the site of infection are transformed by infection. Under conditions that continue to be investigated, infected cells are induced to release mediators some of which can promote parasite persistence, whereas others contribute to the control of infection. A few recent reviews have discussed our current understanding of the role of several protein kinases, such as the mitogen-activated protein kinases (MAPK), kinases in the PI3-kinase signaling pathway, and kinases in the NF-κB-signaling pathway (2–5), in the transformation of infected cells. In this review, the current understanding of the role of the Abl family of protein kinases that play an early role in transduction of signals will be discussed. Interactions with macrophage surface molecules, including the complement receptor, also plays a role in the activation of protein kinase regulated by RNA (PKR), which induces the release of IFNα, IL-27, and IL-10 among other cytokines that also modulate the host response to infection.

PARASITE INTERNALIZATION

There has been long-standing interest in the molecular interactions that mediate *Leishmania* parasite entry into mammalian cells. Several phagocytic receptors, including the mannose receptor, scavenger receptor, complement receptors, and Fc receptors (6), have all been shown to be suitable internalization receptors of *Leishmania* parasites. However, in light of the fact that parasites in mammalian hosts are bathed in serum that contains opsonins including complement components and parasite specific antibodies, it is most likely that opsonin-dependent receptors are the preferred receptors that mediate parasite uptake. The critical importance of antibodies as opsonins for parasite internalization *in vivo* had been suggested by studies in animals that were genetically modified to be defective in circulating antibodies (7, 8). Those mice developed much smaller lesions as compared to wild-type mice when they were infected with *L. mexicana* parasites. Small lesions were proposed to be the result of not only reduced parasite uptake but also to be due to a skewed cytokine response (8). A few recent reports have revisited this topic and have explored the contributions of the opsonin-dependent receptors in mediating *Leishmania* parasite uptake by cells including neutrophils. We initially review the results of those studies to set the appropriate frame of reference for the discussion of the role of the Abl family kinases and PKR.

Numerous reports had assessed the role of the phagocytic receptor in the internalization of *Leishmania* parasites into macrophages. However, a recent study evaluated uptake of *L. donovani* parasites either *via* CR3 or the Fc receptors in the context of their effect on the maturation of the parasitophorous vacuole (PV). These studies were informed by Desjardins and Descoteaux (9) who had shown that upon internalization of *L. donovani* promastigote forms, the nascent PV undergoes a delay in its maturation. Acquisition of late endocytic pathway characteristics, characterized by the loss of early endosome autoantigen 1 (EEA1) and display of the lysosome associate membrane protein (LAMP-1) on the PV membrane, is delayed as compared to internalization of amastigotes that displays LAMP-1 within an hour of infection. Polando et al. (10) found that the phagocytic receptor that is engaged for parasite internalization affects PV maturation. Specifically, promastigote opsonization with C3-containing serum reduced the PV maturation delay by 2 h, whereas opsonization with *Leishmania* immune serum reduced the PV maturation delay by 3 h. In a follow paper, Ricardo-Carter et al. (11) showed that there are other biological consequences to the choice of entry receptor. Their contribution to the well-known phenomenon that *Leishmania*-infected cells do not secrete IL-12 and other inflammatory cytokines in response to lipopolysaccharide was that the underlying suppression mechanism is initiated upon the CR3-mediated uptake of

parasites. They ruled out a role for signaling intermediates NFκB p65, MAPK, IRF-1, or IRF-8 in cytokine suppression induced by parasite uptake *via* CR3. Together, the studies described above are recent contributions to the long held appreciation of the role of the phagocytic receptor in parasite entry into mammalian cells.

As alluded to above, parasites are engulfed by phagocytes that recruited to the site of infection. Among these phagocytes are neutrophils that have been implicated in the "Trojan horse" infection strategy where they serve as a sanctuary for promastigote forms until promastigotes transform into amastigotes and are released or infected neutrophils in distress are engulfed by macrophages. Some of the dynamics of parasite uptake by neutrophils specifically with respect to the phagocytic receptors that mediate parasite entry were investigated by Soong and colleagues (12). When infections were performed in standard cell medium, Carlsen et al. found that although a greater proportion of promastigotes are killed after neutrophil internalization, both promastigote and amastigote forms of *Leishmania amazonensis* infect neutrophils comparably. Incubation of parasites in heat-inactivated serum obtained from *L. amazonensis*-infected mice significantly increased uptake by neutrophils of both promastigote and amastigote forms. Interestingly, tissue-derived amastigotes infected neutrophils at a slightly higher rate. Carlsen et al. then observed differences in the quality of neutrophil activation when infections were initiated by either promastigote or amastigote forms. Uptake of promastigote forms by neutrophils elicited reactive oxygen species (ROS) as well as the production of TNFα. By contrast, internalization of amastigotes by neutrophils resulted in the preferential release of IL-10 and ROS as well. Other consequences of this differential activation of neutrophils included the increase in the lifespan of infected neutrophils and their quality of death. Death by necrosis as compared to apoptotic death by neutrophils elicits dramatically different host responses (13). Differential engagement of the opsonin-dependent receptors by *Leishmania* parasites during their entry into neutrophils appears to result in similar outcomes as compared to macrophages.

ROLE OF Abl FAMILY KINASES IN INTERNALIZATION OF *LEISHMANIA* PARASITES

Of course, phagocytic receptors are associated with receptor linked kinases that transduce signals into the cell. No recent studies have evaluated the roles of these kinases in *Leishmania* infections. It should also be acknowledged that in addition to kinases, small GTPases, including Cdc42, Rac1, and Rho, play distinguishable roles in the uptake of particles as well as *Leishmania* parasites (14). That said, a report by Wetzel et al. (15) uncovered the important role of the Abl family of kinases in the uptake of *Leishmania* parasites *via* either the CR3 or the Fc receptor. The Abl family of protein kinases (first discovered as the oncogene in the Abelson leukemia virus) are non-receptor kinases that transduce signals from diverse extracellular stimuli that can result in cytoskeletal rearrangement during phagocytosis, cell to cell contact, and cell motility (16). They are composed of two members, Abl1 and Abl2

(Arg). Using bone marrow-derived macrophages obtained from mice genetically modified to be deficient in Abl1 (Abl$^{-/-}$), Wetzel et al. found that there was an up to 42% reduction in the uptake of C3bi-opsonized *L. amazonensis* promastigotes as compared to the uptake of these parasites by cells from wild-type mice. Uptake of antibody-opsonized amastigotes was unaffected in Abl$^{-/-}$ cells. They then showed that Imatinib (an Abl family kinase inhibitor) reduced C3bi-opsonized promastigotes uptake down to comparable levels as the Abl$^{-/-}$. In addition, cells from mice that were genetically modified to be deficient in both Abl and Arg [because of embryonic lethality of double knockouts, this was achieved by using a conditional knockout strategy to inactivate the *abl* allele in an arg$^{-/-}$ genetic background (henceforth called dbKO Abl/Arg)] were also shown to have reduced uptake C3bi-opsonized promastigotes. The role of the Abl kinase in mediating the entry of parasites *via* the CR3 receptor was supported by additional experiments in the mouse macrophage cell line RAW264.7. They showed that C3bi-opsonized promastigotes entry into RAW264.7 was significantly reduced when infections of these cells were performed in the presence of M1/70 (a CR3 blocking antibody) and not F16/32 (an FcR blocking antibody).

Parallel experiments were performed to evaluate the role of Arg kinases (other Abl family member) that has also been implicated in interactions with the cell cytoskeleton. Bone marrow-derived macrophages from mice that were engineered to be genetically deficient in Arg (Arg$^{-/-}$) were found to take up C3bi-opsonized promastigotes at comparable levels with cells from wild-type mice. However, in these cells, uptake of IgG-opsonized amastigotes was reduced by 46%. Comparable reductions of opsonized amastigotes were observed in dbKO Abl/Arg. In addition, Imatinib also resulted in significantly reduced uptake of opsonized amastigotes. Although the uptake of C3bi-opsonized amastigotes was also reduced in Arg$^{-/-}$ cells, these cells took up C3bi-coated beads as efficiently as did cells from wild-type mice. This latter observation with *Leishmania* amastigotes underscores the complexity of the internalization schemes employed by these parasites and their capacity to engage other receptors when needed.

The most remarkable part of the Wetzel et al. study was the *in vivo* experiments. It should be noted that in the *in vitro* studies, the experimental design called for short-term incubations of parasites with macrophages (20 min in the case of amastigote infections and 90 min in the case of promastigote infections), as these parasites can apparently employ alternate receptors to ensure their uptake. Remarkably, there was significant reduction in the course of infection in Arg$^{-/-}$ mice. This was consistent with the *in vitro* observations that showed that Arg kinase played a significant role in the uptake of IgG-opsonized amastigotes. Further confirmation of the critical role of parasite uptake mediated by Arg kinase was obtained in dbKO Abl/Arg mice and in mice treated with Imatinib. Reduced lesion sizes in these mice corresponded with significant reductions in the parasite burden that was determined at the end of the experiment. Imatinib had been shown not to have a direct effect on parasite viability. The authors of the study concluded that the reduction in parasite burden and by consequence the limitation of the course of infection was due to the role of the Abl family kinases in the uptake of these

parasites. Although Abl family kinases play important roles in T cell functions, T cell responses in the Abl family kinase knockout mice appeared as well as in the drug-treated mice appeared normal, which ruled out the possibility that alterations in the course of infections were the result of T cell abnormalities.

The Abl family of protein kinases might play a greater role in *Leishmania* infections than is presently appreciated. The observations with Imatinib suggest that this drug or a derivative could be useful in combination with other drugs that target the parasite to control *Leishmania* infections. Along these lines, the uptake of several bacterial pathogens, including *Shigella flexneri*, *Chlamydia trachomatis*, and *Mycobacterium* spp., has also been shown to utilize the Abl family of tyrosine kinases during their entry into cells (17–19). In infections by *Mycobacterium tuberculosis* and *Mycobacterium marinum*, Napier et al. showed that administration of Imatinib to infected mice reduced their bacterial load and associated pathology. Taken together, these studies lend support to the proposition that identification of cellular host genes that are exploited by pathogens could be targeted to control a significant group of pathogens (20).

PROTEIN KINASE REGULATED BY RNA

The PKR is an important antiviral kinase that promotes many cellular processes, including cytokine production. During some viral infections, the viral dsRNA binds to the N-terminal of PKR, which results in dimerization and autophosphorylation of PKRs. Once the PKR is activated, it phosphorylates and inhibits the eukaryotic translation initiation factor 2α (elf2α) to reduce overall translation levels of the host cell proteins (21). Besides the inhibition of translation, PKR can also activate NF-κB, which acts to increase the production of cytokines (such as IL-10 and type 1 interferons) from the host cell. PKR is also activated in bacterial infections; however, its role appears to be controversial. In some bacterial infections, PKR activation has been shown to be induced upon the interactions of their cell wall components with toll-like receptors (TLRs). PKRs in turn activate NF-κB signaling that culminates in the production of inflammatory cytokines, including TNF-α and IL-6 (22). However, as we discuss below, in some infections, activation of PKRs is disease promoting.

ROLE OF PKR IN *L. AMAZONENSIS*

In several recent publications, Lopes and colleagues have provided evidence of an infection enhancing role of PKR in *L. amazonensis* infections. Pereira et al. (23) showed that *L. amazonensis* infection of cultured human or mouse macrophage cell lines induced the activation of PKR as monitored by the time-dependent increase in the phosphorylated form of PKR (pPKR). Supportive evidence of the increased activation of PKR by *L. amazonensis* infection was obtained by the observation of a time-dependent increase in the pelf2α. They then showed interestingly that infections performed in cells expressing a dominant negative mutant of PKR had lesser parasites than in controls, suggesting that pPKR induction augments the *L. amazonensis* parasite burden. Although activation of PKR by other stimuli such as poly(I:C) had been shown to be associated with induction of NO production, they found that *L. amazonensis* infection inhibited PKR-dependent NO production by a mechanism that involved aberrant induction of NF-κB; specifically, infection induced the translocation of the inhibitory p50/p50 homodimer into the nucleus. In a subsequent study by Barreto-de-Souza et al. (24), they showed that PKR augmentation of *L. amazonensis* infections occurs subsequent to its induction of IL-27, which is a cytokine that is structurally related to IL-12 (25). IL-27 that is produced in a time-dependent manner by *L. amazonensis* infection can be inhibited by expression of a dominant negative form of PKR. Addition of exogenous IL-27 enhanced the *L. amazonensis* load in infected cultures. Augmentation of *L. amazonensis* in cells was inhibitable by addition of anti-IL-27 to the infection cultures. Given that some reports have shown that IL-27 induces IL-10 production, which can in turn participate in the inhibition of *L. amazonensis* proliferation, Barreto-de-Souza et al. performed infections in the presence of exogenous IL-27 and antibodies to IL-10 receptor. Inhibition of IL-10 uptake by cells reduced the IL-27 enhancement of *L. amazonensis* infection.

These observations are consistent with studies that evaluated *M. tuberculosis* infections in mice that were deficient in PKR. PKR$^{-/-}$ mice were found to contain fewer viable bacteria than wild-type mice after infection with Mtb (26). In addition, PKR$^{-/-}$ mice exhibited less pulmonary pathology than wild-type mice. It was then shown that in the absence of PKR, infected cells were more prone to undergo apoptosis in response to Mtb infection and exhibited enhanced activation in response to IFN-γ. They reasoned that PKR promotes most likely induces a constitutive low level of IL-10 production that restrains macrophage activation and by so doing promotes pathogen persistence. Administration of a PKR inhibitor is therefore a plausible approach for pathogen control.

ROLE OF PKR IN *L. MAJOR* INFECTIONS

An opposite role for PKR has been observed in *L. major* infections. Here, PKR activation is associated with increased *L. major* death within infected cells. This was observed acutely in infection studies in RAW264.7 macrophages where at 3 h post-infection the parasite burden in wild-type cells (RAW-Bla, transfected with empty plasmid) was comparable to the burden in cells expressing a dominant negative variant of PKR (RAW-DN-PKR) (27). However, at 24 h post-infection, the burden of *L. major* parasites in RAW-Bla cells was significantly less than in RAW-DN-PKR. Poly(I:C) had no effect on the survival of *L. major* in RAW-DN-PKR cells. It had previously been observed that unlike in infections with *L. amazonensis*, in which infection induces the translocation of the p50/p50 NF-κB homodimers into the nucleus, even upon poly(I:C) stimulation, *L. major* infection resulted in translocation of the stimulatory p65/p50 heterodimers. Evidently, *L. major* and *L. amazonensis* must express different infection promoting mechanisms. In light of

previous studies that had characterized the inhibitor of parasite elastase (ISP) in *L. major* and that had shown that ISPs bind to neutrophil elastase (NE) on the surface of macrophages, the role of ISPs as potential regulators of macrophage responses to *L. major* was evaluated. Parasites that were deficient in ISPs (Δisp2/isp3) were derived and evaluated in infections of RAW-Bla and RAW-DN-PKR cells (27, 28). Interestingly, ISP-deficient cells were internalized more efficiently by RAW-Bla macrophages. They also survived better in RAW-DN-PKR macrophages, which suggested that PKR plays a role in reduced survival of *L. major* in macrophages. This observation was confirmed in infections of primary cells from PKR$^{-/-}$ mice and wild-type mice (129Sv); *L. major* parasite burdens at 24 h post-infection were not different from those in cells for PKR$^{-/-}$ mice. This was different from the observations in infections with *L. amazonensis* where a deficiency of PKR eliminated the augmentation of parasite burdens.

Additional studies to identify the cellular receptors that mediate the activation of PKR during *L. major* infections found that TLR2, TLR4, and CR3 most likely work in concert with neutrophil elastase to activate PKR. In light of differences in susceptibility of *L. major* to the induction of PKR, which is in contrast to *L. amazonensis* that is induced to replicate, Faria et al. (27) proposed that ISP characteristics of these species could be the significant difference. *L. major* parasites express higher levels of ISPs as compared to *L. amazonensis*. These ISPs then interact differentially with TLR2 and TLR4 that form a complex with neutrophil elastase and CR3. *L. amazonensis* appears to exhibit a greater preference for interactions with TLR2. Taken together, the parasite's interactions with surface receptors, including CR3, set in motion the eventual activation of PKR that controls parasite replication, survival, as well as the release of critical cytokines by infected cells.

CONCLUDING REMARKS

An understanding of how mammalian cells cope with or are transformed by *Leishmania* infection will necessitate a complete understanding of changes in activation of many host cell protein kinases. Parasite interactions with surface molecules on mammalian host cells initiates host cell responses that significantly influence the progress of the infection. The Abl family of protein kinases and also PKR are non-receptor protein kinases that play important roles in determining the outcome of *Leishmania* infections. Detailed studies of responses elicited to each *Leishmania* species underscore the differences between these parasites.

AUTHOR CONTRIBUTIONS

All authors listed, have made substantial, direct, and intellectual contribution to the work, and approved it for publication.

FUNDING

Supported by NIH grant # R21AI115218-01A.

REFERENCES

1. Ribeiro-Gomes FL, Sacks D. The influence of early neutrophil-*Leishmania* interactions on the host immune response to infection. *Front Cell Infect Microbiol* (2012) 4(2):59. doi:10.3389/fcimb.2012.00059

2. Martinez PA, Petersen CA. Chronic infection by *Leishmania amazonensis* mediated through MAPK ERK mechanisms. *Immunol Res* (2014) 59(1–3): 153–65. doi:10.1007/s12026-014-8535-y

3. Rotella DP. Recent results in protein kinase inhibition for tropical diseases. *Bioorg Med Chem Lett* (2012) 22:6788–93. doi:10.1016/j.bmcl.2012.09.044

4. Reinhard K, Huber M, Lohoff M, Visekruna A. The role of NF-κB activation during protection against *Leishmania* infection. *Int J Med Microbiol* (2012) 302:230–5. doi:10.1016/j.ijmm.2012.07.006

5. Mol M, Patole MS, Singh S. Immune signal transduction in leishmaniasis from natural to artificial systems: role of feedback loop insertion. *Biochim Biophys Acta* (2014) 1840:71–9. doi:10.1016/j.bbagen.2013.08.018

6. Mosser DM. Receptors on phagocytic cells involved in microbial recognition. *Immunol Ser* (1994) 60:99–114.

7. Kima PE, Constant SL, Hannum L, Colmenares M, Lee KS, Haberman AM, et al. Internalization of *Leishmania mexicana* complex amastigotes via the Fc receptor is required to sustain infection in murine cutaneous leishmaniasis. *J Exp Med* (2000) 191:1063–8. doi:10.1084/jem.191.6.1063

8. Miles SA, Conrad SM, Alves RG, Jeronimo SM, Mosser DM. A role for IgG immune complexes during infection with the intracellular pathogen *Leishmania*. *J Exp Med* (2005) 201:747–54. doi:10.1084/jem.20041470

9. Desjardins M, Descoteaux A. Inhibition of phagolysosomal biogenesis by the *Leishmania* lipophosphoglycan. *J Exp Med* (1997) 185:2061–8. doi:10.1084/jem.185.12.2061

10. Polando R, Dixit UG, Carter CR, Jones B, Whitcomb JP, Ballhorn W, et al. The roles of complement receptor 3 and Fcγ receptors during *Leishmania* phagosome maturation. *J Leukoc Biol* (2013) 93:921–32. doi:10.1189/jlb.0212086

11. Ricardo-Carter C, Favila M, Polando RE, Cotton RN, Bogard Horner K, Condon D, et al. *Leishmania major* inhibits IL-12 in macrophages by signalling through CR3 (CD11b/CD18) and down-regulation of ETS-mediated transcription. *Parasite Immunol* (2013) 35:409–20. doi:10.1111/pim.12049

12. Carlsen ED, Hay C, Henard CA, Popov V, Garg NJ, Soong L. *Leishmania amazonensis* amastigotes trigger neutrophil activation but resist neutrophil microbicidal mechanisms. *Infect Immun* (2013) 81:3966–74. doi:10.1128/IAI.00770-13

13. Carlsen ED, Liang Y, Shelite TR, Walker DH, Melby PC, Soong L. Permissive and protective roles for neutrophils in leishmaniasis. *Clin Exp Immunol* (2015) 182:109–18. doi:10.1111/cei.12674

14. Morehead J, Coppens I, Andrews NW. Opsonization modulates Rac-1 activation during cell entry by *Leishmania amazonensis*. *Infect Immun* (2002) 70:4571–80. doi:10.1128/IAI.70.8.4571-4580.2002

15. Wetzel DM, McMahon-Pratt D, Koleske AJ. The Abl and Arg kinases mediate distinct modes of phagocytosis and are required for maximal *Leishmania* infection. *Mol Cell Biol* (2012) 32:3176–86. doi:10.1128/MCB.00086-12

16. Bradley WD, Koleske AJ. Regulation of cell migration and morphogenesis by Abl-family kinases: emerging mechanisms and physiological contexts. *J Cell Sci* (2009) 122:3441–54. doi:10.1242/jcs.039859

23. Pereira RM, Teixeira KL, Barreto-de-Souza V, Calegari-Silva TC, De-Melo LD, Soares DC, et al. Novel role for the double-stranded RNA-activated protein kinase PKR: modulation of macrophage infection by the protozoan parasite *Leishmania FASEB J* (2010) 24:617–26. doi:10.1096/fj.09-140053

24. Barreto-de-Souza V, Ferreira PL, Vivarini Ade C, Calegari-Silva T, Soares DC, Regis EG, et al. IL-27 enhances *Leishmania amazonensis* infection via ds-RNA dependent kinase (PKR) and IL-10 signaling. *Immunobiology* (2015) 220:437–44. doi:10.1016/j.imbio.2014.11.006

25. Iwasaki Y, Fujio K, Okamura T, Yamamoto K. Interleukin-27 in T cell immunity. *Int J Mol Sci* (2015) 16:2851–63. doi:10.3390/ijms16022851

26. Wu K, Koo J, Jiang X, Chen R, Cohen SN, Nathan C. Improved control of tuberculosis and activation of macrophages in mice lacking protein kinase R. *PLoS One* (2012) 7(2):e30512. doi:10.1371/journal.pone.0030512

27. Faria MS, Calegari-Silva TC, de Carvalho Vivarini A, Mottram JC, Lopes UG, Lima AP. Role of protein kinase R in the killing of *Leishmania major* by macrophages in response to neutrophil elastase and TLR4 via TNFα and IFNβ. *FASEB J* (2014) 28:3050–63. doi:10.1096/fj.13-245126

28. Eschenlauer SC, Faria MS, Morrison LS, Bland N, Ribeiro-Gomes FL, DosReis GA, et al. Influence of parasite encoded inhibitors of serine peptidases in early infection of macrophages with *Leishmania major*. *Cell Microbiol* (2009) 11:106–20. doi:10.1111/j.1462-5822.2008.01243.x

17. Elwell CA, Ceesay A, Kim JH, Kalman D, Engel JN. RNA interference screen identifies Abl kinase and PDGFR signaling in *Chlamydia trachomatis* entry. *PLoS Pathog* (2008) 4(3):e1000021. doi:10.1371/journal.ppat.1000021

18. Burton EA, Plattner R, Pendergast AM. Abl tyrosine kinases are required for infection by *Shigella flexneri*. *EMBO J* (2003) 22:5471–9. doi:10.1093/emboj/cdg512

19. Napier RJ, Rafi W, Cheruvu M, Powell KR, Zaunbrecher MA, Bornmann W, et al. Imatinib-sensitive tyrosine kinases regulate mycobacterial pathogenesis and represent therapeutic targets against tuberculosis. *Cell Host Microbe* (2011) 10:475–85. doi:10.1016/j.chom.2011.09.010

20. Schwegmann A, Brombacher F. Host-directed drug targeting of factors hijacked by pathogens. *Sci Signal* (2008) 1(29):re8. doi:10.1126/scisignal.129re8

21. Pfaller CK, Li Z, George CX, Samuel CE. Protein kinase PKR and RNA adenosine deaminase ADAR1: new roles for old players as modulators of the interferon response. *Curr Opin Immunol* (2011) 23:573–82. doi:10.1016/j.coi.2011.08.009

22. Cabanski M, Steinmüller M, Marsh LM, Surdziel E, Seeger W, Lohmeyer J. PKR regulates TLR2/TLR4-dependent signaling in murine alveolar macrophages. *Am J Respir Cell Mol Biol* (2008) 38:26–31. doi:10.1165/rcmb.2007-0010OC

6

The Prophylactic Effect of Probiotic *Enterococcus lactis* IW5 against Different Human Cancer Cells

Yousef Nami[1], Babak Haghshenas[1], Minoo Haghshenas[2], Norhafizah Abdullah[3] and
Ahmad Yari Khosroushahi[4,5]*

[1] Institute of Biosciences, Universiti Putra Malaysia, Selangor, Malaysia, [2] School of Medicine, Shahid Beheshti University of Medical Sciences, Tehran, Iran, [3] Chemical and Environmental Engineering Department, Faculty of Engineering, Universiti Putra Malaysia, Selangor, Malaysia, [4] Drug Applied Research Center, Tabriz University of Medical Sciences, Tabriz, Iran, [5] Department of Pharmacognosy, Faculty of Pharmacy, Tabriz University of Medical Sciences, Tabriz, Iran

Edited by:
Diana Bahia,
Universidade Federal de Minas
Gerais, Brazil

Reviewed by:
Amit Kumar Tyagi,
The University of Texas MD Anderson
Cancer Center, USA
Sahdeo Prasad,
The University of Texas MD Anderson
Cancer Center, USA

*Correspondence:
Ahmad Yari Khosroushahi
Yarikhosroushahia@tbzmed.ac.ir

Enterococcus lactis IW5 was obtained from human gut and the potential probiotic characteristics of this organism were then evaluated. Results showed that this strain was highly resistant to low pH and high bile salt and adhered strongly to Caco-2 human epithelial colorectal cell lines. The supernatant of *E. lactis* IW5 strongly inhibited the growth of several pathogenic bacteria and decreased the viability of different cancer cells, such as HeLa, MCF-7, AGS, HT-29, and Caco-2. Conversely, *E. lactis* IW5 did not inhibit the viability of normal FHs-74 cells. This strain did not generate toxic enzymes, including β-glucosidase, β-glucuronidase, and *N*-acetyl--glucosaminidase and was highly susceptible to ampicillin, gentamycin, penicillin, vancomycin, clindamycin, sulfamethoxazol, and chloramphenicol but resistant to erythromycin and tetracyclin. This study provided evidence for the effect of *E. lactis* IW5 on cancer cells. Therefore, *E. lactis* IW5, as a bioactive therapeutics, should be subjected to other relevant tests to verify the therapeutic suitability of this strain for clinical applications.

Keywords: anticancer, enzyme activity, antibiotic susceptibility, apoptosis, cytotoxicity

INTRODUCTION

Probiotics are non-pathogenic live microorganisms that provide health benefits when these organisms are consumed in sufficient amounts (FAO/WHO, 2001; Mehra et al., 2012; Howarth and Wang, 2013; Haghshenas et al., 2014a; Nami et al., 2014a). Probiotics have been utilized to prevent bacterial infections (Forsyth et al., 2009) and treat cancer (Baldwin et al., 2010; Haghshenas et al., 2014b, 2015a,b; Nami et al., 2014b,c,d). These organisms can also create an acidic environment in the colon by producing short-chain fatty acids. Furthermore, probiotic bacteria can inhibit the occurrence of cancer by (i) lowering pH, (ii) reducing the level of pro-carcinogenic enzymes (Donaldson, 2004), (iii) enhancing cell proliferation by inhibiting normal cell apoptosis and by promoting cell differentiation and cytoprotective activities (Lin et al., 2008), and (iv) suppressing inflammation-induced cell apoptosis (Prisciandaro et al., 2011) caused by lactic acid bacteria (LAB), including *Lactobacillus*, *Enterococcus*, *Streptococcus*, and *Bifidobacterium*. Among these LABs, the genus *Enterococcus* has gained considerable interest in environmental, food, and clinical research (Sharma et al., 2012).

Enterococcus is ubiquitous in nature and considered as the most controversial LAB genus because of unclear functions (Galvez et al., 2009). Enterococci have been utilized as adjutants to treat human and animal diseases. Enterococci have also been used in the food industry as probiotics (Franz et al., 2003) or as starter cultures because these microorganisms produce useful bacteriocins (Fisher and Phillips, 2009). Although *Enterococcus* comprises many species, only a few species are recognized as probiotics, such as *E. faecalis*, *E. faecium*, and *E. lactis*. Probiotics should exhibit important characteristics, such as tolerant to gastrointestinal conditions (acid and bile) and non-pathogenic; probiotics should also display competitive exclusion of pathogens (Collins et al., 1998; Ouwehand et al., 2002). Thus, the selection criteria of probiotic bacteria for clinical applications should be carefully evaluated. This study aimed to determine the probiotic properties (bile tolerance, antimicrobial activity, and antibiotic susceptibility) and antitumor activities of *E. lactis* isolated from the human gut.

MATERIALS AND METHODS

Bacterial Strain and Culture Condition

Enterococcus lactis IW5 was isolated from human fecal samples using streak plate method previously described by Shin et al. (2015) and this strain was maintained at $-70°C$ in de Man Rogosa broth (MRS, Merck, Germany) containing 25% (v/v) glycerol. *E. lactis* IITRHR1 isolated from cheese was used as a control strain. Working cultures were anaerobically incubated at 37°C for 24 h in an anaerobic jar (Mitsubishi Inc. USA) that contains anaerobic gas generation kits (AnaeroPack).

Tolerance to Artificial Gastric Juice and Artificial Bile Acid

Tolerance to artificial gastric juice and bile acid were determined according to previously described method with slight modification (Lee et al., 2014). *E. lactis* was suspended in MRS containing 0.1% pepsin (Sigma, St. Louis, MO, USA) and adjusted to a pH of 2.0 with 0.1 M HCl, and then incubated for 3 h at 37°C. Artificial bile acid tolerance was measured by cultivating cells treated with artificial gastric juice. The cells were incubated at 37°C for 24 h in artificial bile acid consisting of MRS containing 0.3% oxgall (Becton Dickinson, Sparks, MD, USA). The numbers of viable cells were measured by incubating aliquots for 24 h on MRS agar plates at 37°C. The survival rate was calculated using the formulation:

Survival rate (%) = (Log CFU after reaction/Log CFU at 0 h) × 100

Antimicrobial Susceptibility Assay

Thirteen pathogenic organisms from the Persian Type Culture Collection (**Table 1**) were selected to detect antagonistic substances. Well diffusion was performed to detect inhibitory substances produced in the supernatant fluid of the isolate. For this purpose, an overnight culture of the indicator strains was used to inoculate appropriate agar growth media (Dimitonova et al., 2007) at 37°C. Wells with a diameter of 5 mm were cut into agar plates; afterward, 50 μL of filtered cell-free supernatant obtained from the third subculture of the microorganisms grown in MRS broth (cell density 10^8 cfu/mL) was added to each well. The supernatant was obtained by growing inhibitory producer strains overnight in MRS broth at 37°C. The cells were removed through centrifugation; the supernatant was placed in the wells and allowed to diffuse in agar for 2 h at room temperature. The plates were incubated at optimum growth temperature of the indicator strains and examined after 24 h to determine inhibition zone areola diameter (Nowroozi et al., 2004; Maldonado et al., 2012).

Enzyme Activity

Enzyme activity was evaluated using an API ZYM kit (BioMerieux, Paris, France). *E. lactis* IW5 was suspended in sterile saline (0.85% NaCl) at 10^5 CFU/mL and added to each cupule. After inoculation was performed, the cultures were incubated at 37°C for 4 h. One drop of ZYM B reagent was added and a drop of surface-active agent (ZYM reagent) was added to each cupule. ZYM A was introduced to facilitate ZYM B solubilization in the medium. The resulting color was observed for at least 5 min. Values ranging from 0 to 5 were assigned on the basis of color strength to determine the approximate amount (in nmol) of hydrolyzed substrate.

Cell Cultures

Five human cancer lines, namely, Caco-2 (human colorectal carcinoma cell), AGS (human gastric carcinoma cell), MCF-7 (human breast carcinoma cell), HeLa (human cervical carcinoma cell), and HT-29 (human colon carcinoma cell), and one normal cell line, namely, FHs-74 (human intestinal epithelial cells) – obtained from cell resource center of Pasteur institute of Iran (Tehran, Iran) – were used to investigate the anticancer effects of *E. lactis* IW5. The cells were grown in RPMI-1640 medium supplemented with 10% heat-inactivated fetal bovine serum and a 1% penicillin–streptomycin mixture. The cultures were maintained at 37 °C in an atmosphere of 95% O_2 and 5% CO_2 with relative humidity (Merghoub et al., 2009).

Cell-free Culture Supernatant Preparation

The liquid culture of *E. lactis* at the end of the exponential growth phase was centrifuged at 4000 × *g* for 10 min to obtain cell precipitates. The supernatant was collected; pH was adjusted to 7.2 with 1 N NaOH and subjected to lyophilization. Endogenous proteases were inactivated by heat at 100 C for 3–5 min. The desired concentrations of lyophilized culture supernatant (10–50 μg/mL) were prepared in RPMI media by diluting from stock solution (10 mg lyophilized supernatant/mL RPMI media) and sterilized by filtering the supernatant through a 0.22 μm bacterial filter (Millipore); the prepared supernatant was then used to treat cancer cells.

Adhesion to Caco-2 Cells

Enterococcus lactis IW5 was assessed for its adhesion ability to the human colon carcinoma cell line, Caco-2. The cells were seeded in RPMI-1640 medium supplemented with 10% heat-inactivated

TABLE 1 | The inhibitory effect of *Enterococcus lactis* IW5 against pathogenic bacteria.

Test organisms	Growth conditions	Origin	Susceptibility
Salmonella typhimurium	MPA, 37°C	ATCC 14028	R
Escherichia coli O26	LB, 37°C	Native strain	S
E. coli O157	LB, 37°C	PTCC 1276	R
Staphylococcus aureus	Blood agar, 37°C	ATCC 25923	S
Bacillus cereus	MPA, 37°C	PTCC 1539 (ATCC 11778)	S
Listeria monocytogenes	BHI, 37°C	PTCC 1163	ES
Klebsiella pneumoniae	MPA, 37°C	PTCC 1053 (ATCC 10031)	S
Shigella flexneri	MHA, 37°C	PTCC 1234 (NCTC 8516)	S
Pseudomonas aeroginosa	MPA, 37°C	PTCC 1181	R
Candida albicans	MHA, 28°C	PTCC 5027 (ATCC 10231)	R
Serratia marcesens	MHA, 37°C	PTCC 1187 (Native strain)	R
Streptococcus mutans	MHA, 37°C	PTCC 1683 (ATCC 35668)	SS
Staphylococcus saprophyticus	Blood agar, 37°C	PTCC 1440 (CIP 76.125)	R

R: 0 mm; SR: 0–4 mm; SS: 4–8 mm; S: 8–12 mm; ES: >12 mm.

CIP, Collection of Bacteries de l'Institute Pasteur, Paris, France; ATCC, American Type Culture Collection, Virginia, USA; NCTC, National Collection of Type Cultures, London, UK; PTCC, Persian Type Culture Collection, Tehran, Iran.

MPA, mycophonolic acid; LB, Lysogeny broth; BHI, Brain-heart infusion medium; MHA, Mueller Hinton Agar.

fetal bovine serum and 1% penicillin/streptomycin mixture. Cells were seeded on 24-well tissue culture plates and incubated at 37°C in 5% CO_2 in a relatively humid atmosphere until a confluent monolayer was achieved. Adherence assay was carried out by adding 1 mL of the bacterial strain, suspended in RPMI-1640 medium, at a concentration of about 1×10^7 CFU/well and was incubated for 3 h at 37°C in an atmosphere of 5% (v/v) CO_2. Before the adhesion assay, the media in the wells containing a Caco-2 cell monolayer were removed and replaced once with fresh antibiotic-free RPMI.

To remove non-attached bacterial cells, the wells were washed three times with a sterile, pre-warmed PBS solution. To detach the cells from the wells, 1 mL of trypsin/EDTA solution (0.5% porcine trypsin and 0.2% EDTA in PBS; Sigma) was added to each well and the mixture was gently stirred for 5 min. To measure the viable Caco-2 cell count, the cells were counted by the pure plate method onto MRS agar medium and incubated at 37°C under anaerobic conditions. Bacterial adhesion was expressed as the total number of bacteria attached to viable Caco-2 cells.

Cytotoxicity against Different Cancer Cells

The cytotoxicity of the isolated *E. lactis* on tumor/normal cells was evaluated through a microculture tetrazolium [MTT, 3-(4, 5-dimethylthiazol-2-yl)-2,5-diphenyltetrazolium bromide] assay (Mosmann, 1983). In brief, HeLa, AGS, MCF-7, HT-29, Caco-2, and FHs 74 cells (1.2×10^4 cells/well) were seeded in each well of a 96-well microplate with RPMI growth medium. Once 50% confluence was reached 24 h after the cells were seeded, the cells were treated with the filtered supernatant of the isolated strain at different time points (12, 24, and 48 h). After treatment was administered, the medium was replaced with 200 μl of fresh medium containing 50 μl of MTT solution (2 mg/mL in PBS) and incubated for another 4 h at 37 C. After incubation was completed, the MTT mixture was carefully removed, and 200 μl of dimethyl sulfoxide and 25 μl of Sorenson's glycine buffer (0.1 M glycine and 0.1 M NaCl at pH 10.5) were added

to each well and incubated for 30 min. The absorbance of each well was determined after 30 s of shaking by using a microplate reader (Biotek, ELx 800, USA) at 570 nm. The cells treated with MRS (bacterial culture medium) and Taxol (anticancer drug as a reference) served as negative and positive controls, respectively.

Apoptotic Cells Detection
4′,6-diamidino-2-phenylindole (DAPI) staining

All of the cultured cells (treated/untreated groups) were evaluated through 4′,6-diamidino-2-phenylindole (DAPI) staining to detect apoptotic cells. For this purpose, sterile cover slips were placed in each of the six wells of the culture plate. Cancer cells (120×10^4 cells/well) were added to each well and maintained under the desired standard culture condition. At 24 h after the cells were seeded, all of the cultured cells were subjected to *E. lactis* secretion, MRS medium, and Taxol (IC_{50} concentration) treatments. The treated and untreated control groups were incubated for another 24 h and prepared for apoptosis assay. Afterward, 4% paraformaldehyde was added to each well to stain cells with DAPI dye. The cells were fixed and permeabilized with 0.1% Triton-X100 for 5 min. The permeabilized cells were stained with 50 μl of DAPI dye (1:2000 dilutions) and incubated for 3 min at room temperature. The processed cells with cover slips were washed thrice with PBS (pH 7.2) and utilized to assess apoptosis by using a fluorescent microscope (BX64, Olympus, Japan) equipped with a U-MWU2 fluorescence filter (excitation filter BP 330-385, dichromatic mirror DM 400, and emission filter LP 420; Paolillo et al., 2009).

Flow cytometry

The fraction of apoptotic cells was quantitatively measured via flow cytometry using the Annexin V-FITC apoptosis detection kit (eBioscience, San Diego, CA, USA). HeLa cell line (1.2×10^5 cells/well) was seeded into a six-well culture plate and the treatment of cells were similar to DAPI staining. After treatment time point (24 h), the treated/untreated control cells were detached by trypsin, the supernatant was discarded after centrifugation at 900 rpm for 10 min at 28 C, and the cell pellet was resuspended in 500 μl of 1× binding buffer and transferred into a new 5 ml tube. The tubes were centrifuged again and the supernatants were replaced with100 μl binding buffer (1×). Afterward, the tubes were added with 5 μl of FITC-conjugated Annexin V then were incubated for 15 min at room temperature under dark conditions. The incubated cells were centrifuged and the cell plates were resuspended in 500 μl of binding buffer (1×). Finally, 5 μl of propidium iodide solution was added to the cells, and quadrant settings were fixed with untreated, single-stained controls, and copied to dot plots of the treated cells. Quadrant statistic calculations were performed using CELLQuest Pro software (BD Biosciences, San Jose, CA, USA). The experiment was repeated two times with triplicate samples for each experiment. Analyses were accomplished using 10000 cells at a rate of 450 cells/s.

Quantitative Real Time PCR

For RNA analysis, HeLa cells were lysed using TRI Reagent®(Sigma Chemical Co., Poole, UK) according to manufacture guidelines. 24 h post-treatment or untreated control monolayer cells were lysed by adding desired amount of TRI Reagent® (2 mL per 25 cm2 T-flask) accordingly were homogenized and transferred to RNAse/DNAse-free microtubes. Chloroform (0.2 mL per each mL of TRI Reagent^{TM} used for lysing) was added to each sample, and the mixture was vortexed. After maintaining at room temperature for 5 min, the samples were centrifuged at 12000 × g, 4°C and 10 min and the colorless supper aqueous phase was carefully separated and mixed with ice-cold isopropanol (0.5 mL per each mL of TRI Reagent® used initially). The mixture was centrifuged at12000 × g, 4°C for 10 min, yielding total RNA pellet that was washed with 75% ethanol (×3). The air dried samples were dissolved in DEPC treated water and tested qualitatively and quantitatively prior to its use for RT-PCR experiments.

The isolated RNA was reverse transcribed to cDNA using Moloney- murine leukemia virus (MMLV) reverse transcriptase (Bethesda Research Laboratories, Gaithersburg, MD, USA). For RT reaction, 1 μL RNA (1 μg/μL) was mixed with master mix [DEPC treated water 13 μL, dNTP's (10 μM) 2 μL, MMLV buffer with DTT 2 μL, random hexamer primer (pdN6; 400 ng/μL) 0.5 μL], and denatured at 95°C for 5 min. The sample was then cooled down to 4°C for 5 min using ice-bath. Then 1 μL MMLV (200 U/μL) and 0.5 μL RNase in (40 U/μL) were added to the sample and the mixture was incubated using following thermocycling program: 10 min at 25°C, 42 min at 42°C, and 5 min at 95°C. The prepared cDNA templates were used for real time PCR experiments.

Primers were designed from published Gene Bank sequences using Beacon Designer 5.01 (Premier Biosoft International, http://www.premierbiosoft.com) and listed in **Table 3**. All amplification reactions were performed in a total volume of 25 μL using iQ5 Optical System (Bio-Rad Laboratories Inc., Hercules, CA, USA). Each well contained: 1 μL cDNA, 1 μL primer (100 nM each primer), 12.5 μL 2× Power SYBR Green PCR Master Mix (Applied Biosystems, Foster City, CA, USA), and

10.5 μL RNAse/DNAse free water. Thermal cycling conditions were as follow: 1 cycle at 94°C for 10 min, 40 cycles at 95°C for 15 s, 56–62°C (annealing temperature) for 30 s, and 72°C for 25 s. Interpretation of the result was performed using the Pfaffle method and the threshold cycle (C_t) values were normalized to the expression rate of GAPDH as a housekeeping gene. All reactions were performed in triplicate and negative controls were included in each experiment.

Statistical Analysis

Data were analyzed by one-way ANOVA. Significant differences of means ($p < 0.05$) were compared through Duncan's test by using SPSS 19.0. Graphs were prepared using Microsoft Office Excel (Rahmati, 2011).

RESULTS AND DISCUSSION

Isolation and Identification

The bacterial strain was isolated from the human gut. The strain was initially identified by phenotypic methods; the Gram reaction of the isolates was determined by observation under a light microscope after Gram staining by using a Gram staining kit. LAB were considered Gram positive when they appeared blue–purple upon Gram staining. The isolates did not produce gas bubbles when hydrogen peroxide solution (3%) drops (Sigma–Aldrich, USA) were added to bacterial cells to determine catalase positive/negative strains; hence, the result confirmed that this strain is a Gram-positive and catalase-negative bacterium. A total of 45 Gram-positive and catalase-negative strains were obtained. Based on 16S rRNA identification results, the 45 isolated bacteria were classified into three major groups of LAB: enterococci, lactobacilli, and lactococci. After sequencing was performed, the strains belonging to *Enterococcus* genus were categorized into nine different species: one *E. lactis*, two *E. pseudoavium*, four *E. hirae*, two *E. gilvus*, four *E. avium*, three *E. durans*, eight *E. faecalis*, five *E. malodoratus*, and seven *E. faecium*. Moreover, lactobacilli were classified into three diverse species: one *L. casei*, three *L. acidophilus*, and one *L. plantarum*. Lactococci were classified into one species: three *Lactococcus lactis*, with two subspecies, namely, *L. lactis* ssp. *lactis* and *L. lactis* ssp. *cremoris*.

Probiotics have been extensively investigated because these organisms provide health benefits when such probiotics are consumed in sufficient amounts. In this study, LAB species with probiotic and antitumor activities were isolated; the strains that could grow in 5% CO_2 atmosphere. An *E. lactis* strain (Accession number: HF562969.1) resistant to pH 2.0 and 0.3% bile salt was isolated from the human gut and then identified.

Acid and Bile Tolerance

The survival of *E. lactis* IW5 and *E. lactis* IITRHR1 in artificial gastric juice (pH 2.0,0.1% pepsin, for 3 h) and artificial bile salt (0.3% oxgall, for 24 h)was evaluated (**Table 1**). The cells of *E. lactis* IW5 and *E. lactis* IITRHR1 were strongly maintained, with 94.60 and 92.27% survival rate in artificial gastric juice, respectively. In artificial bile salt, the cells of *E. lactis* IW5 and *E. lactis* IITRHR1demonstrated 95.46 and 94.14% survival rate, respectively. Our findings are similar to those of previous studies, which revealed that the survival rates of *Enterococcus* bacteria treated with acid and bile range from 63 to 100% (Haghshenas et al., 2014b; Nami et al., 2014d). Similarly, it has been revealed that *Enterococcus* bacteria were very stable in acidic conditions (pH 2 for 3 h) and high bile salt (0.3% oxgall for 4 h; Bhardwaj et al., 2010).

Antimicrobial Susceptibility Assay

The antimicrobial susceptibility spectrum of *E. lactis* IW5 is shown in **Table 2**. This strain inhibited the growth of pathogenic bacteria, including *Escherichia coli* O26, *Staphylococcus aureus*, *Bacillus cereus*, *Klebsiella pneumoniae*, *Shigella flexneri*, and *Streptococcus mutans*. Moreover, *E. lactis* exhibited strong activity against *Listeria monocytogenes*. No significant activity was observed against *Serratia marcesens*, *Pseudomonas aeruginosa*, *Candida albicans*, *Staphylococcus saprophyticus*, *Escherichia coli* O157, and *Salmonella typhimurium*.

The 50% inhibitory concentration (IC_{50}) of isolated strain metabolites was determined as an index of antagonistic activity from the antimicrobial time and dose-dependent curves. After 24 h of incubation, IC_{50} values were only observed in *E. coli* O26, *S. aureus*, *B. cereus*, *K. pneumoniae*, *S. flexneri*, *S. mutans*, and *L. monocytogenes* cells treated with *E. lactis* secretions. The IC_{50} for *E. lactis* secretions on *E. coli* O26, *S. aureus*, and *B. cereus* cells was 47, 28 and 32 μg/mL, respectively. The IC_{50} values of *E. lactis* secretions on *K. pneumoniae*, *S. flexneri*, and *S. mutans* cells was 31, 26 and 22 μg/mL, respectively. The IC_{50} value of *E. lactis* secretions on *L. monocytogenes* cells showed the lowest value (13 μg/mL). Our results showed that the *E. lactis* IW5 strain obtained from the human gut exhibited good probiotic properties, such as low pH and bile salt resistance. This strain was capable to inhibit several pathogenic bacteria.

Enzyme Activity

Certain enzymes are characteristically produced by probiotics to provide protection from toxic substances. β-glucosidase, *N*-acetyl-β-glucosaminidas, and β-glucuronidase have been associated with certain health disorders (Chen et al., 2014). β-glucuronidase increases the risk of carcinogenesis by secreting toxins and mutagens (Delgado et al., 2007; Dabek et al., 2008). These toxic enzymes could be produced by microorganisms.

TABLE 2 | Tolerance of *E. lactis* IW5 and *E. lactis* IITRHR1 against artificial gastric and bile conditions.

Treatment	Log CFU/mL
E. lactis IW5	
Initial cell no.	8.15 ± 0.26
pH 2.0, 0.1% pepsin, 2 h	7.71 ± 0.12
0.3% oxgall, 24 h	7.78 ± 0.36
E. lactis IITRHR1	
Initial cell no.	8.36 ± 0.18
pH 2.0, 0.1% pepsin, 2 h	7.71 ± 0.21
0.3% oxgall, 24 h	7.87 ± 0.19

The results are represented as mean ± SD.

Our data demonstrated that *E. lactis* did not produce toxic enzymes, including β-glucosidase, *N*-acetyl-β-glucosaminidase, and β-glucuronidase. Conversely, *E. lactis* produced various enzymes, including esterase (20 nmol), acid and alkaline phosphatase (5 nmol), and esterase lipase (≥25 nmol).

Adhesion Ability to Colon Endothelial Cells

Several investigations have implicated a number of factors in the attachment of probiotic bacterial cells to epithelial cells. Such factors include: passive entrapment of the bacterial cells by fimbrial cell matrix material (Sarem et al., 1996), bacterial cell surface-associated lipoteichoic acid (Granato et al., 1999), proteinaceous extracellular adhesins (Conway and Kjelleberg, 1989), and bacterial cell surface-associated proteinaceous factors (Adlerberth et al., 1996). Adhesion of *E. lactis* IW5 and *E. lactis* IITRHR1 was confirmed by using the plating technique. When *E. lactis* IW5 was plated at a concentration of 8.35 ± 0.06 log CFU/well, we found that 8.16 ± 0.04 log CFU/well of the bacteria adhered to the Caco-2 cells. Conversely, when *E. lactis* IITRHR1 was plated at a concentration of 8.13 ± 0.05 log CFU/well, it was found that only 6.45 ± 0.03 log CFU/well of the bacteria adhered to the Caco-2 cells. It has been reported previously that *E. lactis* IITRHR1 can strongly adhere to intestinal epithelial cells, which promote its survival and show a broad range of antimicrobial activity (Sharma et al., 2011). Similar to our findings, these data demonstrated that the bacterial concentration was reduced by 1.68 log CFU/well, following removal of the non-adhered cells.

Toxicity Assay

Microculture tetrazolium assay was performed to determine the cytotoxicity effects of the metabolites secreted by *E. lactis* IW5 on various cancer cell lines, particularly HeLa, Caco-2, AGS, and HT-29. The cytotoxicity potential of the metabolites produced by *E. lactis* IW5 on various cancer cells was determined (**Figures 1** and **2A–D,F**). After 24 h of incubation, the metabolites inhibited all cancer cell lines. Approximately 38, 36, 28, 40, and 30% of MCF-7, HeLa, HT-29, AGS, and Caco-2 cells, respectively, remained viable after these cells were incubated with the metabolites for 24 h. The antiproliferative effect of the metabolites on all of the evaluated cancer cells significantly differed from that of the un-treated and reference strain-treated groups. The effect of the metabolites on FHs 74 normal cells was also examined (**Figure 2E**). *E. lactis* IW5 secretions exhibited no toxic effect on normal cells; more than 95% of the cells grew well. These results indicated that *E. lactis* IW5 is a potential candidate for cancer treatment.

The anticancer activity of probiotic bacteria has been demonstrated by *in vivo* and *in vitro* systems (Ouwehand, 2007). Probiotic organisms inhibit mammalian cell proliferation in primary leukocyte cultures and cell lines. The induction of apoptotic cells by conjugated linoleic acid produced by various probiotic strains has been established in Caco-2 and HT-29 mammalian cancer cell lines. In this study, four human cancer cell lines, namely, Caco-2 (colorectal cancer), AGS (gastric cancer), HeLa (cervical cancer), and HT-29 (colon cancer), and

TABLE 3 | Real time PCR genes and their forward/reverse primers.

Primer	Forward and reverse primer	Sequence	Amplicon size	length
BAX	F	5′-CCCGAGAGGTCTTTTTCCGAG-3′	155	21
	R	5′-CCAGCCCATGATGGTTCTGAT-3′	155	21
BCL2	F	5′-GGTGGGGTCATGTGTGTGG-3′	130	19
	R	5′-CGGTTCAGGTACTCAGTCATCC-3′	130	22
CASPAS 9	F	5′-CTCAGACCAGAGATTCGCAAAC-3′	116	22
	R	5′-GCATTTCCCCTCAAACTCTCAA-3′	116	22
CASPAS 8	F	5′-GACAGAGCTTCTTCGAGACAC-3′	116	21
	R	5′-GCTCGGGCATACAGGCAAAT-3′	116	20
ErbB2	F	5′-TGTGACTGCCTGTCCCTACAA-3′	152	21
	R	5′-CCAGACCATAGCACACTCGG-3′	152	20
ErbB3	F	5′-GACCCAGGTCTACGATGGGAA-3′	99	21
	R	5′-GTGAGCTGAGTCAAGCGGAG-3′	99	20
BCL-XL	F	5′-GAGCTGGTGGTTGACTTTCTC-3′	101	21
	R	5′-TCCATCTCCGATTCAGTCCCT-3′	101	21

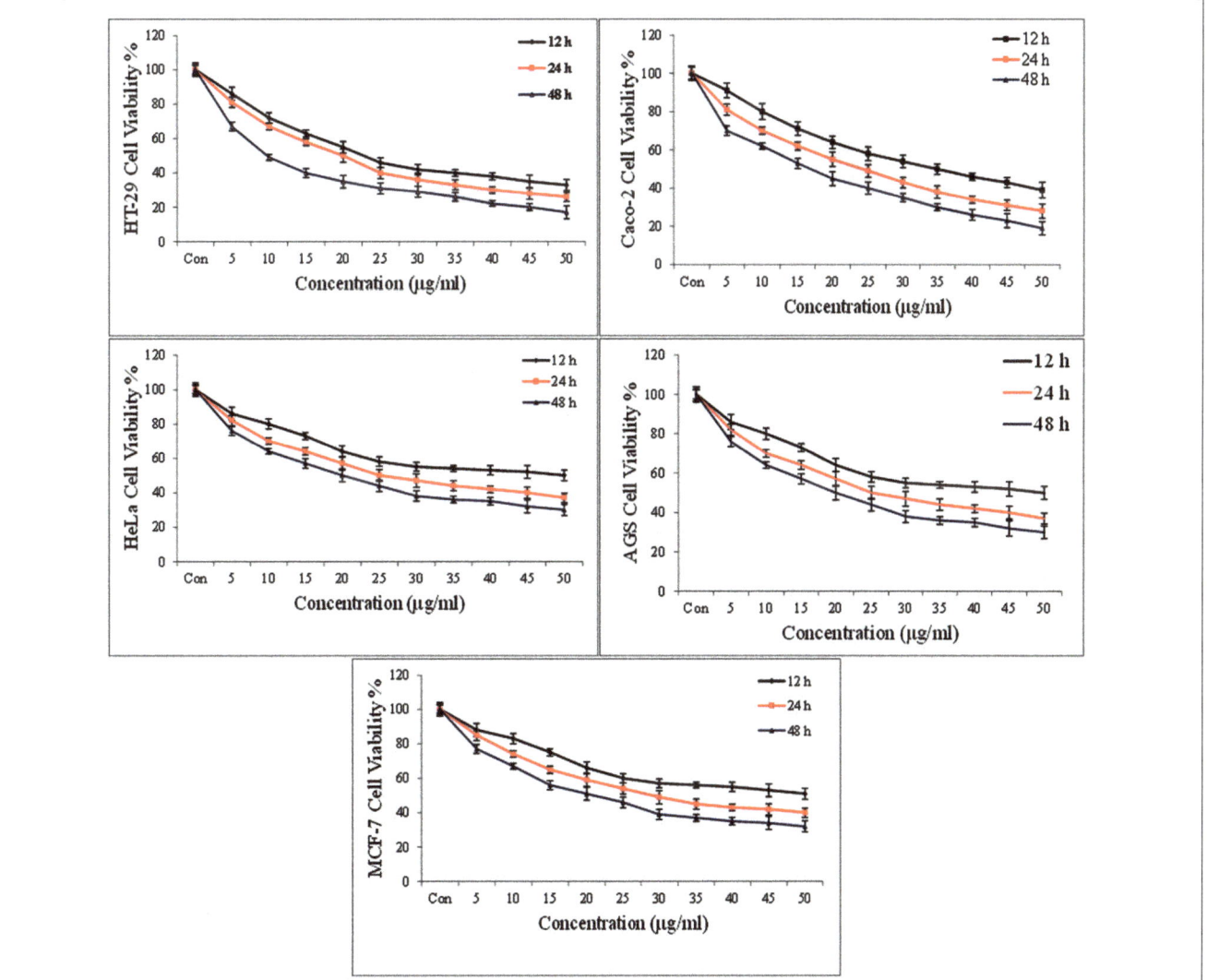

FIGURE 1 | The cytotoxic effects of isolated *Entreococcus lactis* IW5 secretion on different cancer cell lines at three time points 12, 24, and 48 h. Error bars represent the standard deviation of the each mean.

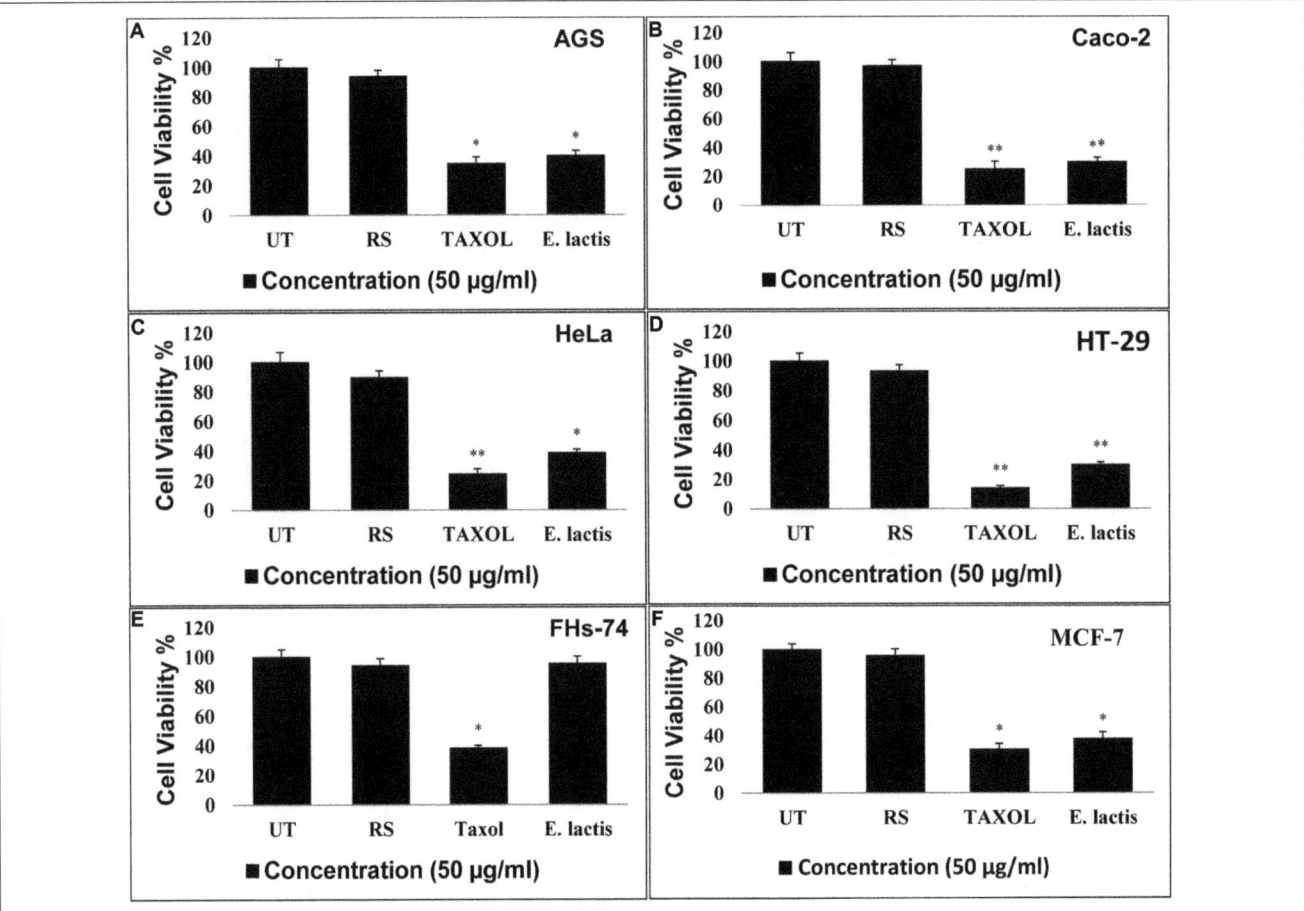

FIGURE 2 | Effect of *E. lactis* secretions on the proliferation of cancerous MCF-7, Caco-2, HT-29, HeLa, AGS, and FHs-74 normal cell lines. *E. lactis* secretions final concentration: 50 μg/mL, Taxol concentration: IC_{50} for each cell line. Incubation time is 24 h. Data are expressed as mean and error bars represent standard deviation of means. UT: Untreated media were used as control. *E. lactis* IITRHR1 was used as Reference Strain (RS) for comparison. Asterisks denote statistically significant differences ($*p < 0.05$; $**p < 0.01$).

one normal cell line, namely, FHs-74, were utilized. The results of this study demonstrated that the metabolites secreted by *E. lactis* IW5 strain significantly inhibited the growth of the four cancer cell lines. *E. lactis* IW5 secretions decreased the proliferation and viability of all cancer cell lines but did not adversely affect FHs-74 normal cells. Therefore, this strain was considered non-toxic. Different cancer cells were treated with 10^6 CFU/well of *E. lactis* IW5; this treatment strongly inhibited the proliferation of cancer cells. In *E. lactis* IW5 treatment, the proliferation of MCF-7, HeLa, HT-29, AGS, and Caco-2 cells was particularly inhibited by 38, 36, 28, 40, and 30%, respectively. Thus, *E. lactis* IW5 can inhibit the proliferation of cancer cells; however, *E. lactis* IITRHR1could not inhibit the proliferation of cancer cells.

Apoptosis Assay

HeLa cells were treated with 50 μg/mL of the filtered secretion after these cells were incubated for 24 h; the treated HeLa cells were stained with DAPI and analyzed through fluorescent microscopy to analyze the effect of *E. lactis* secretions on HeLa cell viability. The intact viable cells displayed completely

healthy nuclei (**Figure 3a**); by contrast, the apoptotic cells were characterized by shrunk cells with condensed (early apoptosis) or fragmented (late apoptosis) nuclei. Other morphological and apoptotic changes, such as membrane blebbing and apoptotic body formation, were observed in the treated cells. This result suggested that apoptosis is the main cytotoxic mechanism of bacterial metabolites (**Figure 3b**).The newly identified *E. lactis* IW5 strain obtained from the human gut exhibited appropriate probiotic properties, such as high tolerance to low pH, resistance to high bile salt concentration, and anti-pathogenic activity against several pathogenic bacteria. Cytotoxic findings indicated that *E. lactis* IW5 secreted metabolites that possessed high anticancer activity against all of the examined cancer cell lines (AGS, Caco-2, HeLa, and HT-29). Therefore, the metabolites produced by *E. lactis* IW5 strain may be used as an alternative nutraceutical with promising therapeutic index because these metabolites are non-cytotoxic to normal mammalian cells.

Compared with the control cells that exhibited natural cell death (**Figure 3c**), the HeLa cells treated with 50 μL/mL of filtered *E. lactis* IW5 secretions demonstrated significant amounts

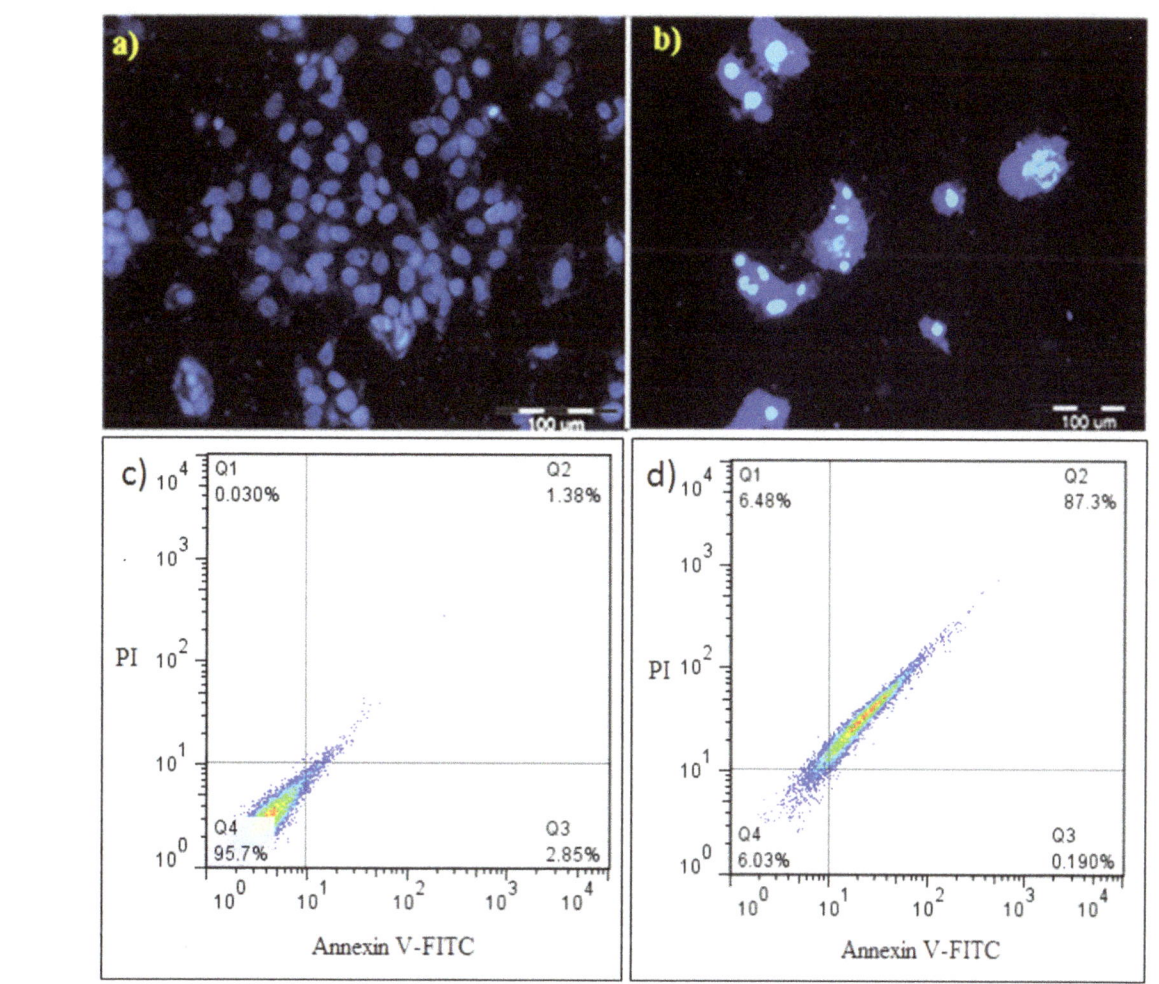

FIGURE 3 | 4′,6-diamidino-2-phenylindole (DAPI) staining and flow cytometric analysis of treated/untreated HeLa cancer cells. (a,b) Untreated and treated DAPI-stained cells; **(c,d)** flow cytometric analysis of untreated and treated cells with 50 µg/mL *E. lactis* secretion metabolites after 24 h incubation. Dots with Annexin V⁻/PI⁺ (Q1), Annexin V⁺/PI⁺ (Q2), Annexin V⁺/PI⁻ (Q3), and Annexin V⁻/PI⁻ (Q4) and feature represent necrotic, late apoptotic, early apoptotic, and viable intact cells, respectively.

($p \leq 0.05$) of annexin V$^+$/PI$^+$ (late apoptotic cells) after incubating for 24 h (**Figure 3d**). In the treated HeLa cells 87.3 and 6.48% were observed in late apoptosis and necrosis, respectively. Based on the flow cytometry findings, *E. lactis* IW5 secretions can inhibit the proliferation of cancer cells and the main mechanism of this prophylactic effect was related to apoptosis induction in cancer cells.

Quantitative Real Time PCR

As shown in **Figure 4**, the expression of anti-apoptotic genes (ERBB 2 and ERBB3), intrinsic apoptosis blocker genes (BCL-2 and BCL-XL), and CASP 8 gene (starter gene in TNF-α apoptosis pathway) were significantly down-regulated by *E. lactis* IW5compared to untreated control group. The down-regulation in the mentioned genes by *E. lactis* IW5 was similar to Taxol® but the expression of CASP 9 (starter gene in intrinsic apoptosis pathway) and BAX (crucial gene in extrinsic IL-3 mediated apoptosis pathway) genes was

significantly different in *E. lactis* IW5 and Taxol treated groups (**Figure 4**). *E. lactis* IW5 up-regulated the expression of BAX gene whereas Taxol up-regulated the expression of CASP9 indicating different inducing pathways of apoptosis. *Lactobacillus paracasei* M5L can induce apoptosis in HT-29 cells through reactive oxygen species generation followed by CRT accompanied endoplasmic reticulum stress and S phase arrest (Hu et al., 2015). The molecular mechanisms of pro-apoptotic effects of human-derived *Lactobacillus reuteri* ATCC PTA 6475 has been previously investigated on myeloid leukemia-derived cells and findings have shown the down-regulation of nuclear factor-kappaB (NF-kappaB)-dependent gene products that mediate cell survival (Bcl-2 and Bcl-xL) related genes (Iyer et al., 2008). Findings of antitumor effects of cell-bound exopolysaccharides (cb-EPS) isolated from *Lactobacillus acidophilus* 606 on HT-29 colon cancer cells have shown the antitumourigenic effects through the induction of BAX gene (Kim et al., 2010). In addition, the human

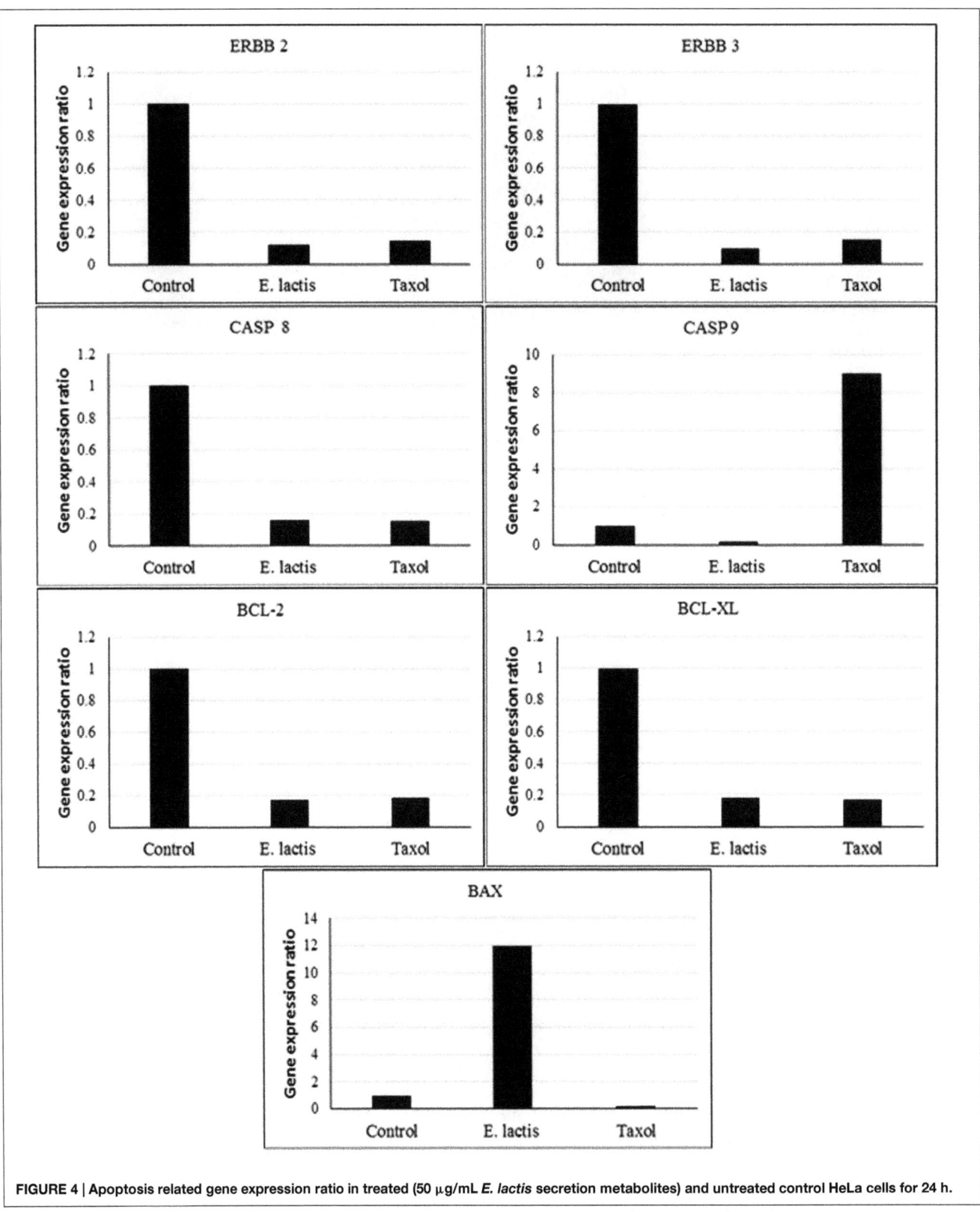

FIGURE 4 | Apoptosis related gene expression ratio in treated (50 μg/mL *E. lactis* secretion metabolites) and untreated control HeLa cells for 24 h.

probiotic *Propionibacterium freudenreichii* could kill HT-29 colorectal adenocarcinoma cells through apoptosis *in vitro* via its metabolites (the short chain fatty acids, acetate and propionate; Lan et al., 2007). Furthermore, the investigation results of the effect of probiotic *Bacillus polyfermenticus* on the growth of human colon cancer cells including HT-29,

DLD-1, and Caco-2 cells have illustrated that *B. polyfermenticus* can inhibit tumor growth and its anticancer activity occurs through suppressing ErbB2 and ErbB3 genes (Ma et al., 2010). Based on our findings, the induction of apoptosis by *E. lactis* IW5 is related to extrinsic IL-3 receptor pathway and it is deferent from Taxol's apoptosis induction (intrinsic mitochondria apoptosis pathway).

REFERENCES

ACKNOWLEDGMENTS

The financial support of the University Putra Malaysia (Putra Grant research no. 9443200), Tabriz University of Medical Sciences, and the moral patronage of Abolfazl Barzegari and Simin Sharifi are gratefully acknowledged.

Adlerberth, I., Ahrne, S., Johansson, M. L., Molin, G., Hanson, L. A., and Wold, A. E. (1996). A mannose-specific adherence mechanism in *Lactobacillus plantarum* conferring binding to the human colonic cell line HT-29. *Appl. Environ. Microbiol.* 62, 2244–2251.

Baldwin, C., Millette, M., Oth, D., Ruiz, M. T., Luquet, F. M., and Lacroix, M. (2010). Probiotic *Lactobacillus acidophilus* and *L. casei* mix sensitize colorectal tumoral cells to 5-fuorouracil-induced apoptosis. *Nutr. Cancer* 62, 371–378. doi: 10.1080/01635580903407197

Bhardwaj, A., Gupta, H., Kapila, S., Kaur, G., Vij, S., and Malik, R. K. (2010). Safety assessment and evaluation of probiotic potential of bacteriocinogenic *Enterococcus faecium* KH 24 strain under in vitro and in vivo conditions. *Int. J. Food Microbiol.* 141, 156–164. doi: 10.1016/j.ijfoodmicro.2010.05.001

Chen, P., Zhang, Q., Dang, H., Liu, X., Tian, F., Zhao, J., et al. (2014). Screening for potential new probiotic based on probiotic properties and α-glucosidase inhibitory activity. *Food Conrol* 35, 65–72. doi: 10.1016/j.foodcont.2013.06.027

Collins, J. K., Thornton, G., and Sullivan, G. O. (1998). Selection of probiotic strains for human applications. *Int. Dairy J.* 8, 487–490. doi: 10.1016/S0958-6946(98)00073-9

Conway, P. L., and Kjelleberg, S. (1989). Protein-mediated adhesion of *Lactobacillus fermentum* strain 737 to mouse stomach squamous epithelium. *J. Gen. Microbiol.* 135, 1175–1186.

Dabek, M., Mccrae, S. I., Stevens, V. J., Duncan, S. H., and Louis, P. (2008). Distribution of beta-glucosidase and beta-glucuronidase activity and of beta-glucuronidase gene gus in human colonic bacteria. *FEMS Microbiol. Ecol.* 66, 487–495. doi: 10.1111/j.1574-6941.2008.00520.x

Delgado, S., O'sullivan, E., Fitzgerald, G., and Mayo, B. (2007). Subtractive screening for probiotic properties of *Lactobacillus* species from the human gastrointestinal tract in the search for new probiotics. *J. Food Sci.* 72, M310–M315. doi: 10.1111/j.1750-3841.2007.00479.x

Dimitonova, S. P., Danova, S. T., Serkedjieva, J. P., and Bakalov, B. V. (2007). Antimicrobial activity and protective properties of vaginal lactobacilli from healthy Bulgarian women. *Anaerobe* 13, 178–184. doi: 10.1016/j.anaerobe.2007.08.003

Donaldson, M. S. (2004). Nutrition and cancer: a review of the evidence for an anti-cancer diet. *Nutr. J.* 3:19. doi: 10.1186/1475-2891-3-19

FAO/WHO (2001). *Food and Agriculture Organization of the United Nations, World Health Organization. Report of a Joint FAO-WHO Expert Consultation on Evaluation of Health and Nutritional Properties of Probiotics in Food Including Powder Milk with Live Lactic Acid Bacteria. Córdoba, 2001.* Available at: http://www.who.int/foddsafety/publications/fs_manage-ment/probiotics.pdf

Fisher, K., and Phillips, C. (2009). The ecology, epidemiology and virulence of *Enterococcus. Microbiology* 155, 1749–1757. doi: 10.1099/mic.0.026385-0

Forsyth, C. B., Farhadi, A., Jakate, S. M., Tang, Y., Shaikh, M., and Keshavarzian, A. (2009). *Lactobacillus* GG treatment ameliorates alcohol-induced intestinal oxidative stress, gut leakiness, and liver injury in a rat model of alcoholic steatohepatitis. *Alcohol* 43, 163–172. doi: 10.1016/j.alcohol.2008.12.009

Franz, C. M., Stiles, M. E., Schleifer, K. H., and Holzapfel, W. H. (2003). Enterococci in foods–a conundrum for food safety. *Int. J. Food Microbiol.* 88, 105–122. doi: 10.1016/S0168-1605(03)00174-0

Galvez, A. A., Dubois-Dauphin, R., Ghalfi, H., Campos, D., and Thonart, P. (2009). Description of two Enterococcus strains isolated from traditional Peruvian artisanal-produced cheeses with a bacteriocin-like inhibitory activity. *Biotechnol. Agron. Soc.* 13, 349–356.

Granato, D., Perotti, F., Masserey, I., Rouvet, M., Golliard, M., Servin, A., et al. (1999). Cell surface-associated lipoteichoic acid acts as an adhesion factor for attachment of *Lactobacillus johnsonii* La1 to human enterocyte-like Caco-2 cells. *Appl. Environ. Microbiol.* 65, 1071–1077.

Haghshenas, B., Abdullah, N., Nami, Y., Radiah, D., Rosli, R., and Khosroushahi, A. Y. (2014a). Different effects of two newly-isolated probiotic *Lactobacillus plantarum* 15HN and *Lactococcus lactis* subsp. *lactis* 44Lac strains from traditional dairy products on cancer cell lines. *Anaerobe* 30, 51–59. doi: 10.1016/j.anaerobe.2014.08.009

Haghshenas, B., Nami, Y., Abdullah, N., Radiah, D., Rosli, R., and Khosroushahi, A. Y. (2014b). Anti-proliferative effects of *Enterococcus* strains isolated from fermented dairy products on different cancer cell lines. *J. Funct. Foods* 11, 363–374. doi: 10.1016/j.jff.2014.10.002

Haghshenas, B., Nami, Y., Abdullah, N., Radiah, D., Rosli, R., Barzegari, A., et al. (2015a). Potentially probiotic acetic acid bacteria isolation and identification from traditional dairies microbiota. *Int. J. Food Sci. Technol.* 50, 1056–1064. doi: 10.1111/ijfs.12718

Haghshenas, B., Nami, Y., Abdullah, N., Radiah, D., Rosli, R., and Khosroushahi, A. Y. (2015b). Anticancer impacts of potentially probiotic acetic acid bacteria isolated from traditional dairy microbiota. *LWT-Food Sci. Technol.* 60, 690–697. doi: 10.1016/j.lwt.2014.09.058

Howarth, G. S., and Wang, H. (2013). Role of endogenous microbiota, probiotics and their biological products in human health. *Nutrients* 5, 58–81. doi: 10.3390/nu5010058

Hu, P., Song, W., Shan, Y., Du, M., Huang, M., Song, C., et al. (2015). *Lactobacillus paracasei* subsp. *paracasei* M5L induces cell cycle arrest and calreticulin translocation via the generation of reactive oxygen species in HT-29 cell apoptosis. *Food Funct.* 6, 2257–2265. doi: 10.1039/c5fo00248f

Iyer, C., Kosters, A., Sethi, G., Kunnumakkara, A. B., Aggarwal, B. B., and Versalovic, J. (2008). Probiotic *Lactobacillus* reuteri promotes TNF-induced apoptosis in human myeloid leukemia-derived cells by modulation of NF-kappaB and MAPK signalling. *Cell. Microbiol.* 10, 1442–1452. doi: 10.1111/j.1462-5822.2008.01137.x

Kim, Y., Oh, S., Yun, H. S., Oh, S., and Kim, S. H. (2010). Cell-bound exopolysaccharide from probiotic bacteria induces autophagic cell death of tumour cells. *Lett. Appl. Microbiol.* 51, 123–130. doi: 10.1111/j.1472-765X.2010.02859.x

Lan, A., Lagadic-Gossmann, D., Lemaire, C., Brenner, C., and Jan, G. (2007). Acidic extracellular pH shifts colorectal cancer cell death from apoptosis to necrosis upon exposure to propionate and acetate, major end-products of the human probiotic propionibacteria. *Apoptosis* 12, 573–591. doi: 10.1007/s10495-006-0010-3

Lee, N. K., Kim, S. Y., Han, K. J., Eom, S. J., and Paik, H. D. (2014). Probiotic potential of *Lactobacillus* strains with anti-allergic effects from kimchi for yogurt starters. *LWT-Food Sci. Technol.* 58, 130–134. doi: 10.1016/j.lwt.2014.02.028

Lin, P. W., Nasr, T. R., Berardinelli, A. J., Kumar, A., and Neish, A. S. (2008). The probiotic *Lactobacillus* GG may augment intestinal host defense by regulating apoptosis and promoting cytoprotective responses in the developing murine gut. *Pediatr. Res.* 64, 511–516. doi: 10.1203/PDR.0b013e3181827c0f

Ma, E. L., Choi, Y. J., Choi, J., Pothoulakis, C., Rhee, S. H., and Im, E. (2010). The anticancer effect of probiotic *Bacillus* polyfermenticus on human colon cancer cells is mediated through ErbB2 and ErbB3 inhibition. *Int. J. Cancer* 127, 780–790. doi: 10.1002/ijc.25011

Maldonado, N. C., De Ruiz, C. S., Otero, M. C., Sesma, F., and Nader-Macias, M. E. (2012). Lactic acid bacteria isolated from young calves: characterization and potential as probiotics. *Res. Vet. Sci.* 92, 342–349. doi: 10.1016/j.rvsc.2011.03.017

Mehra, N., Majumdar, R. S., Kumar, S., and Dhewa, T. (2012). Probiotics: preventive and clinical applications. *Biotechnol. Res. Bull.* 1, 15–20.

Merghoub, N., Benbacer, L., Amzazi, S., Morjani, H., and El Mzibri, M. (2009). Cytotoxic effect of some Moroccan medicinal plant extracts on human cervical cell lines. *J. Med. Plants Res.* 3, 1045–1050.

Mosmann, T. (1983). Rapid colorimetric assay for cellular growth and survival: application to proliferation and cytotoxicity assays. *J. Immunol. Methods* 65, 55–63. doi: 10.1016/0022-1759(83)90303-4

Nami, Y., Abdullah, N., Haghshenas, B., Radiah, D., Rosli, R., and Khosroushahi, A. Y. (2014a). Probiotic assessment of *Enterococcus* durans 6HL and Lactococcus lactis 2HL isolated from vaginal microflora. *J. Med. Microbiol.* 63, 1044–1051. doi: 10.1099/jmm.0.074161-0

Nami, Y., Abdullah, N., Haghshenas, B., Radiah, D., Rosli, R., and Khosroushahi, A. Y. (2014b). Assessment of probiotic potential and anticancer activity of newly isolated vaginal bacterium *Lactobacillus plantarum* 5BL. *Microbiol. Immunol.* 58, 492–502. doi: 10.1111/1348-0421.12175

Nami, Y., Abdullah, N., Haghshenas, B., Radiah, D., Rosli, R., and Khosroushahi, A. Y. (2014c). Probiotic potential and biotherapeutic effects of newly isolated vaginal *Lactobacillus acidophilus* 36YL strain on cancer cells. *Anaerobe* 28, 29–36. doi: 10.1016/j.anaerobe.2014.04.012

Nami, Y., Abdullah, N., Haghshenas, B., Radiah, D., Rosli, R., and Yari Khosroushahi, A. (2014d). A newly isolated probiotic *Enterococcus faecalis* strain from vagina microbiota enhances apoptosis of human cancer cells. *J. Appl. Microbiol.* 117, 498–508. doi: 10.1111/jam.12531

Nowroozi, J., Mirzaii, M., and Norouzi, M. (2004). Study of *Lactobacillus* as probiotic bacteria. *Iranian J. Public Health* 33, 1–7.

Ouwehand, A. C. (2007). Antiallergic effects of probiotics. *J. Nutr.* 137, 794S–797S.

Ouwehand, A. C., Salminen, S., and Isolauri, E. (2002). Probiotics: an overview of beneficial effects. *Antonie Van Leeuwenhoek* 82, 279–289. doi: 10.1023/A:1020620607611

Paolillo, R., Romano Carratelli, C., Sorrentino, S., Mazzola, N., and Rizzo, A. (2009). Immunomodulatory effects of *Lactobacillus plantarum* on human colon cancer cells. *Int. Immunopharmacol.* 9, 1265–1271. doi: 10.1016/j.intimp.2009.07.008

Prisciandaro, L. D., Geier, M. S., Butler, R. N., Cummins, A. G., and Howarth, G. S. (2011). Evidence supporting the use of probiotics for the prevention and treatment of chemotherapy-induced intestinal mucositis. *Crit. Rev. Food Sci.* 51, 239–247. doi: 10.1080/10408390903551747

Rahmati, M. (2011). The apoptotic and cytotoxic effects of *Polygonum avicular* extract on Hela-S cervical cancer cell line. *Afr. J. Biochem. Res.* 5, 373–378.

Sarem, F., Sarem-Damerdji, L. O., and Nicolas, J. P. (1996). Comparison of the adherence of three *Lactobacillus* strains to Caco-2 and Int-407 human intestinal cell lines. *Lett. Appl. Microbiol.* 22, 439–442. doi: 10.1111/j.1472-765X.1996.tb01198.x

Sharma, S., Chaturvedi, J., Chaudhari, B. P., Singh, R. L., and Kakkar, P. (2012). Probiotic *Enterococcus lactis* IITRHR1 protects against acetaminophen-induced hepatotoxicity. *Nutrition* 28, 173–181. doi: 10.1016/j.nut.2011.02.012

Sharma, S., Singh, R. L., and Kakkar, P. (2011). Modulation of Bax/Bcl-2 and caspases by probiotics during acetaminophen induced apoptosis in primary hepatocytes. *Food Chem. Toxicol.* 49, 770–779. doi: 10.1016/j.fct.2010.11.041

Shin, N. R., Moon, J. S., Shin, S. Y., Li, L., Lee, Y. B., Kim, T. J., et al. (2015). Isolation and characterization of human intestinal Enterococcus avium EFEL009 converting rutin to quercetin. *Lett. Appl. Microbiol.* doi: 10.1111/lam.12512 [Epub ahead of print]

7

Inflammasome/IL-1β Responses to Streptococcal Pathogens

Christopher N. LaRock[1]* and Victor Nizet[1,2]*

[1] *Department of Pediatrics, University of California San Diego, La Jolla, CA, USA,* [2] *Skaggs School of Medicine and Pharmaceutical Sciences, University of California San Diego, La Jolla, CA, USA*

Edited by:
Diana Bahia,
Universidade Federal de Minas
Gerais, Brazil

Reviewed by:
Denise Monack,
Stanford University School of
Medicine, USA
Edward A. Miao,
Institute for Systems Biology, USA

*Correspondence:
Christopher N. LaRock
clarock@ucsd.edu;
Victor Nizet
vnizet@ucsd.edu

Inflammation mediated by the inflammasome and the cytokine IL-1β are some of the earliest and most important alarms to infection. These pathways are responsive to the virulence factors that pathogens use to subvert immune processes, and thus are typically activated only by microbes with potential to cause severe disease. Among the most serious human infections are those caused by the pathogenic streptococci, in part because these species numerous strategies for immune evasion. Since the virulence factor armament of each pathogen is unique, the role of IL-1β and the pathways leading to its activation varies for each infection. This review summarizes the role of IL-1β during infections caused by streptococcal pathogens, with emphasis on emergent mechanisms and concepts countering paradigms determined for other organisms.

Keywords: inflammasome, caspase-1, IL-1β, pyroptosis, *Streptococcus*

INTRODUCTION

Humans are frequently colonized by pathogenic species of streptococcal bacteria: the throat and skin by *Streptococcus pyogenes* (group A *Streptococcus*; GAS), the upper respiratory tract by *Streptococcus pneumoniae* (pneumococcus, SPN), and the lower intestine and genital tract by *Streptococcus agalactiae* (group B *Streptococcus*; GBS). This microbial–host association usually occurs in the context of asymptomatic colonization or superficial mucosal infection, but each of these pathogens can also be associated with severe, invasive, even life-threatening, diseases. GAS causes a wide range of diseases, including pharyngitis, cellulitis, puerperal sepsis, necrotizing fasciitis, streptococcal toxic shock syndrome, and rheumatic heart disease, making it one of the top 10 causes of infectious mortality (1). SPN is a similarly prevalent human pathogen responsible for greater than one million annual deaths by pneumonia and meningitis, mostly in young children (2). Lastly, GBS is a common cause of neonatal sepsis and meningitis, making it an important cause of infectious morbidity and mortality among infants in many countries throughout the world (3).

Inflammation is a key component of the immune response during infections with all of the pathogenic streptococci. Inflammation can be protective by preventing bacterial colonization, replication, invasion, and dissemination. Insufficient inflammation commonly leads to a greater infection susceptibly or more prolonged disease. Conversely, excessive inflammation is a driver of several autoimmune diseases and of host tissue injury complicating many severe infectious diseases. Inflammation must therefore be carefully regulated for an optimal immune response, and pathogens can exploit the regulatory processes deployed by the host innate immune system. For example, inflammation in the upper respiratory tract increases the risk of systemic dissemination of SPN, even though it is critical for combating the localized infection at that site. For SPN as well as GBS, inflammation helps break down the blood–brain barrier (BBB) to cause meningitis. In these deadly infections, the tissue damage resulting from inflammation can lead to acute complications, and even if the pathogen is successfully cleared, can be associated with post-infectious sequelae.

The IL-1 and inflammasome pathways in particular exemplify the complex role of inflammation during streptococcal infection. Indeed, GAS is classically defined as a "pyogenic" pathogen, exemplified by pus formation elicited by the robust inflammatory response to its tissue invasion. IL-1β is a highly inflammatory cytokine commonly key in eliciting protective immunity. Caspase-1 and its canonical regulator the inflammasome were first discovered for their ability to activate IL-1β. The inflammasome pathway has since been found to regulate numerous other inflammatory and antimicrobial activities, which in several instances contribute more to the functional immunity than does IL-1β. Activation of IL-1β is also not fully dependent on the inflammasome, but instead requires cooperation between several pathways, many of which also can be activated along redundant routes. When distinctions can be made based on the available literature, we attempt to disambiguate the contribution of each of these signaling and immune effector pathways.

BIOLOGY OF IL-1β AND THE INFLAMMASOME

IL-1

The IL-1 receptor (IL-1R) is widely expressed, which allows IL-1 signaling to induce a variety of cellular effector mechanisms locally as well as systemically. Two cytokines, IL-1α and IL-1β, are recognized by IL-1R to similar effects. The major distinction between these cytokines is that IL-1β is soluble, while IL-1α is typically membrane bound, spatially limiting its function to the activation of neighboring cells. By contrast, IL-1β is free to also act as a chemokine and mediate systemic signaling events. IL-1R1$^{-/-}$ mice, deficient for cell signaling in response to both IL-1α and IL-1β, are more susceptible to most infections, including those caused by GAS (4), GBS (5), and SPN (6–9).

IL-1α is a key mediator of the sterile inflammatory response (10), but is not generally critical for the response to bacterial infection (11). Nevertheless, IL-1α is stimulated during infections by SPN (12, 13), GBS (14), and GAS (15). Genome-wide linkage studies in mice identified a correlation between IL-1α levels and mortality during GAS sepsis (15), suggestive that IL-1α contributes to cytokine storm during sepsis. However, this link was not found in human studies focused on skin infections (16), perhaps because IL-1α might be more beneficial than detrimental in this context. IL-1α probably plays at most a minor role during streptococcal infections, as IL-1β$^{-/-}$ mice phenocopy IL-1R$^{-/-}$ mice in their resistance to GBS (5, 17). The role of IL-1α during experimental GAS and SPN infections is not yet clear.

IL-1β is critical in defense against GAS (4, 18), GBS (19), and SPN (6, 9, 20,21). IL-1β is a major chemoattractant of neutrophils (10), and neutrophil recruitment is largely mediated by IL-1β during GAS (4) and GBS infections (17). This neutrophil influx to the site of infection contributes to GAS and GBS killing, since neutrophil ablated and IL-1R$^{-/-}$ mice have a similar susceptibility to these pathogens (4, 17). SPN is largely resistant to recruited neutrophils during pneumonia, but rather succumbs to the wave

of activated macrophage that follows, which is also largely IL-1β dependent (6, 9, 20, 22). IL-1β also induces fibrinogen expression and localized coagulation, which help to limit dissemination of SPN from the lung (8). It is not clear if this occurs during other streptococcal infections, but if so, the effects may not always benefit the host, as both GAS and GBS have surface-expressed virulence factors that bind fibrinogen and interfere with complement activation and phagocytosis (1, 3).

By controlling early bacterial infection before it becomes this severe, IL-1 can help prevent a pathogen from reaching immune-privileged or vulnerable sites, such as the central nervous system (CNS). Consistent with this notion, IL-1 signaling-deficient mice develop meningitis as a complication of respiratory tract infections at a higher frequency (7). However, once a pathogen reaches the BBB, inflammation is often more harmful then beneficial. GBS crosses the brain microvascular endothelial cells comprising the BBB by direct intracellular invasion (23) without inducing IL-1 (24). SPN can similarly invade the cerebral endothelial cells to gain access to the CNS without barrier damage or disruption (25). Despite these non-inflammatory mechanisms for gaining CNS entry, bacterial CNS infections are inherently inflammatory. Bacterial growth and damage to the initially infected CNS cells greatly induces IL-1 (26), which further breakdowns the BBB to allow more bacterial invasion (27). IL-1 also recruits and activates neutrophils, which are overtly injurious in murine meningitis models (28, 29) and may correlate with poor patient prognosis (30). Neutrophils in the CNS are ineffectual against SPN (31), so there is unfortunately little obvious benefit to this inflammation. Moreover, IL-1 contributes to the pathogenesis of numerous neurodegenerative diseases, and likely has direct role in neurological sequelae common among survivors of streptococcal CNS infection (2, 3, 30, 32).

Interleukin-18

Interleukin-18 (IL-18) is another inflammasome-regulated proinflammatory cytokine. The largest contribution of IL-18 to immunity lies in stimulation of natural killer (NK) cells and induction of interferon-γ (IFN-γ) signaling (33, 34). IL-18 activation is seen during GAS (18, 35), GBS (19), and SPN infections (36). IL-18$^{-/-}$ mice are more susceptible to SPN pneumonia (37). However, in a SPN meningitis model, IL-18$^{-/-}$ mice actually survived longer than WT controls, suggesting that inflammation induced by IL-18 may be more pathological than beneficial in CNS infection, as is the case for IL-1 (38). In GBS infection of neonatal mice, an IL-18 neutralizing antibody increased GBS burden and mortality; conversely, administration of recombinant IL-18 reduced GBS counts (39).

Pyroptosis

In addition to cytokine signaling, activation of inflammasomes initiates programed cell death by pyroptosis (**Figure 1**). This form of cell death releases numerous endogenous damage-associated molecular patterns (DAMPs), including ATP, DNA, HMGB1, and histones, which further amply the inflammatory response through the recruitment and activation of neutrophils and other immune cells (34). Due to the abundance of DAMPs released

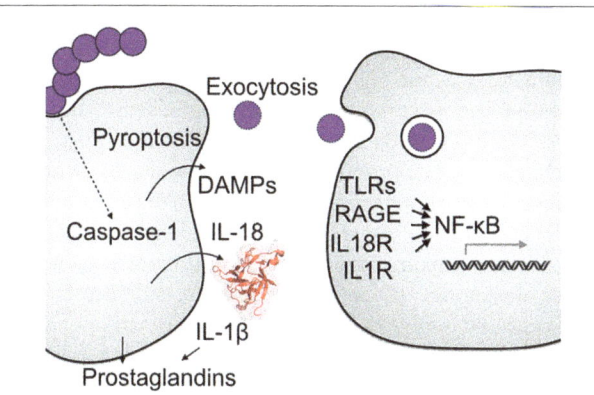

FIGURE 1 | Major effector mechanisms of the inflammasome. A cell containing active inflammasomes releases several inflammatory signals to other cells, including prostaglandins/eicosanoids, IL-1β, and IL-18. The other major cell process activated is programed cell death by pyroptosis, whereupon the released cellular contents can be detected by a number of pattern-recognition receptors to further inflammatory signaling. Pyroptotic cell death also releases any intracellular pathogens, exposing them to direct killing by complement or antimicrobial peptides or phagocytosis by neighboring cells.

during pyroptosis, much of the inflammasome-driven inflammatory response during infection can progress in an IL-1- and IL-18-independent manner (11). In the instance of pneumococcal meningitis, neutralization of IL-1 and IL-18 ameliorate a remarkable amount of the inflammation, yet not all of it (29). A DAMP released during pyroptosis that strongly induces inflammation is HMGB1, a chromatin protein recognized by TLR4 and RAGE receptors. Extracellular HMGB1 is abundant during SPN meningitis, with the levels correlating to severity of disease in both mice and humans (40).

In addition to inflammation, pyroptotic cell death plays an important role in immunity by depriving intracellular pathogens of a replicative niche. Intracellular bacteria are protected from many innate and cellular immune defenses; lysis releases the bacteria where they are exposed to immune cells that are primed and better able to combat the pathogen (41). Though they are commonly treated as exclusively extracellular pathogens, the streptococci can specifically remodel the cellular antimicrobial response to allow intracellular replication (42, 43). It is not yet clear how protective pyroptosis might be for the host during streptococcal infection, but GAS is able to use it to its own advantage. Compared to other cell death programs, pyroptosis occurs relatively rapidly. GAS induction of cell death can be so rapid that IL-1 production is limited, since the cell does not have time to synthesize and convert much cytokine (44).

Other Mechanisms

Several emergent inflammasome effector pathways may also play a role in combating streptococcal infection. The inflammasome can induce secretion of prostaglandin E2, both directly and through IL-1β-induced cell signaling (45). Prostaglandin E2 is markedly induced during GAS (46), GBS (47), and SPN infection (48). This induction has been observed in several infection models

including sepsis (15), necrotizing fasciitis (49), and puerperal infection (50). *In vitro*, prostaglandin E2 is immunosuppressive and impairs killing of GAS (49) due to repression of phagocytosis, reactive oxygen species, and inflammatory cytokines like TNF-α (50). Consistent with these observations, COX-2$^{-/-}$ mice, deficient in prostaglandin E2, had greater GAS resistance (49). However, COX-2-targeting non-steroidal antiinflammatory drugs have long been thought to exacerbate GAS infection and be a risk factor for developing invasive infections (51); therefore, the role of prostaglandin E2 in the anti-GAS immune response is not entirely clear.

Inflammasome activation might also act against intracellular bacteria by mechanisms that do not require death of the host cell. Caspase-1 promotes greater acidification of the phagolysome in GBS-infected cells (52). This mechanism appears to be inactive during infections with Gram-negative bacteria, but operates in response to the Gram-positive bacteria tested, so would likely act against GAS and SPN as well. IL-1β signaling provides another route for killing of several species of intracellular bacteria, including GAS (18). This effect is mediated through autocrine induction of IL-1R-regulated pathways, but which antimicrobial effectors are ultimately involved is not yet known.

THE INFLAMMASOMES

Caspase-1

The inflammasome is a scaffold nucleotide-binding domain and leucine-rich repeat containing receptor (NLR) family of proteins that serves to activate a component conserved between inflammasomes: the cysteine protease Caspase-1. Caspase-1$^{-/-}$ and IL-1β$^{-/-}$IL-18$^{-/-}$ mice often exhibit similar infection response phenotypes (11). The immune contributions of pyroptosis and other cytokine-independent inflammasome effector mechanisms can make the role of Caspase-1 more prominent in certain infections. Alternatively, inflammasome-independent mechanisms for IL-1β secretion can shift this balance in the other direction (34). Consistent with inflammasomes playing a protective role during streptococcal infection, Capase-1$^{-/-}$ mice are more susceptible to GAS (18) and GBS (19). The importance of Caspase-1 in defense against SPN varies greatly depending on model, mirroring the variable role of IL-1 in these infections. In a SPN pneumonia model, Caspase-1 had little effect (18, 53), but in a SPN meningitis model, Capase-1-driven inflammation led to great intracranial pressure and disruption of the BBB (26).

NLRP3 Detection of Pore-Forming Toxins

Several different NLRs can form inflammasomes, but NLRP3 has the most prominent contribution for detection of streptococci (**Figure 2**). Streptococcal pathogens deploy secreted pore-forming toxins, which are well documented to activate the NLRP3 inflammasome (13, 19, 21, 29, 30, 54–57). The precise mechanisms by which NLRP3 senses diverse toxins from a number of bacterial species, as well as numerous other PAMPs-like crystals of uric acid, cholesterol, or amyloid proteins, is not entirely clear. Given the disparate nature of these molecules, and no known binding interactions, NLRP3 does not appear to directly detect these

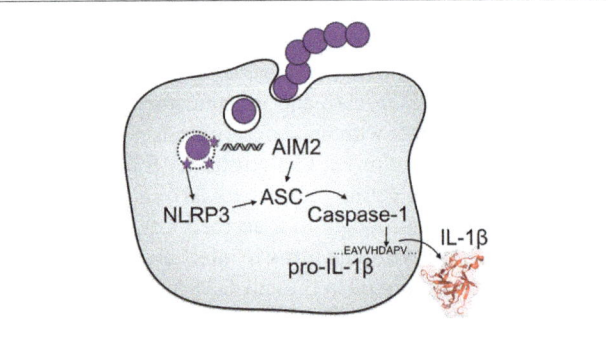

FIGURE 2 | Inflammasome activation by the streptococci. Two primary NLRs form inflammasomes during streptococcal infections, NLRP3 and AIM2. NLRP3 detects membrane disruption by the pore-forming toxins encoded by all the major streptococcal species. These pores also allow bacterial DNA in to the cytosol for detection by AIM2. Either NLR can form an inflammasome scaffold for the activation of caspase-1, the primary protease responsible for the hydrolysis of IL-1β into its mature form.

PAMPs and DAMPs. Several models have been put forward describing a mechanism for NLRP3 activation in response to perturbations in cellular homeostasis. This concept requires a secondary molecule commonly altered by these PAMPs and DAMPs. While the identity of this molecule is not agreed upon, a unifying theme is the disruption of either the outer membrane or endosomal membranes and consequent induction of ER stress (58). As not all NLRP3 stimuli are membrane acting, upstream detection pathways may still be involved in some circumstances. Streptococcal pore-forming toxins directly induce membrane disruption and ER stress (59), so their detection will likely follow whatever paradigm emerges to integrate the different models of NLRP3 activation.

Major pore-forming toxins of GAS and SPN are the cholesterol-dependent cytolysins streptolysin O (SLO) and pneumolysin (PLY), which use cholesterol and glycans as cell surface receptors (60). Both toxins form very large pores in many cell types. In immune cells, pore formation contributes to virulence by killing the cell or inactivating its effector mechanisms, but concurrently activates IL-1β secretion through the NLRP3 inflammasome (12, 13, 44, 54, 61, 62). GAS expresses a second membrane-active pore-forming toxin, streptolysin S (SLS), which is responsible for the classical β-hemolytic phenotype of GAS (63, 64). SLS does not contribute strongly to NLRP3 inflammasome activation (44). This may be due to a dominant role of SLO or the less potent lytic activity of SLS against non-erythrocytes (65), though a toxin's ability to form pores and to activate the inflammasome do not always correlate strictly (62).

The major pore-forming toxin of GBS, β-hemolysin, is highly dissimilar to PLY, SLO, and SLS. This toxin stays tightly associated with the cell surface and plays a key role in the progression from colonization to invasive infection (66). β-hemolysin mutant GBS induce less pyroptosis and IL-1β secretion through the NLRP3 inflammasome (19). The mutation involved, *cylE*, also disrupts synthesis of the characteristic pigment of GBS granadaene (67). Granadaene itself is sufficient to activate the NLRP3 inflammasome (56), and production of granadaene is also linked to the hemolytic activity of *Propionibacterium jensenii* (68). While

suggestive that granadaene is itself the β-hemolysin, CylE expression in *Escherichia coli* confers hemolytic activity but not pigmentation (67), and certain media conditions induce GBS pigmentation without a commensurate increase in hemolytic activity (69). An additional GBS toxin, CAMP factor, also forms pores and delivers bacterial products into the cytosol (70). While this activates several immune detection pathways, the inflammasome does not appear to be one of them for unknown reasons (19).

Pore-forming toxins also activate cell death processes that have features of osmotic lysis, apoptosis, necrosis, and oncosis, which can be confused for pyroptosis and complicate analysis of inflammasome activation (59, 71–74). Since maintaining cell membrane integrity is essential for viability and continued cytokine production, pore-forming toxins can, somewhat paradoxically, actually limit IL-1β by inducing these cell death pathways. The pore-forming toxins of GAS (44), GBS (75), and pneumococcus (74) each can induce the cell to lyse before much IL-1β can be synthesized and processed. Detection of pore-forming toxins, through both caspase-1-dependent and -independent pathways, can also induce membrane-healing mechanisms that limit toxin potency and cell death (34, 76). Therefore, the effect of toxins on the inflammasome appears to be highly concentration dependent: low doses promote cell activation and repair mechanisms, moderate doses activation of the NLRP3 inflammasome, and high doses a rapid cell death that limits IL-1β-driven inflammation.

Alternative NLRP3 PAMPs

Some of the earliest results on the detection of pore-forming toxins by NLRP3 suggested that SLO is not sufficient for inflammasome activation (77). One explanation for this observation is that the NLRP3 inflammasome requires co-stimulatory signals for activation (78). Another explanation for this finding is that low concentrations of pore-forming toxin, themselves insufficient for inflammasome activation, can still mediate the delivery of inflammasome-activating PAMPs and DAMPs, such as bacterial RNA, CpG DNA, Pam$_3$CSK$_4$, zymosan, muramyl dipeptide, and lysozyme-digested peptidoglycan (13, 57, 79–81). Even in circumstances where toxin pore formation is sufficient for inflammasome activation, delivery of these additional PAMPs may provide for a stronger inflammasome stimulus and may allow activation of additional inflammasomes beyond the NLRP3.

Another GAS virulence factor, SpyA, can activate the NLRP3 inflammasome (18). SpyA is delivered in to host cells where it transfers ADP-ribose from nicotinamide adenine dinucleotide (NAD) onto host proteins to modify their activity (82). ADP-ribosylating toxins from *Pseudomonas aeruginosa* and *Mycoplasma pneumoniae* also activate the NLRP3 inflammasome (83), but the precise mechanism underlying the detection of these toxins is unclear. An ADP-ribosyltransferase toxin from *Clostridium botulinum* instead activates a pyrin inflammasome (84), suggesting the target of the toxin dictates which inflammasome is involved. Consistent with this hypothesis, other toxins that target Rho-GTPases like the Clostridial toxin are also detected via pyrin (84). One target of the *M. pneumoniae* toxin is NLRP3 (83), suggesting this could be a target of SpyA and other NLRP3 activating microbial enzymes. Alternatively, SpyA targets vimentin (85), which might de-repress the NLRP3 inflammasome (86). Additionally, ADP-ribosylating toxin depletion of NAD might activate the NLRP3 inflammasome (87); SpyA has very potent NAD-glycohydrolase activity (82). This suggests that another NAD-glycohydrolase of GAS, Nga can activate the inflammasome. Consistent with this hypothesis, Nga does induce cell death, but whether it is morphologically similar to pyroptosis and occurs through the inflammasome has not yet been determined (88).

Alternative Inflammasome and IL-1β Pathways

A second inflammasome pathway activated during streptococcal infection proceeds through AIM2 in response to cytosolic double-stranded DNA from lysed bacteria (**Figure 2**). This PAMP is introduced into the cytosol upon the disruption of the phagosomal membrane by pore-forming toxins, such as PLY (89–91). The AIM2 inflammasome is important in the resistance to SPN (89, 91), but not GAS or GBS (19, 57). Since GAS and GBS are readily detected by other intracellular nucleic acid receptors (57, 70, 92–97), the mechanism underlying AIM2's unresponsiveness is unclear.

The other well-studied inflammasomes, formed via NLRC4, NLRP1, or caspase-11, are not known to be involved in streptococcal infection. They have not been rigorously tested in the context of streptococcal infection, because streptococci do not possess PAMPs similar to those classically known to be detected by these receptors. NLRC4 is exclusively responsive to the flagellin and type III secretion rod proteins of Gram-negative bacteria (98), so expectedly, is unresponsive toward GAS (54). The best established PAMPs for the NLRP1 inflammasome are the *Bacillus anthracis* lethal toxin and an unknown factor of *Toxoplasma gondii* (99). Lastly, caspase-11 can form "non-canonical" inflammasome in response to the lipopolysaccharide of Gram-negative pathogens, but is felt to be non-responsive toward Gram-positive bacteria in general (98).

Group B *Streptococcus* and SPN similarly stimulate multiple pathways for inflammasome activation, and NLRP3⁻/⁻ mice are more susceptible to infection by these pathogens (19, 55, 89). However, there are very likely additional mechanisms allowing for IL-1β activation during streptococcal infection, either by alternative inflammasome or by inflammasome-independent mechanisms. The most telling evidence for this is that all the known inflammasome-activation PAMPs of GAS are detected by NLRP3 (18, 44), but NLRP3 does not contribute to resistance against GAS (54). IL-1β is nonetheless important in the immune response to GAS (4), but the source of its activation remains unclear.

The lack of a phenotype in NLRP3⁻/⁻ mice could be due to redundancy with AIM2, or with another, uncharacterized, inflammasome receptor that detects GAS. The NLR family of pattern-recognition receptors contains dozens of members with unassigned function, so many conventional inflammasomes may yet to be discovered. Alternatively, there may be inflammasome-independent pathways providing for IL-1β signaling. The GAS secreted protease SpeB cleaves and inactivates important immune factors such as immunoglobulins and antimicrobial peptides, making it important in several virulence models (1). In a biochemical assay, SpeB was found to cleave IL-1β (100). However, the pro-domain of IL-1β might just be intrinsically protease labile since it can also be cleaved by proteases from *Candida albicans* (101), *Entamoeba histolytica*, (102) *Staphylococcus aureus* (103), and *Treponema denticola* (104). *In vivo* activation of pro-IL-1β appears nevertheless to be quite specific, as caspase-11 is similar to caspase-1 and presumably cleaves some of the same substrates in order to activate pyroptosis, yet it does not process IL-1β (105). It further remains unclear whether cleavage by proteases other than caspase-1 can occur during infection, or whether it would promote or inhibit IL-1 signaling.

PRIMING OF THE INFLAMMASOME AND IL-1

Induction of IL-1 and the Inflammasome

At several points, the inflammasome and IL-1β signaling pathways intersect with the NF-κB pathway. First, most cells do not constitutively express IL-1β, which is transcriptionally regulated by NF-κB (106). Therefore, most TLR pattern-recognition receptors, acting through MyD88, as well as the subset of NOD receptors that signal through RIP2, can activate NF-κB to induce synthesis of pro-IL-1β (79). IL-1β will also positively regulate itself, since the IL-1R also activates NF-κB (106). Second, both the NLRP3 and AIM2 inflammasome require priming. This priming can occur through TLRs, IL-1R, or TNFR (36, 78, 107, 108). The AIM2 inflammasome is additionally primed by Type I IFN signaling (109), which simultaneously represses the NLRP3 inflammasome (110). GAS, GBS, and SPN can all induce IFN (70, 91–93, 111–113), which could therefore lead to switching of which inflammasomes can form, and consequently, which bacterial factors are detected.

Since the NLRP3 and AIM2 inflammasomes are the only ones known to respond to streptococci (**Figure 2**), stimulatory pathways, such as TLRs, are critical not only for the induction of pro-IL-1β but also its maturation. We will therefore next discuss which of these pathways are known to detect streptococci, and how this detection promotes inflammasome/IL-1 signaling (**Figure 3**). Due to the large number of streptococcal PAMPs contributing to functional redundancies among TLRs, it might be expected that there would often be no immune susceptibility phenotype for any single TLR knockout (114). Nonetheless, through the use of streptococcal and host mutants several specific pathways have been identified. Of further note, a receptor may be found to be essential in one study and dispensable in another; when possible we note how streptococcal genotype, host genotype, and cell or infection model may impact these observations.

Pattern-Recognition Receptor Detection of Streptococcal Pathogens

TLR2 activates NF-κB upon detection of bacterial lipopeptides, lipoteichoic acid, and peptidoglycan (115). These are ubiquitous cell surface components of Gram-positive bacteria, so TLR2 readily detects GAS (95, 116), GBS (117–120), and SPN (121–123). TLR6 and TLR1 cooperate with TLR2 to dictate which PAMPs stimulate signaling. TLR6 contribute to detection of GBS (118, 124) and SPN (53). For GAS, TLR6 is suggested to be dispensable,

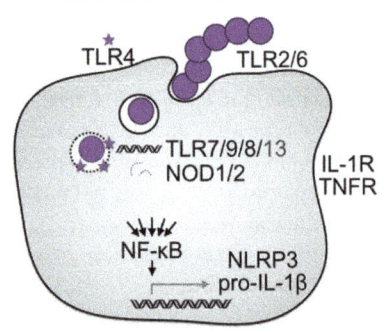

FIGURE 3 | IL-1β/inflammasome licensing pathways. IL-1 and the NLR proteins responsive to the streptococci require induction. Cell–cell signaling can provide this priming signal in the form of IL-1β or TNF-α. More commonly during bacterial infection, bacterial factors detected as pathogen-associated molecular patterns by TLRs provide this signal. TLR2 detects bacterial lipoproteins and is broadly sensitive to Gram-positive pathogens. TLR4 is able to detect the pore-forming toxins of several species of streptococci. When the streptococci are intracellular and the phagosome is disrupted, several additional receptors are involved. TLRs 7, 8, 9, and 13 detect bacterial nucleic acids, while NOD1 and NOD2 detect bacterial cell wall fragments.

but through a dendritic cell model where TLR2 was also dispensable, in contrast to findings with other cell types (114). Even less is certain about TLR1, but it appears to have an overall lesser role upstream of inflammasome activation (118). GBS mutants unable to decorate their cell surface with lipoproteins induce less TLR2 signaling, but the contribution of any particular lipoprotein is unknown (120). The most abundant protein on the GAS surface, M protein, is also detected by TLR2 to stimulate production of several cytokines including IL-1β (125, 126). Lipoteichoic acids may also be detected by TLR2, though GBS lipoteichoic acid is not (115, 120). On possible explanation is that the streptococci post-translationally modify their lipoteichoic acid structure (127); however, since lipoproteins also commonly contaminate lipoteichoic acid preparations (115), this scientific question remains somewhat controversial.

TLR2 activation is specifically connected to the model of inflammasome licensing. Induction of *il1a*, *il1b* (9, 122, 128), and *nlrp3* (21) during SPN infection occurs through TLR2, which was required for normal levels of IL-1β signaling (55). TLR2⁻/⁻ mice are not as attenuated to in their cytokine responses to GBS or SPN infection as MyD88⁻/⁻ mice that are broadly deficient in TLR signaling (117, 128). This finding illustrates that while TLR2 is the canonical receptor for Gram-positive pathogens, additional receptors are activating NF-κB in parallel. Several TLRs more commonly appreciated for their role Gram-negative bacterial and viral infections have also been found to detect streptococci, suggesting their agonist range is broader than commonly appreciated.

TLR4 is the established receptor for lipopolysaccharide, a potent PAMP decorating the surface of Gram-negative bacteria, analogous to the broad importance of TLR2 for detection of Gram-positive bacteria. However, TLR4 is also able to detect PLY (129) through direct binding (130) independent of pore-forming activity (131). Consequently, TLR4 can compensate for TLR2 deficiency (122) to provide resistance to SPN pneumonia

(123, 130, 132). PLY-deficient SPN induce less inflammasome-dependent cytokines IL-1α, IL-1β, and IL-18, with only a modest decrease in other cytokines such as TNF-α, IL-6, and IL-12 (12). The transcription of *il1b* is not greatly impacted by PLY (55), suggesting the toxin is more important for inducing NLRP3 than TLR4. This likely reflects a greater redundancy in the number of activating PAMPs for TLRs relative to NLRs leading to induction of their respective pathways (13). Nonetheless, TLR4 significantly potentiates caspase-dependent death induced by purified PLY (130). TLR4 detection of toxins may be a general mechanism since it has also been shown to mediate responses against several toxins including SLO from GAS (131). TLR4 is not important for detecting GBS (75), possibly due to TLR redundancy or because the GBS pore-forming β-hemolysin lacks homology with other pore-forming toxins (56, 67).

Several nucleic acid receptors are also known to recognize streptococci. ssRNA is recognized by TLR7 and contributes to the detection of GBS (92) but not GAS (93). Unmethylated bacterial DNA can be detected by TLR9, which leads to cell activation in response to SPN (53), GBS (92), and GAS (97). In one study, TLR7 and TLR9 were found to be much more important for the detection the detection of GAS and GBS than was TLR4 (92). For controlling SPN infection, TLR1, TLR2, TLR4, and TLR6 were functionally redundant but TLR9 was essential (53). In more recent studies, TLR7 and TLR9, as well as TLR2, TLR3, and TLR4, had minor roles in the detection of GAS and GBS compared to TLR8 (133). Like TLR7, TLR8 recognizes ssRNA, but this receptor is only present in humans, possibly leading to an overestimation of the relative importance of other TLRs in studies utilizing murine models. Mice instead express TLR13, not found in humans, which recognizes rRNA from several species including GAS (95). While some variation between studies is no doubt due to infection model differences, bacterial genetics can also be contributing variable. Hypervirulent M1T1 strains of GAS secrete a phage-encoded nuclease, Sda1, which degrades their own CpG-rich DNA to evade this detection by TLR9 (96). Similar mechanisms may allow the other streptococcal pathogens to evade TLR9, as well as other nucleic acid-sensing TLRs or NLRs.

NOD1 and NOD2 are related to NLRP3 and NLRC4 but activate NF-κB instead of the inflammasome. Both NOD proteins recognize muramyl dipeptide, a cleavage product of the peptidoglycan that comprises the bacterial cell wall (134) that can be introduced into the cytosol by pore-forming toxins (135). SPN is recognized by NOD1 (136) and NOD2 (137) through a process that requires PLY (136, 138) and bacterial cell wall degradation by lysozyme (81). Macrophages are the major cell recognizing SPN by NOD2 in a pneumonia model (138) with microglia and astrocytes-mediated detection during meningitis (139). NOD2 also is responsive to the GAS cell wall fragments, a commonly used inducer of inflammation in arthritis models (140). It is unknown whether NOD2 detects GAS during infection, and only a minimal role in GBS infection was detected (70, 75, 141). This result could be due to redundancy with other activation pathways since, even for SPN, NOD2 is largely redundant with TLR2 (138). Alternatively, streptococci might evade NOD detection through the same cell wall modifications that prevent detection by other PRRs and confer resistance to lysozyme (127).

Integration of Additional Signaling Pathways

Several of the endogenous DAMPs released during pyroptosis may further amplify the local inflammatory response (10). This second phase of the response could provide for stimulation of TLRs that do not recognize the pathogen directly, which may be particularly important during infection with pathogens adept at evading TLR recognition. Given the multitude of TLR receptors identified to recognize streptococci and their components, pyroptosis might not be essential for initiating an immune response to these pathogens, but would nonetheless amplify inflammation during these infections. Pyroptotic release of DAMPs can also provide an alternative pathway to NF-κB activation in individuals with IRAK-4 deficiencies, who cannot signal via most TLRs with the exception of TLR3, and have an increased susceptibility to SPN and other pathogens (142).

CONCLUSION AND PERSPECTIVE

A growing body of evidence suggests that there is more depth and complexity to IL-1β signaling than previously appreciated. For one, the inflammasome has been found to regulate several pathways in addition to IL-1β, including additional inflammatory signaling cascades, programed cell death, and antimicrobial effector mechanisms. Conversely, the number of pathways that can result in IL-1β activation is also increasing. As the inflammasome field grows, these new discoveries will provide greater insight on the molecular pathogenesis and host response to streptococcal infections. In a complementary fashion, experimental observations made using the streptococci and their unique suite of virulence mechanisms for altering the host response can help shape our understanding of the IL-1β/inflammasome pathway(s), which are so broadly impactful in clinical medicine.

How do alterations in the IL-1β/inflammasome response alter the incidence and outcome of streptococcal infections? Many streptococcal infections disproportionately affect the very young and the very old – and this pattern is mirrored in the quality of the inflammasome response. Neonates and newborns have a diminished ability to produce inflammatory cytokines, such as IL-1β (143). Several mechanisms are at play, including immune system immaturity (144) and active suppression of innate immunity (145), and future work is required to better define the role of the inflammasome in these processes. A different mechanism may be at play in older populations, wherein TLR expression deficiency has been reported to mute cytokine activation in aged mice (146). Local lymphoid tissue responses are aberrant in aged mice, with baseline inflammation and high IL-1β levels already present in the lymphoid tissue of the upper respiratory tract in naive elderly mice, which then failed to upregulate NLRP3 and IL-1β in response to SPN colonization (147). Host genetics also plays a role – MyD88 and IRAK-4 are important for the IL-1β/inflammasome response, and mutations in these genes lead to susceptibility to pyogenic infections similar to those caused by the streptococci (148, 149). Other underlying conditions associated with severe streptococcal infections are inflammatory diseases including diabetes and super-infection by other pathogens, either of which can alter inflammasome responses.

Can pharmacologic targeting of the inflammasome provide a therapeutic benefit during streptococcal infection? Knockout mice deficient in inflammasome factors or inflammasome-regulated cytokines are generally more susceptible to experimental infection. Restoration with exogenous IL-1β is protective in models of GBS septicemia (5) and SPN nasopharyngeal colonization (9, 20). Exogenous IL-18 was also protective in models of GBS sepsis and neonatal infection (39). SPN isolates that do not induce hemolysis or inflammasome activation induce less IL-1β and cause more invasive disease (30, 55, 150). Correspondingly, PLY-mutant SPN bacteria that induce less IL-1 and inflammasome activation (81) are better able to establish chronic infection (151). This mechanism of "flying under the radar" by avoiding inflammasome activation, even at the consequence of losing an important virulence factor, is becoming a paradigm in the field of bacterial pathogenesis. Future therapeutics that take into account the inflammasome pathway when targeting bacterial pathogens may hold promise for better outcomes in treatment of serious bacterial infections.

ACKNOWLEDGMENTS

Streptococcal research in the Nizet lab is supported by NIH grants AI077780 and AI096837. CL received fellowship support from the A.P. Giannini foundation.

REFERENCES

1. Walker MJ, Barnett TC, Mcarthur JD, Cole JN, Gillen CM, Henningham A, et al. Disease manifestations and pathogenic mechanisms of group A Streptococcus. Clin Microbiol Rev (2014) 27:264–301. doi:10.1128/CMR.00101-13

2. Mook-Kanamori BB, Geldhoff M, Van Der Poll T, Van De Beek D. Pathogenesis and pathophysiology of pneumococcal meningitis. Clin Microbiol Rev (2011) 24:557–91. doi:10.1128/CMR.00008-11

3. Nizet V, Doran KS. Group B Streptococcus meningitis. Cell Mol Basis (2013) 26:118. doi:10.1079/9781780641621.0118

4. Hsu L-C, Enzler T, Seita J, Timmer AM, Lee C-Y, Lai T-Y, et al. IL-1β-driven neutrophilia preserves antibacterial defense in the absence of the kinase IKKβ. Nat Immunol (2011) 12:144–50. doi:10.1038/ni.1976

5. Biondo C, Mancuso G, Midiri A, Signorino G, Domina M, Lanza Cariccio V, et al. Essential role of IL-1 signaling in host defenses against group B Streptococcus. MBio (2014) 5:e1428–1414. doi:10.1128/mBio.01428-14

6. Rijneveld AW, Florquin S, Branger J, Speelman P, Van Deventer SJH, Van Der Poll T. TNF-α compensates for the impaired host defense of IL-1 type I receptor-deficient mice during pneumococcal pneumonia. J Immunol (2001) 167:5240–6. doi:10.4049/jimmunol.167.9.5240

7. Zwijnenburg PJG, Van Der Poll T, Florquin S, Roord JJ, Van Furth AM. IL-1 receptor type 1 gene-deficient mice demonstrate an impaired host defense against pneumococcal meningitis. J Immunol (2003) 170:4724–30. doi:10.4049/jimmunol.170.9.4724

8. Yang H, Ko H-J, Yang J-Y, Kim J-J, Seo S-U, Park SG, et al. Interleukin-1 promotes coagulation, which is necessary for protective immunity in the lung against Streptococcus pneumoniae infection. J Infect Dis (2013) 207:50–60. doi:10.1093/infdis/jis651

9. Lemon JK, Miller MR, Weiser JN. Sensing of IL-1 cytokines during Streptococcus pneumoniae colonization contributes to macrophage recruitment and bacterial clearance. Infect Immun (2015) 83:3204–12. doi:10.1128/IAI.00224-15

10. Chen C-J, Kono H, Golenbock D, Reed G, Akira S, Rock KL. Identification of a key pathway required for the sterile inflammatory response triggered by dying cells. Nat Med (2007) 13:851–6. doi:10.1038/nm1603

11. Von Moltke J, Ayres JS, Kofoed EM, Chavarría-Smith J, Vance RE. Recognition of bacteria by inflammasomes. *Annu Rev Immunol* (2013) **31**:73–106. doi:10.1146/annurev-immunol-032712-095944

12. Shoma S, Tsuchiya K, Kawamura I, Nomura T, Hara H, Uchiyama R, et al. Critical involvement of pneumolysin in production of interleukin-1α and caspase-1-dependent cytokines in infection with *Streptococcus pneumoniae* in vitro: a novel function of pneumolysin in caspase-1 activation. *Infect Immun* (2008) **76**:1547–57. doi:10.1128/iai.01269-07

13. McNeela EA, Burke A, Neill DR, Baxter C, Fernandes VE, Ferreira D, et al. Pneumolysin activates the NLRP3 inflammasome and promotes proinflammatory cytokines independently of TLR4. *PLoS Pathog* (2010) **6**:e1001191. doi:10.1371/journal.ppat.1001191

14. Ulett GC, Webb RI, Ulett KB, Cui X, Benjamin WH, Crowley M, et al. Group B *Streptococcus* (GBS) urinary tract infection involves binding of GBS to bladder uroepithelium and potent but GBS-specific induction of interleukin 1α. *J Infect Dis* (2010) **201**:866–70. doi:10.1086/650696

15. Abdeltawab NF, Aziz RK, Kansal R, Rowe SL, Su Y, Gardner L, et al. An unbiased systems genetics approach to mapping genetic loci modulating susceptibility to severe streptococcal sepsis. *PLoS Pathog* (2008) **4**:e1000042. doi:10.1371/journal.ppat.1000042

16. Hannula-Jouppi K, Massinen S, Siljander T, Mäkelä S, Kivinen K, Leinonen R, et al. Genetic susceptibility to non-necrotizing erysipelas/cellulitis. *PLoS One* (2013) **8**:e56225. doi:10.1371/journal.pone.0056225

17. Biondo C, Mancuso G, Midiri A, Signorino G, Domina M, Lanza Cariccio V, et al. The IL-1β/CXCL1/2/neutrophil axis mediates host protection against group B streptococcal infection. *Infect Immun* (2014) **82**:4508–17. doi:10.1128/iai.02104-14

18. Lin AE, Beasley FC, Keller N, Hollands A, Urbano R, Troemel ER, et al. A group A *Streptococcus* ADP-ribosyltransferase toxin stimulates a protective interleukin 1β-dependent macrophage immune response. *MBio* (2015) **6**:e133–115. doi:10.1128/mBio.00133-15

19. Costa A, Gupta R, Signorino G, Malara A, Cardile F, Biondo C, et al. Activation of the NLRP3 inflammasome by group B streptococci. *J Immunol* (2012) **188**:1953–60. doi:10.4049/jimmunol.1102543

20. Kafka D, Ling E, Feldman G, Benharroch D, Voronov E, Givon-Lavi N, et al. Contribution of IL-1 to resistance to *Streptococcus pneumoniae* infection. *Int Immunol* (2008) **20**:1139–46. doi:10.1093/intimm/dxn071

21. Karmakar M, Katsnelson M, Malak HA, Greene NG, Howell SJ, Hise AG, et al. Neutrophil IL-1β processing induced by pneumolysin is mediated by the NLRP3/ASC inflammasome and caspase-1 activation and is dependent on K+ efflux. *J Immunol* (2015) **194**:1763–75. doi:10.4049/jimmunol.1401624

22. Zhang Z, Clarke TB, Weiser JN. Cellular effectors mediating Th17-dependent clearance of pneumococcal colonization in mice. *J Clin Invest* (2009) **119**:1899–909. doi:10.1172/JCI36731

23. Nizet V, Kim KS, Stins M, Jonas M, Chi EY, Nguyen D, et al. Invasion of brain microvascular endothelial cells by group B streptococci. *Infect Immun* (1997) **65**:5074–81.

24. Doran KS, Liu GY, Nizet V. Group B streptococcal β-hemolysin/cytolysin activates neutrophil signaling pathways in brain endothelium and contributes to development of meningitis. *J Clin Invest* (2003) **112**:736. doi:10.1172/JCI200317335

25. Ring A, Weiser JN, Tuomanen EI. Pneumococcal trafficking across the blood-brain barrier. Molecular analysis of a novel bidirectional pathway. *J Clin Invest* (1998) **102**:347–60. doi:10.1172/JCI2406

26. Koedel U, Winkler F, Angele B, Fontana A, Flavell RA, Pfister H-W. Role of caspase-1 in experimental pneumococcal meningitis: evidence from pharmacologic caspase inhibition and caspase-1-deficient mice. *Ann Neurol* (2002) **51**:319–29. doi:10.1002/ana.10103

27. De Vries HE, Blom-Roosemalen MCM, Oosten MV, De Boer AG, Van Berkel TJC, Breimer DD, et al. The influence of cytokines on the integrity of the blood-brain barrier in vitro. *J Neuroimmunol* (1996) **64**:37–43. doi:10.1016/0165-5728(95)00148-4

28. Koedel U, Frankenberg T, Kirschnek S, Obermaier B, Häcker H, Paul R, et al. Apoptosis is essential for neutrophil functional shutdown and determines tissue damage in experimental pneumococcal meningitis. *PLoS Pathog* (2009) **5**:e1000461. doi:10.1371/journal.ppat.1000461

29. Hoegen T, Tremel N, Klein M, Angele B, Wagner H, Kirschning C, et al. The NLRP3 inflammasome contributes to brain injury in pneumococcal meningitis and is activated through ATP-dependent lysosomal cathepsin B release. *J Immunol* (2011) **187**:5440–51. doi:10.4049/jimmunol.1100790

30. Geldhoff M, Mook-Kanamori BB, Brouwer MC, Troost D, Leemans JC, Flavell RA, et al. Inflammasome activation mediates inflammation and outcome in humans and mice with pneumococcal meningitis. *BMC Infect Dis* (2013) **13**:358. doi:10.1186/1471-2334-13-358

31. Ernst JD, Decazes JM, Sande MA. Experimental pneumococcal meningitis: role of leukocytes in pathogenesis. *Infect Immun* (1983) **41**:275–9.

32. Van Der Poll T, Opal SM. Pathogenesis, treatment, and prevention of pneumococcal pneumonia. *Lancet* (2009) **374**:1543–56. doi:10.1016/S0140-6736(09)61114-4

33. Mitchell AJ, Yau B, Mcquillan JA, Ball HJ, Too LK, Abtin A, et al. Inflammasome-dependent IFN-γ drives pathogenesis in *Streptococcus pneumoniae* meningitis. *J Immunol* (2012) **189**:4970–80. doi:10.4049/jimmunol.1201687

34. LaRock CN, Cookson BT. Burning down the house: cellular actions during pyroptosis. *PLoS Pathog* (2013) **9**:e1003793. doi:10.1371/journal.ppat.1003793

35. Miettinen M, Matikainen S, Vuopio-Varkila J, Pirhonen J, Varkila K, Kurimoto M, et al. Lactobacilli and sreptococci induce interleukin-12 (IL-12), IL-18, and gamma interferon production in human peripheral blood mononuclear cells. *Infect Immun* (1998) **66**:6058–62.

36. Fang R, Hara H, Sakai S, Hernandez-Cuellar E, Mitsuyama M, Kawamura I, et al. Type I interferon signaling regulates activation of the absent in melanoma 2 inflammasome during *Streptococcus pneumoniae* infection. *Infect Immun* (2014) **82**:2310–7. doi:10.1128/iai.01572-14

37. Lauw FN, Branger J, Florquin S, Speelman P, Van Deventer SJH, Akira S, et al. IL-18 improves the early antimicrobial host response to pneumococcal pneumonia. *J Immunol* (2002) **168**:372–8. doi:10.4049/jimmunol.168.1.372

38. Zwijnenburg PJG, Van Der Poll T, Florquin S, Akira S, Takeda K, Roord JJ, et al. Interleukin-18 gene-deficient mice show enhanced defense and reduced inflammation during pneumococcal meningitis. *J Neuroimmunol* (2003) **138**:31–7. doi:10.1016/S0165-5728(03)00088-2

39. Cusumano V, Midiri A, Cusumano VV, Bellantoni A, De Sossi G, Teti G, et al. Interleukin-18 is an essential element in host resistance to experimental group B streptococcal disease in neonates. *Infect Immun* (2004) **72**:295–300. doi:10.1128/IAI.72.1.295-300.2004

40. Höhne C, Wenzel M, Angele B, Hammerschmidt S, Häcker H, Klein M, et al. High mobility group box 1 prolongs inflammation and worsens disease in pneumococcal meningitis. *Brain* (2013) **136**:1746–59. doi:10.1093/brain/awt064

41. Miao EA, Leaf IA, Treuting PM, Mao DP, Dors M, Sarkar A, et al. Caspase-1-induced pyroptosis is an innate immune effector mechanism against intracellular bacteria. *Nat Immunol* (2010) **11**:1136–42. doi:10.1038/ni.1960

42. Sakurai A, Maruyama F, Funao J, Nozawa T, Aikawa C, Okahashi N, et al. Specific behavior of intracellular *Streptococcus pyogenes* that has undergone autophagic degradation is associated with bacterial streptolysin O and host small G proteins Rab5 and Rab7. *J Biol Chem* (2010) **285**:22666–75. doi:10.1074/jbc.M109.100131

43. Barnett TC, Liebl D, Seymour LM, Gillen CM, Lim JY, Larock CN, et al. The globally disseminated M1T1 clone of group A *Streptococcus* evades autophagy for intracellular replication. *Cell Host Microbe* (2013) **14**:675–82. doi:10.1016/j.chom.2013.11.003

44. Timmer AM, Timmer JC, Pence MA, Hsu L-C, Ghochani M, Frey TG, et al. Streptolysin O promotes group A *Streptococcus* immune evasion by accelerated macrophage apoptosis. *J Biol Chem* (2009) **284**:862–71. doi:10.1074/jbc.M804632200

45. Von Moltke J, Trinidad NJ, Moayeri M, Kintzer AF, Wang SB, Van Rooijen N, et al. Rapid induction of inflammatory lipid mediators by the inflammasome in vivo. *Nature* (2012) **490**:107–11. doi:10.1038/nature11351

46. Rius J, Guma M, Schachtrup C, Akassoglou K, Zinkernagel AS, Nizet V, et al. NF-kappaB links innate immunity to the hypoxic response through transcriptional regulation of HIF-1α. *Nature* (2008) **453**:807–11. doi:10.1038/nature06905

47. Maloney CG, Thompson SD, Hill HR, Bohnsack JF, Mcintyre TM, Zimmerman GA. Induction of cyclooxygenase-2 by human monocytes exposed to group B streptococci. *J Leukoc Biol* (2000) **67**:615–21.

48. N'Guessan PD, Hippenstiel S, Etouem MO, Zahlten J, Beermann W, Lindner D, et al. *Streptococcus pneumoniae* induced p38 MAPK- and NF-κB-dependent

COX-2 expression in human lung epithelium. *Am J Physiol Lung Cell Mol Physiol* (2006) **290**:L1131–8. doi:10.1152/ajplung.00383.2005

49. Goldmann O, Hertzén E, Hecht A, Schmidt H, Lehne S, Norrby-Teglund A, et al. Inducible cyclooxygenase released prostaglandin E2 modulates the severity of infection caused by *Streptococcus pyogenes*. *J Immunol* (2010) **185**:2372–81. doi:10.4049/jimmunol.1000838

50. Mason KL, Rogers LM, Soares EM, Bani-Hashemi T, Erb Downward J, Agnew D, et al. Intrauterine group A streptococcal infections are exacerbated by prostaglandin E2. *J Immunol* (2013) **191**:2457–65. doi:10.4049/jimmunol.1300786

51. Aronoff DM, Bloch KC. Assessing the relationship between the use of nonsteroidal antiinflammatory drugs and necrotizing fasciitis caused by group A *Streptococcus*. *Medicine* (2003) **82**:225–35. doi:10.1097/00005792-200307000-00001

52. Sokolovska A, Becker CE, Ip WK, Rathinam VA, Brudner M, Paquette N, et al. Activation of caspase-1 by the NLRP3 inflammasome regulates the NADPH oxidase NOX2 to control phagosome function. *Nat Immunol* (2013) **14**:543–53. doi:10.1038/ni.2595

53. Albiger B, Dahlberg S, Sandgren A, Wartha F, Beiter K, Katsuragi H, et al. Toll-like receptor 9 acts at an early stage in host defence against pneumococcal infection. *Cell Microbiol* (2007) **9**:633–44. doi:10.1111/j.1462-5822.2006.00814.x

54. Harder J, Franchi L, Muñoz-Planillo R, Park J-H, Reimer T, Núñez G. Activation of the NLRP3 inflammasome by *Streptococcus pyogenes* requires streptolysin O and NF-κB activation but proceeds independently of TLR signaling and P2X7 receptor. *J Immunol* (2009) **183**:5823–9. doi:10.4049/jimmunol.0900444

55. Witzenrath M, Pache F, Lorenz D, Koppe U, Gutbier B, Tabeling C, et al. The NLRP3 inflammasome is differentially activated by pneumolysin variants and contributes to host defense in pneumococcal pneumonia. *J Immunol* (2011) **187**:434–40. doi:10.4049/jimmunol.1003143

56. Whidbey C, Harrell MI, Burnside K, Ngo L, Becraft AK, Iyer LM, et al. A hemolytic pigment of Group B *Streptococcus* allows bacterial penetration of human placenta. *J Exp Med* (2013) **210**:1265–81. doi:10.1084/jem.20122753

57. Gupta R, Ghosh S, Monks B, Deoliveira RB, Tzeng T-C, Kalantari P, et al. RNA and β-hemolysin of group B *Streptococcus* induce interleukin-1β (IL-1β) by activating NLRP3 inflammasomes in mouse macrophages. *J Biol Chem* (2014) **289**:13701–5. doi:10.1371/journal.pone.0056225

58. Menu P, Mayor A, Zhou R, Tardivel A, Ichijo H, Mori K, et al. ER stress activates the NLRP3 inflammasome via an UPR-independent pathway. *Cell Death Dis* (2012) **3**:e261. doi:10.1038/cddis.2011.132

59. Bentley CC, Hakansson A, Christianson J, Wessels MR. Extracellular group A *Streptococcus* induces keratinocyte apoptosis by dysregulating calcium signalling. *Cell Microbiol* (2005) **7**:945–55. doi:10.1111/j.1462-5822.2005.00525.x

60. Shewell LK, Harvey RM, Higgins MA, Day CJ, Hartley-Tassell LE, Chen AY, et al. The cholesterol-dependent cytolysins pneumolysin and streptolysin O require binding to red blood cell glycans for hemolytic activity. *Proc Natl Acad Sci U S A* (2014) **111**:E5312–20. doi:10.1073/pnas.1412703111

61. Houldsworth S, Andrew PW, Mitchell TJ. Pneumolysin stimulates production of tumor necrosis factor alpha and interleukin-1β by human mononuclear phagocytes. *Infect Immun* (1994) **62**:1501–3.

62. Keyel PA, Roth R, Yokoyama WM, Heuser JE, Salter RD. Reduction of streptolysin O (SLO) pore-forming activity enhances inflammasome activation. *Toxins (Basel)* (2013) **5**:1105–18. doi:10.3390/toxins5061105

63. Betschel SD, Borgia SM, Barg NL, Low DE, De Azavedo JCS. Reduced virulence of group A streptococcal Tn916 mutants that do not produce streptolysin S. *Infect Immun* (1998) **66**:1671–9.

64. Nizet V, Beall B, Bast DJ, Datta V, Kilburn L, Low DE, et al. Genetic locus for streptolysin S production by group A *Streptococcus*. *Infect Immun* (2000) **68**:4245–54. doi:10.1128/IAI.68.7.4245-4254.2000

65. Sierig G, Cywes C, Wessels MR, Ashbaugh CD. Cytotoxic effects of streptolysin O and streptolysin S enhance the virulence of poorly encapsulated group A streptococci. *Infect Immun* (2003) **71**:446–55. doi:10.1128/IAI.71.1.446-455.2003

66. Randis TM, Gelber SE, Hooven TA, Abellar RG, Akabas LH, Lewis EL, et al. Group B *Streptococcus* β-hemolysin/cytolysin breaches maternal-fetal barriers to cause preterm birth and intrauterine fetal demise in vivo. *J Infect Dis* (2014) **210**:265–73. doi:10.1093/infdis/jiu067

67. Pritzlaff CA, Chang JCW, Kuo SP, Tamura GS, Rubens CE, Nizet V. Genetic basis for the β-haemolytic/cytolytic activity of group B *Streptococcus*. *Mol Microbiol* (2001) **39**:236–48. doi:10.1046/j.1365-2958.2001.02211.x

68. Vanberg C, Lutnaes BF, Langsrud T, Nes IF, Holo H. *Propionibacterium jensenii* produces the polyene pigment granadaene and has hemolytic properties similar to those of *Streptococcus agalactiae*. *Appl Environ Microbiol* (2007) **73**:5501–6. doi:10.1128/AEM.00545-07

69. Tapsall JW. Relationship between pigment production and haemolysin formation by Lancefield group B streptococci. *J Med Microbiol* (1987) **24**:83–7. doi:10.1099/00222615-24-1-83

70. Charrel-Dennis M, Latz E, Halmen KA, Trieu-Cuot P, Fitzgerald KA, Kasper DL, et al. TLR-independent type I interferon induction in response to an extracellular bacterial pathogen via intracellular recognition of its DNA. *Cell Host Microbe* (2008) **4**:543–54. doi:10.1016/j.chom.2008.11.002

71. Fink SL, Cookson BT. Apoptosis, pyroptosis, and necrosis: mechanistic description of dead and dying eukaryotic cells. *Infect Immun* (2005) **73**:1907–16. doi:10.1128/iai.73.4.1907-1916.2005

72. Nilsson M, Sørensen OE, Mörgelin M, Weineisen M, Sjöbring U, Herwald H. Activation of human polymorphonuclear neutrophils by streptolysin O from *Streptococcus pyogenes* leads to the release of proinflammatory mediators. *Thromb Haemost* (2006) **95**:982–90. doi:10.1160/TH05-08-0572

73. Goldmann O, Sastalla I, Wos-Oxley M, Rohde M, Medina E. *Streptococcus pyogenes* induces oncosis in macrophages through the activation of an inflammatory programmed cell death pathway. *Cell Microbiol* (2009) **11**:138–55. doi:10.1111/j.1462-5822.2008.01245.x

74. Littmann M, Albiger B, Frentzen A, Normark S, Henriques-Normark B, Plant L. *Streptococcus pneumoniae* evades human dendritic cell surveillance by pneumolysin expression. *EMBO Mol Med* (2009) **1**:211–22. doi:10.1002/emmm.200900025

75. Bebien M, Hensler ME, Davanture S, Hsu L-C, Karin M, Park JM, et al. The pore-forming toxin β-hemolysin/cytolysin triggers p38 MAPK-dependent IL-10 production in macrophages and inhibits innate immunity. *PLoS Pathog* (2012) **8**:e1002812. doi:10.1371/journal.ppat.1002812

76. Keyel PA, Loultcheva L, Roth R, Salter RD, Watkins SC, Yokoyama WM, et al. Streptolysin O clearance through sequestration into blebs that bud passively from the plasma membrane. *J Cell Sci* (2011) **124**:2414–23. doi:10.1242/jcs.076182

77. Kanneganti T-D, Lamkanfi M, Kim Y-G, Chen G, Park J-H, Franchi L, et al. Pannexin-1-mediated recognition of bacterial molecules activates the cryopyrin inflammasome independent of toll-like receptor signaling. *Immunity* (2007) **26**:433–43. doi:10.1016/j.immuni.2007.03.008

78. Bauernfeind FG, Horvath G, Stutz A, Alnemri ES, Macdonald K, Speert D, et al. NF-κB activating pattern recognition and cytokine receptors license NLRP3 inflammasome activation by regulating NLRP3 expression. *J Immunol* (2009) **183**:787–91. doi:10.4049/jimmunol.0901363

79. Kanneganti T-D, Ozoren N, Body-Malapel M, Amer A, Park J-H, Franchi L, et al. Bacterial RNA and small antiviral compounds activate caspase-1 through cryopyrin/Nalp3. *Nature* (2006) **440**:233–6. doi:10.1038/nature04517

80. Shimada T, Park BG, Wolf AJ, Brikos C, Goodridge HS, Becker CA, et al. *Staphylococcus aureus* evades lysozyme-based peptidoglycan digestion that links phagocytosis, inflammasome activation, and IL-1β secretion. *Cell Host Microbe* (2010) **7**:38–49. doi:10.1016/j.chom.2009.12.008

81. Lemon JK, Weiser JN. Degradation products of the extracellular pathogen *Streptococcus pneumoniae* access the cytosol via its pore-forming toxin. *MBio* (2015) **6**:e2110–4. doi:10.1128/mBio.02110-14

82. Coye LH, Collins CM. Identification of SpyA, a novel ADP-ribosyltransferase of *Streptococcus pyogenes*. *Mol Microbiol* (2004) **54**:89–98. doi:10.1111/j.1365-2958.2004.04262.x

83. Bose S, Segovia JA, Somarajan SR, Chang T-H, Kannan TR, Baseman JB. ADP-ribosylation of NLRP3 by *Mycoplasma pneumoniae* CARDS toxin regulates inflammasome activity. *MBio* (2014) **5**:e2186–2114. doi:10.1128/mBio.02186-14

84. Xu H, Yang J, Gao W, Li L, Li P, Zhang L, et al. Innate immune sensing of bacterial modifications of Rho GTPases by the Pyrin inflammasome. *Nature* (2014) **513**:237–41. doi:10.1038/nature13449

85. Icenogle LM, Hengel SM, Coye LH, Streifel A, Collins CM, Goodlett DR, et al. Molecular and biological characterization of streptococcal SpyA-mediated ADP-ribosylation of intermediate filament protein vimentin. *J Biol Chem* (2012) **287**:21481–91. doi:10.1074/jbc.M112.370791

86. Dos Santos G, Rogel MR, Baker MA, Troken JR, Urich D, Morales-Nebreda L, et al. Vimentin regulates activation of the NLRP3 inflammasome. *Nat Commun* (2015) **6**:6574. doi:10.1038/ncomms7574

87. Misawa T, Takahama M, Kozaki T, Lee H, Zou J, Saitoh T, et al. Microtubule-driven spatial arrangement of mitochondria promotes activation of the NLRP3 inflammasome. *Nat Immunol* (2013) 14:454–60. doi:10.1038/ni.2550

88. Bastiat-Sempe B, Love JF, Lomayesva N, Wessels MR. Streptolysin O and NAD-glycohydrolase prevent phagolysosome acidification and promote group A *Streptococcus* survival in macrophages. *MBio* (2014) 5:e1690–1614. doi:10.1128/mBio.01690-14

89. Fang R, Tsuchiya K, Kawamura I, Shen Y, Hara H, Sakai S, et al. Critical roles of ASC inflammasomes in caspase-1 activation and host innate resistance to *Streptococcus pneumoniae* infection. *J Immunol* (2011) 187:4890–9. doi:10.4049/jimmunol.1100381

90. Parker D, Martin FJ, Soong G, Harfenist BS, Aguilar JL, Ratner AJ, et al. *Streptococcus pneumoniae* DNA initiates type I interferon signaling in the respiratory tract. *MBio* (2011) 2:16–11. doi:10.1128/mBio.00016-11

91. Koppe U, Högner K, Doehn J-M, Müller HC, Witzenrath M, Gutbier B, et al. *Streptococcus pneumoniae* stimulates a STING- and IFN regulatory factor 3-dependent type I IFN production in macrophages, which regulates RANTES production in macrophages, cocultured alveolar epithelial cells, and mouse lungs. *J Immunol* (2012) 188:811–7. doi:10.4049/jimmunol.1004143

92. Mancuso G, Gambuzza M, Midiri A, Biondo C, Papasergi S, Akira S, et al. Bacterial recognition by TLR7 in the lysosomes of conventional dendritic cells. *Nat Immunol* (2009) 10:587–94. doi:10.1038/ni.1733

93. Gratz N, Hartweger H, Matt U, Kratochvill F, Janos M, Sigel S, et al. Type I interferon production induced by *Streptococcus pyogenes*-derived nucleic acids is required for host protection. *PLoS Pathog* (2011) 7:e1001345. doi:10.1016/j.immuni.2011.02.006

94. Eigenbrod T, Franchi L, Muñoz-Planillo R, Kirschning CJ, Freudenberg MA, Núñez G, et al. Bacterial RNA mediates activation of caspase-1 and IL-1β release independently of TLRs 3, 7, 9 and TRIF but Is dependent on UNC93B. *J Immunol* (2012) 189:328–36. doi:10.4049/jimmunol.1103258

95. Fieber C, Janos M, Koestler T, Gratz N, Li X-D, Castiglia V, et al. Innate immune response to *Streptococcus pyogenes* depends on the combined activation of TLR13 and TLR2. *PLoS One* (2015) 10:e0119727. doi:10.1371/journal.pone.0119727

96. Uchiyama S, Andreoni F, Schuepbach RA, Nizet V, Zinkernagel AS. DNase Sda1 allows invasive M1T1 group A *Streptococcus* to prevent TLR9-dependent recognition. *PLoS Pathog* (2012) 8:e1002736. doi:10.1371/journal.ppat.1002736

97. Zinkernagel AS, Hruz P, Uchiyama S, Schuepbach RA, Hayashi T, Carson DA, et al. Importance of toll-like receptor 9 in host defense against M1T1 group A *Streptococcus* infections. *J Innate Immun* (2012) 4:213–8. doi:10.1159/000329550

98. Jorgensen I, Miao EA. Pyroptotic cell death defends against intracellular pathogens. *Immunol Rev* (2015) 265:130–42. doi:10.1111/imr.12287

99. Chavarría-Smith J, Vance RE. The NLRP1 inflammasomes. *Immunol Rev* (2015) 265:22–34. doi:10.1111/imr.12283

100. Kapur V, Majesky MW, Li L-L, Black RA, Musser JM. Cleavage of interleukin-1β (IL-1β) precursor to produce active IL-1β by a conserved extracellular cysteine protease from *Streptococcus pyogenes*. *Proc Natl Acad Sci U S A* (1993) 90:7676–80.

101. Beauséjour A, Grenier D, Goulet J-P, Deslauriers N. Proteolytic activation of the interleukin-1β precursor by *Candida albicans*. *Infect Immun* (1998) 66:676–81.

102. Zhang Z, Wang L, Seydel KB, Li E, Ankri S, Mirelman D, et al. *Entamoeba histolytica* cysteine proteinases with interleukin-1β converting enzyme (ICE) activity cause intestinal inflammation and tissue damage in amoebiasis. *Mol Microbiol* (2000) 37:542–8. doi:10.1046/j.1365-2958.2000.02037.x

103. Black RA, Kronheim SR, Cantrell M, Deeley MC, March CJ, Prickett KS, et al. Generation of biologically active interleukin-1 beta by proteolytic cleavage of the inactive precursor. *J Biol Chem* (1988) 263:9437–42.

104. Beausejour A, Deslauriers N, Grenier D. Activation of the interleukin-1β precursor by *Treponema denticola*: a potential role in chronic inflammatory periodontal diseases. *Infect Immun* (1997) 65:3199–202.

105. Kayagaki N, Warming S, Lamkanfi M, Walle LV, Louie S, Dong J, et al. Non-canonical inflammasome activation targets caspase-11. *Nature* (2011) 479:117–21. doi:10.1038/nature10558

106. Hiscott J, Marois J, Garoufalis J, D'Addario M, Roulston A, Kwan I, et al. Characterization of a functional NF-κB site in the human interleukin 1β promoter: evidence for a positive autoregulatory loop. *Mol Cell Biol* (1993) 13:6231–40. doi:10.1128/mcb.13.10.6231

107. Franchi L, Eigenbrod T, Núñez G. Cutting edge: TNF-α mediates sensitization to ATP and silica via the NLRP3 inflammasome in the absence of microbial stimulation. *J Immunol* (2009) 83:792–6. doi:10.4049/jimmunol.0900173

108. Wu J, Fernandes-Alnemri T, Alnemri ES. Involvement of the AIM2, NLRC4, and NLRP3 inflammasomes in caspase-1 activation by *Listeria monocytogenes*. *J Clin Immunol* (2010) 30:693–702. doi:10.1007/s10875-010-9425-2

109. Fernandes-Alnemri T, Yu JW, Datta P, Wu J, Alnemri ES. AIM2 activates the inflammasome and cell death in response to cytoplasmic DNA. *Nature* (2009) 458:509–13. doi:10.1038/nature07710

110. Guarda G, Braun M, Staehli F, Tardivel A, Mattmann C, Förster I, et al. Type I interferon inhibits interleukin-1 production and inflammasome activation. *Immunity* (2011) 34:213–23. doi:10.1016/j.immuni.2011.02.006

111. Weigent DA, Beachey EH, Huff T, Peterson JW, Stanton GJ, Baron S. Induction of human gamma interferon by structurally defined polypeptide fragments of group A streptococcal M protein. *Infect Immun* (1984) 43:122–6.

112. Mancuso G, Midiri A, Biondo C, Beninati C, Zummo S, Galbo R, et al. Type I IFN signaling is crucial for host resistance against different species of pathogenic bacteria. *J Immunol* (2007) 178:3126–33. doi:10.4049/jimmunol.178.5.3126

113. Gratz N, Siller M, Schaljo B, Pirzada ZA, Gattermeier I, Vojtek I, et al. Group A *Streptococcus* activates type I interferon production and MyD88-dependent signaling without involvement of TLR2, TLR4, and TLR9. *J Biol Chem* (2008) 283:19879–87. doi:10.1074/jbc.M802848200

114. Loof TG, Goldmann O, Medina E. Immune recognition of *Streptococcus pyogenes* by dendritic cells. *Infect Immun* (2008) 76:2785–92. doi:10.1128/IAI.01680-07

115. Oliveira-Nascimento L, Massari P, Wetzler LM. The role of TLR2 in infection and immunity. *Front Immunol* (2012) 3:79. doi:10.3389/fimmu.2012.00079

116. Joosten LA, Koenders MI, Smeets RL, Heuvelmans-Jacobs M, Helsen MM, Takeda K, et al. Toll-like receptor 2 pathway drives streptococcal cell wall-induced joint inflammation: critical role of myeloid differentiation factor 88. *J Immunol* (2003) 171:6145–53. doi:10.4049/jimmunol.171.11.6145

117. Mancuso G, Midiri A, Beninati C, Biondo C, Galbo R, Akira S, et al. Dual role of TLR2 and myeloid differentiation factor 88 in a mouse model of invasive group B streptococcal disease. *J Immunol* (2004) 172:6324–9. doi:10.4049/jimmunol.172.10.6324

118. Henneke P, Morath S, Uematsu S, Weichert S, Pfitzenmaier M, Takeuchi O, et al. Role of lipoteichoic acid in the phagocyte response to group B *Streptococcus*. *J Immunol* (2005) 174:6449–55. doi:10.4049/jimmunol.174.10.6449

119. Lehnardt S, Henneke P, Lien E, Kasper DL, Volpe JJ, Bechmann I, et al. A mechanism for neurodegeneration induced by group B streptococci through activation of the TLR2/MyD88 pathway in microglia. *J Immunol* (2006) 177:583–92. doi:10.4049/jimmunol.177.1.583

120. Henneke P, Dramsi S, Mancuso G, Chraibi K, Pellegrini E, Theilacker C, et al. Lipoproteins are critical TLR2 activating toxins in group B streptococcal sepsis. *J Immunol* (2008) 180:6149–58. doi:10.4049/jimmunol.180.9.6149

121. Yoshimura A, Lien E, Ingalls RR, Tuomanen E, Dziarski R, Golenbock D. Cutting edge: recognition of Gram-positive bacterial cell wall components by the innate immune system occurs via toll-like receptor 2. *J Immunol* (1999) 163:1–5.

122. Dessing MC, Florquin S, Paton JC, Van Der Poll T. Toll-like receptor 2 contributes to antibacterial defence against pneumolysin-deficient pneumococci. *Cell Microbiol* (2008) 10:237–46. doi:10.1111/j.1462-5822.2007.01035.x

123. Klein M, Obermaier B, Angele B, Pfister H-W, Wagner H, Koedel U, et al. Innate immunity to pneumococcal infection of the central nervous system depends on toll-like receptor (TLR) 2 and TLR4. *J Infect Dis* (2008) 198:1028–36. doi:10.1086/591626

124. Henneke P, Takeuchi O, Van Strijp JA, Guttormsen H-K, Smith JA, Schromm AB, et al. Novel engagement of CD14 and multiple toll-like receptors by group B streptococci. *J Immunol* (2001) 167:7069–76. doi:10.4049/jimmunol.167.12.7069

125. Påhlman LI, Mörgelin M, Eckert J, Johansson L, Russell W, Riesbeck K, et al. Streptococcal M protein: a multipotent and powerful inducer of inflammation. *J Immunol* (2006) 177:1221–8. doi:10.4049/jimmunol.177.2.1221

126. Severin A, Nickbarg E, Wooters J, Quazi SA, Matsuka YV, Murphy E, et al. Proteomic analysis and identification of *Streptococcus pyogenes*

surface-associated proteins. *J Bacteriol* (2007) **189**:1514–22. doi:10.1128/jb.01132-06

127. LaRock CN, Nizet V. Cationic antimicrobial peptide resistance mechanisms of streptococcal pathogens. *Biochem Biophys Acta* (2015). doi:10.1016/j.bbamem.2015.02.010

128. Lee KS, Scanga CA, Bachelder EM, Chen Q, Snapper CM. TLR2 synergizes with both TLR4 and TLR9 for induction of the MyD88-dependent splenic cytokine and chemokine response to *Streptococcus pneumoniae*. *Cell Immunol* (2007) **245**:103–10. doi:10.1016/j.cellimm.2007.04.003

129. Malley R, Henneke P, Morse SC, Cieslewicz MJ, Lipsitch M, Thompson CM, et al. Recognition of pneumolysin by toll-like receptor 4 confers resistance to pneumococcal infection. *Proc Natl Acad Sci U S A* (2003) **100**:1966–71. doi:10.1073/pnas.0435928100

130. Srivastava A, Henneke P, Visintin A, Morse SC, Martin V, Watkins C, et al. The apoptotic response to pneumolysin is toll-like receptor 4 dependent and protects against pneumococcal disease. *Infect Immun* (2005) **73**:6479–87. doi:10.1128/IAI.73.10.6479-6487.2005

131. Park JM, Ng VH, Maeda S, Rest RF, Karin M. Anthrolysin O and other gram-positive cytolysins are toll-like receptor 4 agonists. *J Exp Med* (2004) **200**:1647–55. doi:10.1084/jem.20041215

132. Branger J, Knapp S, Weijer S, Leemans JC, Pater JM, Speelman P, et al. Role of toll-like receptor 4 in Gram-positive and Gram-negative pneumonia in mice. *Infect Immun* (2004) **72**:788–94. doi:10.1128/iai.72.2.788-794.2004

133. Eigenbrod T, Pelka K, Latz E, Kreikemeyer B, Dalpke AH. TLR8 senses bacterial RNA in human monocytes and plays a nonredundant role for recognition of *Streptococcus pyogenes*. *J Immunol* (2015) **195**:1092–9. doi:10.4049/jimmunol.1403173

134. Girardin SE, Boneca IG, Viala J, Chamaillard M, Labigne A, Thomas G, et al. Nod2 is a general sensor of peptidoglycan through muramyl dipeptide (MDP) detection. *J Biol Chem* (2003) **278**:8869–72. doi:10.1074/jbc.C200651200

135. Hruz P, Zinkernagel AS, Jenikova G, Botwin GJ, Hugot J-P, Karin M, et al. NOD2 contributes to cutaneous defense against *Staphylococcus aureus* through α-toxin-dependent innate immune activation. *Proc Natl Acad Sci U S A* (2009) **106**:12873–8. doi:10.1073/pnas.0904958106

136. Ratner AJ, Aguilar JL, Shchepetov M, Lysenko ES, Weiser JN. Nod1 mediates cytoplasmic sensing of combinations of extracellular bacteria. *Cell Microbiol* (2007) **9**:1343–51. doi:10.1111/j.1462-5822.2006.00878.x

137. Opitz B, Püschel A, Schmeck B, Hocke AC, Rosseau S, Hammerschmidt S, et al. Nucleotide-binding oligomerization domain proteins are innate immune receptors for internalized *Streptococcus pneumoniae*. *J Biol Chem* (2004) **279**:36426–32. doi:10.1074/jbc.M403861200

138. Davis KM, Nakamura S, Weiser JN. Nod2 sensing of lysozyme-digested peptidoglycan promotes macrophage recruitment and clearance of *S. pneumoniae* colonization in mice. *J Clin Invest* (2011) **121**:3666–76. doi:10.1172/JCI57761

139. Liu X, Chauhan VS, Young AB, Marriott I. NOD2 mediates inflammatory responses of primary murine glia to *Streptococcus pneumoniae*. *Glia* (2010) **58**:839–47. doi:10.1002/glia.20968

140. Heinhuis B, Koenders MI, Van De Loo FA, Van Lent PL, Kim SH, Dinarello CA, et al. IL-32γ and *Streptococcus pyogenes* cell wall fragments synergise for IL-1-dependent destructive arthritis via upregulation of TLR-2 and NOD2. *Ann Rheum Dis* (2010) **69**:1866–72. doi:10.1136/ard.2009.127399

141. Lemire P, Calzas C, Segura M. The NOD2 receptor does not play a major role in the pathogenesis of group B *Streptococcus* in mice. *Microb Pathog* (2013) **65**:41–7. doi:10.1016/j.micpath.2013.09.006

142. Ku C-L, Von Bernuth H, Picard C, Zhang S-Y, Chang H-H, Yang K, et al. Selective predisposition to bacterial infections in IRAK-4-deficient children: IRAK-4-dependent TLRs are otherwise redundant in protective immunity. *J Exp Med* (2007) **204**:2407–22. doi:10.1084/jem.20070628

143. Kollmann TR, Levy O, Montgomery RR, Goriely S. Innate immune function by toll-like receptors: distinct responses in newborns and the elderly. *Immunity* (2012) **37**:771–83. doi:10.1016/j.immuni.2012.10.014

144. Sadeghi K, Berger A, Langgartner M, Prusa A-R, Hayde M, Herkner K, et al. Immaturity of infection control in preterm and term newborns is associated with impaired toll-like receptor signaling. *J Infect Dis* (2007) **195**:296–302. doi:10.1086/509892

145. Elahi S, Ertelt JM, Kinder JM, Jiang TT, Zhang X, Xin L, et al. Immunosuppressive CD71+ erythroid cells compromise neonatal host defense against infection. *Nature* (2013) **504**:158–62. doi:10.1038/nature12675

146. Hinojosa E, Boyd AR, Orihuela CJ. Age-associated inflammation and toll-like receptor dysfunction prime the lungs for pneumococcal pneumonia. *J Infect Dis* (2009) **200**:546–54. doi:10.1086/600870

147. Krone CL, Trzciński K, Zborowski T, Sanders EA, Bogaert D. Impaired innate mucosal immunity in aged mice permits prolonged *Streptococcus pneumoniae* colonization. *Infect Immun* (2013) **81**:4615–25. doi:10.1128/IAI.00618-13

148. Picard C, Puel A, Bonnet M, Ku C-L, Bustamante J, Yang K, et al. Pyogenic bacterial infections in humans with IRAK-4 deficiency. *Science* (2003) **299**:2076–9. doi:10.1126/science.1081902

149. Von Bernuth H, Picard C, Jin Z, Pankla R, Xiao H, Ku C-L, et al. Pyogenic bacterial infections in humans with MyD88 deficiency. *Science* (2008) **321**:691–6. doi:10.1126/science.1158298

150. Weinberger DM, Harboe ZB, Sanders EA, Ndiritu M, Klugman KP, Rückinger S, et al. Association of serotype with risk of death due to pneumococcal pneumonia: a meta-analysis. *Clin Infect Dis* (2010) **51**:692–9. doi:10.1086/655828

151. Benton KA, Everson MP, Briles DE. A pneumolysin-negative mutant of *Streptococcus pneumoniae* causes chronic bacteremia rather than acute sepsis in mice. *Infect Immun* (1995) **63**:448–55.

8

Infection Strategies of Intestinal Parasite Pathogens and Host Cell Responses

Bruno M. Di Genova[1] and Renata R. Tonelli[1,2]*

[1] *Departamento de Microbiologia e Imunologia, Universidade Federal de São Paulo, São Paulo, Brazil,* [2] *Instituto de Ciências Ambientais, Químicas e Farmacêuticas, Departamento de Ciências Biológicas, Universidade Federal de São Paulo, Diadema, Brazil*

Edited by:
Olivier Dussurget,
University Paris Diderot, Institut
Pasteur, INSERM, INRA, France

Reviewed by:
Thomas Dandekar,
University of Wuerzburg, Germany
Alexandre Morrot,
Federal University of Rio de Janeiro,
Brazil

*Correspondence:
Renata R. Tonelli
r.tonelli@unifesp.br

Giardia lamblia, Cryptosporidium sp., and *Entamoeba histolytica* are important pathogenic intestinal parasites and are amongst the leading causes worldwide of diarrheal illness in humans. Diseases caused by these organisms, giardiasis, cryptosporidiosis, and amoebiasis, respectively, are characterized by self-limited diarrhea but can evolve to long-term complications. The cellular and molecular mechanisms underlying the pathogenesis of diarrhea associated with these three pathogens are being unraveled, with knowledge of both the strategies explored by the parasites to establish infection and the methods evolved by hosts to avoid it. Special attention is being given to molecules participating in parasite–host interaction and in the mechanisms implicated in the diseases' pathophysiologic processes. This review focuses on cell mechanisms that are modulated during infection, including gene transcription, cytoskeleton rearrangements, signal transduction pathways, and cell death.

Keywords: protozoan parasites, intestinal infection, diarrhea, gastrointestinal tract, intestinal epithelial barrier, parasite–host interaction

INTRODUCTION

Intestinal infection is the most common cause of diarrhea in humans worldwide and although presenting low mortality rates, complications are not uncommon, with some cases requiring hospital care. Diarrhea may be caused by viruses, bacteria, helminths and protozoa, most of which are disseminated with feces-contaminated water and food. Amongst protozoan parasites, *Giardia lamblia, Cryptosporidium parvum,* and *Entamoeba histolytica* are the three most common etiological agents of diarrhea and other related diseases (giardiasis, cryptosporidiosis, and amoebiasis, respectively) characterized as acute and self-limited dysentery. Nevertheless, in some patients disease may become chronic with long-term effects such as malnutrition, growth delays, and cognitive impairment.

Diarrhea is an increase in the volume or liquidity of stool and it may or may not be accompanied by frequent evacuations (Viswanathan et al., 2009). Disorders of both the small and large intestines can result in diarrhea which, based on the duration, may be classified as acute ($\leq$14 days), persistent (from 15 to 29 days) or chronic ($\geq$30 days). This classification is clinically important to determine the etiological agent for diagnostic and treatment purposes (Guerrant et al., 2001). Diarrhea may also be classified into five categories based on the pathophysiological mechanisms as osmotic, secretory, exudative, inflammatory and resulting from motility disturbances (Field, 2003).

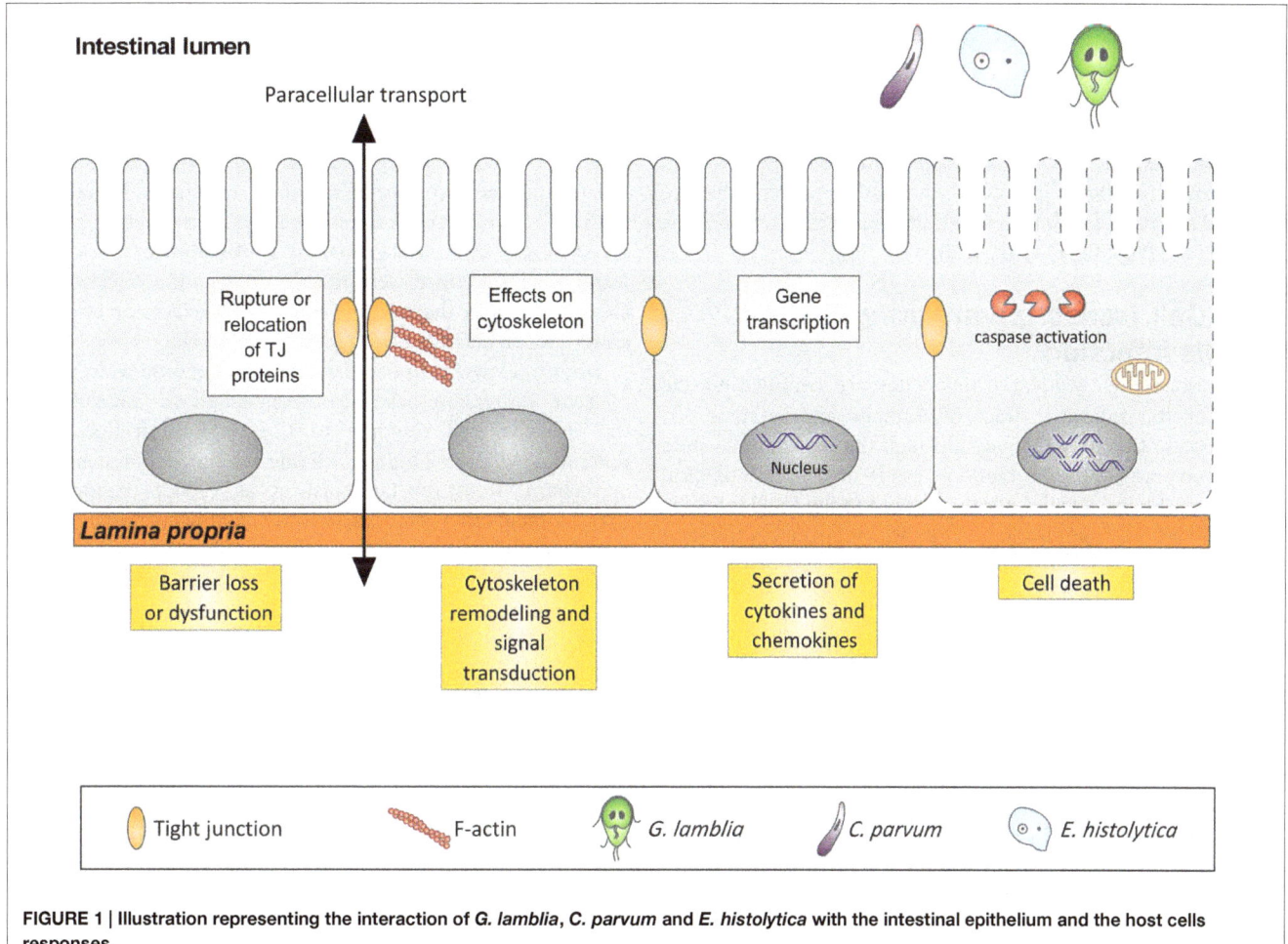

FIGURE 1 | Illustration representing the interaction of *G. lamblia*, *C. parvum* and *E. histolytica* with the intestinal epithelium and the host cells responses.

Osmotic diarrhea is triggered when healthy individuals (with normal gut functions) ingest large amounts of poorly absorbed substrates, usually carbohydrates (polyethylene glycol, mannitol, lactulose) and divalent ions ($MgSO_4$, $MgOH_2$; Hammer et al., 1989; Izzo et al., 1994). An increase in intraluminal unabsorbed nutrients associated with epithelial damage and reduction of the intestinal absorptive surface also characterizes osmotic diarrhea. Secretory diarrhea results from overstimulation of intestinal tract secretory capacity. Exposure to enterotoxins from several types of bacteria (e.g., *Escherichia coli* heat-labile toxin, *Cholera* toxin), excessive bile acid synthesis, low levels of short-chain fatty acids and intestinal inflammation (seen in autoimmune diseases like inflammatory bowel disease and celiac disease) can trigger this type of diarrhea (Sullivan et al., 2009; Walters et al., 2009; Binder, 2010). When the intestinal barrier is compromised due to loss of epithelial cells or disruption of tight junctions (TJs), diarrhea is referred as exudative. Finally, increased or decreased propulsion of stools relates to diarrhea caused by motility problems (Field, 2003).

In many gastrointestinal infectious diseases, more than one of the five pathophysiological mechanisms is involved in the development of diarrhea. This is the case for giardiasis, cryptosporidiosis, and amoebiasis that, in spite of sharing similar pathophysiological mechanisms of diarrhea, have different initiating events. The early events triggered by the interaction of these three protozoans with their respective hosts are the focus of this review, with special attention to gene transcription, signal transduction pathways, cytoskeleton rearrangements, and cell death in host cells (**Figure 1**).

Giardia lamblia AND GIARDIASIS

The genus *Giardia* comprises many species that inhabit the intestinal tract of a series of vertebrate hosts including domestic animals, rodents, dogs, cats, livestock, and wildlife. However, one species, *G. lamblia* (synonyms *G. duodenalis* and *G. intestinalis*), is known to infect and cause giardiasis in humans and mammals, suggesting a zoonotic transmission (Ryan and Cacciò, 2013).

Giardia lamblia has a simple lifecycle comprising two morphogenetic stages, the infectious and environmentally resistant cyst stage and the vegetative trophozoite stage, which colonizes the small intestine epithelium and causes the disease. Infection initiates when a host ingests viable cysts directly or with contaminated water and food (the infective dose for a symptomatic infection is about 10–100 cysts). After passing through the stomach, cysts begin excysting (excystation process), releasing two trophozoites in the upper part of the small intestine where they adhere to the cells lining the intestinal lumen (enterocytes) through an adhesive ventral disk and multiply by binary fission. Under suitable environmental conditions (i.e., increased bile salt concentration and cholesterol deprivation), trophozoites transform into cysts (encystation process) that are excreted and passed with the feces, thus completing their lifecycle (Gillin et al., 1988; Luján et al., 1996).

Host Cell Transcription during *Giardia lamblia* Infection

As mentioned, *G. lamblia* trophozoites are confined to the lumen of the intestinal tract of humans and animals. To cause disease they must "swim" through the intraluminal fluid flow, overcome peristaltic motions, evade host immunological defense mechanisms and replicate attached to the small intestine mucosal surface. Colonization of the host intestinal epithelium by trophozoites is mediated by an adhesive, microtubule-based organelle denominated the ventral disk (Holberton, 1973; Adam, 2001; Schwartz et al., 2012). Specific molecular mechanisms may be involved, with a range of *Giardia* cell-surface constituents, including lectins and saccharides, being described as ligands for host cell attachment (Farthing et al., 1986; Inge et al., 1988; Ward et al., 1988; Céu Sousa et al., 2001). Membrane rafts present at the surface of trophozoites are also implicated in trophozoite adhesion to human enterocyte-like cells, as parasite treatment with methyl-β-cyclodextrin (a lipid raft disorganizing agent) resulted in diminished attachment of *G. lamblia* to Caco-2/TC7 cells (Humen et al., 2011). Independent of whether adhesion results from a mere mechanical adhesion through the ventral disk or involves ligand–receptor interactions, the host cells are not passive recipients for *Giardia* attachment but are active participants, having evolved specialized strategies to resist infection. These include, for example, the amplification and regulation of the expression of genes, many of which are involved in the immunological defense of host cells to *Giardia* (Ferella et al., 2014). The importance of the early transcription of genes coding for components of intestinal mucosal immunity can be appreciated by studies using animal models. For example, it has been shown that infection of C57BL/6 mice with *G. muris* induces upregulation of interleukin 17A starting 1 week post-infection (Dreesen et al., 2014). The immune-modulating cytokine interleukin-6 (IL-6) is also involved in the control of *G. lamblia* infection, as IL-6-deficient mice were not able to handle the acute phase of the disease and developed chronic giardiasis (Zhou et al., 2003). Importantly, reverse transcription-PCR-based quantitation of cytokine mRNA levels in peripheral lymph node cells exhibited a short-term upregulation of IL-4 expression in IL-6-deficient mice that seemed to be associated with failure to control the parasite population (Zhou et al., 2003).

Recently, host gene expression of mice whole small intestinal tissue following *G. lamblia* infection has been analyzed using oligonucleotide arrays (Tako et al., 2013). The results from this analysis indicated that genes associated with antibodies, mast cell proteases and matrix metalloprotease 7 (Mmp7) were upregulated (Tako et al., 2013). The role of Mmp7 was then confirmed *in vivo*, as Mmp7-deficient mice presented increased numbers of trophozoites in the small intestine when compared to control animals (Tako et al., 2013). In mice, the *Mmp7* gene encodes the processing proteinase of murine Paneth cell defensins, a class of antimicrobial peptides important in the innate immune response of the small intestines (Ostaff et al., 2013). *G. lamblia* trophozoites are lyzed by α-defensin peptides *in vitro* (Aley et al., 1994), making it plausible to consider that Mmp7-deficient mice were unable to clear the infection with *Giardia* probably due to their inability to release or to process antimicrobial defensins from intestinal epithelia.

In vitro as well as *in vivo* models for genome-wide analysis of gene expression have also contributed to understanding the epithelial cell response to *G. lamblia* infection. Using differentiated Caco-2 human cell line as a model of the intestinal epithelium, Roxström-Lindquist et al. (2005) demonstrated that co-incubation of cells with *G. lamblia* resulted in the upregulation of genes coding for chemokines (CCL2, CCL20, CXCL1, CXCL2, CXCL3) and stress-induced genes like *c-Fos*, *c-Jun* and immediate-early response 3 (*IER3*), to cite a few examples. In addition, genes involved in cellular proliferation (*G0S2*, *PCNA*, *ORC5L*, *MCM2*, *MCM3*) had their expression reduced after 6 and 18 h of co-incubation of host cells with trophozoites. Therefore, it appears that infection with *Giardia* interferes with the transcription of host genes involved in innate immunity and prevents cell turnover, probably to maintain a stable niche for colonization as suggested by Stadelmann et al. (2012).

Giardia lamblia Infection and Host Cell Death

At a certain point, the physical attachment of trophozoites to epithelial cells may target specific signaling networks, provoking downstream events that impair normal organ function and lead to associated signs and symptoms of giardiasis. In this scenario, the most striking outcome of *Giardia*–host interaction may be considered the activation of cell death mechanisms such as apoptosis. Apoptosis is described as a regulated and controlled process of autonomous cell death that avoids eliciting inflammation (Elmore, 2007). It can be activated extrinsically by receptor-mediated signaling (through death receptor ligation and DISK assembly) or intrinsically by disruption of the mitochondria. In the first case, the major effectors of apoptosis are a family of aspartic acid-specific proteases known as caspases (Thornberry and Lazebnik, 1998). These are normally synthesized as inactive precursors, but become activated at the onset of apoptosis by activation signals. At the initiation of apoptosis, different sets of caspases are activated depending on whether cell death was triggered extrinsically (caspases 8 and 10) or intrinsically (caspases 9 and 2). However, propagation of the apoptosis signal relies on the direct cleavage of a downstream effector caspase such as caspase-3 that, in turn, cleaves key substrates, producing many of the cellular and biochemical events of apoptosis (Thornberry and Lazebnik, 1998; Slee et al., 2001; Elmore, 2007).

In the gastrointestinal tract, epithelial cells are organized as a single cell layer that covers the entire tissue. They are originated at the crypt and migrate up to the villous tip where they are constantly renewed by extrusion to the lumen. At this point, the turnover of cells is extremely fast (5–7 days) and crucial for the maintenance of normal organ morphology and function. A number of studies have demonstrated that this process is highly regulated and can only be maintained by balancing the levels of cell death and proliferation (Günther et al., 2013). In the gut, cell death occurs through the activation of apoptosis and, although important for gut homeostasis, excessive enterocyte death has been associated with different disorders of the gastrointestinal tract including ulcerative colitis, celiac disease and Crohn's disease (Ciccocioppo et al., 2001; Di Sabatino et al., 2003; Turner, 2009). Infection of human ileocecal adenocarcinoma cell line HCT-8 with G. lamblia can also induce host cell apoptosis. In this case, signs of chromatin condensation were observed within the nuclei of intestinal cell monolayers exposed to different G. lamblia assemblages (A, B, and E) or with a combination of these assemblages (Koh et al., 2013). In addition, caspase-3 activation was found to occur in G. lamblia-induced apoptosis, as nuclear fragmentation and cell death was effectively suppressed by a caspase-3 inhibitor (Koh et al., 2013). Interestingly, the apoptotic response elicited by Giardia was not dependent on co-incubation of host cells with live trophozoites, as sonicated parasites induced caspase-3-dependent apoptosis of non-transformed human duodenal epithelial cell line (SCBN; Chin et al., 2002). Confirming these data, morphological changes consistent with apoptosis and activation of caspase-3 were also observed by others (Panaro et al., 2007). The results from this work indicated that initiation of apoptosis occurred through the activation of caspase-8 and caspase-9, demonstrating that both the intrinsic and extrinsic pathways were triggered during Giardia infection (Panaro et al., 2007). At the same time, a significant downregulation of Bcl-2 and increased expression of the pro-apoptotic protein Bax were observed, being the first demonstration of the participation of members from the Bcl-2 family in the induction of enterocyte apoptosis during Giardia–host cell interaction (Panaro et al., 2007). Bcl-2 family proteins, which have either pro-apoptotic (Bax and Bak) or anti-apoptotic activities (Bcl-2, Bcl-X$_L$, and induced myeloid leukemia cell differentiation protein 1 or Mcl-1), are critical regulators of caspase activation and apoptosis (Adams and Cory, 1998, 2001). In mammalian cells, overexpression of Bax was associated with loss of mitochondrial membrane potential, cytosolic accumulation of cytochrome c, caspase activation, cleavage of poly(ADP-ribose)-polymerase (PARP), DNA fragmentation, and cell death (Oliver et al., 1998). Interestingly, during host cell infections with G. lamblia, cleavage of PARP occurred upon activation of caspase-3 (Panaro et al., 2007). Taken together, these data indicate that downregulation of Bcl-2 and upregulation of Bax is associated with the activation of both the intrinsic and extrinsic apoptotic pathways during G. lamblia infection. In this scenario, the activated initiators caspase-8 and caspase-9 trigger executioner caspase-3, leading to the proteolytic cleavage of PARP and induction of apoptosis (Panaro et al., 2007).

While the critical role of apoptosis in the pathophysiology of giardiasis is well-documented, host responses to prevent cell death during Giardia infection remain poorly understood. In an attempt to shed light on this issue, some authors postulated that inhibition of apoptosis in Giardia-infected cells would involve the upregulation of sodium-dependent glucose cotransporter (SGLT)-1 as demonstrated by bacteria. In this work, Yu et al. (2005) described that activation and enhanced glucose uptake into enterocytes rescued cells from lipopolysaccharide (LPS)-induced apoptosis via SGLT-1. In G. lamblia infections, it was shown that, when SGLT-1-transfected Caco-2 cells were exposed to trophozoite products in high (25 mM) glucose media, host cell apoptosis was abolished (Yu et al., 2008). In addition, a soluble proteolytic fraction of G. lamblia was found to upregulate SGLT-1-mediated glucose uptake in association with increased apical SGLT-1 expression in epithelial cells (Yu et al., 2008). These findings indicated that SGLT-1-dependent glucose uptake might represent a novel epithelial cell rescue mechanism against G. lamblia-induced apoptosis.

Disassembly of Tight Junctions and Cytoskeleton Reorganization during Giardia lamblia Infection

The intestinal epithelium is formed by a single layer of epithelial cells that function as a physical barrier between the lumen and the subepithelial tissue. It prevents the entrance of microorganisms, luminal antigens, and toxins to the mucosal tissue; controls the paracellular movement (transport in the space between epithelial cells) of water, solutes, and macromolecules; and regulates cell proliferation, polarization, and differentiation (Turner, 2009). Within the barrier, epithelial cells are held together by complex structures of tetraspan (claudins, occludins, and tricellulin) and single-span transmembrane proteins known as TJs or zonulae occludentes (ZOs; Balda and Matter, 2008; Turner, 2009). TJs form a continuous belt-like structure that completely encircles the cell at the apical region of the plasma membrane. They also work as signaling platforms, as junctional proteins like ZO-1 interact with cytoskeleton actin through a plaque of cytoplasmic proteins localized under the junction (Matter and Balda, 2003; Turner, 2009). Due to their importance in providing adhesive contacts between neighboring cells and in controlling the permeability of the intestinal barrier, it is not surprising that disruption or reduced expression of TJ proteins and loss of epithelial barrier function have been associated with many intestinal disorders including giardiasis (Teoh et al., 2000; Buret et al., 2002; Chin et al., 2002; Müller and von Allmen, 2005; Buret, 2007; Troeger et al., 2007; Koh et al., 2013). Indeed, some in vitro studies demonstrated that co-incubation of different cell lines (SCBN and HCT-8) with trophozoites results in the disruption of the TJ protein ZO-1 which, in some cases, leads to increased cell permeability (Buret et al., 2002; Chin et al., 2002; Koh et al., 2013). Disruption of ZO-1 by Giardia was associated with caspase-3-dependent apoptosis, as the loss of this protein was abolished in cells treated with caspase-3 inhibitors prior to infection (Chin et al., 2002; Koh et al., 2013). In line with these observations, apoptosis has been observed in many diseases involving TJ disruption (Zeissig et al., 2007; Su et al., 2013). However, whereas in some of these diseases apoptosis is a downstream response to loss of junctional proteins, in giardiasis this fact has yet to be determined.

In recent years, a body of evidence has indicated that loss of TJs may result not only from reduced expression of junctional components but also as a consequence of their relocation/reorganization within cells. For example, in Alzheimer's disease, it was shown that endothelial cells exposed to β-amyloid peptide (Aβ) display a disrupted plasma membrane pattern of claudin-5 and ZO-2, which are relocated to the cytoplasm (Marco and Skaper, 2006). During enteropathogenic E. coli infections, loss of occludin association with claudin-1 and ZO-1 was observed to occur due to the translocation of apically localized TJ proteins to the lateral membrane (Muza-Moons et al., 2004). In this case, the E. coli-induced reorganization of junctional complexes resulted in decreased transepithelial electrical resistance (TEER) and disruption of the intestinal barrier (Muza-Moons et al., 2004). In G. lamblia infections, relocation of TJ proteins ZO-1 and claudin-1 from the cell–cell contact region to the cytoplasm were shown to occur during co-incubation of Caco-2 cells with trophozoites. Moreover, F-actin was retracted and concentrated near cellular contacts, resulting in microvillous atrophy as observed by scanning electron microscopy (Maia-Brigagão et al., 2012). This is in accordance with previous observations describing that infections of colonic cells with Giardia induced localized condensation of F-actin, loss of perijunctional α-actinin and increased cell permeability (Teoh et al., 2000). The mechanism by which epithelial TJs and cytoskeleton were disassembled involved the post-translational modification of the myosin light chain (MLC) of myosin II by MLC kinase (MLCK), as exposition of cells with Giardia triggered MLC phosphorylation (Scott et al., 2002). The importance of MLCK in TJ and cytoskeleton disassembly was further reinforced by the observation that co-incubation of cells with a specific MLCK inhibitor blocked the effects of Giardia on epithelial permeability, F-actin, and ZO-1 (Scott et al., 2002).

The hypothesis that TJ disruption, either by reduced expression or relocalization of junctional components, has an important role in barrier function in giardiasis is supported by in vivo data. In this case, analysis of duodenal biopsy specimens from patients with chronic giardiasis has shown reduced claudin-1 expression and serious villous shortening (Troeger et al., 2007). These were accompanied by a decrease in the absorptive capacity of the duodenum and active anion secretion as evidenced by reduced Na^{2+}-coupled D-glucose absorption and electrophysiological measurements, respectively (Troeger et al., 2007).

Therefore, it can be concluded that the activation of the MLCK signaling pathway during giardiasis relates to loss of TJ proteins, cytoskeleton rearrangement and barrier dysfunction, which can contribute to the pathophysiological mechanisms underlying diarrhea such as electrolyte secretion and malabsorption. A summary of the studies about the pathophysiology of Giardia infections is listed in **Table 1**.

Cryptosporidium parvum

Cryptosporidium parvum is an obligatory intracellular intestinal parasite of humans and animals and is responsible for many cases of cryptosporidiosis worldwide. Cryptosporidiosis is characterized as a watery diarrhea and is a potentially life-threatening disease in both immunocompetent and immunosuppressed hosts. The parasite exists in the environment as an oocyst that contains four sporozoites. When ingested by a host (by fecal-oral contact or by contaminated drinking or recreational water), the oocyst travels through the gut lumen to the small intestine where the sporozoites are released by excystation. Motile sporozoites then attach to the intestinal epithelium and are enveloped by the host cell apical membrane forming an extracytoplasmic parasitophorous vacuole inside which the parasite undergoes asexual multiplication. The sporozoites then enter a sexual reproductive stage and develop into female macrogamonts and male microgamonts. After fertilization, the zygote can develop into two types of oocysts: (a) a thick-walled oocyst that is excreted into the environment or (b) a thin-walled oocyst that can auto-infect the host (Petersen, 1993).

Cryptosporidiosis is characterized by watery diarrhea, malabsorption and wasting. The pathophysiology of cryptosporidiosis is multifactorial, and three pathophysiological mechanisms have been proposed to occur during infection: first, infiltration of the lamina propria by host immune cells (lymphocytes, macrophages, and neutrophils), responsible for inflammatory diarrhea; increased transepithelial permeability, villous atrophy, crypt hyperplasia and cell death, characteristic of exudative diarrhea; and malabsorption due to loss of the intestinal architecture relating to osmotic diarrhea. In the following section, the events preceding the development of diarrhea during C. parvum infection are described.

Host Cell Transcription during Cryptosporidium Infection

As a member of the phylum Apicomplexa, which also includes Plasmodium and Toxoplasma sp., C. parvum is equipped with a specialized apical apparatus named the apical complex. This complex is composed of secretory organelles such as rhoptries, micronemes and dense granules, which play distinct roles during Cryptosporidium–host cell interaction. As an obligatory intracellular parasite, invasion of cells by Cryptosporidium is mandatory and is preceded by the adhesion of parasites to intestinal epithelial cells. Adherence allows the parasite to anchor itself to the epithelial layer, and this process has been shown to be mediated by the presence of adhesins like thrombospondins (TSPs) and thrombospondin-related adhesive protein (TRAP)-C1, and glycoproteins such as mucins and mucin-like proteins (GP900), to cite a few examples (Wanyri and Ward, 2006). Following adherence, parasites are internalized, resulting in subtle changes of host cell gene expression such as the development of a mechanism for evading the host cell immune response. Indeed, several studies have described changes in certain selected host cellular genes due to Cryptosporidium

TABLE 1 | Summary on studies describing the effect of *Giardia lamblia* infection on host cell responses.

Target on host cells	Effect	Reference
Gene transcription		
Induction of IL-17	Protective immune response against *G. muris* infection in C57BL/6J mice	Dreesen et al., 2014
Induction of IL-6	Early control of acute *G. lamblia* infection in C57BL/6J mice	Zhou et al., 2003
Induction of matrix metalloprotease 7 (Mmp7)	Production of mature α-defensins in C57BL/6J mice and control of *G. lamblia* infection	Tako et al., 2013
Induction of chemokines (CCL2, CCL20, CXCL1, CXCL2, and CXCL3)	Recruitment of host immune cells to the site of infection (?)	Roxström-Lindquist et al., 2005
Induction of stress-induced genes (c-Fos, c-Jun, and IER3)	Regulation of cell stress during *G. lamblia* infections in Caco-2 cells (?)	Roxström-Lindquist et al., 2005
Reduction of cell proliferation genes (G0S2, PCNA, ORC5L, MCM2, MCM3)	Response to NO production in Caco-2 cells infected with *G. lamblia*	Roxström-Lindquist et al., 2005
Cell viability		
Activation of caspase-3-dependent apoptosis	Assembled-specific induction of apoptosis by *G. lamblia*-infected HCT-8 cells and strain-dependent apoptosis of human duodenal epithelial cell line	Chin et al., 2002; Koh et al., 2013
Activation of caspase-3, caspase-8, and caspase-9-dependent apoptosis	Activation of both the intrinsic and the extrinsic apoptotic pathways of HCT-8 cells infected with *G. lamblia*	Panaro et al., 2007
Activation of sodium-dependent glucose cotransporter (SGLT)-1	Protection against *G. lamblia*-induced apoptosis in Caco-2 cell cultured in high glucose media	Yu et al., 2008
TJs and cytoskeleton		
Disruption of ZO-1	Increased permeability in HCT-8 and human duodenal epithelial cell line infected with *G. lamblia*	Buret et al., 2002; Chin et al., 2002; Koh et al., 2013
Reduced claudin-1 expression	Decreased absorption, increased ion secretion and villous shortening in duodenal biopsies from *G. lamblia* infected patients	Troeger et al., 2007
Relocation of claudin-1 and F-actin retraction	Increased paracellular permeability and microvilli atrophy in *G. lamblia*-infected Caco-2 cells	Maia-Brigagão et al., 2012
F-actin condensation and loss of perijunctional α-actinin	Increased permeability of Caco-2 and non-transformed human epithelial cell line (SCBN) infected with *G. lamblia*	Teoh et al., 2000
F-actin and ZO-1 reorganization	Myosin-light chain kinase (MLCK)-dependent increased cell permeability in *G. lamblia* infections	Scott et al., 2002

infection (Castellanos-Gonzalez et al., 2008; Zhou et al., 2009). For example, microarray analysis of human ileal mucosa explants infected with *C. parvum* or *C. hominis* demonstrated increased expression of osteoprotegerin (OPG) mRNA compared to uninfected cells (Castellanos-Gonzalez et al., 2008). The relevance of this finding was further extended, as jejunal biopsy specimens obtained from a volunteer (before and after experimental infection with *C. meleagridis*) displayed a 1281-fold increase in OPG mRNA post-infection (Castellanos-Gonzalez et al., 2008). OPG is a soluble glycoprotein produced by osteoblasts, intestine cells, hematopoietic and immune cells (dendritic cells and lymphocytes; Simonet et al., 1997; Vidal et al., 2004). It is a member of the TNF superfamily which includes proteins such as TNF-α, Fas/FasL and TRAIL, which are involved in cell differentiation, proliferation, survival, apoptosis and in immune responses (Simonet et al., 1997). Various human intestinal epithelial cell lines were reported to constitutively express OPG, especially during inflammation, suggesting that it may play an important role as a mucosal immunoregulatory factor (Vidal et al., 2004). Whether OPG exerts its biological effect in response to cell inflammation during *C. parvum* infection remains to be elucidated. Nonetheless, data obtained by Castellanos-Gonzalez et al. (2008) demonstrated that OPG is

produced during the early stages of *C. parvum* infection blocking TRAIL-mediated apoptosis of host cells, indicating OPG as a protective factor in cryptosporidiosis.

Cytokines and chemokines are proteins that regulate inflammation and modulate cellular activities such as growth, survival and differentiation (Zlotnik, 2000; Dinarello, 2007). They may act as pro- or anti-inflammatory factors (cytokines) or as chemotactic attractants (chemokines) to leukocytes and in trafficking of immune cells (Dinarello, 2007). They can be produced by a series of cells such as T helper cells (Th) and macrophages but also by intestinal epithelial cells (Stadnyk, 2002). Reports on the pathogenesis of cryptosporidiosis have shown increased mRNA levels for cytokines (IL-1β, IL-4, IL-8, IL-14, IL-15, IFN-γ, TGF-β) and chemokines (C-C and C-X-C subfamilies) in human and murine intestinal cells and xenografts infected with *C. parvum* (Laurent et al., 1997; Seydel et al., 1998b; Robinson et al., 2000, 2001; Lacroix-Lamandé et al., 2002; Deng et al., 2004; Tessema et al., 2009). These results are consistent with previous observations showing the recruitment of effector cells to the site of inflammation in the intestinal lamina propria from *Cryptosporidium*-infected patients (Robinson et al., 2001), as chemokines may control the localization of immune cells throughout the body (Griffith et al., 2014). Therefore, these data suggest that expression of multiple cytokines and chemokines by host cells may play an important role in the control of inflammation during *Cryptosporidium* infections. More recently, fractalkine or CX3CL1, a membrane-bound chemokine of the CX3C family, was shown to be upregulated in human biliary epithelial cells following *C. parvum* infection. Induction of CX3CL1 expression involved downregulation of microRNAs (miR-424 and miR-503) both known to target the CX3CL1 3′ UTR, suppressing its translation and inducing RNA degradation (Zhou et al., 2013). MicroRNAs are non-coding, single-stranded RNAs that negatively regulate gene expression through interactions with 3′ UTRs of the target mRNA (Bartel, 2009).

Defensins are a family of antimicrobial peptides expressed by different cells including Paneth cells in the epithelium of the small intestine. They are subdivided into two families, the α-defensins (also known as cryptdins in mice) and the β-defensins, both displaying microbicidal activity (Bevins and Salzman, 2011). Experiments with human colonic (HT29) and murine rectal adenocarcinoma (CMT-93) cell lines infected with *C. parvum* have shown differential β-defensin gene expression (Zaalouk et al., 2004). Indeed, using reverse transcription-PCR, a reduction in human-defensin-1 (hBD-1) and induction of hBD-2 were observed in *Cryptosporidium*-infected colonic cells. Furthermore, enterocytes infected with *C. parvum* and treated with recombinant hBD-1 and hBD-2 showed a reduction in the percentage of viable sporozoites, indicating that these peptides may have an important role in the host's innate response against infection (Zaalouk et al., 2004). Upregulation of inducible nitric oxide synthase (iNOS) was also shown to occur in neonatal piglets during acute *C. parvum* infection. Curiously, expression of iNOS was not restricted to infected cells, possibly indicating a non-specific response against *Cryptosporidium* infection, although the importance of iNOS in the control of tissue parasitism was further confirmed, as inhibition of iNOS activity resulted in increased parasite burden in intestinal epithelial cells (Gookin et al., 2006). Finally, iNOS induction was shown to be NF-κB dependent, as iNOS activity was abolished when infected cells were incubated with lactacystin (a proteasome inhibitor that prevents degradation of IκBα; Gookin et al., 2006). These data, together with the demonstration that iNOS expression by macrophages and other cell types occurs in tissues from patients with a wide variety of infectious diseases (Bogdan, 2001), may suggest a protective role for nitric oxide in cryptosporidiosis. Further experiments are needed to confirm this hypothesis in the human disease.

Cryptosporidium parvum Infection and Host Cell Death

The first report on the occurrence of apoptosis in *C. parvum*-infected human biliary epithelial cells (H69 cells) was issued by Chen et al. (1998). Later, nuclear condensation and DNA fragmentation (as markers of apoptosis) during *C. parvum* infection were shown to be caspase-dependent and induced by Fas/FasL, as caspase inhibitors or neutralizing antibodies to either the Fas receptor (Fas) or Fas ligand (FasL) blocked these events (Chen et al., 1999; Ojcius et al., 1999). *Cryptosporidium*-induced apoptosis was documented to occur independent of cell line (CaCo-2, MDBK, and HCT-8 cells) and resulted in impaired *C. parvum* development *in vitro*, suggesting a host–cell mechanism to control the spread of infection (Widmer et al., 2000). Further studies, however, demonstrated that apoptotic changes of intestinal epithelial cells were modulated by the *C. parvum* development stages and displayed biphasic activation with early inhibition (at the trophozoite stage) and late moderate promotion (at the sporozoite and merozoite stages; Mele et al., 2004; Liu et al., 2009). On the basis of these data, it has been suggested that *Cryptosporidium* precisely regulates host cell apoptosis to favor its growth and development at initial stages of infection, and to promote its propagation later on.

In line with this hypothesis, namely that *Cryptosporidium* can exert control on the processes that regulate apoptosis in the host, Chen et al. (2001) have shown that *C. parvum*-infected biliary cells activated the NF-κB signaling cascade, leading to secretion of the pro-inflammatory cytokine IL-8 and inhibition of cell apoptosis. Moreover, these events were restricted to infected cells given that *C. parvum*-induced apoptosis was limited to bystander uninfected cells (Chen et al., 2001). Inhibition of host cell apoptosis during *Cryptosporidium* infection has also been reported to involve the expression of members of the IAP family (inhibitors of apoptosis proteins) such as c-IAP1, c-IAP2, XIAP, and survivin. In this case, it was demonstrated that knockdown of survivin (but not that of c-IAP1, c-IAP2, or XIAP) by siRNA enhanced caspase-3/7 activity and resulted in increased host cell apoptosis and decreased *C. parvum* infection (Liu et al., 2008). The role of IAPs in the control of cell death in cryptosporidiosis is reinforced by a study showing that XIAP mediated proteasome-dependent inhibition of activated caspase-3 in *C. parvum* infection (Foster et al., 2012).

Disassembly of Tight Junctions during *Cryptosporidium* Infection

Dysfunction of the epithelial barrier during *in vivo* intestinal infections with *Cryptosporidium* has been documented in humans and animals. Villous atrophy, hyperplasia of the crypt epithelium and increased transepithelial permeability are some of the abnormalities reported in cryptosporidiosis (Genta et al., 1993; Adams et al., 1994; Gookin et al., 2002). *In vitro* studies on *Cryptosporidium andersoni*-infected human (Caco-2) and bovine (MDBK and NBL-1) epithelial cells reported disruption of ZO-1 and nuclear fragmentation during infection (Buret et al., 2003). Interestingly, both events were reversed by pretreatment of host cells with recombinant human epidermal growth factor (rhEGF), and significantly reduced infection rates in bovine and human enterocytes (Buret et al., 2003). The relationship, if any, between *C. andersoni*-induced ZO-1 disruption and loss of barrier function is still unknown. Moreover, how EGF exerts its biological effect during *Cryptosporidium* infection deserves further investigation. **Table 2** summarizes work on *Cryptosporidium*.

TABLE 2 | Summary on studies describing the effect of *Cryptosporidium* infection on host cell responses.

Target on host cells	Effect	Reference
Gene transcription		
Induction of osteoprotegerin (OPG)	Immune modulation of host cell response to *C. parvum* and *C. hominis* infection of human ileal explants	Castellanos-Gonzalez et al., 2008
Induction of cytokines (IL-1β, IL-4, IL-8, IL-14, IL-15, IFN-γ, TGF-β)	Control of inflammation in human intestinal xenografts, jejunal biopsies, C57BL/6 mice and HCT-8 cells infected with *C. parvum*	Laurent et al., 1997; Seydel et al., 1998b; Robinson et al., 2000, 2001; Lacroix-Lamandé et al., 2002; Deng et al., 2004; Tessema et al., 2009
Induction of C-X-C chemokines	Recruitment of immune cells to the *lamina propria* of *C. parvum*-infected cells	Laurent et al., 1997
Induction of fractalkine or CX3CL1 chemokine	Donwregulation of microRNAs and activation of mucosal antimicrobial defense against *C. parvum* in human biliary epithelial cells	Zhou et al., 2013
Modulation of β-defensin expression	Control of host innate immune response in *C. parvum*-infected HT29 cells	Zaalouk et al., 2004
Induction of nitric oxide synthase (iNOS)	Control of tissue parasitism in neonatal piglets and human epithelial cells infected with *C. parvum*	Gookin et al., 2006
Cell viability		
Activation of caspase-3-dependent signaling cascade	Induced Fas/FasL-dependent apoptosis in *C. parvum*-infected biliary epithelial cells	Chen et al., 1999; Ojcius et al., 1999
Expression of survivin	Protection against *C-parvum*-induced caspase 3/7 apoptosis in HCT-8 cells	Liu et al., 2008
Activation of XIAP	Proteasome-dependent inhibition of activated caspase-3 and cell apoptosis in piglets infected with *C. parvum*	Foster et al., 2012
TJs and cytoskeleton		
Disruption of ZO-1	Unknown effect in *C. andersoni*-infected Caco-2 and MDBK cells	Buret et al., 2003

Entamoeba histolytica

Entamoeba histolytica is a protozoan parasite that colonizes the large intestine of humans causing amoebiasis. Although most infections with *E. histolytica* are asymptomatic, some patients may experience clinical manifestations of invasive amoebiasis such as amoebic colitis and amoebic liver abscess (Marie and Petri, 2014).

Entamoeba histolytica has a simple lifecycle that involves two distinct morphogenetic stages, the amoeboid and proliferative trophozoite, and the infectious cyst form. Human infections begin with ingestion of the viable cysts in food or water that has been contaminated by feces. Excystation occurs in the small intestine, and released trophozoites migrate to the colon where they multiply by binary fission. In the end, trophozoites encyst, completing the lifecycle when they are excreted into the environment in stool (Marie and Petri, 2014). Trophozoite adhesion to colonic mucus and epithelial cells is a critical step in the colonization of the large intestine by *E. histolytica* and a Gal/GalNAc lectin (260 kDa) expressed at the parasite surface was shown to mediate its binding to host mucins and cell surface carbohydrates (Frederick and Petri, 2005).

Host Cell Transcription during *Entamoeba histolytica* Infection

It is well-known that infection with the protozoan parasite *E. histolytica* results in significant inflammatory responses that contribute to tissue damage and invasion. *In vitro*

studies using co-culture of human epithelial and stromal cells and cell lines (HeLa, HT29, and T84) demonstrated an upregulation of IL-8 transcripts during *E. histolytica* infections that correlated with increased secretion of this pro-inflammatory cytokine and others such as GROα, GM-CSF, IL-6, and IL-1α (Eckmann et al., 1995). Increased mRNA for both IL-1β and IL-8 were also reported to occur *in vivo* when mouse–human intestinal xenografts (SCID-HU-INT) were infected with *E. histolytica* trophozoites (Seydel et al., 1997). The relevance of these findings was further reinforced when intraluminal administration of an antisense oligonucleotide (to block the production of IL-1β and IL-8) inhibited the gut inflammatory response to *E. histolytica* infection (Seydel et al., 1998a).

In human patients with acute or convalescent amoebiasis, gene expression profiles obtained by microarray analysis of intestinal biopsies clearly demonstrated upregulation of *REG1A* and *REG1B* genes (Peterson et al., 2011). *REG1A* and *REG1B* belong to the regenerating islet-derived (REG) gene family encoding for C-type lectin-like proteins (Parikh et al., 2012). They are involved in the proliferation and differentiation of diverse cell types and are well-known to be highly expressed in some pathologies as inflammatory diseases, cancer and diabetes (Parikh et al., 2012). One member of this family, the REG1α protein, was also shown to mediate the anti-apoptotic effect of STAT3 in cancer cells (Sekikawa et al., 2008). During *E. histolytica* infections, expression of REG1α and REG1β were shown to inhibit parasite-induced apoptosis *in vitro* as REG1−/− mice were found to be more susceptible to cell death.

Entamoeba histolytica Infection and Host Cell Death

As the name suggests, *E. histolytica* (*histo*: tissue and *lytica*: destroyer) is a tissue-destroying amoeba (Pinilla et al., 2008) and host cell apoptosis is one of the most common events associated with infections with this parasite (Huston et al., 2000; Christy and Petri, 2011; Marie and Petri, 2014). The mechanisms leading to host cell death in *E. histolytica* infections are not completely understood. However, apoptosis and trogocytosis have been reported to occur in amoebiasis without parasite penetration within host cells. Apoptosis was first suggested to occur during *E. histolytica* infections as DNA fragmentation was observed after trophozoite adhesion to a murine myeloid cell line (FDC-P1; Ragland et al., 1994). Cell killing by *E. histolytica* was further shown to occur via a Bcl-2-independent mechanism, as FDC-P1 cells transfected with a retrovirus construct to express the Bcl-2 protein were susceptible to amoeba contact-dependent killing (Ragland et al., 1994). Later, using Jurkat cells it was demonstrated that infection with *E. histolytica* rapidly activated caspase-3, independently of caspase-8 and -9 activation (Huston et al., 2000). Interestingly, *E. histolytica* activation of caspase-3 was followed by phagocytosis of host cells, suggesting that cell killing precedes ingestion by trophozoites (Huston et al., 2003). Over the past years, diverse studies have tried to elucidate the pathways explored by *E. histolytica* to trigger host cell apoptosis. For example, in hepatocytes, live *E. histolytica* was shown to induce an apoptosis-like death without the participation of both Fas and TNF-α pathways (Seydel and Stanley, 1998). Teixeira and Mann have observed that adhesion of *E. histolytica* trophozoites to Jurkat cells induced a contact-dependent protein dephosphorylation by host cell protein tyrosine phosphatases (PTPs) such as SHP-1 and SHP-2 (Kim et al., 2010), since pretreatment of cells with a PTP inhibitor inhibited amoeba-induced dephosphorylation and cell apoptosis (Teixeira and Mann, 2002). Activation of host cell PTP occurred through a calcium-dependent calpain protease responsible for PTP1B cleavage that led, at last, to cell death (Teixeira and Mann, 2002). Reinforcing these data, activation of host cell calpain by *E. histolytica* was also observed by others (Kim et al., 2007; Jang et al., 2011) and was shown to modulate the degradation of STAT proteins (STAT3 and STAT5) and NF-κB (p65) in Caco-2 cells (Kim et al., 2014). Furthermore, pretreatment of Caco-2 cells with calpeptin (a calpain inhibitor) or calpain silencing partially reduced *Entamoeba*-induced DNA fragmentation (Kim et al., 2014).

In recent years, a number of studies have shown that oxidative stress could cause cellular apoptosis via both the extrinsic and intrinsic pathways in health and pathological conditions (Lin and Beal, 2006). In amoebiasis, for example, incubation of human neutrophils with *E. histolytica* trophozoites triggered NADPH oxidase-dependent production of reactive oxygen species (ROS) and cell apoptosis (Sim et al., 2005). The mechanism involved in *Entamoeba*-induced ROS generation and apoptosis was associated with ERK1/2 activation, possibly through β2-integrin, as cells pretreated with a MEK1 inhibitor (PD98059) and with a monoclonal antibody to CD18 (anti-integrin β2 subunit) prevented cell apoptosis (Sim et al., 2005, 2007). Phosphatidylinositol-3-kinase (PI-3-kinase) was also involved in ROS production and apoptosis during *Entamoeba* infection, suggesting that signaling molecules may be key factors in *E. histolytica*-induced, ROS-dependent apoptosis (Sim et al., 2007). Similar results were reported in colonic Caco-2 and HT-29 cells, as increased levels of intracellular ROS were reported to occur through NOX1 oxidase after cell exposure to trophozoites (Kim et al., 2011, 2013). In this case, cell death was shown to be caspase-independent and the signaling cascade activated during this event is still unknown.

The historical concept that *E. histolytica* kills cells by apoptosis was recently challenged by Ralston et al. (2014). Using both Caco-2 and Jurkat cells, the authors demonstrated that, immediately after contact with human cells, *E. histolytica* ingests small fragments of the cell membrane, some containing cellular components like cell cytoplasm and mitochondria (Ralston et al., 2014). Surprisingly, host cells were alive when ingestion of fragments was initiated, and resulted in the elevation of the intracellular amount of calcium before the eventual death of cells as trophozoites detached from corpses. The internalization of cell fragments by *E. histolytica* was named as amoebic trogocytosis (from the Greek *trogo*, for nibble) and only occurred with live cells as pre-killed cells are ingested intact (Ralston et al., 2014).

On the whole, the results of these studies demonstrated that *E. histolytica* infections might result in cell death both by apoptosis and trogocytosis, and that these events might contribute to tissue invasion by the parasite.

Disassembly of Tight Junctions during *Entamoeba* Infection

It is widely accepted that tissue invasion by *E. histolytica* is preceded by the interaction of trophozoites with intestinal epithelial cells, and a series of studies have shown that this interaction impacts cell morphology, intercellular contacts and regulation of paracellular transport of molecules across the intestinal epithelium. Indeed, *in vitro* studies have shown a rapid decrease in transepithelial resistance (TER) and increased mannitol flux during trophozoite interaction with polarized human intestinal Caco-2 and T84 cells (Martinez-Palomo et al., 1985; Li et al., 1994; Leroy et al., 2000). Apical injury of host cells such as loss of brush border in regions of contact between epithelial cells and amoebae was also reported. Importantly, these changes were only observed when Caco-2 cells were co-incubated with live trophozoites but not with amoeba lysates or conditioned medium, indicating that they were not mediated by soluble amoebic cytotoxins (Li et al., 1994; Leroy et al., 2000). In human enteric T84 cells co-cultured with amoebae, decreased TER was associated with changes in TJ proteins like release of ZO-1 from ZO-2, degradation of ZO-1 and dephosphorylation of ZO-2 (Leroy et al., 2000).

Besides *E. histolytica*-host cell contact, amoebic products also have been shown to be crucial for cellular barrier dysfunctions during parasite infections. For example, prostaglandin E_2 (PGE_2) secreted by *E. histolytica* was shown to alter the spatial localization of claudin-4 that resulted in increased sodium ion permeability through TJs (Lejeune et al., 2011). EhCP112, an *E. histolytica*-secreted cysteine protease, has been shown to digest gelatin, collagen type I, fibronectin, hemoglobin and, most importantly, to destroy MDCK cell monolayers (Ocádiz et al., 2005). When complexed with EhADH112 adhesin, the formed EhCPADH112 complex was shown to co-localize with claudin-1 and occludin at the TJs after the incubation of epithelial MDCK cells with trophozoite extracts. Furthermore, EhCPADH112 induced progressive disruption of the paracellular barrier as measured by TER. Importantly, these effects were reversed when co-cultures were incubated with a protease inhibitor cocktail or a monoclonal antibody against the EhCPADH112 complex (Betanzos et al., 2013). Cysteine proteases are important virulence factors in *E. histolytica*, and 20 genes encoding for these proteases have been identified on the genome (Bruchhaus et al., 2003). Their role in parasite loss of host cell integrity is highlighted in a study showing that calpain (a calcium-dependent cysteine protease) activation resulted in degradation of paxillin, Cas, vimentin, vinculin, talin, and α- or β-spectrin in Jurkat T cells infected with

E. histolytica (Lee et al., 2011). In addition to proteases, Goplen et al. (2013) have shown that *E. histolytica* expressed a cognate "occludin-like" protein of the host, as revealed by confocal microscopy using antibodies for human occludin. Apical administration of "occluding-like" protein to T84 human colonic epithelial cells resulted in epithelial disruption and decreased TER, suggesting the involvement of this protein in the pathophysiology of amoebiasis. The exact mechanism by which "occluding-like" protein exerts its effects is not completely understood but the authors suggested that it might compete for epithelial occludin–occludin interactions, a hypothesis that needs further investigation. Studies on *E. histolytica* are summarized in **Table 3**.

GIARDIASIS, CRYPTOSPORIDIOSIS, AND AMOEBIASIS: MECHANISTIC SIMILARITIES AND DIFFERENCES

The pathophysiology of diarrhea caused by *G. lamblia*, *Cryptosporidium* sp., and *E. histolytica* is multifactorial and,

TABLE 3 | Summary on studies describing the effect of *Entamoeba histolytica* infection on host cell responses.

Target on host cells	Effect	Reference
Gene transcription		
Induction of cytokines (GROα, GM-CSF, IL-6, IL-8, IL-1α, and IL-1β)	Control of inflammation in *E. histolytica*-infected human cells (HeLa, HT29, and T84) and mouse-human intestinal xenografts	Eckmann et al., 1995; Seydel et al., 1997, 1998a
Induction of REG1A and REG1B	Inhibition of parasite-induced apoptosis in colonic biopsies of *E. histolytica*-infected patients	Peterson et al., 2011
Cell viability		
Activation of caspase-3 signaling pathway	Induction of caspase-8 and caspase-9 independent apoptosis of Jurkat cells	Huston et al., 2003
Induction of "apoptosis-like" mechanisms	Death of hepatocytes in Fas and TNF-α independent pathways	Seydel and Stanley, 1998
Dephosphorylation of host cell proteins by PTPs	Induction of calcium-dependent calpain protease and apoptosis of *E. histolytica*-infected Jurkat cells	Teixeira and Mann, 2002
Activation of calpain	Cell death of HT-29 and Jurkat cells infected with *E. histolytica* and modulation of STAT proteins and NF-κB DNA fragmentation	Kim et al., 2007, 2014; Jang et al., 2011
Activation of NADPH-oxidase	Induction of ERK1/2 pathways and ROS-dependent apoptosis of human neutrophils infected with *E. histolytica*	Sim et al., 2005, 2007
Activation of NOX1 oxidase	Production of ROS and caspase-independent apoptosis of Caco-2 and HT-29 cells infected with *E. histolytica*	Kim et al., 2011, 2013
Activation of PI-3-K	ROS-mediated neutrophil apoptosis induced by *E. histolytica*	Sim et al., 2007
Ingestion of host cell membrane fragments by trophozoites	Elevation of intracellular Ca^{2+} and death of cells by trogocytosis	Ralston et al., 2014
TJs and cytoskeleton		
Trophozoite interaction with polarized cells	Reduction in transepithelial resistance and increased mannitol flux in *E. histolytica*-infected Caco-2 and T84 cells	Li et al., 1994; Leroy et al., 2000
Degradation of ZO-1, release of ZO-1 from ZO-2, and dephosphorylation of ZO-2	Reduction in transepithelial resistance and increased mannitol flux in *E. histolytica*-infected T84 cells	Leroy et al., 2000
Relocalization of claudin-4	Increased sodium ion permeability in amoeba infected T84 cells	Lejeune et al., 2011
Secretion of an "occludin-like" molecule by trophozoites	Disruption of epithelial barrier and reduction in transepithelial resistance in *E. histolytica*-infected T84 cells	Goplen et al., 2013

despite depending on the microbiological agent causing it, these diseases share some mechanistic features (**Table 4**).

The exact mechanism leading to giardiasis is unknown, although research points to a combination between osmosis, active secretion, exudation, inflammation and altered motility as drivers of *Giardia*-induced diarrhea. In molecular terms, disruption, reduced expression and/or relocation of TJ and cytoskeleton proteins (such as ZO-1, claudin-1, F-actin, and α-actinin) were shown to result in increased intestinal permeability and a drop in TER, indicating that infection can cause paracellular leakage (exudative diarrhea). The events leading to this class of diarrhea are similar to cryptosporidiosis and amoebiasis. Accordingly, disruption of ZO-1 was reported in colonic cells infected with *Cryptosporidium* (Buret et al., 2003) while in cells infected with *E. histolytica*, contact-dependent degradation of TJ proteins ZO-1 and ZO-2, dephosphorylation of ZO-2, relocation of claudin-4 and reduction in TER were shown to underlie exudative diarrhea (Leroy et al., 2000; Lejeune et al., 2011). However, while *Giardia* causes TJ disruption without penetrating the epithelium, *E. histolytica* kills (through apoptosis and trogocytosis), invades and destroys host tissues. In cryptosporidiosis, further studies are needed to assess whether cell invasion or parasitic products initiate these alterations.

Tight junction alterations were also observed to indirectly increase the luminal Cl⁻ concentration (secretory diarrhea) as a consequence of the loss of absorptive function (villous shortening, microvilli atrophy and increased cell death) and/or increased secretion (destruction of the epithelial barrier) in *Giardia*-infected cells (Troeger et al., 2007; Maia-Brigagão et al., 2012). Similarly, damage to the absorptive villi and enhanced fluid secretion from the crypts have been documented in cryptosporidiosis, supporting diarrhea by active secretion (Guarino et al., 1995, 1997). In amoebiasis, increased mannitol flux and movement of sodium ions into the intestinal lumen were reported (Leroy et al., 2000; Lejeune et al., 2011).

As digestion of nutrients in the small intestine depends on hydrolytic enzymes (disaccharidases such as sucrose, maltase, lactase, and peptidase) produced by the brush border membrane of microvilli, dysfunctional microvilli may interfere significantly with the absorption of nutrients. In giardiasis, loss of microvilli brush border, combined with villous atrophy, is responsible for disaccharidase insufficiencies and malabsorption of nutrients, ultimately causing osmotic diarrhea (Buret, 2007, 2008; Troeger et al., 2007). Likewise, in enteric cryptosporidiosis, villous atrophy and crypt hyperplasia were shown to account for impaired monosaccharide and glucose-Na⁺ absorption while lactose malabsorption was described in individuals infected with *E. histolytica* (Rana et al., 2004).

In some diarrheal infections, the association between impaired absorption and increased secretion may contribute to accelerated intestinal transit. Indeed, in giardiasis and cryptosporidiosis, increased motility was reported, which in turn may contribute to the exacerbation of weight loss observed in *Giardia*-infected patients. On the contrary, whether motility dysfunction occurs and its importance on the development of amoebiasis it are

TABLE 4 | Pathophysiological mechanisms implicated in diarrhea caused by *G. lamblia*, *Cryptosporidium* sp., and *E. histolytica*.

Pathophysiological mechanism	Giardiasis	Cryptosporidiosis	Amoebiasis
Osmotic diarrhea	Malabsorption of nutrients was described to occur in response to reduced disaccharidase activity in the gut (Troeger et al., 2007)	Impaired absorptive function was shown to result in reduced absorption of both monosaccharides and co-transport of glucose-Na⁺ (Argenzio et al., 1990; Farthing, 2000)	Lactose malabsorption was reported in amoeba-infected patients (Rana et al., 2004)
Secretory diarrhea	Loss of epithelial absorptive surface (villous and microvilli atrophy) and chloride secretion were reported in colonic cells *in vitro*, in animal models and human patients (Gorowara et al., 1992; Scott et al., 2000; Troeger et al., 2007)	Damage to the absorptive villous and unbalanced secretory crypts were involved in electrolyte secretion. An unknown cryptosporidial enterotoxin was suggested to trigger net secretion (Guarino et al., 1995, 1997)	Increased mannitol flux and movement of Na⁺ ions into the intestinal lumen (Leroy et al., 2000; Lejeune et al., 2011)
Exudative diarrhea	Disruption or relocation of tight junctions proteins and dysfunctional epithelial barrier were associated with leak flux diarrhea (Troeger et al., 2007; Maia-Brigagão et al., 2012)	Disruption of epithelial tight junction, loss of intestinal barrier, dysregulated influx of immune and inflammatory cells and cell death by apoptosis were related to increased flux into the lumen (White, 2010)	Dysregulation of the TJ protein complex, decreased transepithelial resistance and cell apoptosis were associated with water flow (Ragland et al., 1994; Leroy et al., 2000; Betanzos et al., 2013)
Inflammatory diarrhea	Inflammation was rarely observed in chronically infected patients (Hanevik et al., 2007; Kohli et al., 2008)	Parasite products and infiltration of host immune cells in the *lamina propria* were associated to pathogenesis (Laurent et al., 1999)	Production of inflammatory mediators were correlated to tissue damage in amoebic diarrhea (Seydel et al., 1997, 1998a,b)
Motility problems	Malabsorption of nutrients, water-impaired absorption and electrolyte secretion were suggested to contribute to increased intestinal transit and peristalsis (Cotton et al., 2011)	Intestinal epithelial cell layer breakdown was shown to result in increased intestinal transit (Sharpstone et al., 1999; Brantley et al., 2003)	Not reported

unclear. However, it cannot be ruled out, as increased secretion and malabsorption are triggered by *E. histolytica* infection.

Despite the similarities in the events leading to osmotic, secretory and exudative diarrhea, there are some differences between giardiasis, cryptosporidiosis, and amoebiasis when considering the immunological and inflammatory response of the host (inflammatory diarrhea). For example, while several lines of evidence support the hypothesis that infections with *Giardia* are rarely accompanied by inflammation (Hanevik et al., 2007; Morken et al., 2008), a parasite extract was shown to be a poor cytokine inducer (inducing only small amounts of IL-6 and TNF-α; Zhou et al., 2003, 2007). On the contrary, a hallmark of amoebiasis and cryptosporidiosis is acute intestinal inflammation dominated by NF-κB-mediated secretion of inflammatory cytokines produced by host cells (Eckmann et al., 1995; Seydel et al., 1997; McCole et al., 2000; Chen et al., 2001; Hou et al., 2010). For example, IL-1β, IL-6, IL-8, TNF-α, and IFN-γ are key factors in the inflammatory response elicited by host cells after contact with amoebae (Eckmann et al., 1995; Seydel et al., 1997; Hou et al., 2010). However, whether production of pro-inflammatory cytokines influences the permeability of epithelial cell TJs and gut absorption is not known. Similar to amoebiasis, upon *Cryptosporidium* infection, epithelial cells release pro-inflammatory cytokines (IL-1β, IL-8, TNF-α, IFN-γ) and chemokines (C-X-C and fractalkine) to the site of infection, which in turn may contribute to increased epithelial permeability, impaired intestinal absorption and enhanced secretion (Seydel et al., 1998a; Farthing, 2000; Lacroix-Lamandé et al., 2002).

Collectively, these observations suggest that malabsorption, secretion of electrolytes and impairment of TJs may underlie luminal fluid accumulation during *G. lamblia* infection. Marked mucosal inflammation, decreased absorptive surface and malabsorption are thought to contribute to the pathogenesis of *Cryptosporidium*-induced diarrhea, while in *E. histolytica*-infected cells, epithelial destruction and inflammation infection appears to be the basis of the disease.

CONCLUDING REMARKS

Intestinal parasitism is extremely common, with *G. lamblia*, *C. parvum*, and *E. histolytica* being the most important intestinal protozoan parasites of humans worldwide. Infections begin when a person ingests the infective stage of the parasite with contaminated food or water. Once inside the host, parasites lodge in the intestinal tract causing acute and self-limited diarrhea. However, in some patients, the disease can progress to chronic diarrhea and related complications such as malnutrition, growth delays and cognitive impairment.

Significant progress has been made in understanding the processes by which *G. lamblia*, *C. parvum*, and *E. histolytica* trigger diarrhea and how the host cell responds to infection. Disruption of TJ barrier function, alterations of host cell architecture, and transcription of genes involved in host immunity and cell death are some of the events elicited in the host cell when interacting with these parasites.

Future elucidation of the processes that integrate these events and eliminate the disease may lead to novel therapeutic approaches for diarrhea caused by enteropathogenic parasites.

AUTHOR CONTRIBUTIONS

Conceived and wrote the paper RT. Wrote the paper BG.

FUNDING

The work herein mentioned was supported by grants of the Fundação de Amparo à Pesquisa do Estado de São Paulo (RT 2010/15042-2; BDG 2013/04272-5).

REFERENCES

Adam, R. D. (2001). Biology of *Giardia lamblia*. *Clin. Microbiol. Rev.* 14, 447–475. doi: 10.1128/CMR.14.3.447-475.2001

Adams, J. M., and Cory, S. (1998). The Bcl-2 protein family: arbiters of cell survival. *Science* 281, 1322–1326. doi: 10.1126/science.281.5381.1322

Adams, J. M., and Cory, S. (2001). Life-or-death decisions by the Bcl-2 protein family. *Trends Biochem. Sci.* 26, 61–66. doi: 10.1016/S0968-0004(00)01740-0

Adams, R. B., Guerrant, R. L., Zu, S., Fang, G., and Roche, J. K. (1994). *Cryptosporidium parvum* infection of intestinal epithelium: morphologic and functional studies in an in vitro model. *J. Infect. Dis.* 169, 170–177. doi: 10.1093/infdis/169.1.170

Aley, S. B., Zimmerman, M., Hetsko, M., Selsted, M. E., and Gillin, F. D. (1994). Killing of *Giardia lamblia* by cryptdins and cationic neutrophil peptides. *Infect. Immun.* 62, 5397–5403.

Argenzio, R. A., Liacos, J. A., Levy, M. L., Meuten, D. J., Lecce, J. G., and Powell, D. W. (1990). Villous atrophy, crypt hyperplasia, cellular infiltration, and impaired glucose-Na absorption in enteric cryptosporidiosis of pigs. *Gastroenterology* 98, 1129–1140.

Balda, M. S., and Matter, K. (2008). Tight junctions at a glance. *J. Cell Sci.* 121, 3677–3682. doi: 10.1242/jcs.023887

Bartel, D. P. (2009). MicroRNAs: target recognition and regulatory functions. *Cell* 136, 215–233. doi: 10.1016/j.cell.2009.01.002

Betanzos, A., Javier-Reyna, R., García-Rivera, G., Bañuelos, C., González-Mariscal, L., Schnoor, M., et al. (2013). The EhCPADH112 complex of *Entamoeba histolytica* interacts with tight junction proteins occludin and claudin-1 to produce epithelial damage. *PLoS ONE* 8:e65100. doi: 10.1371/journal.pone.0065100

Bevins, C. L., and Salzman, N. H. (2011). Paneth cells, antimicrobial peptides and maintenance of intestinal homeostasis. *Nat. Rev. Microbiol.* 9, 356–368. doi: 10.1038/nrmicro2546

Binder, H. J. (2010). Role of colonic short-chain fatty acid transport in diarrhea. *Annu. Rev. Physiol.* 72, 297–313. doi: 10.1146/annurev-physiol-021909-135817

Bogdan, C. (2001). Nitric oxide and the immune response. *Nat. Immunol.* 2, 907–916. doi: 10.1038/ni1001-907

Brantley, R. K., Williams, K. R., Silva, T. M., Sistrom, M., Thielman, N. M., Ward, H., et al. (2003). AIDS-associated diarrhea and wasting in Northeast Brazil is associated with subtherapeutic plasma levels of antiretroviral medications and with both bovine and human subtypes of *Cryptosporidium parvum*. *Braz. J. Infect. Dis.* 7, 16–22. doi: 10.1590/S1413-86702003000100003

Bruchhaus, I., Loftus, B. J., Hall, N., and Tannich, E. (2003). The intestinal protozoan parasite *Entamoeba histolytica* contains 20 cysteine protease genes, of which only a small subset is expressed during in vitro cultivation. *Eukaryot. Cell* 2, 501–509. doi: 10.1128/EC.2.3.501-509.2003

Buret, A. G. (2007). Mechanisms of epithelial dysfunction in giardiasis. *Gut* 56, 316–317. doi: 10.1136/gut.2006.107771

Buret, A. G. (2008). Pathophysiology of enteric infections with *Giardia duodenalis*. *Parasite* 15, 262–265. doi: 10.1051/parasite/2008153261

Buret, A. G., Chin, A. C., and Scott, K. G. E. (2003). Infection of human and bovine epithelial cells with *Cryptosporidium andersoni* induces apoptosis and disrupts tight junctional ZO-1: effects of epidermal growth factor. *Int. J. Parasitol.* 33, 1363–1371. doi: 10.1016/S0020-7519(03)00138-3

Buret, A. G., Mitchell, K., Muench, D. G., and Scott, K. G. (2002). *Giardia lamblia* disrupts tight junctional ZO-1 and increases permeability in non-transformed human small intestinal epithelial monolayers: effects of epidermal growth factor. *Parasitology* 125, 11–19. doi: 10.1017/S0031182002001853

Castellanos-Gonzalez, A., Yancey, L. S., Wang, H. C., Pantenburg, B., Liscum, K. R., and Lewis, D. (2008). *Cryptosporidium* infection of human intestinal epithelial cells increases expression of osteoprotegerin: a novel mechanism for evasion of host defenses. *J. Infect. Dis.* 197, 916–923. doi: 10.1086/528374

Céu Sousa, M., Gonçalves, C. A., Bairos, V. A., and Poiares-Da-Silva, J. (2001). Adherence of *Giardia lamblia* trophozoites to Int-407 human intestinal cells. *Clin. Diagn. Lab. Immunol.* 8, 258–265.

Chen, X. M., Gores, G. J., Paya, C. V., and Larusso, N. F. (1999). *Cryptosporidium parvum* induces apoptosis in biliary epithelia by a Fas/Fas ligand-dependent mechanism. *Am. J. Physiol.* 277, G599–G608.

Chen, X. M., Levine, S. A., Splinter, P. L., Tietz, P. S., Ganong, A. L., Jobin, C., et al. (2001). *Cryptosporidium parvum* activaes nuclear factor κB in biliary epithelia preventing epithelial cell apoptosis. *Gastroenterology* 120, 1774–1783. doi: 10.1053/gast.2001.24850

Chen, X. M., Levine, S. A., Tietz, P., Krueger, E., McNiven, M. A., Jefferson, D. M., et al. (1998). *Cryptosporidium parvum* is cytopathic for cultured human biliary epithelia via an apoptotic mechanism. *Hepatology* 28, 906–913. doi: 10.1002/hep.510280402

Chin, A. C., Teoh, D. A., Scott, K. G.-E., Meddings, J. B., MacNaughton, W. K., and Buret, A. G. (2002). Strain-dependent induction of enterocyte apoptosis by *Giardia lamblia* disrupts epithelial barrier function in a caspase-3-dependent manner. *Infect. Immun.* 70, 3673–3680. doi: 10.1128/IAI.70.7.3673-3680.2002

Christy, N. C., and Petri, W. A. Jr. (2011). Mechanisms of adherence, cytotoxicity and phagocytosis modulate the pathogenesis of *Entamoeba histolytica*. *Future Microbiol.* 6, 1501–1519. doi: 10.2217/fmb.11.120

Ciccocioppo, R., Di Sabatino, A., Parroni, R., Muzi, P., D'Alò, S., Ventura, T., et al. (2001). Increased enterocyte apoptosis and Fas-Fas ligand system in celiac disease. *Am. J. Clin. Pathol.* 115, 494–503. doi: 10.1309/UV54-BHP3-A66B-0QUD

Cotton, J. A., Beatty, J. K., and Buret, A. G. (2011). Host parasite interactions and pathophysiology in *Giardia* infections. *Int. J. Parasitol.* 41, 925–933. doi: 10.1016/j.ijpara.2011.05.002

Deng, M., Lancto, C. A., and Abrahamsen, M. S. (2004). *Cryptosporidium parvum* regulation of human epithelial cell gene expression. *Int. J. Parasitol.* 34, 73–82. doi: 10.1016/j.ijpara.2003.10.001

Di Sabatino, A., Ciccocioppo, R., Luinetti, O., Ricevuti, L., Morera, R., Cifone, M. G., et al. (2003). Increased enterocyte apoptosis in inflamed areas of Crohn's disease. *Dis. Colon Rectum* 46, 1498–1507. doi: 10.1007/s10350-004-6802-z

Dinarello, C. A. (2007). Historical review of cytokines. *Eur. J. Immunol.* 37, S34–S45. doi: 10.1002/eji.200737772

Dreesen, L., Bosscher, K. D., Grit, G., Staels, B., Lubberts, E., Bauge, E., et al. (2014). *Giardia muris* infection in mice is associated with a protective interleukin 17A response and induction of peroxisome proliferator-activated receptor alpha. *Infect. Immun.* 82, 3333–3340. doi: 10.1128/IAI.01536-14

Eckmann, L., Reed, S. L., Smith, J. R., and Kagnoff, M. F. (1995). *Entamoeba histolytica* trophozoites induce an inflammatory cytokine response by cultured human cells through the paracrine action of cytolytically released interleukin-1 alpha. *J. Clin. Invest.* 96, 1269–1279. doi: 10.1172/JCI118161

Elmore, S. (2007). Apoptosis: a review of programmed cell death. *Toxicol. Pathol.* 35, 495–516. doi: 10.1080/01926230701320337

Farthing, M. J., Pereira, M. E., and Keusch, G. T. (1986). Description and characterization of a surface lectin from *Giardia lamblia*. *Infect. Immun.* 51, 661–667.

Farthing, M. J. G. (2000). "Clinical aspects of human cryptosporidiosis," in *Cryptosporidiosis and Microsporidiosis*, ed. F. Petry (Basel: Karger), 1–268.

Ferella, M., Davids, B. J., Cipriano, M. J., Birkeland, S. R., Palm, D., Gillin, F. D., et al. (2014). Gene expression changes during *Giardia*–host cell interactions in serum-free medium. *Mol. Biochem. Parasitol.* 197, 21–23. doi: 10.1016/j.molbiopara.2014.09.007

Field, M. (2003). Intestinal ion transport and the pathophysiology of diarrhea. *J. Clin. Invest.* 111, 931–943. doi: 10.1172/JCI200318326

Foster, D. M., Stauffer, S. H., Stone, M. R., and Gookin, J. L. (2012). Proteasome inhibition of pathologic shedding of enterocytes to defend barrier function requires X-linked inhibitor of apoptosis protein and nuclear factor κB. *Gastroenterology* 143, 133–144. doi: 10.1053/j.gastro.2012.03.030

Frederick, J. R., and Petri, W. A. Jr. (2005). Roles for the galactose-/N-acetylgalactosamine-binding lectin of *Entamoeba* in parasite virulence and differentiation. *Glycobiology* 15, 53R–59R. doi: 10.1093/glycob/cwj007

Genta, R. M., Chappell, C. L., White, A. C. Jr., Kimball, K. T., and Goodgame, R. W. (1993). Duodenal morphology and intensity of infection in AIDS-related intestinal cryptosporidiosis. *Gastroenterology* 105, 1769–1775.

Gillin, F. D., Reiner, D. S., and Boucher, S. E. (1988). Small-intestinal factors promote encystation of *Giardia lamblia* in vitro. *Infect. Immun.* 56, 705–707.

Gookin, J. L., Chiang, S., Allen, J., Armstrong, M. U., Satuffer, S. H., Finnegan, C., et al. (2006). NF-κB-mediated expression of iNOS promotes epithelial defense against infection by *Cryptosporidium parvum* in neonatal piglets. *Am. J. Physiol. Gastrointest. Liver Physiol.* 290, G164–G174. doi: 10.1152/ajpgi.00460.2004

Gookin, J. L., Nordone, S. K., and Argenzio, R. A. (2002). Host response to *Cryptosporidium* infection. *J. Vet. Intern. Med.* 16, 12–21. doi: 10.1111/j.1939-1676.2002.tb01602.x

Goplen, M., Lejeune, M., Cornick, S., Moreau, F., and Chadee, K. (2013). *Entamoeba histolytica* contains an occludin-like protein that can alter colonic epithelial barrier function. *PLoS ONE* 8:e73339. doi: 10.1371/journal.pone.0073339

Gorowara, S., Ganguly, N. K., Mahajan, R. C., and Walia, B. N. (1992). Study on the mechanism of *Giardia lamblia* induced diarrhoea in mice. *Biochim. Biophys. Acta* 1138, 122–126. doi: 10.1016/0925-4439(92)90051-N

Griffith, J. W., Sokol, C. L., and Luster, A. D. (2014). Chemokines and chemokine receptors: positioning cells for host defense and immunity. *Annu. Rev. Immunol.* 32, 659–702. doi: 10.1146/annurev-immunol-032713-120145

Guarino, A., Canani, R. B., Casola, A., Pozio, E., Russo, R., Bruzzese, E., et al. (1995). Human intestinal cryptosporidiosis: secretory diarrhea and enterotoxic activity in Caco-2 cells. *J. Infect. Dis.* 171, 976–983. doi: 10.1093/infdis/171.4.976

Guarino, A., Castaldo, A., Russo, S., Spagnuolo, M. I., Canani, R. B., Tarallo, L., et al. (1997). Enteric cryptosporidiosis in pediatric HIV infection. *J. Pediatr. Gastroenterol. Nutr.* 25, 187–192. doi: 10.1097/00005176-199708000-00009

Guerrant, R. L., Van Gilder, T., Steiner, T. S., Thielman, N. M., Slutsker, L., Tauxe, R. V., et al. (2001). Practice guidelines for the management of infectious diarrhea. *Clin. Infect. Dis.* 32, 331–351. doi: 10.1086/318514

Günther, C., Neumann, H., Neurath, M. F., and Becker, C. (2013). Apoptosis, necrosis and necroptosis: cell death regulation in the intestinal epithelium. *Gut* 62, 1062–1071. doi: 10.1136/gutjnl-2011-301364

Hammer, H. F., Santa Ana, C. A., Schiller, L. R., and Fordtran, J. S. (1989). Studies of osmotic diarrhea induced in normal subjects by ingestion of polyethylene glycol and lactulose. *J. Clin. Invest.* 84, 1056–1062. doi: 10.1172/JCI114267

Hanevik, K., Hausken, T., Morken, M., Strand, E., Morch, K., Coll, P., et al. (2007). Persisting symptoms and duodenal inflammation related to *Giardia duodenalis* infection. *J. Infect.* 55, 524–530. doi: 10.1016/j.jinf.2007.09.004

Holberton, D. V. (1973). Fine structure of the ventral disk apparatus and the mechanism of attachment in the flagellate *Giardia muris*. *J. Cell Sci.* 13, 11–41.

Hou, Y., Mortimer, L., and Chadee, K. (2010). *Entamoeba histolytica* cysteine proteinase 5 binds integrin on colonic cells and stimulates NFkappaB-mediated pro-inflammatory responses. *J. Biol. Chem.* 285, 35497–35504. doi: 10.1074/jbc.M109.066035

Humen, M. A., Pérez, P. F., and Moal, V. L. L. (2011). Lipid raft-dependent adhesion of *Giardia intestinalis* trophozoites to a cultured human enterocyte-like Caco-2/TC7 cell monolayer leads to cytoskeleton-dependent functional injuries. *Cell. Microbiol.* 13, 1683–1702. doi: 10.1111/j.1462-5822.2011.01647.x

Huston, C. D., Boettner, D. R., Miller-Sims, V., and Petri, W. A. Jr. (2003). Apoptotic killing and phagocytosis of host cells by the parasite *Entamoeba histolytica*. *Infect. Immun.* 71, 964–972. doi: 10.1128/IAI.71.2.964-972.2003

Huston, C. D., Houpt, E. R., Mann, B. J., Hahn, C. S., and Petri, W. A. Jr. (2000). Caspase 3-dependent killing of host cells by the parasite *Entamoeba histolytica*. *Cell. Microbiol.* 2, 617–625. doi: 10.1046/j.1462-5822.2000.00085.x

Inge, P. M., Edson, C. M., and Farthing, M. J. (1988). Attachment of *Giardia lamblia* to rat intestinal epithelial cells. *Gut* 29, 795–801. doi: 10.1136/gut.29.6.795

Izzo, A. A., Gaginella, T. S., Mascolo, N., and Capasso, F. (1994). Nitric oxide as a mediator of the laxative action of magnesium sulphate. *Br. J. Pharmacol.* 113, 228–232. doi: 10.1111/j.1476-5381.1994.tb16198.x

Jang, Y. S., Song, K. J., Kim, J. Y., Lee, Y. A., Kim, K. A., Lee, S. K., et al. (2011). Calpains are involved in *Entamoeba histolytica*-induced death of HT-29 colonic epithelial cells. *Korean J. Parasitol.* 49, 177–180. doi: 10.3347/kjp.2011.49.2.177

Kim, K. A., Kim, J. Y., Lee, Y. A., Min, A., Bahk, Y. Y., and Shin, M. H. (2013). *Entamoeba histolytica* induces cell death of HT29 colonic epithelial cells via NOX1-derived ROS. *Korean J. Parasitol.* 51, 61–68. doi: 10.3347/kjp.2013.51.1.61

Kim, K. A., Kim, J. Y., Lee, Y. A., Song, K. J., Min, D., and Shin, M. H. (2011). NOX1 participates in ROS-dependent cell death of colon epithelial Caco2 cells induced by *Entamoeba histolytica*. *Microbes Infect.* 13, 1052–1061. doi: 10.1016/j.micinf.2011.06.001

Kim, K. A., Lee, Y. A., and Shin, M. H. (2007). Calpain-dependent calpastatin cleavage regulates caspase-3 activation during apoptosis of Jurkat T cells induced by *Entamoeba histolytica*. *Int. J. Parasitol.* 37, 1209–1219. doi: 10.1016/j.ijpara.2007.03.011

Kim, K. A., Lee, Y. A., and Shin, M. H. (2010). Calpain-dependent cleavage of SHP-1 and SHP-2 is involved in the dephosphorylation of Jurkat T cells induced by *Entamoeba histolytica*. *Parasite Immunol.* 32, 176–183. doi: 10.1111/j.1365-3024.2009.01175.x

Kim, K. A., Min, A., Lee, Y. A., and Shin, M. H. (2014). Degradation of the transcription factor NF-κB, STAT3, and STAT5 is involved in *Entamoeba histolytica*-induced cell death in Caco-2 colonic epithelial cells. *Korean J. Parasitol.* 52, 459–469. doi: 10.3347/kjp.2014.52.5.459

Koh, W. H., Geurden, T., Paget, T., O'Handley, R., Steuart, R. F., Thompson, R. C. A., et al. (2013). *Giardia duodenalis* assemblage-specific induction of apoptosis and tight junction disruption in human intestinal epithelial cells: effects of mixed infections. *J. Parasitol.* 99, 353–358. doi: 10.1645/GE-3021.1

Kohli, A., Bushen, O. Y., Pinkerton, R. C., Houpt, E., Newman, R. D., Sears, C. L., et al. (2008). *Giardia duodenalis* assemblage, clinical presentation and markers of intestinal inflammation in Brazilian children. *Trans. R. Soc. Trop. Med. Hyg.* 102, 718–725. doi: 10.1016/j.trstmh.2008.03.002

Lacroix-Lamandé, S., Mancassola, R., Naciri, M., and Laurent, F. (2002). Role of gamma interferon in chemokine expression in the ileum of mice and in a murine intestinal epithelial cell line after *Cryptosporidium parvum* infection. *Infect. Immun.* 70, 2090–2099. doi: 10.1128/IAI.70.4.2090-2099.2002

Laurent, F., Eckmann, L., Savidge, T. C., Morgan, G., Theodos, C., Naciri, M., et al. (1997). *Cryptosporidium parvum* infection of human intestinal epithelial cells induces the polarized secretion of C-X-C chemokines. *Infect. Immun.* 65, 5067–5073.

Laurent, F., McCole, D., Eckmann, L., and Kagnoff, M. F. (1999). Pathogenesis of *Cryptosporidium parvum* infection. *Microbes Infect.* 2, 141–148. doi: 10.1016/S1286-4579(99)80005-7

Lee, Y. A., Kim, K. A., and Shin, M. H. (2011). Calpain mediates degradation of cytoskeletal proteins during Jurkat T-cell death induced by *Entamoeba histolytica*. *Parasite Immunol.* 33, 349–356. doi: 10.1111/j.1365-3024.2011.01290.x

Lejeune, M., Moreau, F., and Chadee, K. (2011). Prostaglandin E2 produced by *Entamoeba histolytica* signals via EP4 receptor and alters claudin-4 to increase ion permeability of tight junctions. *Am. J. Pathol.* 179, 807–818. doi: 10.1016/j.ajpath.2011.05.001

Leroy, A., Lauwaet, T., De Bruyne, G., Cornelissen, M., and Mareel, M. (2000). *Entamoeba histolytica* disturbs the tight junction complex in human enteric T84 cell layers. *FASEB J.* 14, 1139–1146.

Li, W., Stenson, W. F., Kunz-Jenkins, C., Swanson, P. E., Duncan, R., and Stanley, S. L. Jr. (1994). *Entamoeba histolytica* interactions with polarized human intestinal Caco-2 epithelial cells. *Infect. Immun.* 62, 5112–5119.

Lin, M. T., and Beal, M. F. (2006). Mitochondrial dysfunction and oxidative stress in neurodegenerative diseases. *Nature* 443, 787–795. doi: 10.1038/nature05292

Liu, J., Deng, M., Lancto, C. A., Abrahamsen, M. S., Rutherford, M. S., and Enomoto, S. (2009). Biphasic modulation of apoptotic pathways in *Cryptosporidium parvum*-infected human intestinal epithelial cells. *Infect. Immun.* 77, 837–849. doi: 10.1128/IAI.00955-08

Liu, J., Enomoto, S., Lancto, C. A., Abrahamsen, M. S., and Rutherford, M. S. (2008). Inhibition of apoptosis in *Cryptosporidium parvum*-infected intestinal epithelial cells is dependent on survivin. *Infect. Immun.* 76, 3784–3792. doi: 10.1128/IAI.00308-08

Luján, H. D., Mowatt, M. R., Byrd, L. G., and Nash, T. E. (1996). Cholesterol starvation induces differentiation of the intestinal parasite *Giardia lamblia*. *Proc. Natl. Acad. Sci. U.S.A.* 93, 7628–7633. doi: 10.1073/pnas.93.15.7628

Maia-Brigagão, C., Morgado-Díaz, J. A., and De Souza, W. (2012). *Giardia* disrupts the arrangement of tight, adherens and desmosomal junction proteins of intestinal cells. *Parasit. Int.* 61, 280–287. doi: 10.1016/j.parint.2011.11.002

Marco, S., and Skaper, S. D. (2006). Amyloid β-peptide1–42 alters tight junction protein distribution and expression in brain microvessel endothelial cells. *Neurosci. Lett.* 401, 219–224. doi: 10.1016/j.neulet.2006.03.047

Marie, C., and Petri, W. A. Jr. (2014). Regulation of virulence of *Entamoeba histolytica*. *Annu. Rev. Microbiol.* 68, 493–520. doi: 10.1146/annurev-micro-091313-103550

Martinez-Palomo, A., Gonzalez-Robles, A., Chavez, B., Orozco, E., Fernandez-Castelo, S., and Cervantes, A. (1985). Structural bases of the cytolytic mechanisms of *Entamoeba histolytica*. *J. Protozool.* 32, 166–175. doi: 10.1111/j.1550-7408.1985.tb03033.x

Matter, K., and Balda, M. (2003). Signalling to and from tight junctions. *Nat. Rev. Mol. Cell Biol.* 4, 225–236. doi: 10.1038/nrm1055

McCole, D. F., Eckmann, L., Laurent, F., and Kagnoff, M. F. (2000). Intestinal epithelial cell apoptosis following *Cryptosporidium parvum* infection. *Infect. Immun.* 68, 1710–1713. doi: 10.1128/IAI.68.3.1710-1713.2000

Mele, R., Gomez Morales, M. A., Tosini, F., and Pozio, E. (2004). *Cryptosporidium parvum* at different developmental stages modulates host cell apoptosis in vitro. *Infect. Immun.* 72, 6061–6067. doi: 10.1128/IAI.72.10.6061-6067.2004

Morken, M. H., Nysaeter, G., Strand, E. A., Hausken, T., and Berstad, A. (2008). Lactulose breath test results in patients with persistent abdominal symptoms following *Giardia lamblia* infection. *Scand. J. Gastroenterol.* 43, 141–145.

Müller, N., and von Allmen, N. (2005). Recent insights into the mucosal reactions associated with *Giardia lamblia* infections. *Int. J. Parasitol.* 35, 1339–1347. doi: 10.1016/j.ijpara.2005.07.008

Muza-Moons, M. M., Schneeberger, E. E., and Hecht, G. A. (2004). Enteropathogenic *Escherichia coli* infection leads to appearance of aberrant tight junctions strands in the lateral membrane of intestinal epithelial cells. *Cell. Microbiol.* 6, 783–793. doi: 10.1111/j.1462-5822.2004.00404.x

Ocádiz, R., Orozco, E., Carrillo, E., Quintas, L. I., Ortega-López, J., García-Pérez, R. M., et al. (2005). EhCP112 is an *Entamoeba histolytica* secreted cysteine protease that may be involved in the parasite-virulence. *Cell. Microbiol.* 7, 221–232. doi: 10.1111/j.1462-5822.2004.00453.x

Ojcius, D. M., Perfettini, J. L., Bonnin, A., and Laurent, F. (1999). Caspase-dependent apoptosis during infection with *Cryptosporidium parvum*. *Microbes Infect.* 1, 1163–1168. doi: 10.1016/S1286-4579(99)00246-4

Oliver, F. J., De La Rubia, G., Rolli, V., Ruiz-Ruiz, M. C., Murcia, G., and Murcia, J. M. (1998). Importance of poly(ADP-ribose) polymerase and its cleavage in apoptosis. Lesson from an uncleavable mutant. *J. Biol. Chem.* 273, 33533–33539.

Ostaff, M. J., Stange, E. F., and Wehkamp, J. (2013). Antimicrobial peptides and gut microbiota in homeostasis and pathology. *EMBO Mol. Med.* 5, 1465–1483. doi: 10.1002/emmm.201201773

Panaro, M. A., Cianciulli, A., Mitolo, V., Mitolo, C. I., Acquafredda, A., Brandonisio, O., et al. (2007). Caspase-dependent apoptosis of the HCT-8 epithelial cell line induced by the parasite *Giardia intestinalis*. *FEMS Immunol. Med. Microbiol.* 51, 302–309. doi: 10.1111/j.1574-695X.2007.00304.x

Parikh, A., Stephan, A. F., and Tzanakakis, E. S. (2012). Regenerating proteins and their expression, regulation and signaling. *Biomol. Concepts* 3, 57–70. doi: 10.1515/bmc.2011.055

Petersen, C. (1993). Cellular biology of *Cryptosporidium parvum*. *Parasitol. Today* 9, 87–91. doi: 10.1016/0169-4758(93)90211-W

Peterson, K. M., Guo, X., Elkahloun, A. G., Mondal, D., Bardhan, P. K., Sugawara, A., et al. (2011). The expression of REG 1A and REG 1B is increased during acute amebic colitis. *Parasit. Int.* 60, 296–300. doi: 10.1016/j.parint.2011.04.005

Pinilla, A. E., Lopez, M. C., and Viasus, D. F. (2008). [History of the *Entamoeba histolytica* protozoan]. *Rev. Med. Chile* 136, 118–124. doi: 10.4067/S0034-98872008000100015

Ragland, B. D., Ashley, L. S., Vaux, D. L., and Petri, W. A. (1994). *Entamoeba histolytica*: target cells killed by trophozoites undergo DNA fragmentation which is not blocked by Bcl-2. *Exp. Parasitol.* 79, 460–467. doi: 10.1006/expr.1994.1107

Ralston, K. S., Solga, M. D., Mackey-Lawrence, N. M., Bhattacharya, A., and Petri, W. A. (2014). Trogocytosis by *Entamoeba histolytica* contributes to cell killing and tissue invasio. *Nature* 508, 526–530. doi: 10.1038/nature13242

Rana, S. V., Bhasin, D. K., and Vinayak, V. K. (2004). Prospective evaluation of lactose malabsorption by lactose hydrogen breath test in individuals infected with *Entamoeba histolytica* and passing cysts. *Br. J. Nutr.* 92, 207–208. doi: 10.1079/BJN20041194

Robinson, P., Okhuysen, P. C., Chappell, C. L., Lewis, D. E., Shahab, I., and Lahoti, S. (2000). Transforming growth factor β1 is expressed in the jejunum after experimental *Cryptosporidium parvum* infection in humans. *Infect. Immun.* 68, 5405–5407. doi: 10.1128/IAI.68.9.5405-5407.2000

Robinson, P., Okhuysen, P. C., Chappell, C. L., Lewis, D. E., Shahab, I., and Lahoti, S. (2001). Expression of IL-15 and IL-4 in IFN-gamma-independent control of experimental human *Cryptosporidium parvum* infection. *Cytokine* 15, 39–46. doi: 10.1006/cyto.2001.0888

Roxström-Lindquist, K., Ringqvist, E., Palm, D., and Svard, S. (2005). *Giardia lamblia*-induced changes in gene expression in differentiated Caco-2 human intestinal epithelial cells. *Infect. Immun.* 73, 8204–8208. doi: 10.1128/IAI.73.12.8204-8208.2005

Ryan, U., and Cacciò, S. M. (2013). Zoonotic potential of *Giardia*. *Int. J. Parasitol.* 43, 943–956. doi: 10.1016/j.ijpara.2013.06.001

Schwartz, C. L., Heumann, J. M., Dawson, S. C., and Hoenger, A. (2012). A detailed, hierarchical study of *Giardia lamblia*'s ventral disc reveals novel microtubule-associated protein complexes. *PLoS ONE* 7:e43783. doi: 10.1371/journal.pone.0043783

Scott, K., Meddings, J., Kirk, D., Lees-Miller, S., and Buret, A. (2002). Intestinal infection with *Giardia* spp. reduces epithelial barrier function in a myosin light chain kinase-dependent fashion. *Gastroenterology* 123, 1179–1190. doi: 10.1053/gast.2002.36002

Scott, K. G.-E., Logan, M. R., Klammer, G. M., Teoh, D. A., and Buret, A. G. (2000). Jejunal brush border microvillous alterations in *Giardia muris*-infected mice: role of T lymphocytes and interleukin-6. *Infect. Immun.* 68, 3412–3418. doi: 10.1128/IAI.68.6.3412-3418.2000

Sekikawa, A., Fukui, H., Fujii, S., Ichikawa, K., Tomita, S., Imura, J., et al. (2008). REG Ialpha protein mediates an anti-apoptotic effect of STAT3 signaling in gastric cancer cells. *Carcinogenesis* 29, 76–83. doi: 10.1093/carcin/bgm250

Seydel, K. B., Li, E., Swanson, P. E., and Stanley, S. L. Jr. (1997). Human intestinal epithelial cells produce proinflammatory cytokines in response to infection in a SCID mouse-human intestinal xenograft model of amebiasis. *Infect. Immun.* 65, 1631–1639.

Seydel, K. B., Li, E., Zhang, Z., and Stanley, S. L. Jr. (1998a). Epithelial cell-initiated inflammation plays a crucial role in early tissue damage in amebic infection of human intestine. *Gastroenterology* 115, 1446–1453. doi: 10.1016/S0016-5085(98)70023-X

Seydel, K. B., and Stanley, S. L. Jr. (1998). *Entamoeba histolytica* induces host cell death in amebic liver abscess by a non-Fas-dependent, non-tumor necrosis factor alpha-dependent pathway of apoptosis. *Infect. Immun.* 66, 2980–2983.

Seydel, K. B., Zhang, T., Champion, G. A., Fichtenbaum, C., Swanson, P. E., Tzipori, S., et al. (1998b). *Cryptosporidium parvum* infection of human intestinal xenografts in SCID mice induces production of human tumor necrosis factor alpha and interleukin-8. *Infect. Immun.* 66, 2379–2382.

Sharpstone, D., Neild, P., Crane, R., Taylor, C., Hodgson, C., Sherwood, R., et al. (1999). Small intestinal transit, absorption, and permeability in patients with AIDS with and without diarrhoea. *Gut* 45, 70–76. doi: 10.1136/gut.45.1.70

Sim, S., Park, S. J., Yong, T. S., Im, K. I., and Shin, M. H. (2007). Involvement of beta 2-integrin in ROS-mediated neutrophil apoptosis induced by *Entamoeba histolytica*. *Microbes Infect.* 9, 1368–1375. doi: 10.1016/j.micinf.2007.06.013

Sim, S., Yong, T. S., Park, S. J., Im, K. I., Kong, Y., Ryu, J. S., et al. (2005). NADPH oxidase-derived reactive oxygen species-mediated activation of ERK1/2 is required for apoptosis of human neutrophils induced by *Entamoeba histolytica*. *J. Immunol.* 174, 4279–4288. doi: 10.4049/jimmunol.174.7.4279

Simonet, W. S., Lacey, D. L., Dunstan, C. R., Kelley, M., Chang, M. S., Lüthy, R., et al. (1997). Osteoprotogerin: a novel secreted protein involved in the regulation of bone density. *Cell* 89, 309–319. doi: 10.1016/S0092-8674(00)80209-3

Slee, E. A., Adrain, C., and Martin, S. J. (2001). Executioner caspase-3, -6, and -7 perform distinct, non-redundant roles during the demolition phase of apoptosis. *J. Biol. Chem.* 276, 7320–7326. doi: 10.1074/jbc.M008363200

Stadelmann, B., Merino, M. C., Persson, L., and Svärd, S. G. (2012). Arginine consumption by the intestinal parasite *Giardia* intestinalis reduces proliferation of intestinal epithelial cells. *PLoS ONE* 7:e45325. doi: 10.1371/journal.pone.0045325

Stadnyk, A. W. (2002). Intestinal epithelial cells as a source of inflammatory cytokines and chemokines. *Can. J. Gastroenterol.* 16, 241–246.

Su, L., Nalle, S. C., Shen, L., Turner, E. S., Singh, G., Breskin, L. A., et al. (2013). TNFR2 activates MLCK-dependent tight junction dysregulation to cause apoptosis-mediated barrier loss and experimental colitis. *Gastroenterology* 145, 407–415. doi: 10.1053/j.gastro.2013.04.011

Sullivan, S., Alex, P., Dassopoulos, T., Zachos, N. C., Iacobuzio-Donahue, C., Donowitz, M., et al. (2009). Down-regulation of sodium transporters and NHERF proteins in IBD patients and mouse colitis models: potential contributors to IBD-associated diarrhea. *Inflamm. Bowel Dis.* 15, 261–274. doi: 10.1002/ibd.20743

Tako, E. A., Hassimi, M. F., and Singer, S. M. (2013). Transcriptomic analysis of the host response to *Giardia duodenalis* infection reveals redundant mechanisms for parasite control. *mBio* 4:e660-13. doi: 10.1128/mBio.00660-13

Teixeira, J. E., and Mann, B. J. (2002). *Entamoeba histolytica*-induced dephosphorylation in host cells. *Infect. Immun.* 70, 1816–1823. doi: 10.1128/IAI.70.4.1816-1823.2002

Teoh, D., Kamieniecki, D., Pang, G., and Buret, A. (2000). *Giardia lamblia* rearranges F-actin and alpha-actinin in human colonic and duodenal monolayers and reduces transepithelial electrical resistance. *J. Parasitol.* 86, 800–806. doi: 10.1645/0022-3395(2000)086[0800:GLRFAA]2.0.CO;2

Tessema, T. S., Schwamb, B., Lochner, M., Förster, I., Jakobi, V., and Petry, F. (2009). Dynamics of gut mucosal and systemic Th1/Th2 cytokine responses in interferon-gamma and interleukin-12p40 knock out mice during primary and challenge *Cryptosporidium parvum* infection. *Immunobiology* 214, 454–466. doi: 10.1016/j.imbio.2008.11.015

Thornberry, N. A., and Lazebnik, Y. (1998). Caspases: enemies within. *Science* 281, 1312–1316. doi: 10.1126/science.281.5381.1312

Troeger, H., Epple, H.-J., Schneider, T., Wahnschaffe, U., Ullrich, R., Burchard, G.-D., et al. (2007). Effect of chronic *Giardia lamblia* infection on epithelial transport and barrier function in human duodenum. *Gut* 56, 328–335. doi: 10.1136/gut.2006.100198

Turner, J. (2009). Intestinal mucosal barrier function in health and disease. *Nat. Rev. Immunol.* 9, 799–809. doi: 10.1038/nri2653

Vidal, K., Serrant, P., Schlosser, B., Van Den Broek, P., Lorget, F., and Donnet-Hughes, A. (2004). Osteoprotegerin production by human intestinal epithelial cells: a potential regulator of mucosal immune responses. *Am. J. Physiol. Gastrointest. Liver Physiol.* 287, G836–G844. doi: 10.1152/ajpgi.004 28.2003

Viswanathan, V. K., Hodges, K., and Hecht, G. (2009). Enteric infection meets intestinal function: how bacterial pathogens cause diarrhoea. *Nat. Rev. Microbiol.* 7, 110–119. doi: 10.1038/nrmicro2053

Walters, J. R. F., Tasleem, A. M., Omer, O. S., Brydon, W. G., Dew, T., and Le Roux, C. W. (2009). A new mechanism for bile acid diarrhea: defective feedback inhibition of bile acid biosynthesis. *Clin. Gastroenterol. Hepatol.* 7, 1189–1194. doi: 10.1016/j.cgh.2009.04.024

Wanyri, J., and Ward, H. (2006). Molecular basis of *Cryptosporidium*-host cell interactions: recent advances and future prospects. *Future Microbiol.* 1, 201–208. doi: 10.2217/17460913.1.2.201

Ward, H. D., Alroy, J., Lev, B. I., Keusch, G. T., and Pereira, M. E. (1988). Biology of *Giardia lamblia*. Detection of N-acetyl-D-glucosamine as the only surface saccharide moiety and identification of two distinct subsets of trophozoites by lectin binding. *J. Exp. Med.* 167, 73–88. doi: 10.1084/jem.167.1.73

White, C. (2010). "*Cryptosporidium* species," in *Principles and Practice of Infectious Diseases*, 7th Edn, eds G. L. Mandell, J. E. Bennett, and R. Dolin (Philadelphia: Churchill Livingstone, Elsevier).

Widmer, G., Corey, E. A., Stein, B., Griffiths, J. K., and Tzipori, S. (2000). Host cell apoptosis impairs *Cryptosporidium parvum* development in vitro. *J. Parasitol.* 86, 922–928. doi: 10.1645/0022-3395(2000)086[0922:HCAICP]2.0.CO;2

Yu, L. C., Flynn, A. N., Turner, J. R., and Buret, A. G. (2005). SGLT-1-mediated glucose uptake protects intestinal epithelial cells against LPS-induced apoptosis and barrier defects: a novel cellular rescue mechanism? *FASEB J.* 19, 1822–1835. doi: 10.1096/fj.05-4226com

Yu, L. C. H., Huang, C. Y., Kuo, W., Sayer, H., Turner, J. R., and Buret, A. G. (2008). SGLT-1-mediated glucose uptake protects human intestinal epithelial cells against *Giardia duodenalis*-induced apoptosis. *Int. J. Parasitol.* 38, 923–934. doi: 10.1016/j.ijpara.2007.12.004

Zaalouk, T. K., Bajaj-Elliott, M., George, J. T., and McDonald, V. (2004). Differential regulation of β-defensin gene expression during *Cryptosporidium parvum* infection. *Infect. Immun.* 72, 2772–2779. doi: 10.1128/IAI.72.5.2772-2779.2004

Zeissig, S., Bürge, N., Günzel, D., Richter, J., Mankertz, J., Wahnschaffe, U., et al. (2007). Changes in expression and distribution of claudin 2, 5 and 8 lead to discontinuous tight junctions and barrier dysfunction in active Crohn's disease. *Gut* 56, 61–72. doi: 10.1136/gut.2006.094375

Zhou, P., Li, E., Robertson, J., Nash, T. E., and Singer, S. M. (2003). Role of interleukin-6 in the control of acute and chronic *Giardia lamblia* infections in mice. *Infect. Immun.* 71, 1566–1568. doi: 10.1128/IAI.71.3.1566-1568.2003

Zhou, P., Li, E., Shea-Donohue, T., and Singer, S. M. (2007). Tumour necrosis factor alpha contributes to protection against *Giardia lamblia* infection in mice. *Parasite Immunol.* 29, 367–374. doi: 10.1111/j.1365-3024.2007.00953.x

Zhou, R., Gong, A., Chen, D., Miller, R. E., Eischeid, A. N., and Chen, X. M. (2013). Histone deacetylases and NF-kB signaling coordinate expression of CX3CL1 in epithelial cells in response to microbial challenge by suppressing miR-424 and miR-503. *PLoS ONE* 8:e65153. doi: 10.1371/journal.pone.0065153

Zhou, R., Hu, G. Z., Liiu, J., Gong, A., Drescher, K. M., and Chen, X. M. (2009). NF-kappaB p65-dependent transactivation of miRNA genes following *Cryptosporidium parvum* infection stimulates epithelial cell immune responses. *PLoS Pathog.* 5:e1000681. doi: 10.1371/journal.ppat.1000681

Zlotnik, A. (2000). Chemokines: a new classification system and their role in immunity. *Immunity* 12, 121–127. doi: 10.1016/S1074-7613(00)80165-X

9

Stochastic Methods for Inferring States of Cell Migration

R. J. Allen[1†], C. Welch[1†], Neha Pankow[1], Klaus M. Hahn[1,2]* and Timothy C. Elston[1,2]*

[1] *Department of Pharmacology, University of North Carolina at Chapel Hill, Chapel Hill, NC, United States,* [2] *Computational Medicine Program, University of North Carolina at Chapel Hill, Chapel Hill, NC, United States*

Edited by:
Zhike Zi,
Max Planck Institute for Molecular
Genetics, Germany

Reviewed by:
David McMillen,
University of Toronto Mississauga,
Canada
Nathan Weinstein,
Universidad Nacional Autónoma
de México, Mexico

*Correspondence:
Klaus M. Hahn
khahn@med.unc.edu
Timothy C. Elston
telston@med.unc.edu

†Present address:
R. J. Allen,
Early Clinical Development,
Quantitative Systems Pharmacology,
Pfizer Inc., Cambridge, MA,
United States
C. Welch,
Otolaryngology/Head and Neck
Surgery, University of North Carolina
School of Medicine, Chapel Hill, NC,
United States

Cell migration refers to the ability of cells to translocate across a substrate or through a matrix. To achieve net movement requires spatiotemporal regulation of the actin cytoskeleton. Computational approaches are necessary to identify and quantify the regulatory mechanisms that generate directed cell movement. To address this need, we developed computational tools, based on stochastic modeling, to analyze time series data for the position of randomly migrating cells. Our approach allows parameters that characterize cell movement to be efficiently estimated from cell track data. We applied our methods to analyze the random migration of Mouse Embryonic Fibroblasts (MEFS) and HeLa cells. Our analysis revealed that MEFs exist in two distinct states of migration characterized by differences in cell speed and persistence, whereas HeLa cells only exhibit a single state. Further analysis revealed that the Rho-family GTPase RhoG plays a role in determining the properties of the two migratory states of MEFs. An important feature of our computational approach is that it provides a method for predicting the current migration state of an individual cell from time series data. Finally, we applied our computational methods to HeLa cells expressing a Rac1 biosensor. The Rac1 biosensor is known to perturb movement when expressed at overly high concentrations; at these expression levels the HeLa cells showed two migratory states, which correlated with differences in the spatial distribution of active Rac1.

Keywords: cell migration, stochastic modeling, RHOG, Rac1, biosenor, migration states

INTRODUCTION

The ability of cells to move is essential to many biological processes, such as tissue development, the immune response and wound healing (Franca-Koh et al., 2007; Petrie et al., 2009; Cain and Ridley, 2012). Anomalous cell migration plays a role in diseases, such as cancer and atherosclerosis (Cain and Ridley, 2012; Hall, 2009; Lemarié et al., 2010; Finney et al., 2017). During cell migration, intracellular signaling networks tightly control the spatiotemporal dynamics of the cytoskeleton. In particular, the Rho family of small GTPases has been implicated in membrane protrusion, adhesion, contraction and de-adhesion, all steps necessary for cell migration (Rottner et al., 1999; Jaffe and Hall, 2005; Goley and Welch, 2006; Ridley, 2006; Iden and Collard, 2008; Ladwein and Rottner, 2008). Rac1, the family member studied here, produces cell protrusions by interacting with effector proteins that modulate actin polymerization, including formins and Paks. A prevailing hypothesis is that Rac1 induces localized actin polymerization to trap random, thermal driven outward movements of the cell edge (Ridley, 2015; Marston et al., 2019; Schaks et al., 2019).

During random cell migration, in which cells do not experience directional environmental cues, cells move in a persistent manner, but with significant variability in their direction and speed. Therefore, methods for quantifying cell movement that take into account the stochastic nature of this phenomenon are needed. Previous studies have analyzed cell migration in terms of quantitative metrics such as the mean squared deviation in cell position, which can be linked to both speed and persistence (Othmer et al., 1988; Dimilla et al., 1992; Rosello et al., 2004; Dieterich et al., 2008). Additionally, it has been suggested that fractional diffusion models are required to accurately describe cell movement (Dieterich et al., 2008). We refer the reader to a recent review which describes these approaches and others (Svensson et al., 2018). We used stochastic modeling to develop tools for quantifying cell migration such that it can be characterized in terms of biologically relevant parameters. In our approach, the motion of cells is assumed to follow a 2D random walk with persistence. A related method that takes into account the probability of turning and contains a parameter related to persistence also has been applied to analyze random cell migration (Arrieumerlou and Meyer, 2005). An important distinction of our approach is that our model allows for the possibility of multiple states of migration, distinguished by differences in speed and persistence. This feature allowed us to determine that Mouse Embryonic Fibroblasts (MEFS) exist in two distinct states during random migration. Knock down of the Rho-GTPase RhoG suggests that this protein plays an important role in establishing the two states. We next demonstrated how our method allows the migration state of a cell to be predicted from time series data. Finally, we applied our method to examine the activation of Rac1, a GTPase known to be important in producing localized protrusions. Interestingly, we found that overexpressed, biosensor induced two states of migration in HeLa cells that correlated with different numbers of active Rac1 foci.

RESULTS

Preliminary Analysis

To develop our methods, we collected data sets that consisted of time series for the x and y coordinates of the cell centroids of randomly migrating MEF cells (**Figures 1A,B**). We chose this cell type because it shows persistent migration in the absence of directional cues. As an initial analysis of the data, we computed the average persistence of cell movement defined as $P = <\cos(\theta)>$, where θ is the change in the direction of cell movement between measurements (**Figure 1C**) and the angular bracket denotes averaging over cell tracks. If θ is uniformly distributed, then the motion of the cell lacks persistence and $P = 0$. This behavior would be consistent with a pure random walk (diffusive motion). For values of P greater than zero, the movement of the cell shows persistence, with a value of 1 indicating motion in a straight line. Combining the cell tracks for individual cells, produced a value of $P = 0.43$. This value is consistent with cells that show persistent motion. We also generated histograms from the Δx and Δy displacements and empirically calculated

cumulative density functions (**Supplementary Figure S1**, top left panel). These distributions were found to show slight deviations from a Gaussian distribution.

A Stochastic Model for Cell Migration

Our preliminary cell track analysis led us to model cell movement as a 2D random walk with persistence (**Figure 1C**). In our model, for each time interval i, the distance, r_i, traveled by a cell and the angle, θ_i, through which the cell moves are considered random variables. The random variable r_i is taken to have a Gaussian distribution characterized by mean μ_R, and variance σ_R^2. We allowed for negative values of r_i to account for the scenario in which a cell maintains its direction of polarization, but its centroid moves in a rearward direction. The directional angle θ_i, is also taken to have a Gaussian distribution with variance σ_s^2, and centered on the value of the previous angle θ_{i-1}. Small values of σ_s^2 correspond to highly persistent migration. For large values of σ_s^2 the new direction becomes uniformly distributed on the interval $[-\pi, \pi]$ and the model represents a purely diffusive process.

It is not possible to tell from cell track data alone if changes in θ_i of magnitude greater than $\pi/2$ resulted from large deviations in orientation or negative r_i. Thus, the probability distribution for these variables cannot be constructed unambiguously from the cell track data. To overcome this difficulty, we performed a change of variables from (r_i, θ_i) to $(\Delta x_i^{\parallel}, \Delta y_i^{\perp})$, where these new variables correspond to changes in the centroid's position during the ith time interval that are parallel and perpendicular to the direction of the previous step (**Figure 1C**). An important feature of the model is that analytical expressions for the probability density functions (pdfs) of $\Delta x_i^{\parallel}$ and $\Delta y_i^{\perp}$ can be found (**Supplementary Information**), allowing estimation of model parameters from experimental data to be performed in a computationally efficient manner, relative to the alternative of estimating probability density functions via repeated simulation of the stochastic model (**Figure 1C**). These co-ordinates explicitly handle the degeneracy in θ_i described above, because in these co-ordinates all possibilities that could have led to a given observation are considered. If cells show persistent motion, $\Delta x_i^{\parallel}$ has a positive mean value. Also, if there are no external cues in the experiments to define a preferred direction of motion, $\Delta y_i^{\perp}$ is symmetric about zero. Therefore, the distribution for $\Delta x_i^{\parallel}$ is more informative, and we use it to compare the experimental results with the model's behavior. It is possible to simultaneously fit the $\Delta x_i^{\parallel}$ and $\Delta y_i^{\perp}$ distributions, but this comes at an increased computational cost. As a consistency check, after performing parameter estimation, we verify that the model accurately captures the $\Delta y_i^{\perp}$ distribution. If the model failed this consistency test, we could repeat the parameter estimation using both distributions. However, this was not required for any of the cases considered here.

We used a Monte Carlo method based on the Metropolis algorithm to perform parameter estimation. This was followed by local optimization algorithms to identify parameters associated with the global minimum error between the model and data (**Supplementary Information**). To test the accuracy and

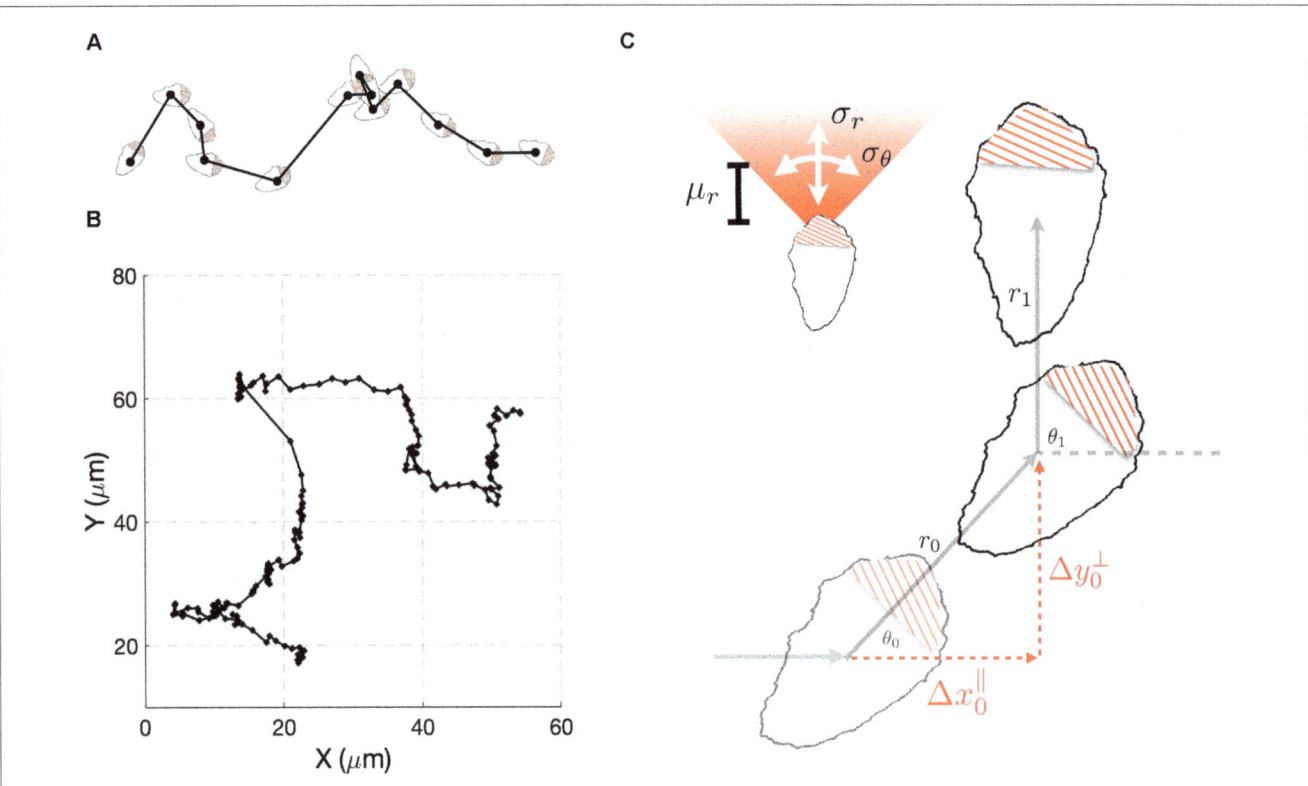

FIGURE 1 | A stochastic model for cell migration. **(A)** Cell tracks are constructed by recording the geometric center of the cell over time. **(B)** Example track resulting from tracking the cell centroid at 5 min intervals (black dots). **(C)** A stochastic model of migration in which during each time interval, a cell moves a distance r through and angle θ with respect to the direction of the previous step. The random variable r is taken to be normally distributed with mean μ_r and variance σ_r^2. The angle θ is also normally distributed with mean zero and variance σ_θ^2. To compare the model to experimental data we change variables to $\Delta x^{\parallel}$ and $\Delta y^{\perp}$, the directions parallel and perpendicular to previous step.

efficiency of this method, we benchmarked our approach using data generated from computational simulations of the stochastic model (**Supplementary Figure S2**). Having validated our computational methods, we next fit the model to the experimentally measured distributions. The model did not generate a good fit to experimental data for MEF cells (**Supplementary Figure S3**, dashed curve). In particular, we found that the model could not capture the second mode observed in the $\Delta x_i^{\parallel}$.

A Multistate Model for Cell Migration

Further inspection of the MEF cell tracks suggested that individual cells might exist in different modes of migration, distinguished by differences in speed and persistence. We therefore expanded our model to allow for different states of migration. That is, we hypothesized that at any given time a migrating cell is in one of n states denoted by S_i, with $i \in \{1 \dots n\}$. Each state is characterized by the parameters μ_r^i, σ_r^i, and σ_θ^i. The additional parameters, α^i, denoting the fraction of time spent in state i, are required to fully specify the model. Since $\sum \alpha^i = 1$, in the two-state case the total number of parameters is seven. Note that if a two-state model is fit to data consisting of only a single state, then we expect our Monte Carlo method to produce parameter sets in which α^1 takes on values of 0 or

1, or $\mu_r^1 = \mu_r^2$, $\sigma_r^1 = \sigma_r^2$, and $\sigma_\theta^1 = \sigma_\theta^2$. The extended model is essentially a mixture model, which is itself a reduced hidden Markov model under the assumption that the probabilities of transitioning between states are independent and identically distributed. We again used simulated data to validate the accuracy and efficiency of our Monte Carlo method when multiple states are considered (**Supplementary Figure S4**).

The multi-state model produced a good fit to the MEF $\Delta x_i^{\parallel}$ distribution (**Figure 2A**). To assess the accuracy of our parameter estimates we used confidence-interval profiling (Raue et al., 2009). To determine acceptable values for the sum of the squared errors (SSE) we boot-strapped the original datasets to assess plausible differences in our observed distributions should we repeat the experiments (**Supplementary Information**). The results of this analysis provide a measure of the confidence that should be placed on each estimated parameter value (**Supplementary Figures S5A,B**). Of particular interest is the parameter α which represents the fraction of time in each state. The best fits were achieved with $\alpha = 0.12$. We confirmed that the model also captured the distributions for $\Delta y_i^{\perp}$ (**Supplementary Figure S6A**). The results of our analysis suggest that randomly migrating MEFs exist in one of two states. About 12% of time these cells are in a state with a well-defined characteristic step of $\sim$3 μm (State 1 – blue distribution in **Figure 2A** left

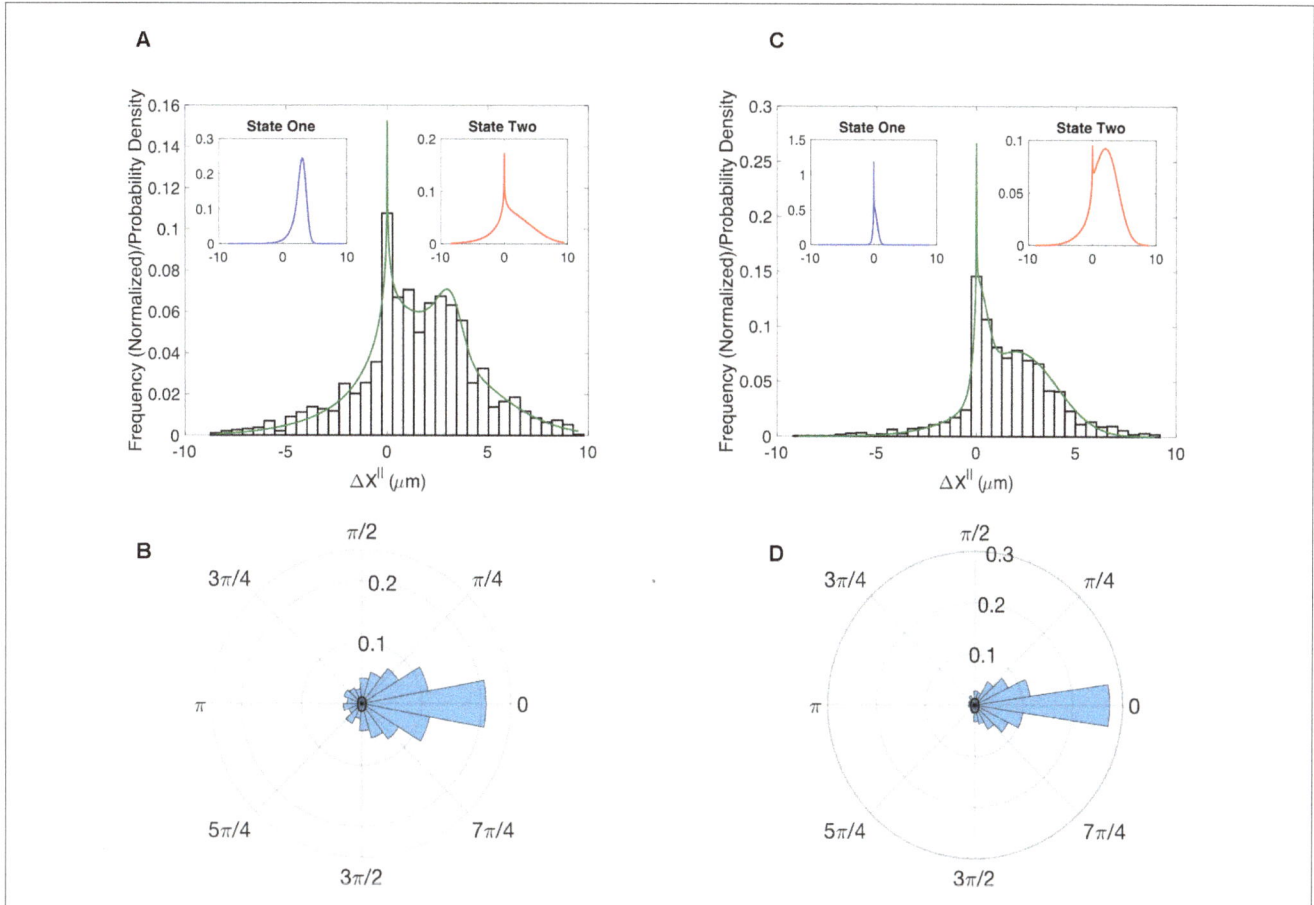

FIGURE 2 | Results for the multistate model of migration. **(A)** Comparison of the experimentally determined distribution of $\Delta x^{\parallel}$ for WT MEF cells (histogram) to the results of a two-state model (green curve). Insets show the distributions for the predicted two states. **(B)** Experimentally determined distribution for the angle θ for WT MEF cells. **(C)** Same as **(A)** except for cells in which RhoG has been knocked down. **(D)** Same as B except for cells in which RhoG has been knocked down.

inset) and an angular distribution with $\sigma_\theta^1 = 0.7$. In the second state, the step size is highly variable (State 2 – red distribution **Figure 2A** right inset) and the motion is less persistent $\sigma_\theta^2 = 1.3$. For completeness, we also show the distribution for the angle θ (**Figure 2B**).

RhoG's Role in Migration

It has long been appreciated that the canonical Rho-GTPases RhoA, Rac1, and Cdc42 play important roles in cell migration. However, the role of RhoG in migration is less well studied. To determine if RhoG plays a role in the random migration of MEFs, we generated time series data for cells in which this protein was knocked down. While the angular distributions for the WT and knockdown do not show clear differences (**Figures 2B,D**), the $\Delta x_{\parallel}$ distributions indicate RhoG does effect migration (**Figures 2A,C**). Moreover, by fitting our simple two-state model we can quantify this effect and ascertain that the persistent state 1 in the MEF control has been converted to a state in which the cells do not show significant movement ($\mu_r^1 = 3.2 \ \mu\text{m}$ for the WT to $\mu_r^1 = 0.25 \ \mu\text{m}$ for the KD). State 2 seems to be preserved by the KD in the sense that the confidence intervals defining state 2 parameters are overlapping in the two

cases (**Supplementary Figure S5**). A putative mechanism for how RhoG activation influences cell migration via recruitment of the DOCK180/ELMO complex (Katoh and Negishi, 2003; Katoh et al., 2006), which acts as GEF for Rac1. However, whether this is the key pathway in this process, and how it is organized spatio-temporally, is a direction of future research.

Inferring States From Time Series Data

We next sought to develop computational tools that could be used to determine if the predicted states of migration correspond to subpopulations of cells with distinct phenotypes or if individual cells could transition between states. To test if individual cells change their migration state, we developed a method to infer migration states from individual cell tracks. Our approach uses a Bayesian prediction method based on the probability that a sequence of k successive steps arises from one of the identified states (see Methods for details). Before applying our state prediction method on the experimental data, we first validated the approach using synthetic data. To generate this data, we performed computational simulations of the stochastic model using the parameters estimated from the experimental data for MEF cells. With these values our state-prediction algorithm correctly identified the states more than 90% of the time, validating the approach (**Supplementary Figure S7**).

Having demonstrated our method's ability to infer cell migration states from simulated track data and demonstrate a role for RhoG, we examined whether the different migration states could be correlated with molecular changes within cells. The Hahn lab has used HeLa cells to develop new biosensors and optogenetic probes. It is well established that these molecular tools must be used at controlled concentrations, below levels where they perturb cell movement (Kraynov et al., 2000; Machacek et al., 2009). Controls in earlier studies have shown that HeLa cells exhibit altered motility when the Rac1 biosensor is expressed at high levels. We decided to investigate if our stochastic modeling approach could quantify the effects of biosensor overexpression. We compared WT HeLa cells without biosensor expression to cells with high levels of Rac1 biosensor. Our analysis revealed that WT cells showed little directed motion and a single migration state was sufficient to capture the distributions of steps sizes (**Figures 3A,B**). In contrast, cells with highest levels of biosensor exhibited two states of migration

(**Figure 3C**). In particular, two states were needed to capture the long tail of the distribution (see **Supplementary Figure S9** for comparison of one-and two state results). In state 1 the cell moves in persistent manner, whereas in state 2 the cell is mostly stationary. To test if the predicted two states are correlated with differences in cell signaling, we ran our state prediction algorithm on the track data. Interestingly, our analysis predicted that individual cells randomly switch between the two states (**Figure 4A**), and qualitative observations indicated that the slow state showed multiple disperse Rac activation events at the edge of the cell, while the fast state showed a single Rac activation at the leading edge (see **Supplementary Movies M1, M2**). To quantify this observation, we identified and counted the number of foci of active Rac1 in each image and grouped these counts by the predicted state (see **Supplementary Information** for details), reasoning that random movement would require more cell protrusions distributed around the cell perimeter. Rac activation is known to be sufficient to generate cell protrusions (Wu et al., 2009; Wang et al., 2016). Cells predicted be in state 1, which corresponds to the fast-persistent state, had fewer Rac1 foci than

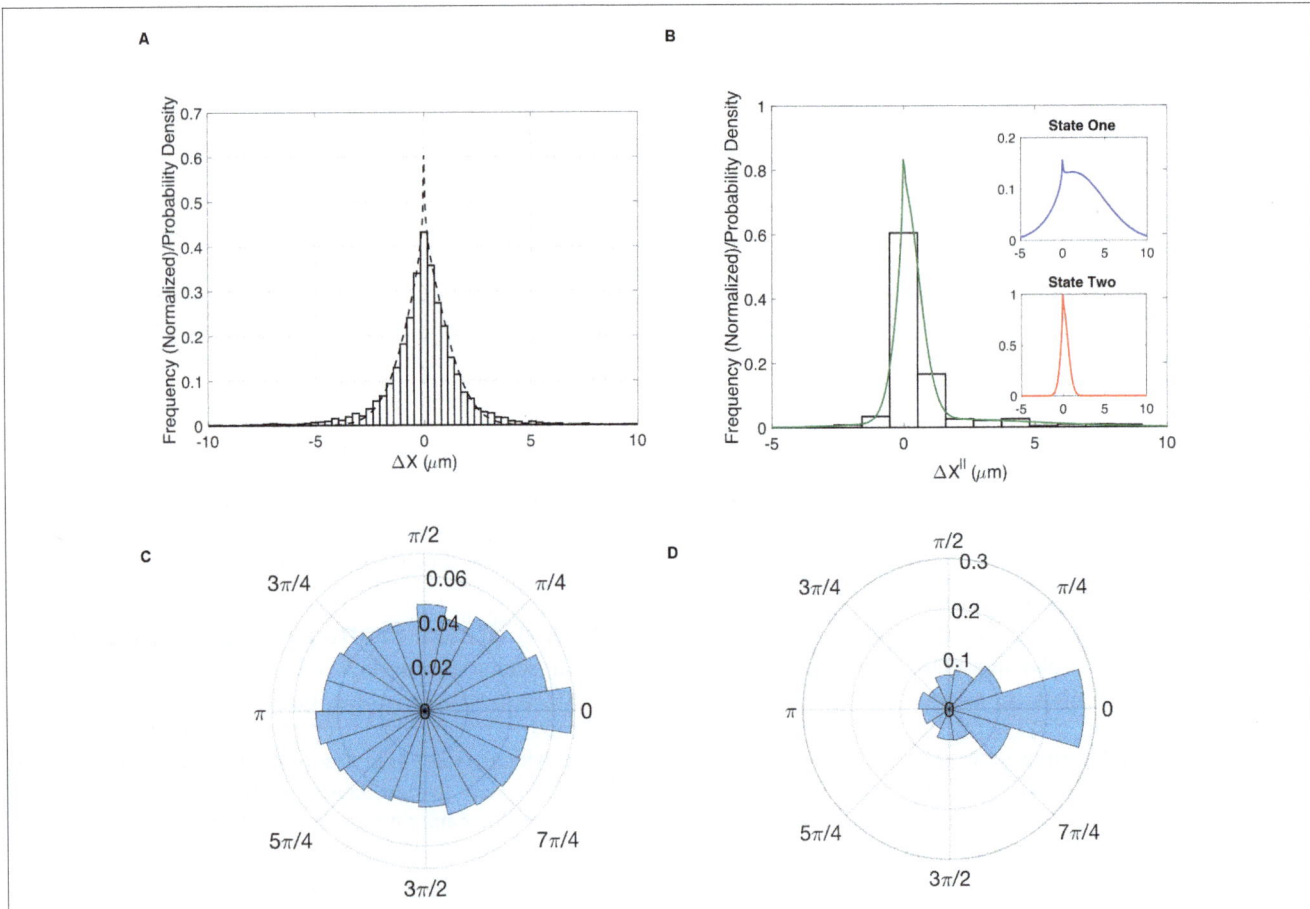

FIGURE 3 | Results for HeLa cells with and without expression of a Rac1 biosensor. **(A)** Comparison of the experimentally determined distribution of $\Delta x^{\parallel}$ for HeLa cells not expressing the biosensor (histogram) to the results of a one-state model (dashed curve). Insets show the distributions for the predicted two states. **(B)** Experimentally determined distribution for the angle θ for HeLa cells not expressing the biosensor. **(C)** Comparison of the experimentally determined distribution of $\Delta x^{\parallel}$ for HeLa cells expressing Rac1 biosensor (histogram) to the results of a two-state model (green curve). Insets show the distributions for the predicted two states. **(D)** Experimentally determined distribution for the angle θ for HeLa cells expressing biosensor.

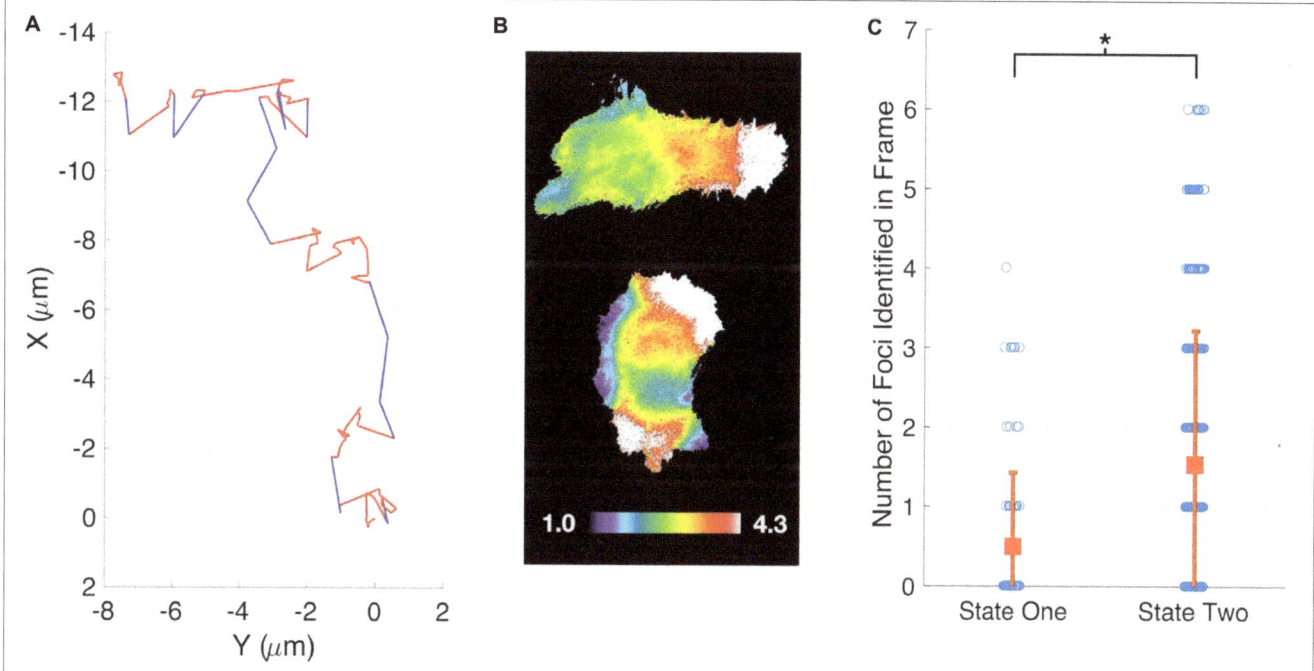

FIGURE 4 | State prediction for HeLa Cells expressing a Rac1 biosensor. **(A)** Example cell track showing switching between a fast-persistent state (blue) and slower-less persistent state (red). **(B)** When in the fast-persistent state (state 1, upper panel) cells have fewer active Rac1 foci as compared to the slower-less persistent state (state 2, lower panel). The upper cell is undergoing persistent motion to the right. The color scale indicates the ratiometric readout of Rac activation, normalized so the lowest 5% of the cell = 1. **(C)** Quantification of the number of foci in each state.

those predicted to be in state 2, which show little net movement (**Figure 4B**). This observation is consistent with highly motile cells typically showing strong polarity.

DISCUSSION

We developed novel computational methods for analyzing the movement of randomly migrating cells. Our approach combines stochastic modeling with statistical inference methods to detect and quantify migratory phenotypes. Migrating cells have a biochemical, morphological, and structural orientation that persists as these cells move. Our model captures this 'memory' by conditioning the cell's movement during the current time interval on its previous direction of motion. An important feature of our model is that analytic expressions for the probability densities for cell displacements parallel and perpendicular to the previous direction of motion can be found. This feature allows us to generate the probability density function for a given set of parameters rather than generating an approximation to this PDF via stochastic simulation of our migration model (**Figure 1C**). In most use cases the analytical PDF is computationally more efficient due to the high number of repeats required to estimate the PDF with sufficent accuracy. We have validated all our approaches using simulated data, and then applied the methodology to study randomly migrating MEF and HeLa cells.

Our modeling approach allows for multiple states of migration. This feature allowed us to demonstrate that migrating cells randomly transition between modes of movement. Crucial to the detection of these states is the quantification of parameter values and the associated confidence in those estimates. This process allowed us to be confident in the existence of two states of migration for MEF and HeLa cells over-expressing a Rac1 biosensor.

The identification of multiple states of migration for MEF cells led us to assess the role of RhoG in establishing these states. To do this we used siRNA to reduce RhoG expression. This perturbation suggests that RhoG plays a role in directed migration, because reducing RhoG eliminated net movement in the first predicted state and shortened the range of step sizes taken in the second state. We next developed a Bayesian approach to predict the current migration state of a cell from time series of the cell's position. Using this method, we demonstrated that individual HeLa cells expressing a Rac1 biosensor switched between migratory states. Importantly, we were able to correlate these two states with differences in the distribution of Rac1 activity.

We believe that our methods provide useful tools for quantifying and characterizing cell migration. Our stochastic model characterizes cell migration using parameters with straightforward biological interpretations. Hence, application of this model can lead to biological insights not apparent in the data from visual inspection or simple quantitative measures. In this case, our analysis suggests a role of RhoG in allowing cells to change direction, which may play a role in the ability of randomly migrating cells to search their environment.

MATERIALS AND METHODS

Computational Methods

The full code and analysis for this paper is available (Allen, 2020).

Coordinate Transformation

We modeled cell migration as a stochastic sequence of steps characterized by the step size r_i and directional angle θ_i (**Figure 1**). Since we assume r_i and θ_i to be realizations of independent random variables R and Θ the probability the cell moves (r, θ) is defined by

$$f(r, \theta|\theta_{i-1}) = g_R(r).g_\Theta(\theta|\theta_{i-1}) \qquad (1)$$

where $g_R(r)$ is the probability density function (pdf) for the step magnitude, which we take to have the normal distribution $\mathcal{N}(\mu_r, \sigma_r^2 \Delta t)$, and $g_\Theta(\theta|\theta_{i-1})$ is the pdf generating the new orientation conditioned on the previous angle, which we take to have the normal distribution $\mathcal{N}(\theta_{i-1}, \sigma_\theta^2 \Delta t)$. The experimental data is collected in Cartesian coordinates (X, Y). In principle we could transform the data into the coordinates R and Θ. However this transformation cannot be completed uniquely, because there is no way to distinguish a backward step in which the cell maintains its direction of polarity ($\theta_i = \theta_{i-1}$) from one in which the front and back of the cell have reversed ($\theta_i = \theta_{i-1} + (2k + 1)\pi$). Furthermore, the value of θ_i cannot be determined if $r_i = 0$. For these reasons, we transform the model to the coordinates $(\Delta x_i^{\parallel}, \Delta y_i^{\perp})$, where these new variables correspond to changes in the centroid's position during the ith time interval that are parallel and perpendicular to the direction of the previous step.

To compare with the model the data needs to be manipulated to generate histograms for steps in the $x^{\parallel}$ and $y^{\perp}$ directions. For each sequential triplet of coordinates $\{(x_{i-1}, y_{i-1}), (x_i, y_i), (x_{i+1}, y_{i+1})\}$, we rotate the steps as a rigid body about (x_{i-1}, y_{i-1}) by a four quadrant inverse tangent based on $\tan^{-1} \frac{y_i - y_{i-1}}{x_i - x_{i-1}}$. The result of this is that all steps are pre-oriented in a positive x-direction and initiated at $(0, 0)$, and can be plotted as histograms of step distance in the x and y direction: $(\Delta x_i^{\parallel}, \Delta y_i^{\perp}) = (x_{i+1}' - x_i', y_{i+1}' - y_i')$.

The pdf for $\Delta x^{\parallel}$ is:

$$f_X(\Delta x^{\parallel}) = \int_{-1}^{1} \frac{f_H(\Delta x^{\parallel}/h, h)}{h} dh \qquad (2)$$

where $h = \cos(\theta)$, and,

$$f_H(\Delta x^{\parallel}/h, h) = g_r(\Delta x^{\parallel}/h) \sum_k \frac{1}{(1 - h^2)^{\frac{1}{2}}} \times$$

$$(g_\Theta(\arccos(h) + 2\pi k) + g_\Theta(-\arccos(h) + 2\pi k)). \qquad (3)$$

The expression for $f_Y(\Delta y^{\perp})$ is similar, however now with $g_\Theta(\arccos(h) + \pi/2 + 2\pi k) + g_\Theta(-\arccos(h) + \pi/2 + 2\pi k)$ in the summation term. A derivation of these results is presented in the **Supplementary Information**.

Parameter Estimation

Parameters were estimated by simulated annealing, which is a Monte Carlo method based on the Metropolis algorithm (24, 25).

Initial choices of parameters generate an analytical solution (Eq. 2), which is scored against the experimental data ($\Delta x_i^{\parallel}$) by the sum of least squared differences. At each step of the algorithm the parameters are updated by a small addition of Gaussian noise, if this update scores better than the current score then these parameters are accepted. If the score is higher, the parameter set is accepted with probability $e^{-\frac{\Delta s}{T}}$, where Δs is the difference between the current and previous scores and T is the current temperature. Over the course of the fitting T, the temperature is reduced. This fixes the parameter choices into a local minimum. Here we choose a geometric cooling regime. Due to the stochastic nature of the simulation, and that there could be many local minima, it is necessary to run this fitting procedure multiple times. The best fit of this routine was then further refined using MATLABs fmincon routine, which was also used to assess the sensitivity of our fit to altering parameter values via confidence-interval profiling (**Supplementary Figure S4**, **Supplementary Material** for details).

The histograms were amalgamated from multiple cell tracks. For the case of two states, the pdf for Δx becomes

$$f_X(\Delta x) = \alpha f_X^1(\Delta x) + (1 - \alpha)f_X^2(\Delta x)$$

where α is the fraction of time spent in state 1 and the distributions $f_X^1(\Delta x)$ and $f_X^2(\Delta x)$ are parameterized by $(\mu_r^1, \sigma_r^1, \sigma_\theta^1)$ and $(\mu_r^2, \sigma_r^2, \sigma_\theta^2)$, respectively.

Parameter sets were identified by multiple simulated annealing runs, followed by local-optimization routines.

Validation of Methods

To validate the pdfs and the parameter estimation algorithm, we simulated cell tracks using the stochastic model (**Figure 1**). Cell tracks were generated using two states, each with distinct parameter sets. At each step a state was chosen at random with probability 0.5. As above, the simulated cell tracks were used to construct the distributions for $\Delta x_i^{\parallel}$ and $\Delta y_i^{\perp}$. We assumed model parameters were not known and used the Monte Carlo method to fit Eq. 2, modified to two states (see below) to the simulated data for $\Delta x_i^{\parallel}$. The Monte Carlo method quickly converged on the correct parameter values (**Supplementary Figure S4**), validating the analytical solution to the model and our fitting procedure. In theory we also could fit the pdf for $\Delta y_i^{\perp}$. However, the pdf for $\Delta y_i^{\perp}$ is symmetric, because there is no preferred direction of migration and therefore less informative than the distribution for $\Delta x_i^{\parallel}$. We found that we could maintain the accuracy of our parameter estimation while improving the computational cost by only considering the $\Delta x_i^{\parallel}$ distribution. As a consistency check, we always verify that the estimated parameters accurately reproduce the pdfs for $\Delta y_i^{\perp}$ (**Supplementary Figure S6**).

State Prediction

To identify which state a cell is in at a given time, we used Bayes' theorem to invert the problem. That is, we calculate the probability that a cell is in state S_i given the experimental data. Note that in calculating this probability, we also get the false positive rate or p-value. To make a reliable prediction of

S_i may require an n-step window, where n is odd, such that, $\{x_{i-n/2}, \ldots, x_{i-1}, x_i, x_{i+1}, \ldots, x_{i+n/2}\}$. Then:

$$P(s_i|X) = \frac{P(X|s_i)P(s_i)}{P(X|s_i)P(s_i) + P(X|s_i^c)P(s_i^c)}$$

where $P(X|s_i)$ is calculated from the model, and we take $P(s_i) = \alpha$. Windows of length one, three and five were tested. For the case presented here, we found that the window of length one produced results similar to the other two window lengths.

Foci Identification

Ratiometric images of the FRET based Rac1 biosensor were analyzed for localized regions of higher Rac1 activity near the periphery of the cell. We call these regions "foci". We used custom application of the image processing toolbox in MATLAB to identify foci, which we define as contiguous regions within the cell that were simultaneously: (1) 60% above the average intensity of the cell, (2) greater than 100 pixels in area, and (3) contained at least one point within 5 pixels of the cell edge. The length of time (or number of frames) that a cell could be followed for varied. So, to not overweight any one cell, the number of image frames analyzed, n, was selected to maximize $n \times m$ where m is the number of cells with at least n images.

Experimental Methods
Cell Culture and Transient Transfections

HeLa cells were maintained in Dulbecco's modified Eagle's Medium (DMEM) (Cellgro) supplemented with 10% fetal bovine serum (FBS) (HyClone), 100 U/mL penicillin and 100 μg/mL streptomycin (Cellgro) and 2 mM L-glutamine (Invitrogen) at 37°C and 5% CO_2. All cDNA constructs were transfected into cells using FuGene6 (Roche) according to the manufacturer's instructions. IA32 Mouse Embryonic Fibroblast (MEF) cells were maintained in Dulbecco's modified Eagle's Medium (DMEM) (Cellgro) supplemented with 10% fetal bovine serum (FBS) (HyClone) and 1× GlutaMAX (Thermo Fisher Scientific).

IA32MEFs were transfected with either RhoG siRNA (CAGGTTTACCTAAGAGGCCAA) or Allstars Negative Control siRNA (Qiagen, United States). 7.5 μL, 10 μM siRNA was added to 250 μL serum-free DMEM. 3 μL lipofectamine RNAimax was added to another 250 μL serum-free DMEM. After 5 min, the two solutions were mixed and incubated for 20 min, followed by dropwise addition to a 35 mm dish. Medium was changed after 24 h and cells were split as required for use in experiments 48–72 h post-transfection, when knock-down efficiency was maximal. Control siRNA cells were incubated with 5 μM CFDA green for 20 min in serum-free DMEM. CFDA-labeled control cells were mixed with unlabeled RhoG siRNA cells immediately prior to the experiment.

Live Cell Imaging

For live cell imaging, cells were plated on fibronectin-coated coverslips (10 μg/ml fibronectin) 4 h before imaging, then transferred to Ham's F12-K imaging medium supplemented with 2% FBS and 15 mM HEPES. Live cell imaging was performed in a closed heated chamber (20/20 Bionomic).

For biosensor imaging, photobleach-corrected time-lapse image stacks were acquired for 18 h at 5 min intervals and processed as previously described (Pertz et al., 2006; Machacek et al., 2009; Hodgson et al., 2010). The following filter sets were used (Chroma Technology Corp.): CFP: D436/20, D470/40; FRET: D436/20, ET535/30; YFP: D500/20, ET535/30. Cells were illuminated with a 100 W Hg arc lamp through a 1.0 neutral density filter.

For RhoG siRNA experiments cell tracks were generated through 10× DIC imaging of cells plated as above, but using Ham's F12K medium supplemented with 5% FBS. Images were acquired for at least 70 frames at 10 min intervals in a closed, heated chamber. This length of track was objectively identified as optimal by maximizing the total number of analyzed frames in the entire data set.

DATA AVAILABILITY STATEMENT

The raw data supporting the conclusions of this article will be made available by the authors, without undue reservation.

AUTHOR CONTRIBUTIONS

RA, TE, and KH devised the research plan and wrote the manuscript. RA performed the mathematical and computational calculations. CW and NP collected the data. All authors contributed to the article and approved the submitted version.

FUNDING

The work was supported by grants R35GM127145 (TE), R35GM122596 (KH), and W911NF-15-1-063 (TE and KH). A previous version of this manuscript has been released as a pre-print at BioRxiv (Allen et al., 2018).

SUPPLEMENTARY MATERIAL

FIGURE S1 | Cumulative distribution functions (CDFs) for x and y step sizes compared to the CDFs for normal distributions.

FIGURE S2 | Validation of analytical approach. Simulated data (histogram) is generated by simulating data (blue bars) stochastically with parameters $(\mu_r, \sigma_r, \sigma_\theta, \sigma_0) = (0.1, 0.1, 0.5, \pi/5)$. Comparison with the analytical PDF (red) gives good agreement.

FIGURE S3 | Fitting the model to MEF WT cells indicates that a one state model of migration cannot adequately capture key features of the data. Observed data (open bars), model fit (green).

FIGURE S4 | Analytics and Fitting Validation. Simulated data (histogram) is generated by applying the model with parameters $(\mu_r^1, \sigma_r^1, \sigma_\theta^1) = (1, 0.3, 3)$, $(\mu_r^2, \sigma_r^2, \sigma_\theta^2) = (5, 1, 0.5)$ and $\alpha = 0.5$. Assuming the underlying parameters were unknown, we used simulated annealing to fit the analytical solution (green line). This validates our analytical solution and fitting procedure (we estimate $(\hat{\mu}_r^1, \hat{\sigma}_r^1, \hat{\sigma}_\theta^1) = (1.02, 0.30, 2.72)$, $(\hat{\mu}_r^2, \hat{\sigma}_r^2, \hat{\sigma}_\theta^2) = (5.08, 0.93, 0.51)$, and $(\alpha = 0.50)$.

FIGURE S5 | Parameter values and associated confidence intervals. For each condition, and each parameter, the best-fit is the minimum of the blue curve. Acceptable parameter values are those which can lead to a value below the threshold (red dots) when the other parameters are re-fit, highlighted by the red-dash lined. Note, that two states were clearly identified in both cases.

FIGURE S6 | Model prediction versus the ΔY distribution from the cases shown in **Figure 2**. **(A)** MEF WT **(B)** MEF RhoG KD. In **(A)** and **(B)**: Observed data (open bars), model fit (green) and individual pdfs for state one and state two (insets).

FIGURE S7 | Testing the accuracy of predicting states using HeLa Rac1 model parameters. Data simulated stochastically, where we know the state a given step was generated from, is compared with our prediction of the state using Bayes Theorem. Overall, we are correct more than 90% of the time. However, steps truly in state one get occasionally mischaracterized are overlapping leading to an accuracy of around 70% in this case.

FIGURE S8 | Bootstrapped Analysis of the Difference in Foci Between States. To ascertain if the difference we observed in the mean foci count in state one versus state two was meaningful or could have been observed by chance, we performed a boostrap analysis. For a given iteration of the bootstrap, we randomly assigned state one or state two to each image frame with probability α and $1 - \alpha$, respectively. Then, for each iteration, calculated Δ mean foci as the difference in the mean number of foci in state one versus state two. Blue bars, bootstrapped distribution (50,000 iterations), dotted red line observed Δ mean foci.

FIGURE S9 | Comparison of 1-state (dashed curves) and 2-state (green curves) model fits for all of the experimental results.

FIGURE S10 | Overview of algorithm to estimate model parameters and confidence intervals.

FILES M1 | M1 (slow_state_hela_rac1.mp4) and M2 (fast_state_hela_rac1.mp4). The predicted "fast state" of migration was correlated with fewer Rac1 foci, most frequently a single focus at the leading edge, and in the slow state multiple brief Rac1 activation foci appear.

REFERENCES

Allen, R. J. (2020). *Cell Tracking Project*. Available at: https://github.com/rallen81/CellTracking (accessed January 17, 2018).

Allen, R. J., Welch, C., Pankow, N., Hahn, K., and Elston, T. C. (2018). Stochastic methods for inferring states of cell migration. *bioRxiv [Preprint]*. doi: 10.1101/249656

Arrieumerlou, C., and Meyer, T. (2005). A local coupling model and compass parameter for eukaryotic chemotaxis. *Dev. Cell* 8, 215–227. doi: 10.1016/j.devcel.2004.12.007

Cain, R. J., and Ridley, A. J. (2012). Phosphoinositide 3-kinases in cell migration. *Biol. Cell* 101, 13–29. doi: 10.1042/bc20080079

Dieterich, P., Klages, R., Preuss, R., and Schwab, A. (2008). Anomalous dynamics of cell migration. *Proc. Natl. Acad. Sci. U.S.A.* 105, 459–463. doi: 10.1073/pnas.0707603105

Dimilla, P. A., Quinn, J. A., Albelda, S. M., and Lauffenburger, D. A. (1992). Measurement of individual cell migration parameters for human tissue cells. *AIChE J.* 38, 1092–1104. doi: 10.1002/aic.690380712

Finney, A. C., Stokes, K. Y., Pattillo, C. B., and Orr, A. W. (2017). Integrin signaling in atherosclerosis. *Cell Mol. Life Sci.* 74, 2263–2282. doi: 10.1007/s00018-017-2490-4

Franca-Koh, J., Kamimura, Y., and Devreotes, P. N. (2007). Leading-edge research: PtdIns(3,4,5)P3 and directed migration. *Nat. Cell Biol.* 15–17. doi: 10.1038/ncb0107-15

Goley, E. D., and Welch, M. D. (2006). The ARP2/3 complex: an actin nucleator comes of age. *Nat. Rev. Mol. Cell Biol.* 7, 713–726. doi: 10.1038/nrm2026

Hall, A. (2009). The cytoskeleton and cancer. *Cancer Metastasis Rev.* 28, 5–14.

Hodgson, L., Shen, F., and Hahn, K. (2010). Biosensors for characterizing the dynamics of rho family GTPases in living cells. *Curr. Protoc. Cell Biol.* 14, Unit 14.11.1–26. doi: 10.1002/0471143030.cb1411s46

Iden, S., and Collard, J. G. (2008). Crosstalk between small GTPases and polarity proteins in cell polarization. *Nat. Rev. Mol. Cell Biol.* 9, 846–859. doi: 10.1038/nrm2521

Jaffe, A. B., and Hall, A. (2005). "Rho GTPases: biochemistry and biology. *Annu. Rev. Cell Dev. Biol.* 21, 247–269.

Katoh, H., Hiramoto, K., and Negishi, M. (2006). Activation of Rac1 by RhoG regulates cell migration. *J. Cell Sci.* 119, 56–65. doi: 10.1242/jcs.02720

Katoh, H., and Negishi, M. (2003). RhoG activates Rac1 by direct interaction with the Dock180-binding protein Elmo. *Nature* 424, 461–464. doi: 10.1038/nature01817

Kraynov, V. S., Chamberlain, C., Bokoch, G. M., Schwartz, M. A., Slabaugh, S., and Hahn, K. M. (2000). Localized Rac activation dynamics visualized in living cells. *Science* 290, 333–337. doi: 10.1126/science.290.5490.333

Ladwein, M., and Rottner, K. (2008). On the Rho'd: the regulation of membrane protrusions by Rho-GTPases. *FEBS Lett.* 582, 2066–2074. doi: 10.1016/j.febslet.2008.04.033

Lemarié, C. A., Tharaux, P. L., and Lehoux, S. (2010). Extracellular matrix alterations in hypertensive vascular remodeling. *J. Mol. Cell. Cardiol.* 48, 433–439. doi: 10.1016/j.yjmcc.2009.09.018

Machacek, M., Hodgson, L., Welch, C., Elliott, H., Pertz, O., Nalbant, P., et al. (2009). Coordination of Rho GTPase activities during cell protrusion. *Nature* 461, 99–103. doi: 10.1038/nature08242

Marston, D. J., Anderson, K. L., Swift, M. F., Rougie, M., Page, C., Hahn, K. M., et al. (2019). High Rac1 activity is functionally translated into cytosolic structures with unique nanoscale cytoskeletal architecture. *Proc. Natl. Acad. Sci. U. S. A.* 116, 1267–1272.

Othmer, H. G., Dunbar, S. R., and Alt, W. (1988). Models of dispersal in biological systems. *J. Math. Biol.* 26, 263–298. doi: 10.1007/bf00277392

Pertz, O., Hodgson, L., Klemke, R. L., and Hahn, K. (2006). Spatiotemporal dynamics of RhoA activity in migrating cells. *Nature* 440, 1069–1072. doi: 10.1038/nature04665

Petrie, R. J., Doyle, A. D., and Yamada, K. M. (2009). Random versus directionally persistent cell migration. *Nat. Rev. Mol. Cell Biol.* 10, 538–549. doi: 10.1038/nrm2729

Raue, A., Kreutz, C., Maiwald, T., Bachmann, J., Schilling, M., Klingmüller, U., et al. (2009). Structural and practical identifiability analysis of partially observed dynamical models by exploiting the profile likelihood. *Bioinformatics* 25, 1923–1929. doi: 10.1093/bioinformatics/btp358

Ridley, A. J. (2006). Rho GTPases and actin dynamics in membrane protrusions and vesicle trafficking. *Trends Cell Biol.* 16, 522–529. doi: 10.1016/j.tcb.2006.08.006

Ridley, A. J. (2015). Rho GTPase signalling in cell migration. *Curr. Opin. Cell Biol.* 36, 103–112. doi: 10.1016/j.ceb.2015.08.005

Rosello, C., Ballet, P., Planus, E., and Tracqui, P. (2004). Model driven quantification of individual and collective cell migration. *Acta Biotheor.* 52, 343–363. doi: 10.1023/b:acbi.0000046602.58202.5e

Rottner, K., Hall, A., and Small, J. V. (1999). Interplay between rac and rho in the control of substrate dynamics. *Curr. Biol.* 9, 640–648.

Schaks, M., Giannone, G., and Rottner, K. (2019). Actin dynamics in cell migration. *Essays. Biochem.* 63, 483–495.

Svensson, C. M., Medyukhina, A., Belyaev, I., Al Zaben, N., and Figge, M. T. (2018). Untangling cell tracks: quantifying cell migration by time lapse image data analysis. *Cytometry* 93, 357–370.

Wang, H., Vilela, M., Winkler, A., Tarnawski, M., Schlichting, I., Yumerefendi, H., et al. (2016). LOVTRAP: an optogenetic system for photoinduced protein dissociation. *Nat. Meth.* 13, 755–758.

Wu, Y. I., Frey, D., Lungu, O. I., Jaehrig, A., Schlichting, I., Kuhlman, B., et al. (2009). A genetically encoded photoactivatable Rac controls the motility of living cells. *Nature* 461, 104–108.

10

H9N2 Avian Influenza Virus Protein PB1 Enhances the Immune Responses of Bone Marrow-Derived Dendritic Cells by Down-Regulating miR375

Jian Lin[1], Jing Xia[1], Chong Z. Tu[2], Ke Y. Zhang[1], Yan Zeng[1] and Qian Yang[1]*

[1] *Department of Zoology, College of Life Science, Nanjing Agricultural University, Jiangsu, China,* [2] *Department of Histoembryology, College of Veterinary Medicine, Nanjing Agricultural University, Jiangsu, China*

Edited by:
Diana Bahia,
Universidade Federal de Minas Gerais,
Brazil

Reviewed by:
Hridayesh Prakash,
University of Hyderabad, India
David MacHugh,
University College Dublin, Ireland

*Correspondence:
Qian Yang
zxbyq@njau.edu.cn

Polymerase basic protein 1 (PB1), the catalytic core of the influenza A virus RNA polymerase complex, is essential for viral transcription and replication. Dendritic cells (DCs) possess important antigen presenting ability and a crucial role in recognizing and clearing virus. MicroRNA (miRNA) influence the development of DCs and their ability to present antigens as well as the ability of avian influenza virus (AIV) to infect host cells and replicate. Here, we studied the molecular mechanism underlying the miRNA-mediated regulation of immune function in mouse DCs. We first screened for and verified the induction of miRNAs in DCs after PB1 transfection. Results showed that the viral protein PB1 down-regulated the expression of miR375, miR146, miR339, and miR679 in DCs, consistent with the results of H9N2 virus treatment; however, the expression of miR222 and miR499, also reduced in the presence of PB1, was in contrast to the results of H9N2 virus treatment. Our results suggest that PB1 enhanced the ability of DCs to present antigens, activate lymphocytes, and secrete cytokines, while miR375 over-expression repressed activation of DC maturation. Nevertheless, PB1 could not promote DC maturation once miR375 was inhibited. Finally, we revealed that PB1 inhibited the P-Jnk/Jnk signaling pathway, but activated the p-Erk/Erk signaling pathway. While inhibition of miR375 -activated the p-Erk/Erk and p-p38/p38 signaling pathway, but repressed the P-Jnk/Jnk signaling pathway. Taken together, results of our studies shed new light on the roles and mechanisms of PB1 and miR375 in regulating DC function and suggest new strategies for combating AIV.

Keywords: H9N2 AIV, PB1, miRNA, dendritic cells, immune regulation

INTRODUCTION

The influenza virus contains eight segments of a single-stranded RNA genome with negative polarity. The H9N2 subtype avian influenza virus (AIV), classified as a low pathogenic AIV, has high genetic variability and has shown both increases in virulence and ability to cross the host barrier (Peiris et al., 1999; Jin et al., 2014; Shaib et al., 2014; Zhou et al., 2014). Since the H7N9 and

H10N8 AIV outbreaks in 2013 resulted from recombination between H9N2 and other influenza subtypes, H9N2 AIV is a subject of intense research (Fang et al., 2013; Pu et al., 2015). The virus polymerase complex of H9N2 AIV, consisting of the polymerase basic protein 1 (PB1), polymerase basic protein 2 (PB2) and polymerase acidic protein (PA) subunits, has been reported to charge for catalyzing both viral RNA genome replication and transcription. Previous studies have shown that PB1 serves as a core subunit to incorporate PA and PB2 into the polymerase complex by directly interacting with PA and PB2 (Hemerka et al., 2009). Recently studies also reveal that PB1 interacted with PA was an attractive target for drug treatment (Massari et al., 2016; Swale et al., 2016).

There is a continuing need for novel anti-influenza therapeutics using new targets or creative strategies and the pathogenicity of a virus is determined not only by its characteristics but also by the host immune response (O'Donnell and Subbarao, 2011). Dendritic cells (DCs) is a professional and effective antigen-presenting cells in the innate immune response (Tucci et al., 2014). Influenza virus is a human pathogen and also naturally infects a large range of animals. Previous studies have found that the mouse model of influenza virus infection is useful for understanding host immune responses and host-pathogen interaction. Zhou found that miR2911, a honeysuckle (HS)-encoded atypical microRNA (miRNA), directly targets Influenza A virus with various subtypes (Zhou et al., 2015). Also, Isakova-Sivak and colleagues have used the mouse model to study the infectivity, immunogenicity and cross-protective efficacy of live attenuated influenza vaccines containing nucleoprotein from cold-adapted or wild-type influenza virus (Isakova-Sivak et al., 2017). Consequently, because human blood and humanized mice were not available for our studies, we developed a suitable animal model to study AIV responses in mammalian DCs. AIV infection affects the maturation, antigen presenting ability, and cytokine secretion of DCs (Lin et al., 2014). The binding of pathogen-associated molecular patterns to receptors expressed by DCs may activate DCs (Lopez et al., 2004; Liang et al., 2013), but it remains unclear how AIVs produce changes in DCs and how DCs respond to AIV infection.

MicroRNA (miRNA) have emerged as key regulators of innate immunity and modulate the ability of DCs to present antigens and secrete cytokines (Gantier et al., 2007; Cheng et al., 2012; Smyth et al., 2015). For example, miR-24 has been shown to be regulated during macrophage and dendritic cell differentiation potentiates innate immunity (Fordham et al., 2015). MicroRNA-146a reported to regulate human dendritic cell apoptosis and cytokine production by targeting TRAF6 and IRAK1 proteins (Park et al., 2015). Furthermore, AIV infection leads to the differential expression of cellular miRNA in chickens and mice, and miR491 and miR654 inhibit the replication of H1N1 virus through binding to PB1 in MDCK cells (Song et al., 2010). Since miRNA have the ability to modulate DC function, and our previous research work demonstrated that H9N2 AIV significantly influenced miRNA expression of DCs, the purpose of our study was to investigate the role miRNA play in regulating the immune response of DCs to PB1 stimulation.

RESULTS

miRNA Expression following H9N2 AIV Infection and Viral Fragment Transfection

To study how H9N2 might control miRNA expression, three segments of H9N2 AIV (PB1, PA, and NP) were cloned into pcDNA3.1 and transfected into bone marrow-derived DCs (BMDCs) (**Supplementary Image 1**). The expression of selected miRNAs was then examined by reverse transcription quantitative real-time PCR (RT-qPCR). Interestingly, for all of the miRNAs up-regulated by H9N2, PB1, PA, and NP significantly repressed their expression. For the miRNAs down-regulated by H9N2, the PB1 segment also mostly reduced their expression, especially for miR339, miR375, and miR146 (**Figures 2A,B**). Segments PB1, PA, and NP are involved in the transcription and replication of the AIV RNA genome (Bouvier and Palese, 2008).

Activation of Mouse BMDCs by PB1

We first examined the phenotypic changes in BMDCs transfected with PB1, PA, and NP. The transfection efficiency of PB1, PA, and NP was detected by RT-qPCR, and results are listed in **Supplementary Data Sheet 1**. Fluorescence-activated cell sorting (FACS) suggested that the mean fluorescence intensity (MFI) of MHCII was significantly enhanced by PB1, as were the co-stimulatory molecules CD40, CD80, and CD86 (**Figures 3A,B**). PA segment up-regulated CD80 and CD86, whilst NP had no effect on them (**Figures 3A,B**). Next, we assessed the ability of DCs to activate T lymphocytes and secrete cytokines. As shown in **Figures 3C,D**, PB1-stimulated DCs showed enhanced stimulation at a ratio of 1:1, and they expressed higher levels of interleukin-6 (IL-6) and tumor necrosis factor-α (TNF-α) than did the pcDNA3.1-transfected controls ($P < 0.05$). Whilst the LPS stimulation (positive control) were demonstrated to enhance the ability of DCs to present antigens, activate T lymphocytes, and secrete cytokines.

The Immune Function of miR375 and miR181b in Regulating Mice BMDCs

Recent studies have shown that miRNA regulate the immune responses of BMDCs (Wu et al., 2012). As H9N2 and PB1 significantly down-regulated the expression of miR375 (**Figure 2**), we examined the functions of miR375 in BMDCs. MiRNA over-expression vectors were constructed and validated by digesting with restriction enzyme Hind III and BamHI (**Supplementary Image 2**). FACS showed that miR375 over-expression decreased the percentage of CD80-, CD86-, CD40-, and MHCII when compared with the pSilencer4.1 group. While inhibition of miR375 greatly down-regulated the MFI of CD80 and CD86, but up-regulated the MFI of CD40 and MHCII when compared with the blank group (**Figures 4A,B**). Moreover, results also shown that the abundance of miR181b significant down-regulated all the surface -markers except MHC-II.). Interestingly, we found that the inhibition of miR181b significant increased the MFI of MHCII and CD-86 when compared with the blank group (**Figures 4A,B**).

Inhibition of Endogenous miR375 and miR181b Blocked PB1-Induced Phenotypic Alterations in BMDCs

Previous studies demonstrated that PB1 and a number of miRNAs, including miR375 and miR181b, can influence the phenotype of BMDCs. Thus, we investigated whether the PB1-induced changes in DCs are mediated by miRNAs. To test this hypothesis, miRNA inhibitors were added to DCs to repress endogenous miRNAs before PB1 transfection. FACS revealed that the inhibition of endogenous miR375 and miR181b decreased the expression of co-stimulatory molecules (CD80/CD86 and CD40) and MHCII, which was induced by PB1 ($P < 0.05$; **Figures 4C,D**).

Effects on Signaling Pathways Stimulated by PB1, miR375 and miR181b

Mitogen-activated protein kinase (MAPK) pathways exist in all eukaryotes and control a wide range of cellular processes, such as proliferation, differentiation, and survival (Kolch, 2000). We previously demonstrated that H9N2 AIV activates interferon (IFN) regulatory factor (IRF)-7. In this study, we first found that PB1 not only significant activated the p-Erk/Erk signaling pathway, but also inhibited the P-Jnk/Jnk signaling pathway when compared with the pcDNA3,1 group. Then, we found that over expressed miR375 significant decreased the expression level of p-IkBa/IkBa and IRF7, whilst the over expression of miR181b significant decreased the expression level of p-IkBa/IkBa, IRF3 and IRF7, but significant increased the expression level of p-Erk/Erk. Furthermore, we found that the inhibition of miR375 hugely up-regulated the expression level of p-p38/p38, p-Erk/Erk and IRF7, but down-regulated the expression level of p-IkBa/IkBa and p-Jnk/Jnk which suggested activation of P38, Erk, NF-κB, and IFN signaling pathway. Finally, the inhibition of miR181b significant repressed the p-Jnk/Jnk signaling pathway when compared with the blank group (**Figures 5A,B**).

DISCUSSION

The interactions between miRNA and DCs are important for AIV infection. In this report, we focused on miR375, which exhibited decreased expression in the PB1 segment stimulated group. DCs play an important role in the generation and maintenance of immune responses (Smyth et al., 2015). There are three standards for evaluating the immune function of BMDCs, including phenotypic alterations, the ability to activate T lymphocytes and the ability to secrete cytokines (Banchereau et al., 2000). Our study suggests that the immune function of DCs, including phenotypic alteration, T lymphocyte activation, and cytokine secretion, was greatly stimulated by PB1. Also, we demonstrated that expression of miR375 was suppressed by PB1.

MiRNAs repress key regulatory components of the innate immune response and markedly affected the capacity of DCs to present antigens and secrete cytokines. MicroRNA-375 was observed to influence cell proliferation, apoptosis and differentiation through the Notch signaling pathway, while microRNA-181b modulated the secretion of TNF-α and IL-1β in macrophages (Zhang et al., 2015; Wang et al., 2016). Our

research suggests that increased expression of miR375 attenuated the DC immune responses induced by PB1. PB1 over-expression increased the levels of CD80-, CD86-, MHCII-, and CD40 in cultured BMDCs, whilst the inhibition of endogenous miR375 had the opposite effect (**Figures 4C,D**). MiR375 also modestly decreased the expression of CD80; this effect was reversed by inhibiting endogenous miR375. Also, miR181b repressed the expression of CD40 and MHCII; this inhibition effect was relieved when endogenous miR181b was silenced. All four surface markers are characteristics of fully mature DCs and represent different functionalities (Geissmann et al., 2010). The major histocompatibility complex class II (MHCII) are family of molecules normally found on antigen-presenting cells such as DCs and mononuclear phagocytes. The MHCII-dependent pathway of antigen presentation is called the exogenous pathway, which has been shown to be regulated by miRNAs (Tomasi et al., 2010). Over-expression of PB1 enhances the ability of DCs to express the surface markers, activate lymphocytes and secrete inflammatory cytokines. Whilst the addition of miR181b decreased the expression of MHCII and CD40, this effect can be repressed by inhibiting expression of miR181b. Thus, miR181b may enhance the function of DC by down-regulating surface maturation molecules MHCII. PB1 increased the expression of the pro-inflammatory cytokines IL-6 and TNF-α, but had no effect on the anti-inflammatory cytokines IL-10 and IL-12, which seem to trigger Th17 programming in the BMDC. DCs also promote Th1 responses via IL-12 (de Jong et al., 2002).

IFN-α, controlled primarily by IFN regulatory factor 3 (IRF-3) and IFN regulatory factor (IRF-7), plays a crucial role in host defense processes against viral infection (Taniguchi and Takaoka, 2002; Yanai and Taniguchi, 2008). H9N2 AIV infection could result in the activation of IRF-7 on DCs (Lin et al., 2014). Our results show that IRF-3 and IRF-7 were all down-regulated in miR375 and miR181b groups, while inhibition of endogenous miR375 and miR181b significantly decreases IRF-3 and IRF7, suggesting that miR375 and miR181b are necessary for the production of IFN-α.

In summary, our results suggest that on the one hand, H9N2 virus protein PB1 can enhance the ability of DC to induce their phenotype, activate lymphocytes and secrete cytokines, and this effect may be accomplished by reducing the Jnk signaling pathway and activating the Erk signaling pathway. On the other hand, our results also suggest that miR375 can inhibit maturation of DC by decreasing expression of surface markers. Here, we demonstrated a previously unidentified role for PB1 in the regulation of murine immune responses of DCs, which was mediated by miR375 and miR181b. We propose that PB1 may enhance the function of DC by down-regulating miR375. Thus, miR375 may have a previously uncharacterised immunomodulatory role that can activate DCs for defense against H9N2 AIV (**Figure 1**).

METHODS

Ethics Statement

SPF C57BL/6 and BALB/c mice were obtained from Comparative Medical Center of Yang Zhou University. This study was

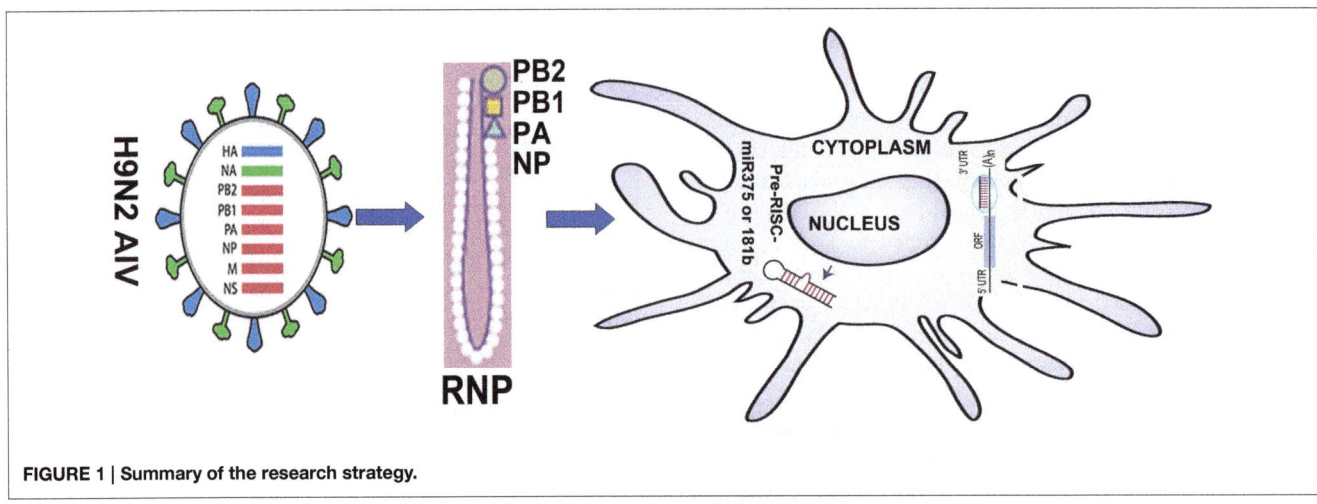

FIGURE 1 | Summary of the research strategy.

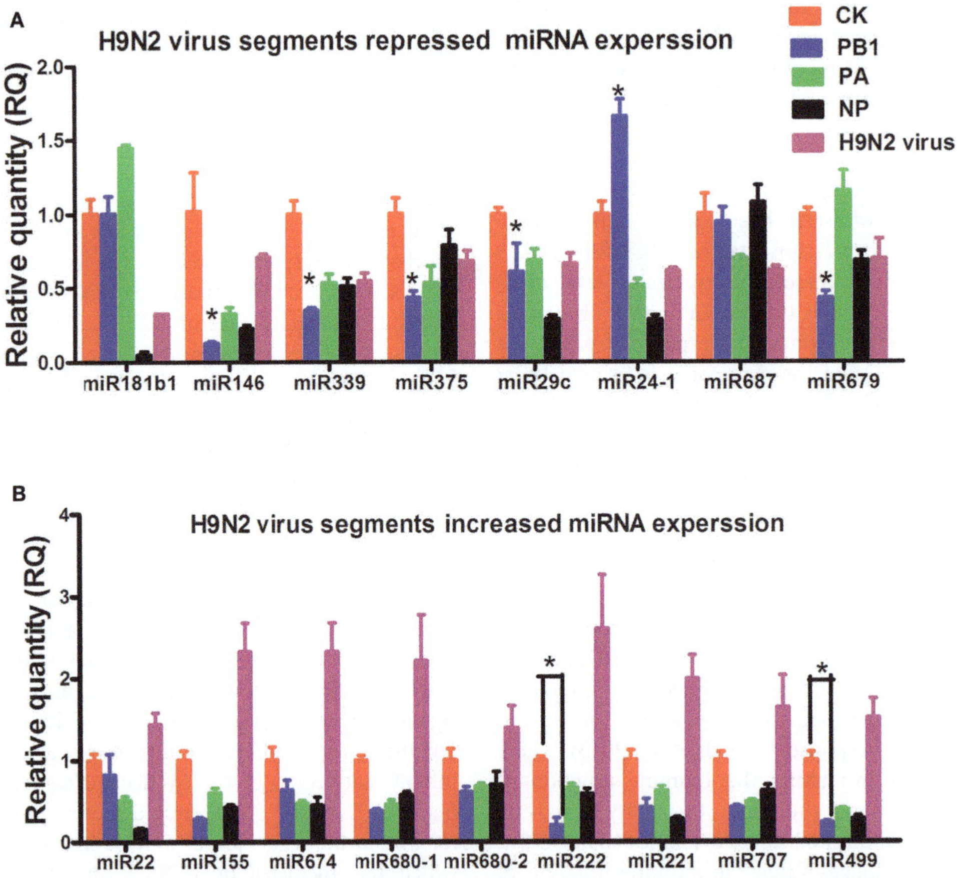

A

H9N2 virus segments repressed miRNA experssion

- CK
- PB1
- PA
- NP
- H9N2 virus

B

H9N2 virus segments increased miRNA experssion

FIGURE 2 | Results of the qPCR analysis of select miRNAs following stimulation by PB1, PA, or NP. (A) The expression levels of down-regulated miRNAs stimulated by PB1, PA, or NP (*$P < 0.01$, the significance of the data was determined by one-way ANOVA with Duncan test) (CK: blank DCs; PB1, PA, and NP: plasmid over-expressed DCs; LPS: 1 μg/ml, each times, three 4–6 week wild-type male C57BL/6 mice were sacrificed to isolated BMDCs and experiments were performed at least in triplicate). **(B)** The expression levels of up-regulated miRNAs stimulated by PB1, PA, or NP (*$P < 0.01$, the significance of the data was determined by one-way ANOVA with Duncan test) (CK: blank DCs; PB1, PA, and NP: plasmid over-expressed DCs; LPS: 1 μg/ml).

approved by the Ethical Committee of Animal Experiments of the College of Veterinary Medicine, Nanjing Agricultural University. All animal care and use were conducted in strict accordance with the Animal Research Committee guidelines of the College of Veterinary Medicine, Nanjing Agricultural University.

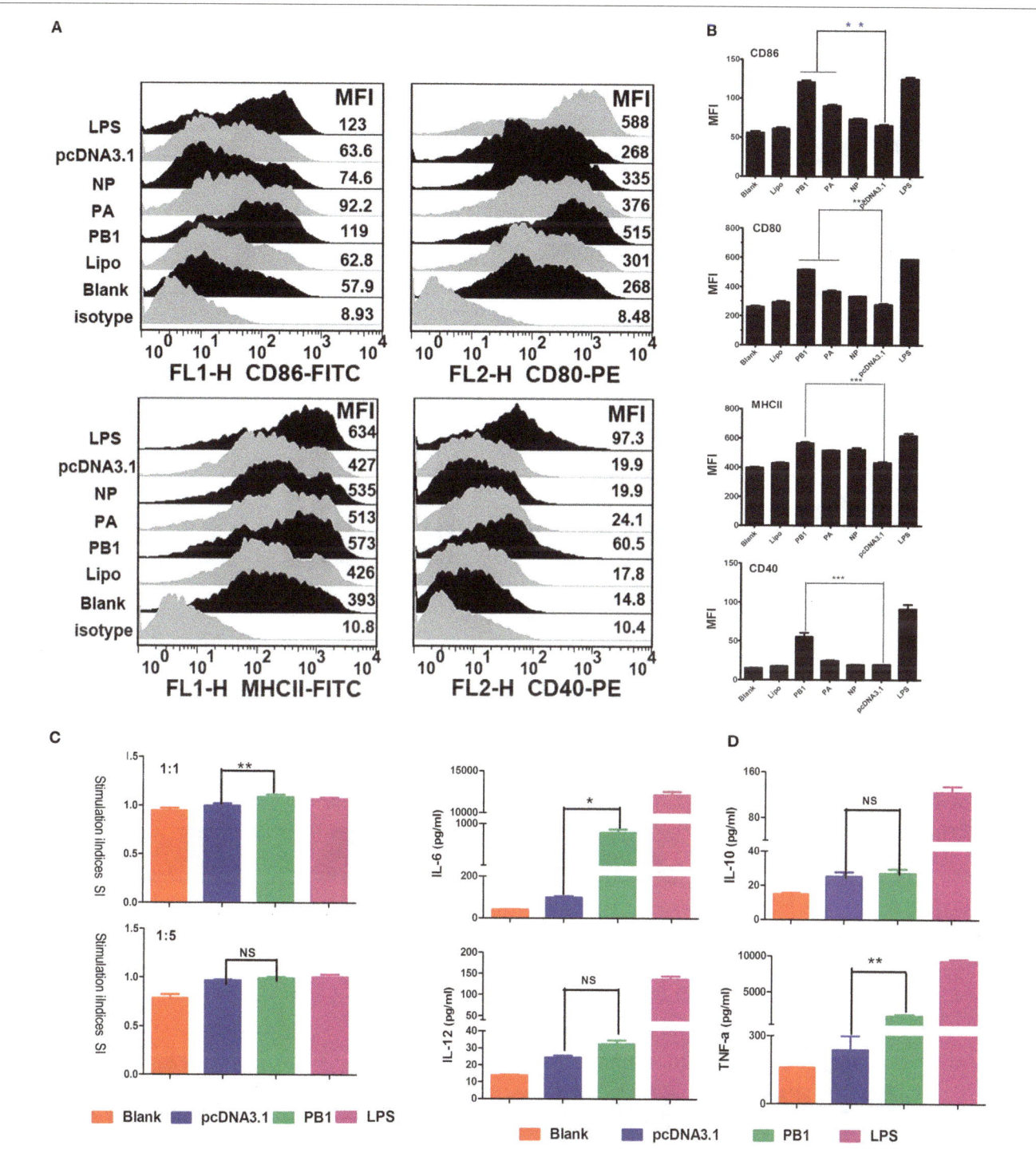

FIGURE 3 | Immune activation of BMDCs stimulated by PB1. (A) Flow cytometric analysis of the phenotypic alterations in DCs stimulated with PB1, PA, or NP (i.e., the expressions of CD40, CD80/86, and MHCII on BMDCs stimulated with PB1, PA, or NP). (Isotype: IgG2a for CD40, IgG1 for CD80, and CD86, IgG2b for MHCII; Blank: DCs without any treatment; Lipo: DCs added the same lipofectame as transfection groups; PB1, PA, and NP: plasmids transfected DCs; pcDNA3.1 group: plasmid pcDNA3.1 transfected DCs; LPS: Positive control, 1 μg/ml LPS, three 4–6 week wild-type male C57BL/6 mice were sacrificed to isolated BMDCs and experiments were performed at least in triplicate). **(B)** The MFI of CD40, CD80/86, and MHCII (*$P < 0.05$, **$P < 0.01$, or ***$P < 0.001$ the significance of the data was determined by one-way ANOVA with Tukey's multiple comparison test). **(C)** PB1-stimulated BMDCs stimulated the proliferation of naive T cells in mixed-lymphocyte reactions (MLR). The stimulator cells were BMDCs stimulated with or without PB1, pcDNA3.1, or LPS at 37°C for 24 h. All experiments were performed at least in triplicate. Significant differences between the treated and pcDNA3.1 groups are expressed as *$P < 0.05$ or **$P < 0.01$. The significance of the data was determined by one-way ANOVA with Tukey's multiple comparison test. **(D)** Cytokine release from PB1-stimulated BMDCs was measured by enzyme-linked immunosorbent assays (ELISAs). Data for IL-6, IL-10, IL-12, and TNF-α are shown as means ± standard deviation (SD) of three samples. Significant differences between the treated and pcDNA3.1 groups are expressed as *$P < 0.05$ or **$P < 0.01$. The significance of the data was determined by one-way ANOVA with Tukey's multiple comparison test.

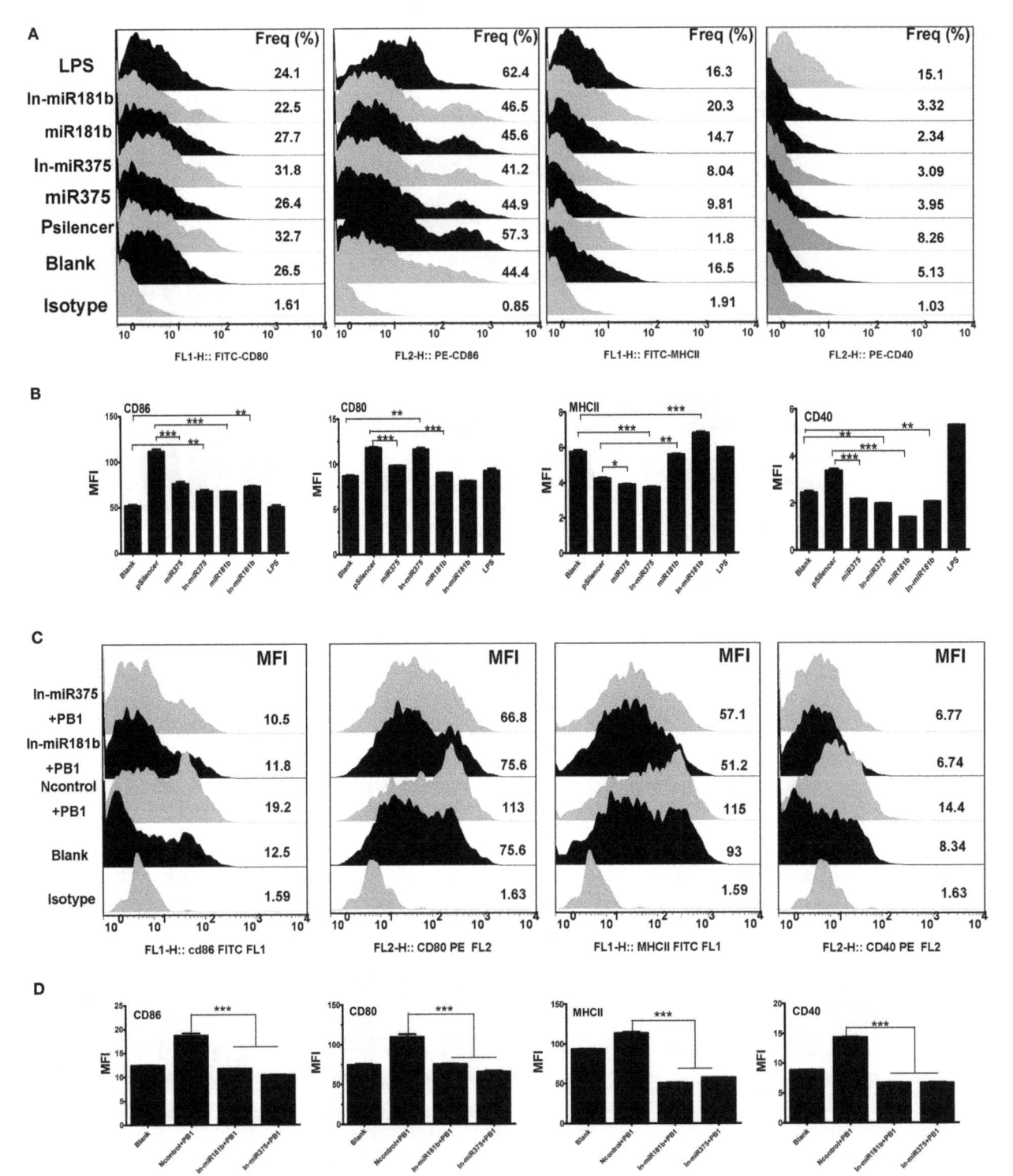

FIGURE 4 | The immune function of BMDCs stimulated by miR375 andmiR181b. (A) Flow cytometric analysis of the phenotypic alterations in DCs stimulated by miR375, miR181b, In-miR375, and In-miR181b (i.e., the expressions of CD40, CD80/86, and MHCII on BMDCs stimulated by miRNAs). (Isotype: IgG2a for CD40, IgG1 for CD80 and CD86, IgG2b for MHCII; Positive control, 1 μg/ml LPS, three 4–6 week wild-type male C57BL/6 mice were sacrificed to isolated BMDCs and experiments were performed at least in triplicate). **(B)** The MFI of CD40, CD80/86, and MHCII. (*$P < 0.05$, **$P < 0.01$, or ***$P < 0.001$ the significance of the data was determined by one-way ANOVA with Tukey's multiple comparison test). **(C)** PB1 mediated BMDCs activation when miR375 and miR181b were inhibited (100 μg/ml inhibitor for each miRNA, Isotype: IgG2a for CD40, IgG1 for CD80 and CD86, IgG2b for MHCII, three 4–6 week wild-type male C57BL/6 mice were sacrificed to isolated BMDCs and experiments were performed at least in triplicate). **(D)** The MFI data for CD40, CD80/86, and MHCII. (*$P < 0.05$, **$P < 0.01$, or ***$P < 0.001$ the significance of the data was determined by one-way ANOVA with Tukey's multiple comparison test).

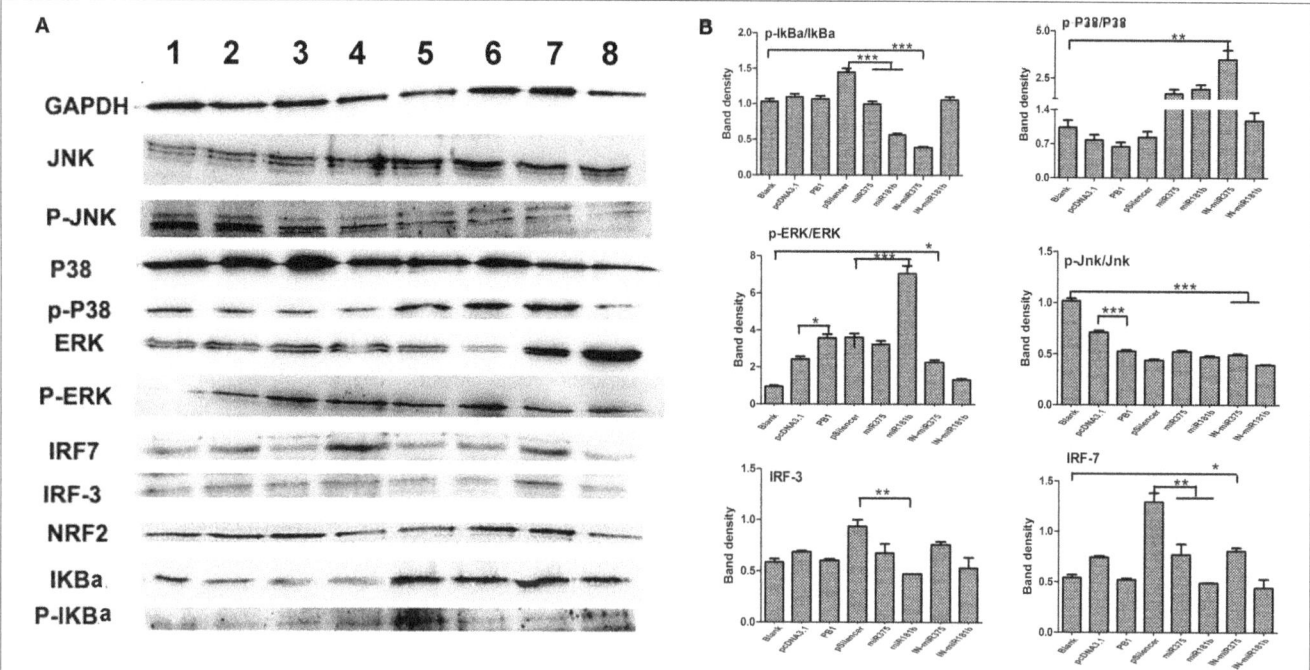

FIGURE 5 | Regulatory protein expression on BMDCs stimulated by PB1 and miRNAs as determined by Western blotting. (A) Western blot results for IkBa, P38, ERK, Jnk, IRF-3, and IRF-7 in cells stimulated by PB1, miR375 and miR181b (lane 1: blank group; lane 2: pcDNA3.1 stimulated group; lane 3: PB1 stimulated group; lane 4: pSilencer4.1 stimulated group; lane 5: miR375 stimulated group; lane 6: miR181b1 stimulated group; lane 7: InmiR375 stimulated group; lane 8: In-miR181b1 stimulated group; three 4–6 week wild-type male C57BL/6 mice were sacrificed to isolated BMDCs and experiments were performed at least in triplicate). **(B)** The expression levels of IkBa, P38, ERK, Jnk, IRF-3, and IRF-7 in cells stimulated by PB1, miR375 and miR181b (the data shown are the means ± standard error of the mean of double wells from three independent experiments). The levels of significance are identified by *$P < 0.05$, **$P < 0.01$, or ***$P < 0.001$.

Plasmids and Cell Culture

Three RNA segments that encode proteins involved in viral replication (PB1, PA, and NP) were amplified from the H9N2 virus and cloned into pcDNA3.1 (Invitrogen). MiRNAs (miR-375 and miR-181) were amplified and cloned into pSilencer4.1 (Invitrogen). Bone marrow-derived dendritic cells (BMDCs) were prepared from the femurs and tibias of sacrificed 4–6 week wild-type male C57BL/6 mice and treated with red blood cell lysis buffer (Beyotime) (Lin et al., 2014). Briefly, bone marrow cells were flushed from the tibias and femurs and cultured in complete medium (RPMI1640 (Invitrogen) with 10% FBS (Hyclone), 1% streptomycin and penicillin, 10 ng/ml recombinant granulocyte-macrophage colony-stimulating factor (GM-CSF) and IL-4 (Peprotech) and plated in 6-well plates. At day 6, the non-adherent, relatively immature DCs (1×10^6 cells/ml) were harvested and centrifuged to remove debris and dead cells, then cultured overnight in complete medium and transfected with different plasmids for subsequent assays. Transferred cell samples (1×10^6 cells) were washed twice with PBS and incubated at 4°C for 30 min with the following monoclonal antibodies (anti-mouse CD11c (N418), anti-mouse CD40 (1C10), anti-mouse CD86 (GL1), anti-mouse MHC (major histocompatibility complex) class II (M5/114.15.2) and anti-mouse CD80 antibody (16-10A1), respectively) (eBioscience). Finally, cells were analyzed using a Fluorescence Active Cell Sorter (FACS) (BD, FACS Aria) after two separate washes.

The Choice of miRNAs and Quantitative PCR Validation

To test which viral protein had the largest effect on expression of miRNAs, we amplified three replication related RNA segments (PB1, PA, and NP) and then cloned these into pcDNA3.1 vector. Primers are listed in **Supplementary Table 1.** Based on the microarray result, 9 up-regulated and 8 down-regulated genes were selected for RT-qPCR verification. Small RNAs were purified using the miRNeasy mini kit (Qiagen) and reverse transcribed to cDNA by miScript Reverse Transcriptase. QuantiTect SYBR Green PCR master mix (Qiagen) was used to perform qPCR according to the manufacturer's instructions. miRNA expression was normalized to the internal control 5S rRNA. Primers for 17 selected miRNAs are listed in **Supplementary Table 2.** All assays were performed in triplicate. Relative expression levels were calculated using the 2-ΔΔCt method (Livak and Schmittgen, 2001).

Immune Response of BMDCs Stimulated by PB1

Surface Marker Alterations of BMDCs

Immature BMDCs were plated into fresh medium (1×10^6 cells/ml) and transfected with one of the three vector constructs(PB1, PA, and NP), pcDNA3.1 (negative control) and LPS (1 µg/ml, positive control) for 48 h. Then cell samples

$(1 \times 10^6$ cells, 1.5 ml tube) were collected, washed twice with PBS and incubated at 4°C for 30 min with the following monoclonal antibodies (anti-mouse CD11c, anti-mouse CD40, anti-mouse CD86, anti-mouse MHC class II and anti-mouse CD80 antibody or the respective isotype controls, respectively). After washing, cells were analyzed with a Fluorescence Activated Cell Sorter (FACS) (BD, FACS Aria)

Allogeneic Mixed Leukocyte Reaction (MLR) Proliferation Assays

The primary T-cell stimulatory capacity of BMDCs was examined in a MLR. Untreated and variously treated BMDCs [pcDNA3.1-stimulated, PB1-stimulated and LPS-stimulated (1 μg /ml)] were used as the stimulator cells. Allogeneic lymphocytes were obtained from BALB/c mice as follows. Leukocytes were isolated from the spleens of 4 to 6 week-old allogeneic BALB/c mice with a T cell isolation kit (Miltenyi, Bergisch Gladbach, Germany) and cultured in complete RPMI 1640 medium supplementaryed with 10% FCS in 96-well plates at 37°C for 48 h. Graded numbers of responder cells $(1 \times 10^5$ cells/well) were added to 96-well round bottomed plates, giving responder: stimulator ratios of 1:1 or 5:1, in a culture volume of 100 μl. Cell proliferation assays was conducted with the Cell Counting Kit-8 (CCK-8, Beyotime). Each well received 20 μl CCK-8 solution and was incubated for a further 2 h at 37°C before absorbance measurement at 450 nm. All experiments were conducted in triplicate. The Stimulation Index was calculated using the formula:

$$SI = (OD_{sample} - OD_{stimulator\ cells\ only})/$$
$$(OD_{responder\ cells\ only} - OD_{blankcontrol}).$$

Cytokine Analysis

Bone marrow-derived dendritic cells (BMDC) culture supernatants were collected at 24 h after treatments (Groups were divided as MLR experiments). Concentrations of TNF-α, IL-6, IL-10, and IL-12p_{70} in the supernatants were measured using the Quantikine Elisa kit (Boster) according to the manufacturer's instructions. The sensitivity of the assay was 2 pg/ml for TNF-α, and 4 pg/ml for IL-6, IL-10, and IL-12p_{70}.

qRT-PCR Validation

BMDCs was cultured and collected at 24 h after treatments with PB1, PA and NP segments. qPCR was conducted to examine expression variation for selected genes. Individual samples were diluted 1:5 and 2 μl was amplified in a 20 μl reaction containing 10 μl of SYBR Premix™ Ex Taq (TaKaRa), 0.4 μl of ROX dye II and 0.4 μM of each of the forward and reverse gene-specific primers using an ABI 7500 instrument (Applied Biosystems, USA). - We also evaluated the transcription efficiency of plasmid pcDNA3.1-PA, NP, and PB1 by qPCR. Primers are- listed in **Supplementary Table 3**.

Immune Response of BMDCs Stimulated with miRNAs

Plasmid Construction and Phenotypic Detection

To confirm that the phenotype alteration induced by PB1 is meditated by miRNA, miRNA over-expression vectors were constructed using the pSilencer4.1 vector (Invitrogen). Four selected miRNAs (miR375 and miR181b) were amplified and then cloned into pSilencer4.1. Primers are listed in **Supplementary Table 4**. The isolation of BMDCs and phenotypic detection were performed as described above Plasmids were transfected with lipofectame2000 reagent (Invitrogen). MiR181b and miR375 inhibitors, which were chemically modified single stranded RNAs, were designed and purchased from RiboBio to evaluate miRNA function (Guangzhou, China). Each 100 nM miRNA inhibitor (micrOFF™ mmu-miR-181b-3p inhibitor, micrOFF™ mmu-miR-375-5p inhibitor, and micrOFF™ inhibitor Negative Control) was transfected into BMDCs for 24 h to analyze their effect on DCs via detection of- phenotypic alteration with FACS.

miRNA Inhibition Experiment

To detect whether the phenotypic alteration of BMDCs induced by PB1 was mediated by miR181b or miR375, miRNAs inhibitors were transfected into BMDCs for 4 h as described above, before PB1 over-expression plasmid was transfected. After another 24 h, BMDCs were collected for phenotypic detection using FACS.

Western Blot Assay

MAPK, NF-kB, and IFN-a signaling pathways control a wide range of cellular processes, especially for the immune response. Thus, we tried to evaluated how PB1 and their induced miRNAs affect the MAPK, NF-kB, and IFN-a signal pathways by western blot. BMDCs were transfect with PB1, miR375, miR181b, In-miR375, and In-miR181b for 48 h. Then cells were collected and washed with PBS three times for the next experiments. Western blot detection was performed as previously described by us. Mouse IkBa, P-IkBa, P38, P-P38, ERK, P-ERK, JUK, P-JUK, IRF-3, and IRF-7 were purchased from Abcam or Cell Signaling Technology and detected according to each manufacturer's protocol. Protein bands were visualized using the Super ECL Plus system. GAPDH was used as a loading control (Abcam).

Statistical Analyses

Data were evaluated by unpaired two-tailed Student's *t*-test using GraphPad Prism 5 (http://www.graphpad.com) (CSSN), with $p < 0.05$ considered to be statistically significant. The significance of the data was also determined by one-way ANOVA, followed by Tukey's multiple comparison tests. FACS data were analyzed by FlowJo software (FlowJo, China). All data are expressed as mean ± standard error of the mean.

AUTHOR CONTRIBUTIONS

JL design and performed all the experiments, analyzed the data and drafted the manuscript, JX and CT developed the dendritic cells and performed flow cytometry analyses, KZ charged for the data analyzed, YZ and QY supervised the experiment and participated in the design. All the authors read and approved the final manuscript.

ACKNOWLEDGMENTS

This work was supported by the National Natural Science Foundation of China (No. 31172302), the Natural Science Foundation of Jiangsu Province (No. BK20150666), the National Natural Science Foundation of China (No. 31570843) and A Project Funded by the Priority Academic Program Development of Jiangsu Higher Education Institutions (PAPD).

SUPPLEMENTARY MATERIAL

Supplementary Image 1 | Identification and construction of pcDNA3.1-PB1, pcDNA3.1-PA, and pcDNA3.1-NP. (A) Identification of pcDNA3.1-PB1 by digestion with *XhoI and KpnI* (M1: DL5000 DNA marker; 1: plasmid pcDNA3.1-PB1; 2: plasmid pcDNA3.1-PB1 digested with *XhoI and KpnI*). **(B)** Identification of pcDNA3.1-NP by digestion with *XhoI and HindIII* (M1: DL5000 DNA marker; 1: plasmid pcDNA3.1-NP; 2: plasmid pcDNA3.1-NP digested with *XhoI and HindIII*). **(C)** Identification of pcDNA3.1-PA by digestion with *XhoI and HindIII* (M1: DL5000 DNA marker; 1-3: plasmid pcDNA3.1-PA digested with *XhoI and HindIII*; 4: plasmid pcDNA3.1-PA).

Supplementary Image 2 | Identification and construction of pSilencer-miR375 and pSilencer-miR181b by digestion with *Bam*HI and *Hind*III. (A) Identification of pSilencer-miR181b by digestion with *Bam*HI and *Hind*III (M1: DL5000 DNA marker; 1: plasmid pSilencer- miR181b; 2: pSilencer-miR181b digested with *Bam*HI and *Hind*III). **(B)** Identification of pSilencer-miR375 by digestion with *Bam*HI and *Hind*III (M1: DL5000 DNA marker; 1: pSilencer-miR375 digested with *Bam*HI and *Hind*III; 2: plasmid pSilencer-miR375).

Supplementary Table 1 | Primers used in amplified PB1, PA and NP.

Supplementary Table 2 | qRT-PCR primers used for detecting miRNAs alteration.

Supplementary Table 3 | qRT-PCR primers used for detecting target genes and viral segments.

Supplementary Table 4 | Primers used in amplified miR375 and miR181b1.

Supplementary Data Sheet 1 | Results of the qPCR analysis of the transfection efficient on BMDCs with plasmid pcDNA3.1-PA, pcDNA3.1-NP and pcDNA3.1-PB1.

REFERENCES

Banchereau, J., Briere, F., Caux, C., Davoust, J., Lebecque, S., Liu, Y. J., et al. (2000). Immunobiology of dendritic cells. *Annu. Rev. Immunol.* 18, 767–811. doi: 10.1146/annurev.immunol.18.1.767

Bouvier, N. M., and Palese, P. (2008). The biology of influenza viruses. *Vaccine* 26(Suppl. 4), D49–D53.

Cheng, J. C., Yeh, Y. J., Tseng, C. P., Hsu, S. D., Chang, Y. L., Sakamoto, N., et al. (2012). Let-7b is a novel regulator of hepatitis C virus replication. *Cell. Mol. Life Sci.* 69, 2621–2633. doi: 10.1007/s00018-012-0940-6

de Jong, E. C., Vieira, P. L., Kalinski, P., Schuitemaker, J. H., Tanaka, Y., Wierenga, E. A., et al. (2002). Microbial compounds selectively induce Th1 cell-promoting or Th2 cell-promoting dendritic cells in vitro with diverse th cell-polarizing signals. *J. Immunol.* 168, 1704–1709. doi: 10.4049/jimmunol.168.4.1704

Fang, L. Q., Li, X. L., Liu, K., Li, Y. J., Yao, H. W., Liang, S., et al. (2013). Mapping spread and risk of avian influenza A (H7N9) in China. *Sci. Rep.* 3:2722. doi: 10.1038/srep02722

Fordham, J. B., Naqvi, A. R., and Nares, S. (2015). Regulation of miR-24, miR-30b, and miR-142-3p during macrophage and dendritic cell differentiation potentiates innate immunity. *J. Leukoc. Biol.* 98, 195–207. doi: 10.1189/jlb.1A1014-519RR

Gantier, M. P., Sadler, A. J., and Williams, B. R. (2007). Fine-tuning of the innate immune response by microRNAs. *Immunol. Cell Biol.* 85, 458–462. doi: 10.1038/sj.icb.7100091

Geissmann, F., Manz, M. G., Jung, S., Sieweke, M. H., Merad, M., and Ley, K. (2010). Development of monocytes, macrophages, and dendritic cells. *Science* 327, 656–661. doi: 10.1126/science.1178331

Hemerka, J. N., Wang, D., Weng, Y., Lu, W., Kaushik, R. S., Jin, J., et al. (2009). Detection and characterization of influenza A virus PA-PB2 interaction through a bimolecular fluorescence complementation assay. *J. Virol.* 83, 3944–3955. doi: 10.1128/JVI.02300-08

Isakova-Sivak, I., Korenkov, D., Smolonogina, T., Tretiak, T., Donina, S., Rekstin, A., et al. (2017). Comparative studies of infectivity, immunogenicity and cross-protective efficacy of live attenuated influenza vaccines containing nucleoprotein from cold-adapted or wild-type influenza virus in a mouse model. *Virology* 500, 209–217. doi: 10.1016/j.virol.2016.10.027

Jin, Y., Yu, D., Ren, H., Yin, Z., Huang, Z., Hu, M., et al. (2014). Phylogeography of Avian influenza A H9N2 in China. *BMC Genomics.* 15:1110. doi: 10.1186/1471-2164-15-1110

Kolch, W. (2000). Meaningful relationships: the regulation of the Ras/Raf/MEK/ERK pathway by protein interactions. *Biochem. J.* 351, 289–305. doi: 10.1042/bj3510289

Liang, J., Fu, J., Kang, H., Lin, J., Yu, Q., and Yang, Q. (2013). The stimulatory effect of TLRs ligands on maturation of chicken bone marrow-derived dendritic cells. *Vet. Immunol. Immunopathol.* 155,205–210. doi: 10.1016/j.vetimm.2013.06.014

Lin, J., Yin, Y. Y., Qin, T., Zhu, L. Q., Yu, Q. H., and Yang, Q. (2014). Enhanced immune response of BMDCs pulsed with H9N2 AIV and CpG. *Vaccine* 32,6783–6790. doi: 10.1016/j.vaccine.2014.10.013

Livak, K. J., and Schmittgen, T. D. (2001). Analysis of relative gene expression data using real-time quantitative PCR and the $2^{-\Delta\Delta C_T}$ Method. *Methods* 25, 402–408. doi: 10.1006/meth.2001.1262

Lopez, C. B., Moltedo, B., Alexopoulou, L., Bonifaz, L., Flavell, R. A., and Moran, T. M. (2004). TLR-independent induction of dendritic cell maturation and adaptive immunity by negative-strand RNA viruses. *J. Immunol.* 173, 6882–6889. doi: 10.4049/jimmunol.173.11.6882

Massari, S., Goracci, L., Desantis, J., and Tabarrini, O. (2016). Polymerase Acidic Protein-Basic Protein 1 (PA-PB1) Protein-Protein Interaction as a Target for Next-Generation Anti-influenza Therapeutics. *J. Med. Chem.* 59, 7699–7718. doi: 10.1021/acs.jmedchem.5b01474

O'Donnell, C. D., and Subbarao, K. (2011). The contribution of animal models to the understanding of the host range and virulence of influenza A viruses. *Microbes Infect.* 13, 502–515. doi: 10.1016/j.micinf.2011.01.014

Park, H., Huang, X., Lu, C., Cairo, M. S., and Zhou, X. (2015). MicroRNA-146a and microRNA-146b regulate human dendritic cell apoptosis and cytokine production by targeting TRAF6 and IRAK1 proteins. *J. Biol. Chem.* 290, 2831–2841. doi: 10.1074/jbc.M114.591420

Peiris, M., Yuen, K. Y., Leung, C. W., Chan, K. H., Ip, P. L., Lai, R. W., et al. (1999). Human infection with influenza H9N2. *Lancet* 354, 916–917.

Pu, J., Wang, S., Yin, Y., Zhang, G., Carter, R. A., Wang, J., et al. (2015). Evolution of the H9N2 influenza genotype that facilitated the genesis of the novel H7N9 virus. *Proc. Natl. Acad. Sci. U.S.A.* 112, 548–553. doi: 10.1073/pnas.1422456112

Shaib, H. A., Cochet, N., Ribeiro, T., Abdel Nour, A. M., Nemer, G., Azhar, E., et al. (2014). Passaging impact of H9N2 avian influenza virus in hamsters on its pathogenicity and genetic variability. *J. Infect. Dev. Ctries.* 8, 570–580. doi: 10.3855/jidc.4023

Smyth, L. A., Boardman, D. A., Tung, S. L., Lechler, R., and Lombardi, G. (2015). MicroRNAs affect dendritic cell function and phenotype. *Immunology* 144, 197–205. doi: 10.1111/imm.12390

Song, L., Liu, H., Gao, S., Jiang, W., and Huang, W. (2010). Cellular microRNAs inhibit replication of the H1N1 influenza A virus in infected cells. *J. Virol.* 84, 8849–8860. doi: 10.1128/JVI.00456-10

Swale, C., Monod, A., Tengo, L., Labaronne, A., Garzoni, F., Bourhis, J. M., et al. (2016). Structural characterization of recombinant IAV polymerase reveals a stable complex between viral PA-PB1 heterodimer and host RanBP5. *Sci. Rep.* 6:24727. doi: 10.1038/srep24727

Taniguchi, T., and Takaoka, A. (2002). The interferon-alpha/beta system in antiviral responses: a multimodal machinery of gene regulation by the IRF family of transcription factors. *Curr. Opin. Immunol.* 14, 111–116. doi: 10.1016/S0952-7915(01)00305-3

Tomasi, T. B., Magner, W. J., Wiesen, J. L., Oshlag, J. Z., Cao, F., Pontikos, A. N., et al. (2010). MHC class II regulation by epigenetic agents and microRNAs. *Immunol. Res.* 46, 45–58. doi: 10.1007/s12026-009-8128-3

Tucci, M., Stucci, S., Passarelli, A., Giudice, G., Dammacco, F., and Silvestris, F. (2014). The immune escape in melanoma: role of the impaired dendritic cell function. *Expert Rev. Clin. Immunol.* 10, 1395–1404. doi: 10.1586/1744666X.2014.955851

Wang, L., Song, G., Liu, M., Chen, B., Chen, Y., Shen, Y., et al. (2016). MicroRNA-375 overexpression influences P19 cell proliferation, apoptosis and

differentiation through the Notch signaling pathway. *Int. J. Mol. Med.* 37, 47–55. doi: 10.3892/ijmm.2015.2399

Wu, C., Gong, Y., Yuan, J., Zhang, W., Zhao, G., Li, H., et al. (2012). microRNA-181a represses ox-LDL-stimulated inflammatory response in dendritic cell by targeting c-Fos. *J. Lipid Res.* 53, 2355–2363. doi: 10.1194/jlr.M028878

Yanai, H., and Taniguchi, T. (2008). [IRF family transcription factors and host defense signaling]. *Tanpakushitsu Kakusan Koso* 53, 1231–1238.

Zhang, Y., Wang, F., Lan, Y., Zhou, D., Ren, X., Zhao, L., et al. (2015). Roles of microRNA-146a and microRNA-181b in regulating the secretion of tumor necrosis factor-α and interleukin-1β in silicon dioxide-induced NR8383 rat macrophages. *Mol. Med. Rep.* 12, 5587–5593. doi: 10.3892/mmr.2015.4083

Zhou, P., Zhu, W., Gu, H., Fu, X., Wang, L., Zheng, Y., et al. (2014). Avian influenza H9N2 seroprevalence among swine farm residents in China. *J. Med. Virol.* 86, 597–600. doi: 10.1002/jmv.23869

Zhou, Z., Li, X., Liu, J., Dong, L., Chen, Q., Kong, H., et al. (2015). Honeysuckle-encoded atypical microRNA2911 directly targets influenza A viruses. *Cell Res.* 25, 39–49. doi: 10.1038/cr.2014.130

11

Bacterial Control of Pores Induced by the Type III Secretion System

Julie Guignot[1,2,3,4] and Guy Tran Van Nhieu[1,2,3,4]*

[1] Equipe Communication Intercellulaire et Infections Microbiennes, Centre de Recherche Interdisciplinaire en Biologie (CIRB), Collège de France, Paris, France, [2] Institut National de la Santé et de la Recherche Médicale U1050, Paris, France, [3] Centre National de la Recherche Scientifique UMR7241, Paris, France, [4] MEMOLIFE Laboratory of Excellence and Paris Sciences et Lettres, Paris, France

Edited by:
Olivier Dussurget,
University Paris Diderot, France

Reviewed by:
Cammie Lesser,
Harvard Medical School, USA
Roland Benz,
Jacobs University Bremen, Germany

*Correspondence:
Guy Tran Van Nhieu
guy.tran-van-nhieu@college-de-france.fr

Type III secretion systems (T3SSs) are specialized secretion apparatus involved in the virulence of many Gram-negative pathogens, enabling the injection of bacterial type III effectors into host cells. The T3SS-dependent injection of effectors requires the insertion into host cell membranes of a pore-forming "translocon," whose effects on cell responses remain ill-defined. As opposed to pore-forming toxins that damage host cell plasma membranes and induce cell survival mechanisms, T3SS-dependent pore formation is transient, being regulated by cell membrane repair mechanisms or bacterial effectors. Here, we review host cell responses to pore formation induced by T3SSs associated with the loss of plasma membrane integrity and regulation of innate immunity. We will particularly focus on recent advances in mechanisms controlling pore formation and the activity of the T3SS linked to type III effectors or bacterial proteases. The implications of the regulation of the T3SS translocon activity during the infectious process will be discussed.

Keywords: SPATE, pore formation, T3SS, membrane repair, cell death

INTRODUCTION

Through secreted proteins, bacterial pathogens have the capacity to induce the formation of pores into eukaryotic host cell membranes. Pore-forming toxins (PFTs) can exert direct cytotoxic effect by irreversibly damaging the plasma membrane, or, at sub-lethal concentrations, can induce cell signaling involved in cytoskeletal reorganization, or in a variety of defense and innate immune responses (1–6). Alternatively, secreted bacterial proteins, such as AB toxins or type III secretion system (T3SS) translocon components, can form transient pores at the plasma membranes to promote the delivery of bacterial virulence factors into the host cytosol. Although host cell responses to various AB toxins have been largely described (7–9), relatively little is known about signaling linked to pore formation mediated by T3SS translocon components.

The T3SS can be viewed as a molecular syringe that upon cell contact, allows the delivery of bacterial effectors directly from the bacterial cytoplasm to the host cytosol [for review, see Ref. (10, 11)]. This system is widely spread among Gram-negative bacterial pathogens and shows conserved structural and functional features. Much of our knowledge has been inferred from extensive studies on the *Shigella*, *Salmonella*, or *Yersinia* T3SSs, and specific characteristics have been reported for other T3SSs, such as those from enteropathogenic (EPEC) and enterohemorrhagic (EHEC) *Escherichia coli.*

As for AB toxins, T3SS-mediated injection of bacterial effectors through eukaryotic cell plasma membranes requires the formation of a "translocation" pore, which occurs upon contact of T3SS with host cell membrane. Cell contact triggers the secretion of translocators proteins through the T3SS: two hydrophobic translocators proteins insert in the host cell membrane to form the so-called translocon, whereas one hydrophilic translocator protein is thought to connect the membrane-inserted translocon and the T3SS needle [for review, see Ref. (12); **Table 1**]. Here, we will review the responses elicited by host cells, linked to the pore-forming activity of the T3SS, and discuss their role during bacterial pathogenesis.

T3 TRANSLOCON AND PORE ACTIVITY

Upon cell contact, two hydrophobic proteins forming the translocon and containing trans-membrane domains insert into the host cell plasma membrane. Membrane insertion is associated with conformational changes, leading to oligomerization occurring through coiled-coil domain interactions, required for pore formation [**Table 1**; (12–16)]. Interestingly, coiled-coil domain of translocator proteins share homology with PFT, suggesting common origins and oligomerization mechanisms (17). Although the hydrophilic protein does not integrate in membranes, it is absolutely required for pore activity, possibly by acting as an assembly platform for proper oligomerization of the translocon components (12). The hydrophilic protein is also presumed to provide a molecular link between the translocon and the T3SS needle, through which type III effectors are channeled to get access to the cell cytosol. It is generally admitted that during type III effector translocation into host cells, the translocon is connected to the needle, forming a sealed conduct that does not allow exchange with the extracellular medium. This view is supported by cryo-EM studies showing a continuum between the T3SS and host cell membranes during bacterial infection (18, 19). However, the *Yersinia* type III effector YopH secreted in the extracellular media was shown to translocate into host cells by hijacking translocon components, suggesting that an alternate AB5-like toxin translocation mechanism could also occur for type III effectors (20). Presumably, only translocons detached from T3SS are expected to form pores opened to the extracellular medium. While such considerations remain speculative,

and such disconnection may occur following the translocation of injected type III effectors. Studies using artificial membranes have illustrated the pore-forming activity of purified translocon components (21). Although there are numerous evidence demonstrating pore-activity linked to T3SS, structures corresponding to pore-forming translocons are yet to be visualized during bacterial infection (13, 22–25).

Red blood cells (RBCs), which lack internal organelles, are unable to reseal membrane injuries and have been used to demonstrate T3SS-mediated pore formation (26). Release of hemoglobin by RBCs provides a metric for membrane damage linked to pore formation, which, in combination with solute size-dependent osmoprotection experiments, allows to estimate the size of membrane pores. Such experiments indicate that the T3SS induces the formation of pores within host cell membranes with an estimated size ranging from 1.2 to 5 nm, depending on the studies and bacterial systems (27–29). This diameter size is comparable to with that estimated for the inner diameter of the T3SS needle, consistent with a continuum between the needle and the membrane-inserted translocon during the injection of type III effectors. The analysis of the effects of mutations in translocator proteins shows a lack of correlation between T3SS-dependent RBCs' hemolysis and translocation of type III effectors in epithelial cells (30–34). This suggests that T3SS-dependent pore formation measured by the RBC's hemolysis assay does not implicate the same requirements as pore formation during translocation of effectors in epithelial cells. These issues are a matter of current debates. Other methods, including the use of fluorescent dyes, have been developed to demonstrate T3SS-dependent pore activity (25, 35).

MECHANISM OF T3SS-DEPENDENT PORE FORMATION

The observations that (i) translocated effectors do not leak into the extracellular medium after injection into cells and (ii) only a minority of cells infected with T3SS-expressing bacteria show dye incorporation assay or K^+ efflux, point to the inefficient capacity of the T3SS to mediate the formation of pore in nucleated cells (36–38). It was generally thought that as opposed to RBCs, membrane repair in nucleated cells was responsible for this relatively low pore-forming activity. As developed further, it is now clear that bacteria also control pore formation to avoid/or counteract detection by host cells.

In a very recent study, Sheahan and Isberg have identified host cell factors required for *Yersinia* T3SS-associated pore activity. Insertion and assembly of the translocon into the host cell membrane is a more complex process than originally thought, as numerous cytoskeletal and membrane trafficking proteins have been involved (39). This study confirms the key role played by actin and the small Rho GTPase in pore formation (40–42). Unexpectedly, Sheahan and Isberg also identified CCR5, a plasma membrane receptor, as playing a major role in T3-pore formation. CCR5 was recently identified to be a receptor for some PFT, emphasizing the functional homology the between T3 translocon and PFT (43).

TABLE 1 | Translocators components in various T3SSs [for review, see Ref. (12)].

	Hydrophilic protein	Hydrophobic protein with 2 TM domain	Hydrophobic protein with 1 TM domain
EPEC/EHEC	EspA	EspB	EspD
Yersinia	LcrV	YopD	YopB
Salmonella	SipD	SipC	SipB
Shigella	IpaD	IpaC	IpaB
Pseudomonas	PcrV	PopD	PopB

HOST CELL RESPONSES TO PORE FORMATION IN PLASMA MEMBRANES

In response to membrane injuries, cells trigger repair mechanisms involving the detection and removal of damaged plasma membranes. Membrane injuries, such as those induced by PFTs, immediately trigger an osmotic stress response, as well as a Ca^{2+} influx and a K^+ efflux that are sensed by host cells (4, 44–46). These responses activate MAP kinase signaling, inflammasomes, and NF-κB activation, which in turn lead to the elicitation of inflammatory and innate immune responses (**Figure 1**). Such signaling also activates membrane repair mechanisms: K^+ efflux triggers NLRP3 activation, leading to the recruitment of Caspase-1 (IL-1-converting enzyme) (47). Caspase-1 has a dual effect; it cleaves pro-IL-1β to generate mature IL-1β and stimulates the sterol regulatory element-binding proteins (SREBPs) to promote membrane biogenesis (48). Fast-acting cortical membrane repair involving exocytic and endocytic processes are also well described (49, 50). Ca^{2+} influx triggered by pore formation is sensed by synaptotagmin, a Ca^{2+} sensor present at the surface of lysosomes. Intracellular Ca^{2+} increase determines the synaptotagmin-dependent fusion of specialized lysosomes, named secretory lysosomes, in large vesicles. These vesicles fuse with wounded membranes, a process that contributes to the patching of

pores at the plasma membranes (26, 49, 50). Fusion of secretory lysosomes with wounded plasma membranes also leads to the release of lysosomal enzymes, such as sphingomyelinases, into the medium. Sphingomyelinases hydrolyze sphingomyelin to form ceramides that induce membrane curvature. This curvature is thought to initiate endocytosis of damage membranes that are subsequently targeted to intracellular degradation. Endocytosis has been proposed as an active repair mechanism of membrane damaged by PFTs (44). Ca^{2+} influx also leads to the binding of cytoplasmic annexins to the plasma membrane, resulting in the connection of the membrane to actin network. Annexin A5 was also shown to form a network limiting diffusion at the site of membrane injury (51). Ca^{2+} influx has also been associated with the annexin-dependent blebbing of the plasma membrane leading to the shedding of vesicles containing pores mediated by PFT in the extracellular milieu (52–54).

CHARACTERIZING SIGNALS LINKED TO MEMBRANE INSERTION AND PORE ACTIVITY OF THE T3 TRANSLOCON

Identifying signals that specifically associated with the T3SS translocon is challenging because it is also required for the translocation

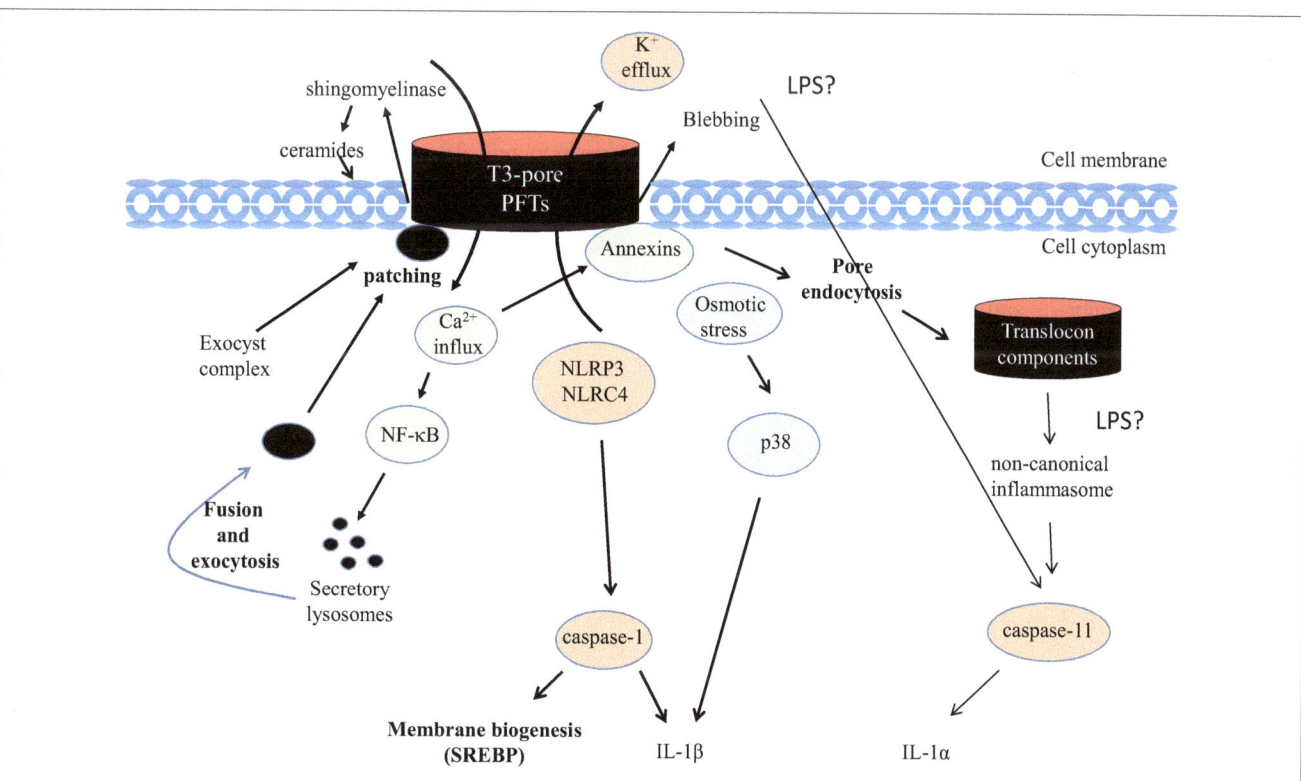

FIGURE 1 | **Membrane repair and inflammasome activation mediated by T3 translocons and PFTs**. Membrane injuries by PFTs or T3 translocon (T3-T) trigger an osmotic stress response, Ca^{2+} influx, and K^+ efflux that are sensed by host cells. These responses activate innate immune responses and membrane repair mechanisms. K^+ efflux, or possibly osmotic stress, associated with PFTs leads to the activation of the p38 MAPK and IL-1β secretion. In response to T3-pore formation, inflammasome and caspase-1 activation are also observed in association with K^+ influx into the translocon component (T3-TC) containing vacuole. Following endocytosis, T3 translocon components can activate caspase-11 through the activation of the non-canonical inflammasome. Membrane repair mechanisms linked to Ca^{2+} influx, lysosomal exocytosis and annexin recruitment are observed. New membrane recruitment to the site of infection by the exocyst complex could also contribute to patch T3-pores.

of type III effectors, many of which being reported to regulate innate immune responses. Furthermore, various microbial structures, including structural components of the T3SA, act as pathogen-associated molecular patterns (PAMPs) and are sensed by host cells to induce innate immune responses that are not directly associated with translocon insertion into host membranes (55, 56). To identify translocon-specific signals, studies have reported the use of bacterial mutants lacking all type III effectors (effectorless strain) and/or using cells lacking the two main TLR adaptor proteins (MyD88$^{-/-}$ and Trif$^{-/-}$) and hence, deficient for TLR signaling downstream of PAMPs. Such studies showed that the insertion of translocon components into the host plasma membrane activates an innate immune response that differs depending on the cell type (42). Insertion of the *Yersinia* translocon is associated with NLRP3 and NLRC4 activation with downstream signaling events leading to caspase-1 activation and IL-1β production (57, 58). The T3SS-dependent activation of NLRC4 has also been observed for *Shigella*, *Salmonella*, and *Pseudomonas* (59, 60). For *Salmonella and Pseudomonas*, such NLRC4 activation was shown to depend on T3SS-dependent pore formation and K$^+$ efflux (37). Activation of the non-canonical caspase-11 (caspase 4 in humans) inflammasome has also been described to be dependent on the T3SS, although recent evidences indicate that bacterial LPS could account for caspase-11 activation (61–64).

The cytosolic presence of translocators, rather than pore formation, has also been described to activate the inflammasome (65). The detection of translocator components in the cytosol has been attributed to the cytoplasmic tail of one of translocators following its insertion in the plasma membrane, or, alternatively, to the endocytosis of the pore-forming translocon complex. In both cases, cytosolic access of T3 translocon components leads to canonical NLRP3 and non-canonical caspase-11 activation, similar to what has been described for cytosolic PAMPs (62, 65). Consistent with a role for translocon endocytosis, Senerovic et al. have described that the purified translocator component IpaB oligomerizes in membrane and forms ion channels promoting K$^+$ influx upon internalization within endosomes, responsible for macrophages cell death. In this case, translocon-dependent K$^+$ influx into vacuoles may affect endolysosomal membranes' integrity, leading to caspase-1 activation downstream of the NLRC4 inflammasome (66).

Perhaps most indicative of T3SS-dependent pore-forming activity, membrane repair mechanisms are also activated upon bacterial infection. In response to Ca^{2+} influx linked to T3SS-dependent pores, synaptotagmin-dependent lysosomal exocytosis has been reported in *Salmonella* and *Yersinia* infected cells (39, 67). During infection, *Salmonella* and EPEC also trigger the recruitment and activation of the Ca^{2+}-sensors annexins at the site of bacterial attachment (68–73).

BACTERIAL MECHANISMS OF AVOIDING CELL DEATH LINKED TO T3SS-MEDIATED PORE FORMATION

Invasive bacteria, such as *Salmonella* or *Shigella*, promote their uptake in vacuole, resulting in a process leading to the removal of membrane-inserted translocons from the plasma membrane. This "self-removal" of membrane-inserted translocons may represent an additional factor contributing to the difficulty in detecting pore formation in epithelial cells infected by these bacteria. To minimize plasma membrane damages linked to T3 translocons, bacteria that multiply extracellularly have developed multiple strategies against inflammatory cell death. Injected type III effectors may down-regulate cell death and inflammatory signals, by interfering with initiator or effector caspases and NLRC4 inflammasome activation (74). The role of these type III effectors has been recently reviewed elsewhere (57, 58, 75–77). Here, we will mostly discuss the bacterial regulation of T3SS-dependent pore formation.

In *Yersinia*, at least three different type III effectors, such as YopK, YopE, and YopT, regulate T3SS-dependent pore formation and effector injection into host cells. The translocon component YopB activates both pro-inflammatory response and the small Rho GTPase, Ras (42, 78). YopB/D translocon insertion, in cooperation with invasin-beta1 integrin signaling, activates multiple Rho GTPases leading to actin polymerization, a step absolutely required for the *Yersinia* T3SS-dependent pore formation in the plasma membrane. The role of actin polymerization in the formation of the *Yersinia* T3SS-dependent pore is not clear but might reflect the importance to affix plasma membrane while translocon is inserted, or the translocon disconnection from the T3SA following effector injection in host cells. Among injected effectors, YopE and YopT display pore inhibition activity through the downregulation of several Rho GTPases (RhoA, RhoG, Rac1, and CDC42), linked to a GAP and protease activity toward these GTPases, respectively. Inhibition of Rho GTPase activity associated with actin depolymerization not only prevents pore formation but also reduces effector translocation. YopK also negatively regulates injection of type III effectors and cytotoxicity. As opposed to YopE and YopT regulating the T3SS activity through their action of Rho GTPases, YopK binds to the translocon and may directly clot it or induce conformational changes leading to translocation blockage (76). Although sharing little primary sequence homology with YopK, the EPEC/EHEC type III effector EspZ displays a similar activity (**Figure 2**). EspZ was shown to interact with the EPEC-translocon component EspD and prevents cell death by preventing the translocation of T3SS effectors into infected cells (79).

PROTEOLYTIC DEGRADATION OF T3-PORES BY A BACTERIAL SERINE PROTEASE AUTOTRANSPORTER OF ENTEROBACTERIACEAE

More recently, our group has reported a novel mechanism controlling T3SS-mediated pore formation and cytotoxicity induced by EPEC and EHEC (38). In addition to the T3SS, EPEC secretes other bacterial toxins involved in virulence. Among these, EspC, a protease belonging to the serine protease autotransporter of enterobacteriaceae (SPATE) family (80, 81), was shown to degrade the T3SS translocon components following contact with epithelial cells, thus downregulating T3SS-dependent pore formation and cytotoxicity. In EPEC, the hydrophilic translocator component EspA polymerizes into a filament connecting the

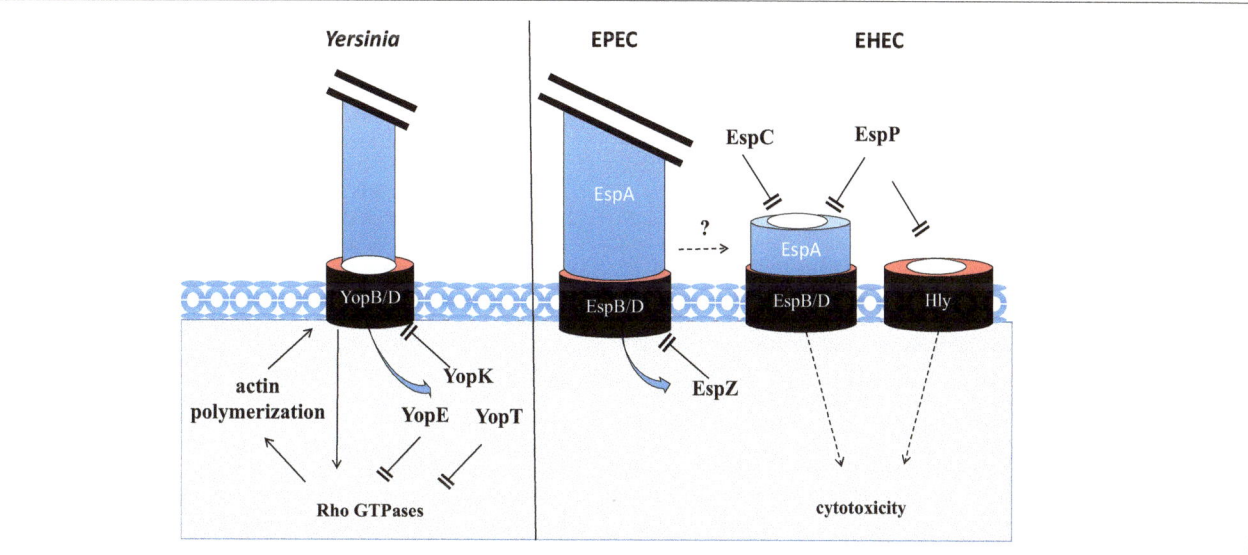

FIGURE 2 | Bacterial effectors regulating T3-pore formation. Upon cell contact and T3SS activation, the *Yersinia* YopB/D translocon components activate Rho GTPases leading to the polymerization of actin and T3-pore formation. The injected T3 effectors, such as YopK, YopE, and YopT, downregulate T3-pore formation and effector translocation. YopK directly acts on the T3 translocon. YopE and YopT inhibit RhoGTPases. EspZ shares an activity related to that of YopK by binding to the EPEC T3 translocon, inhibiting T3-pore formation and effector injection. EspC downregulates T3 pore by degrading the translocator components EspA/D, an activity shared by the EHEC EspP. EspP also downregulate the Hly PFT inserted in plasma membranes.

T3SS needle to the translocon that is composed of the EspB and EspD hydrophobic proteins. EspC appears to preferentially target EspA associated with EspD. Since EspC does not prevent type III effector injection, it may recognize a specific conformation of EspA/D corresponding to a T3SS "by-product" with potential cytotoxic activity. Interestingly, EspP, the EspC hortologue in EHEC has been involved in the proteolytic degradation of the *E. coli* hemolysin Hly, a pore-forming cytolysin (82). The cleavage of Hly by EspP occurs in the region of the hydrophobic domain and lead to the inactivation of its pore-forming activity.

EPITHELIAL CELL DEATH LINKED TO T3SS-PORES

Depending on the cell type and the extent of pore formation, membrane lesions can lead to apoptotic or necrotic cell death. It has been suggested that pores detected in epithelial cells infected with effectorless *Yersinia* or an EPEC *espC* mutant result from unsealed translocons similar to those found in membranes of erythrocytes. With the exception of T3SS-dependent cell death induced by *Yersinia*, which appears to implicate distinct pathways, T3SS-dependent cytotoxicity appears to be caspase independent (38, 79, 83, 84). Epithelial cells dying from T3SS-dependent unregulated pore formation show nuclear shrinkage without signs of nuclear fragmentation, consistent with non-apoptotic cell death (38, 79, 83, 84). The precise mechanism implicated in this T3SS-dependent death is unknown. In unrelated studies, however, nuclear shrinkage and caspase-independent cell death have been linked to the activation of phospholipase A2 (PLA2) (85). Interestingly, PLA2 activation associated with K^+ efflux and/or Ca^{2+} influx triggers IL-1 β secretion (86, 87), as observed for T3SS-dependent pore formation. Nuclear shrinkage may correspond to

a common response to membrane insults induced by PFTs and T3SS-dependent unregulated pore formation (88, 89).

CONCLUDING REMARKS AND PERSPECTIVES

As reviewed here, T3SS-expressing bacteria have developed a diversity of mechanisms to downregulate the formation of pores linked to the activity of T3SS translocon, reflecting the importance of this process in the pathophysiology of bacterial infections. In the absence of such translocon regulatory processes, a variety of inflammatory and death processes can be induced, depending on the bacterial pathogen. Although the insertion of T3SS-translocons during type III effector injection may induce a common canonical response associated with the activation of the NLRC4 inflammasome and eventually, necrotic cell death, these responses may be subsequently further tuned by other bacterial effectors. Deciphering how these signals integrate during the course of the bacterial infectious process represents a challenge needed to be addressed in future studies. Understanding how the T3SS pore formation and injection of effector is regulated could also lead to the development of innovative therapeutic molecules, widening the spectrum of currently studied T3SS inhibitor (90, 91).

AUTHOR CONTRIBUTIONS

JG wrote the manuscript, and GN edited the manuscript.

FUNDING

The work was supported in part by grants from the Labex Memolife and the Idex (ANR-10-IDEX-0001-02 PSL).

REFERENCES

1. Freche B, Reig N, van der Goot FG. The role of the inflammasome in cellular responses to toxins and bacterial effectors. *Semin Immunopathol* (2007) **29**:249–60. doi:10.1007/s00281-007-0085-0

2. Gonzalez MR, Bischofberger M, Pernot L, van der Goot FG, Freche B. Bacterial pore-forming toxins: the (w)hole story? *Cell Mol Life Sci* (2008) **65**:493–507. doi:10.1007/s00018-007-7434-y

3. Garcia-Saez AJ, Buschhorn SB, Keller H, Anderluh G, Simons K, Schwille P. Oligomerization and pore formation by equinatoxin II inhibit endocytosis and lead to plasma membrane reorganization. *J Biol Chem* (2011) **286**:37768–77. doi:10.1074/jbc.M111.281592

4. Los FC, Randis TM, Aroian RV, Ratner AJ. Role of pore-forming toxins in bacterial infectious diseases. *Microbiol Mol Biol Rev* (2013) **77**:173–207. doi:10.1128/MMBR.00052-12

5. Wiles TJ, Mulvey MA. The RTX pore-forming toxin alpha-hemolysin of uropathogenic *Escherichia coli*: progress and perspectives. *Future Microbiol* (2013) **8**:73–84. doi:10.2217/fmb.12.131

6. Diabate M, Munro P, Garcia E, Jacquel A, Michel G, Obba S, et al. *Escherichia coli* alpha-hemolysin counteracts the anti-virulence innate immune response triggered by the Rho GTPase activating toxin CNF1 during bacteremia. *PLoS Pathog* (2015) **11**:e1004732. doi:10.1371/journal.ppat.1004732

7. Beddoe T, Paton AW, Le Nours J, Rossjohn J, Paton JC. Structure, biological functions and applications of the AB5 toxins. *Trends Biochem Sci* (2010) **35**:411–8. doi:10.1016/j.tibs.2010.02.003

8. Odumosu O, Nicholas D, Yano H, Langridge W. AB toxins: a paradigm switch from deadly to desirable. *Toxins (Basel)* (2010) **2**:1612–45. doi:10.3390/toxins2071612

9. Lemichez E, Barbieri JT. General aspects and recent advances on bacterial protein toxins. *Cold Spring Harb Perspect Med* (2013) **3**:a013573. doi:10.1101/cshperspect.a013573

10. Cornelis GR. The type III secretion injectisome. *Nat Rev Microbiol* (2006) **4**:811–25. doi:10.1038/nrmicro1526

11. Puhar A, Sansonetti PJ. Type III secretion system. *Curr Biol* (2014) **24**:R784–91. doi:10.1016/j.cub.2014.07.016

12. Mattei PJ, Faudry E, Job V, Izore T, Attree I, Dessen A. Membrane targeting and pore formation by the type III secretion system translocon. *FEBS J* (2011) **278**:414–26. doi:10.1111/j.1742-4658.2010.07974.x

13. Schoehn G, Di Guilmi AM, Lemaire D, Attree I, Weissenhorn W, Dessen A. Oligomerization of type III secretion proteins PopB and PopD precedes pore formation in *Pseudomonas*. *EMBO J* (2003) **22**:4957–67. doi:10.1093/emboj/cdg499

14. Dickenson NE, Choudhari SP, Adam PR, Kramer RM, Joshi SB, Middaugh CR, et al. Oligomeric states of the *Shigella* translocator protein IpaB provide structural insights into formation of the type III secretion translocon. *Protein Sci* (2013) **22**:614–27. doi:10.1002/pro.2245

15. Myeni SK, Wang L, Zhou D. SipB-SipC complex is essential for translocon formation. *PLoS One* (2013) **8**:e60499. doi:10.1371/journal.pone.0060499

16. Adam PR, Dickenson NE, Greenwood JC II, Picking WL, Picking WD. Influence of oligomerization state on the structural properties of invasion plasmid antigen B from *Shigella flexneri* in the presence and absence of phospholipid membranes. *Proteins* (2014) **82**:3013–22. doi:10.1002/prot.24662

17. Barta ML, Dickenson NE, Patil M, Keightley A, Wyckoff GJ, Picking WD, et al. The structures of coiled-coil domains from type III secretion system translocators reveal homology to pore-forming toxins. *J Mol Biol* (2012) **417**:395–405. doi:10.1016/j.jmb.2012.01.026

18. Hu B, Morado DR, Margolin W, Rohde JR, Arizmendi O, Picking WL, et al. Visualization of the type III secretion sorting platform of *Shigella flexneri*. *Proc Natl Acad Sci U S A* (2015) **112**:1047–52. doi:10.1073/pnas.1411610112

19. Nans A, Kudryashev M, Saibil HR, Hayward RD. Structure of a bacterial type III secretion system in contact with a host membrane in situ. *Nat Commun* (2015) **6**:10114. doi:10.1038/ncomms10114

20. Edgren T, Forsberg A, Rosqvist R, Wolf-Watz H. Type III secretion in *Yersinia*: injectisome or not? *PLoS Pathog* (2012) **8**:e1002669. doi:10.1371/journal.ppat.1002669

21. Wager B, Faudry E, Wills T, Attree I, Delcour AH. Current fluctuation analysis of the PopB and PopD translocon components of the *Pseudomonas aeruginosa* type III secretion system. *Biophys J* (2013) **104**:1445–55. doi:10.1016/j.bpj.2013.02.018

22. Ide T, Laarmann S, Greune L, Schillers H, Oberleithner H, Schmidt MA. Characterization of translocation pores inserted into plasma membranes by type III-secreted Esp proteins of enteropathogenic *Escherichia coli*. *Cell Microbiol* (2001) **3**:669–79. doi:10.1046/j.1462-5822.2001.00146.x

23. Veenendaal AK, Hodgkinson JL, Schwarzer L, Stabat D, Zenk SF, Blocker AJ. The type III secretion system needle tip complex mediates host cell sensing and translocon insertion. *Mol Microbiol* (2007) **63**:1719–30. doi:10.1111/j.1365-2958.2007.05620.x

24. Montagner C, Arquint C, Cornelis GR. Translocators YopB and YopD from *Yersinia enterocolitica* form a multimeric integral membrane complex in eukaryotic cell membranes. *J Bacteriol* (2011) **193**:6923–8. doi:10.1128/JB.05555-11

25. Chatterjee A, Caballero-Franco C, Bakker D, Totten S, Jardim A. Pore-forming activity of the *Escherichia coli* type III secretion system protein EspD. *J Biol Chem* (2015) **290**:25579–94. doi:10.1074/jbc.M115.648204

26. McNeil PL, Kirchhausen T. An emergency response team for membrane repair. *Nat Rev Mol Cell Biol* (2005) **6**:499–505. doi:10.1038/nrm1665

27. Blocker A, Gounon P, Larquet E, Niebuhr K, Cabiaux V, Parsot C, et al. The tripartite type III secretion of *Shigella flexneri* inserts IpaB and IpaC into host membranes. *J Cell Biol* (1999) **147**:683–93. doi:10.1083/jcb.147.3.683

28. Buttner D, Bonas U. Port of entry – the type III secretion translocon. *Trends Microbiol* (2002) **10**:186–92. doi:10.1016/S0966-842X(02)02331-4

29. Mueller CA, Broz P, Cornelis GR. The type III secretion system tip complex and translocon. *Mol Microbiol* (2008) **68**:1085–95. doi:10.1111/j.1365-2958.2008.06237.x

30. Barzu S, Benjelloun-Touimi Z, Phalipon A, Sansonetti P, Parsot C. Functional analysis of the *Shigella flexneri* IpaC invasin by insertional mutagenesis. *Infect Immun* (1997) **65**:1599–605.

31. Olsson J, Edqvist PJ, Broms JE, Forsberg A, Wolf-Watz H, Francis MS. The YopD translocator of *Yersinia pseudotuberculosis* is a multifunctional protein comprised of discrete domains. *J Bacteriol* (2004) **186**:4110–23. doi:10.1128/JB.186.13.4110-4123.2004

32. Picking WL, Nishioka H, Hearn PD, Baxter MA, Harrington AT, Blocker A, et al. IpaD of *Shigella flexneri* is independently required for regulation of Ipa protein secretion and efficient insertion of IpaB and IpaC into host membranes. *Infect Immun* (2005) **73**:1432–40. doi:10.1128/IAI.73.3.1432-1440.2005

33. Luo W, Donnenberg MS. Analysis of the function of enteropathogenic *Escherichia coli* EspB by random mutagenesis. *Infect Immun* (2006) **74**:810–20. doi:10.1128/IAI.74.2.810-820.2006

34. Roehrich AD, Martinez-Argudo I, Johnson S, Blocker AJ, Veenendaal AK. The extreme C terminus of *Shigella flexneri* IpaB is required for regulation of type III secretion, needle tip composition, and binding. *Infect Immun* (2010) **78**:1682–91. doi:10.1128/IAI.00645-09

35. Faudry E, Perdu C, Attree I. Pore formation by T3SS translocators: liposome leakage assay. *Methods Mol Biol* (2013) **966**:173–85. doi:10.1007/978-1-62703-245-2_11

36. Viboud GI, Bliska JB. A bacterial type III secretion system inhibits actin polymerization to prevent pore formation in host cell membranes. *EMBO J* (2001) **20**:5373–82. doi:10.1093/emboj/20.19.5373

37. Arlehamn CS, Petrilli V, Gross O, Tschopp J, Evans TJ. The role of potassium in inflammasome activation by bacteria. *J Biol Chem* (2010) **285**:10508–18. doi:10.1074/jbc.M109.067298

38. Guignot J, Segura A, Tran Van Nhieu G. The serine protease EspC from enteropathogenic *Escherichia coli* regulates pore formation and cytotoxicity mediated by the type III secretion system. *PLoS Pathog* (2015) **11**:e1005013. doi:10.1371/journal.ppat.1005013

39. Sheahan KL, Isberg RR. Identification of mammalian proteins that collaborate with type III secretion system function: involvement of a chemokine receptor in supporting translocon activity. *MBio* (2015) **6**:e2023–2014. doi:10.1128/mBio.02023-14

40. Mounier J, Popoff MR, Enninga J, Frame MC, Sansonetti PJ, Van Nhieu GT. The IpaC carboxyterminal effector domain mediates Src-dependent actin polymerization during *Shigella* invasion of epithelial cells. *PLoS Pathog* (2009) **5**:e1000271. doi:10.1371/journal.ppat.1000271

41. Verove J, Bernarde C, Bohn YS, Boulay F, Rabiet MJ, Attree I, et al. Injection of *Pseudomonas aeruginosa* exo toxins into host cells can be modulated by host factors at the level of translocon assembly and/or activity. *PLoS One* (2012) **7**:e30488. doi:10.1371/journal.pone.0030488

42. Bliska JB, Wang X, Viboud GI, Brodsky IE. Modulation of innate immune responses by *Yersinia* type III secretion system translocators and effectors. *Cell Microbiol* (2013) 15:1622–31. doi:10.1111/cmi.12164

43. DuMont AL, Torres VJ. Cell targeting by the *Staphylococcus aureus* pore-forming toxins: it's not just about lipids. *Trends Microbiol* (2014) 22:21–7. doi:10.1016/j.tim.2013.10.004

44. Andrews NW, Almeida PE, Corrotte M. Damage control: cellular mechanisms of plasma membrane repair. *Trends Cell Biol* (2014) 24:734–42. doi:10.1016/j.tcb.2014.07.008

45. Draeger A, Schoenauer R, Atanassoff AP, Wolfmeier H, Babiychuk EB. Dealing with damage: plasma membrane repair mechanisms. *Biochimie* (2014) 107(Pt A):66–72. doi:10.1016/j.biochi.2014.08.008

46. Babiychuk EB, Draeger A. Defying death: cellular survival strategies following plasmalemmal injury by bacterial toxins. *Semin Cell Dev Biol* (2015) 45:39–47. doi:10.1016/j.semcdb.2015.10.016

47. Munoz-Planillo R, Kuffa P, Martinez-Colon G, Smith BL, Rajendiran TM, Nunez G. K(+) efflux is the common trigger of NLRP3 inflammasome activation by bacterial toxins and particulate matter. *Immunity* (2013) 38:1142–53. doi:10.1016/j.immuni.2013.05.016

48. Gurcel L, Abrami L, Girardin S, Tschopp J, van der Goot FG. Caspase-1 activation of lipid metabolic pathways in response to bacterial pore-forming toxins promotes cell survival. *Cell* (2006) 126:1135–45. doi:10.1016/j.cell.2006.07.033

49. Idone V, Tam C, Andrews NW. Two-way traffic on the road to plasma membrane repair. *Trends Cell Biol* (2008) 18:552–9. doi:10.1016/j.tcb.2008.09.001

50. Idone V, Tam C, Goss JW, Toomre D, Pypaert M, Andrews NW. Repair of injured plasma membrane by rapid Ca^{2+}-dependent endocytosis. *J Cell Biol* (2008) 180:905–14. doi:10.1083/jcb.200708010

51. Bouter A, Carmeille R, Gounou C, Bouvet F, Degrelle SA, Evain-Brion D, et al. Review: annexin-A5 and cell membrane repair. *Placenta* (2015) 36(Suppl 1):S43–9. doi:10.1016/j.placenta.2015.01.193

52. Draeger A, Monastyrskaya K, Babiychuk EB. Plasma membrane repair and cellular damage control: the annexin survival kit. *Biochem Pharmacol* (2011) 81:703–12. doi:10.1016/j.bcp.2010.12.027

53. Cassidy SK, O'Riordan MX. More than a pore: the cellular response to cholesterol-dependent cytolysins. *Toxins (Basel)* (2013) 5:618–36. doi:10.3390/toxins5040618

54. Jaiswal JK, Lauritzen SP, Scheffer L, Sakaguchi M, Bunkenborg J, Simon SM, et al. S100A11 is required for efficient plasma membrane repair and survival of invasive cancer cells. *Nat Commun* (2014) 5:3795. doi:10.1038/ncomms4795

55. Miao EA, Mao DP, Yudkovsky N, Bonneau R, Lorang CG, Warren SE, et al. Innate immune detection of the type III secretion apparatus through the NLRC4 inflammasome. *Proc Natl Acad Sci U S A* (2010) 107:3076–80. doi:10.1073/pnas.0913087107

56. Hwang I, Park S, Hong S, Kim EH, Yu JW. *Salmonella* promotes ASC oligomerization-dependent caspase-1 activation. *Immune Netw* (2012) 12:284–90. doi:10.4110/in.2012.12.6.284

57. Brodsky IE, Palm NW, Sadanand S, Ryndak MB, Sutterwala FS, Flavell RA, et al. A *Yersinia* effector protein promotes virulence by preventing inflammasome recognition of the type III secretion system. *Cell Host Microbe* (2010) 7:376–87. doi:10.1016/j.chom.2010.04.009

58. Kwuan L, Adams W, Auerbuch V. Impact of host membrane pore formation by the *Yersinia pseudotuberculosis* type III secretion system on the macrophage innate immune response. *Infect Immun* (2013) 81:905–14. doi:10.1128/IAI.01014-12

59. Sutterwala FS, Mijares LA, Li L, Ogura Y, Kazmierczak BI, Flavell RA. Immune recognition of *Pseudomonas aeruginosa* mediated by the IPAF/NLRC4 inflammasome. *J Exp Med* (2007) 204:3235–45. doi:10.1084/jem.20071239

60. Abdelaziz DH, Amr K, Amer AO. Nlrc4/Ipaf/CLAN/CARD12: more than a flagellin sensor. *Int J Biochem Cell Biol* (2011) 42:789–91. doi:10.1016/j.biocel.2010.01.003

61. Casson CN, Copenhaver AM, Zwack EE, Nguyen HT, Strowig T, Javdan B, et al. Caspase-11 activation in response to bacterial secretion systems that access the host cytosol. *PLoS Pathog* (2013) 9:e1003400. doi:10.1371/journal.ppat.1003400

62. Shin S, Brodsky IE. Caspase-11: the noncanonical guardian of cytosolic sanctity. *Cell Host Microbe* (2013) 13:243–5. doi:10.1016/j.chom.2013.02.011

63. Rivers-Auty J, Brough D. Potassium efflux fires the canon: potassium efflux as a common trigger for canonical and noncanonical NLRP3 pathways. *Eur J Immunol* (2015) 45:2758–61. doi:10.1002/eji.201545958

64. Stowe I, Lee B, Kayagaki N. Caspase-11: arming the guards against bacterial infection. *Immunol Rev* (2015) 265:75–84. doi:10.1111/imr.12292

65. Zwack EE, Snyder AG, Wynosky-Dolfi MA, Ruthel G, Philip NH, Marketon MM, et al. Inflammasome activation in response to the *Yersinia* type III secretion system requires hyperinjection of translocon proteins YopB and YopD. *MBio* (2015) 6:e2095–2014. doi:10.1128/mBio.02095-14

66. Senerovic L, Tsunoda SP, Goosmann C, Brinkmann V, Zychlinsky A, Meissner F, et al. Spontaneous formation of IpaB ion channels in host cell membranes reveals how *Shigella* induces pyroptosis in macrophages. *Cell Death Dis* (2012) 3:e384. doi:10.1038/cddis.2012.124

67. Roy D, Liston DR, Idone VJ, Di A, Nelson DJ, Pujol C, et al. A process for controlling intracellular bacterial infections induced by membrane injury. *Science* (2004) 304:1515–8. doi:10.1126/science.1098371

68. Zobiack N, Rescher U, Laarmann S, Michgehl S, Schmidt MA, Gerke V. Cell-surface attachment of pedestal-forming enteropathogenic *E. coli* induces a clustering of raft components and a recruitment of annexin 2. *J Cell Sci* (2002) 115:91–8.

69. Rescher U, Ruhe D, Ludwig C, Zobiack N, Gerke V. Annexin 2 is a phosphatidylinositol (4,5)-bisphosphate binding protein recruited to actin assembly sites at cellular membranes. *J Cell Sci* (2004) 117:3473–80. doi:10.1242/jcs.01208

70. Goebeler V, Ruhe D, Gerke V, Rescher U. Annexin A8 displays unique phospholipid and F-actin binding properties. *FEBS Lett* (2006) 580:2430–4. doi:10.1016/j.febslet.2006.03.076

71. Tobe T. Cytoskeleton-modulating effectors of enteropathogenic and enterohemorrhagic *Escherichia coli*: role of EspL2 in adherence and an alternative pathway for modulating cytoskeleton through annexin A2 function. *FEBS J* (2010) 277:2403–8. doi:10.1111/j.1742-4658.2010.07654.x

72. Munera D, Martinez E, Varyukhina S, Mahajan A, Ayala-Sanmartin J, Frankel G. Recruitment and membrane interactions of host cell proteins during attachment of enteropathogenic and enterohaemorrhagic *Escherichia coli*. *Biochem J* (2012) 445:383–92. doi:10.1042/BJ20120533

73. Jolly C, Winfree S, Hansen B, Steele-Mortimer O. The annexin A2/p11 complex is required for efficient invasion of *Salmonella* Typhimurium in epithelial cells. *Cell Microbiol* (2014) 16:64–77. doi:10.1111/cmi.12180

74. Al Moussawi K, Kazmierczak BI. Distinct contributions of interleukin-1alpha (IL-1alpha) and IL-1beta to innate immune recognition of *Pseudomonas aeruginosa* in the lung. *Infect Immun* (2014) 82:4204–11. doi:10.1128/IAI.02218-14

75. Shin H, Cornelis GR. Type III secretion translocation pores of *Yersinia enterocolitica* trigger maturation and release of pro-inflammatory IL-1beta. *Cell Microbiol* (2007) 9:2893–902. doi:10.1111/j.1462-5822.2007.01004.x

76. Dewoody RS, Merritt PM, Marketon MM. Regulation of the *Yersinia* type III secretion system: traffic control. *Front Cell Infect Microbiol* (2013) 3:4. doi:10.3389/fcimb.2013.00004

77. Wong Fok Lung T, Pearson JS, Schuelein R, Hartland EL. The cell death response to enteropathogenic *Escherichia coli* infection. *Cell Microbiol* (2014) 16:1736–45. doi:10.1111/cmi.12371

78. Viboud GI, So SS, Ryndak MB, Bliska JB. Proinflammatory signalling stimulated by the type III translocation factor YopB is counteracted by multiple effectors in epithelial cells infected with *Yersinia pseudotuberculosis*. *Mol Microbiol* (2003) 47:1305–15. doi:10.1046/j.1365-2958.2003.03350.x

79. Berger CN, Crepin VF, Baruch K, Mousnier A, Rosenshine I, Frankel G. EspZ of enteropathogenic and enterohemorrhagic *Escherichia coli* regulates type III secretion system protein translocation. *MBio* (2012) 3:e00317-12. doi:10.1128/mBio.00317-12

80. Dautin N. Serine protease autotransporters of enterobacteriaceae (SPATEs): biogenesis and function. *Toxins (Basel)* (2010) 2:1179–206. doi:10.3390/toxins2061179

81. Ruiz-Perez F, Nataro JP. Bacterial serine proteases secreted by the autotransporter pathway: classification, specificity, and role in virulence. *Cell Mol Life Sci* (2014) 71:745–70. doi:10.1007/s00018-013-1355-8

82. Brockmeyer J, Aldick T, Soltwisch J, Zhang W, Tarr PI, Weiss A, et al. Enterohaemorrhagic *Escherichia coli* haemolysin is cleaved and inactivated by serine protease EspPalpha. *Environ Microbiol* (2011) 13:1327–41. doi:10.1111/j.1462-2920.2011.02431.x

83. Stockbauer KE, Foreman-Wykert AK, Miller JF. *Bordetella* type III secretion induces caspase 1-independent necrosis. *Cell Microbiol* (2003) 5:123–32. doi:10.1046/j.1462-5822.2003.00260.x

84. Zhao Z, Zhang L, Ren C, Zhao J, Chen C, Jiang X, et al. Autophagy is induced by the type III secretion system of *Vibrio alginolyticus* in several mammalian cell lines. *Arch Microbiol* (2011) **193**:53–61. doi:10.1007/s00203-010-0646-9

85. Shinzawa K, Tsujimoto Y. PLA2 activity is required for nuclear shrinkage in caspase-independent cell death. *J Cell Biol* (2003) **163**:1219–30. doi:10.1083/jcb.200306159

86. Walev I, Klein J, Husmann M, Valeva A, Strauch S, Wirtz H, et al. Potassium regulates IL-1 beta processing via calcium-independent phospholipase A2. *J Immunol* (2000) **164**:5120–4. doi:10.4049/jimmunol.164.10.5120

87. Andrei C, Margiocco P, Poggi A, Lotti LV, Torrisi MR, Rubartelli A. Phospholipases C and A2 control lysosome-mediated IL-1 beta secretion: implications for inflammatory processes. *Proc Natl Acad Sci U S A* (2004) **101**:9745–50. doi:10.1073/pnas.0308558101

88. Cockeran R, Theron AJ, Steel HC, Matlola NM, Mitchell TJ, Feldman C, et al. Proinflammatory interactions of pneumolysin with human neutrophils. *J Infect Dis* (2001) **183**:604–11. doi:10.1086/318536

89. Noor S, Goldfine H, Tucker DE, Suram S, Lenz LL, Akira S, et al. Activation of cytosolic phospholipase A2alpha in resident peritoneal macrophages by *Listeria monocytogenes* involves listeriolysin O and TLR2. *J Biol Chem* (2008) **283**:4744–55. doi:10.1074/jbc.M709956200

90. Marshall NC, Finlay BB. Targeting the type III secretion system to treat bacterial infections. *Expert Opin Ther Targets* (2014) **18**:137–52. doi:10.1517/14728222.2014.855199

91. Gu L, Zhou S, Zhu L, Liang C, Chen X. Small-molecule inhibitors of the type III secretion system. *Molecules* (2015) **20**:17659–74. doi:10.3390/molecules200917659

12

A Comparative Study of the Effect of Leukoreduction and Pre-storage Leukodepletion on Red Blood Cells during Storage

Thelma A. Pertinhez[1,2], Emanuela Casali[2], Fabio Baroni[1], Pamela Berni[1], Roberto Baricchi[1] and Alberto Spisni[3]*

[1] Transfusion Medicine Unit, Arcispedale Santa Maria Nuova - IRCCS, Reggio Emilia, Italy, [2] Department of Biomedical, Biotechnological and Translational Sciences, University of Parma, Parma, Italy, [3] Department of Surgical Sciences, University of Parma, Parma, Italy

Edited by:
Lello Zolla,
University of Tuscia, Italy

Reviewed by:
Eleonora Napoli,
University of California, Davis, USA
Angelo D'Alessandro,
University of Colorado Denver, USA

*Correspondence:
Alberto Spisni
alberto.spisni@unipr.it

Blood transfusion is a fundamental therapy in numerous pathological conditions. Regrettably, many clinical reports describe adverse transfusion's drawbacks due to red blood cells alterations during storage. Thus, the possibility for a blood bank to ameliorate the quality of the erythrocyte concentrates units is crucial to improve clinical results and reduce transfusion adverse occurrences. Leukodepletion is a pre-storage treatment recognized to better preserve the quality of red blood cells with respect to leukoreduction. Aim of this work is to unravel the biochemical and biophysical basis that sustain the good clinical outcomes associated to the use of leukodepleted erythrocytes units. Erythrocytes concentrates were prepared as leukoreduced ($n = 8$) and pre-storage leukodepleted ($n = 8$) and then studied during 6 weeks in blood bank conditions. Overall, the data indicate that leukodepletion not only provide red blood cells with an appropriate amount of nutrients for a longer time but also selects red blood cells characterized by a more resilient plasma membrane fit to prolong their viability. We believe these results will stimulate new ideas to further optimize the current storage protocols.

Keywords: metabolomics, leukoreduction, leukodepletion, NMR, blood transfusion, blood bank conditions

INTRODUCTION

In the past years operators in Transfusion Medicine directed particular attention toward the standardization of donor selection, the modality of blood collection, the production of hemocomponents and the development of tests to detect infectious agents. Among the cutting edge themes that are now under discussion (Spitalnik et al., 2015) two of them are particularly relevant: *i.* the identification and quantification of the component of each transfusion product and *ii.* Which criteria must be used to match the available transfusion unit with the patient in specific clinical scenarios. This relevant topic emerges due to the wide clinical variability of transfusion recipients, varying from patients transfused in ambulatory to critical patients requiring multiple transfusions.

Clinical studies identified a correlation between transfusion of red blood cells (RBCs) stored for more than 14 days and the increase of infectious events (Vamvakas and Carven, 1999; Chang et al., 2000; Leal-Noval et al., 2001; Taylor et al., 2002), extension of hospitalization time, prolonged mechanical ventilation, multiple organ failure (Moore et al., 1997) and mortality (Leal-Noval et al., 2003; Gong et al., 2005). The origin of these negative outcomes are attributed to the morphological and biochemical modifications that RBCs undergo during prolonged storage: e.g., the production of lactate leads to a decrease in pH, depletion in adenosine triphosphate and 2,3-diphosphoglycerate. A conclusive solution to those biochemical alterations has not been found yet: storage at low temperature (2–6°C) slows the rate of glycolysis, unfortunately it also decreases the activity of the Na^+-K^+ pump, resulting in the alteration of electrolytes balance. The increased K^+ concentration in the conservation medium augments the risk of arrhythmia in case of rapid transfusion through a central vein. Storage is also known to induce changes in RBCs morphology as well as membrane loss. In this respect, the exact correlation between those storage lesions and the fate of the transfused cells is still unknown and calls the attention of the scientific community (Hess and Grazzini, 2010; D'Alessandro et al., 2012; Dzieciatkowska et al., 2013; Prudent et al., 2015).

A subject of major debate among clinical practitioners is the definition of the criteria for the selection of the transfusion units: the prevalent opinion, at the moment, points to the storage time of the bags. While there are data indicating a correlation between RBCs storage time and morbidity/mortality of the transfused patients (Wang et al., 2012), randomized controlled trials are in progress to verify this fact (Steiner et al., 2010; Lacroix et al., 2011; Alexander et al., 2016).

Consequent to these considerations, reduction of the RBCs storage time would be a simple solution. Unfortunately, it is a practice very difficult to be implemented in a blood bank. Thus, methods to prolong the quality of RBCs during storage are being searched: the removal of leukocytes, before storage, is a promising one.

Beneficial outcomes of leukoreduction and pre-storage leukodepletion, such as the minimization of febrile non-hemolytic transfusion reactions; of cytomegalovirus transmission and anti HLA immunization leading to platelet refractoriness, are well-documented (Eisenfeld et al., 1992; Fischer et al., 1998; Novotny et al., 1995). Other collateral effects, significantly reduced by the introduction of this storage protocol are post-operatory infectious complications, acute respiratory distress syndrome (ARDS), acute lung injury (ALI), transfusion related acute lung injury (TRALI), transfusion-associated circulatory overload (TACO), prolonged mechanical ventilation, hospitalization time and mortality (Bianchi et al., 2015; Silliman et al., 2011).

Recognizing the clinical relevance of these observations, Canada, France and UK adopted universal leukoreduction in 1990, while Germany introduced this pre-treatment in 2001. Until end of 2015, in Italy, RBCs leukodepletion has been used in only few centers and for specific clinical cases. In our hospital, Arcispedale Santa Maria Nuova, Reggio Emilia, Italy, 30% of the RBCs units were leukodepleted and used for critical patients (intensive care unit, cardio surgery, hematology, neonatology).

A recent decree of the Italian Ministry of Health, 2 November 2015, designated as mandatory RBCs pre-storage leukodepletion starting January 2016.

The origin of the beneficial outcomes associated to the introduction of RBCs leukodepletion before storage, are expected to be due to the preservation of RBCs quality/vitality for a longer time. Because the main difference between leukoreduced and leukodepleted RBCs is the significantly reduced number of leukocytes present in the bags of leukodepleted RBCs, to progress in the optimization of RBCs storage, it is necessary to understand how the presence of leukocytes influences RBCs vitality, a condition that, at present, is generally associated to the maintenance of both their proper shape and efficient metabolism.

The introduction of new *"omics"* methodologies such as proteomic and metabolomics (Pertinhez et al., 2014; D'Alessandro et al., 2015; Zolla et al., 2015; Nemkov et al., 2016) has provided important tools to answer these questions. In this frame, we have been prompted to study the variation of some RBCs morphological and biochemical parameters during storage. We report here a comparative study on leukoreduced RBCs and pre-storage leukodepleted RBCs.

MATERIALS AND METHODS

The study was approved by the Arcispedale Santa Maria Nuova (ASMN) Ethics Committee on January 21, 2013. Written informed consent was obtained from all volunteers donors who participated in this study according to the declaration of Helsinki. Blood components were collected from periodical donors of the Transfusion Medicine Unit of ASMN according to the policy of the Italian National Blood Centre Guidelines. Since our experimental conditions did not allow to carried out a paired study, to reduce individual variability we selected 16 males donors aged 30–50 years.

Blood Collection and Processing

Sixteen whole blood units (450 mL ± 10%) were collected using the top-and-bottom system (Fresenius Kabi Medicare Bad Homburg, Germany). Eight were collected into triple bags, and eight into quadruple bags, containing Citrate, Phosphate, Dextrose solution (CPD). All the units were centrifuged by Hettich Roto Silenta 630 RS centrifuge (22°C, 11 min, 4000 × g) therefore, most of the plasma and buffy coat was removed using a Compomat G4 separator (Fresenius Kabi Medicare) and RBCs were stored in 100 mL of saline, adenine, glucose and mannitol (SAGM) additive solution. Those collected into triple bags were prepared as Non-Leukodepleted Erythrocyte Concentrate (NLPEC) while the ones collected in the quadruple bags were prepared as Leukodepleted Prestorage Erythrocyte Concentrate (LPEC) using in-line filters.

After 24 h, each RBC unit was divided in 7 satellites bags of 40 mL each (Fresenius Kabi Medicare Bad Homburg, Germany). Satellite bags were stored under standard conditions (2–6°C) and analyzed at different Day (2, 9, 16, 23, 30, 36, and 42) of storage.

An aliquot from each bag was taken for cell count, hematocrit, mean corpuscular value (MCV) and total hemoglobin determination (Sapphire instrument Abbott diagnostic Illinois,

---

Abbreviations: CPD, citrate, phosphate, dextrose; LPEC, Leukodepleted-Pre-storage Erythrocyte Concentrate; NLPEC, Non-Leukodepleted-Pre-storage Erythrocyte Concentrate; NMR, Nuclear Magnetic Resonance; RBCs, red blood cells; SAGM, Saline, adenine, glucose, mannitol; WBCs, white blood cells; MCV, mean corpuscular value; PCA, principal component analysis.

USA and Cytomix FC 500 Beckman Coulter IL, Indianapolis, USA). A residual number of leukocytes ($<< 1 \times 10^6$ a depletion of log 4) were present in the final LPEC units according to the Italian Blood National System regulatory law.

Definition of NLPEC and LPEC

NLPEC = buffy-coat poor RBCs = leukoreduced RBCs ($0.95 \pm 0.39 \times 10^9$ WBCs per unit, $n = 8$).

LPEC = buffy-coat removal + leukofiltration = highly depleted or leukodepleted RBCs, according to EU guidelines $<< 1 \times 10^6$ WBCs (a depletion of log 4) per unit.

RBCs Supernatant and RBC Lysate Preparation

The RBCs supernatant was collected after centrifugation at 2000 $\times$ g for 10 min and was divided into two aliquots. One aliquot was used without further modifications for biochemical assays (see below) while the other was depleted of proteins by ultra-filtration (5000 Da cut-off) and frozen at -80°C for subsequent ^{1}H-NMR measurements.

RBCs lysate was prepared as follow: red blood cells were washed twice by suspension in 0.9% NaCl in 5 mM phosphate buffer pH 7.4 followed by centrifugation at 2000 g $\times$ 10 min. The collected RBCs were then lysed through two cycles of freezing in liquid nitrogen and thawing at 37°C followed by sonication for 30 s. Proteins and membranes were eliminated by ultra-filtration (cut-off 5000 Da), as described in Pertinhez et al. (2014).

Supernatant Biochemical Assays

Supernatant: Na$^+$ and K$^+$ were measured by an indirect ion-selective electrode method (Hoffmann-La Roche Ltd). Total proteins, lactate and lactate dehydrogenase (LDH), were measured by Cobas© Roche.

RBCs hemolysis was evaluated by the absorption spectrum of free hemoglobin (HbO$_2$), using an extinction coefficient of 512 mM^{-1}cm^{-1} at 415 nm on a spectrophotometer JASCO V-630.

^{1}H-NMR Experiments

Samples were prepared by mixing 570 microliters of the ultra-filtrate either of the RBCs supernatant or of the RBCs lysate, with 30 microliters of TSP (1% in D$_2$O) and 10 microliters of 1 M phosphate buffer pH 7.4 (Pertinhez et al., 2014).

1D ^{1}H-NMR spectra were acquired at 25°C on a Spectrometer Varian Inova 600 MHz (Palo Alto, USA); processing and peaks assignment was performed with Chenomx NMR suite 7.6 (Edmonton, Canada) as previously described in Pertinhez et al. (2014). The ^{1}H-NMR spectra were automatically reduced into consecutive integrated spectral regions (buckets) of an equal width (0.03 ppm). The region containing the water resonance (4.5–5.0 ppm) was not included in the analysis.

Statistical Analysis

Significance has been evaluated by two-way ANOVA tests followed by Fisher LSD *post hoc* tests. *P*-values smaller than 0.05 were considered to be significant.

MestReNova 8.1 software (Santiago de Compostela, Spain) was used to perform the principal component analysis (PCA).

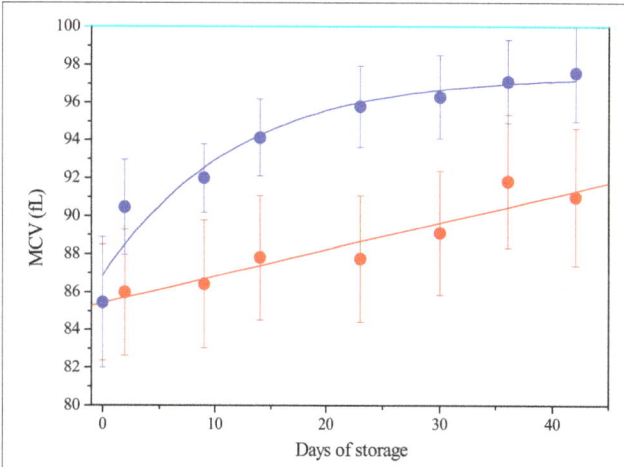

FIGURE 1 | Mean Corpuscular Value (fL) of RBC during storage: LPEC (●) linear fit $r = 0.96$; NLPEC (●) sigmoidal fit $r^2 = 0.95$ Chi$^2 = 0.137$. ANOVA showed a significant effect for both the storage time and preparation ($p < 0.0005$). Interaction between variables was not significant. *Post hoc* tests indicated always significant differences at each storage time between preparations ($p < 0.05$).

Prior to multivariate statistic, pareto scaling, which scales data by dividing each variable by the square root of the standard deviation, was applied.

Note that metabolites concentration were measured starting from the 2nd day of conservation (that therefore is our initial time, t1), after mandatory tests performed by the Transfusion Medicine Unit in accordance with the Transfusion Regulatory Italian law (n° 219, 21 October 2005).

RESULTS

RBCs Mean Corpuscular Value during Storage

Under physiological conditions, RBCs lifespan is 120 days: their aging *in vivo* is associated to a decrease in volume and an increase in cell density (Bosman, 2013). In blood bank conditions, instead, after the removal of plasma and buffy coat, and re-suspension in SAGM, we observe that RBCs, irrespective of being leukodepleted or not, increase their Mean Corpuscular Value (MCV) during storage (**Figure 1**).

Note that, in the 16 whole blood freshly collected units, the RBCs showed an initial similar MCV of 85.5 fL $\pm$ 2.95 and the MCV values remained within the laboratory reference range (80-100 fL), for both preparations, throughout the storage time.

Nonetheless, **Figure 1** shows that NLPEC MCV undergoes a higher increase with respect to LPEC MCV. The increase of red blood cells MCV, which we measure in SAGM, is consistent with the current data (Veale et al., 2011).

The different behavior between NLPEC and LPEC is already evident at time 1, Day 2, ($p = 0.011$) and it becomes more significant at Day 42 ($p = 0.006$). The linear increment ($r = 0.96$) of the MCV values measured for LPEC, as compared to the NLPEC units, that exhibits a sigmoidal behavior ($r^2 = 0.95$), suggests that leukodepletion selects a homogeneous RBCs

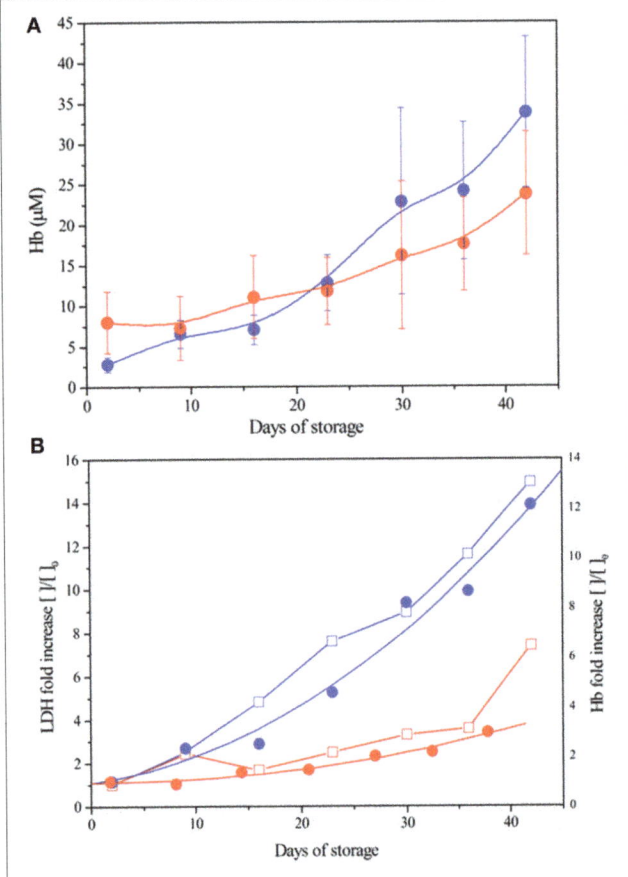

FIGURE 2 | Hb and LDH in the supernatant of LPEC and NLPEC during storage. (A) Changes in Hb concentration: LPEC (●), NLPEC (●). ANOVA showed a significant main effect of storage time ($p < 0.0005$) and significant interaction between storage time and preparation ($p = 0.0002$) **(B)** Fold increase comparison: LPEC LDH (□) and Hb (●), NLPEC LDH (□) and Hb (●).

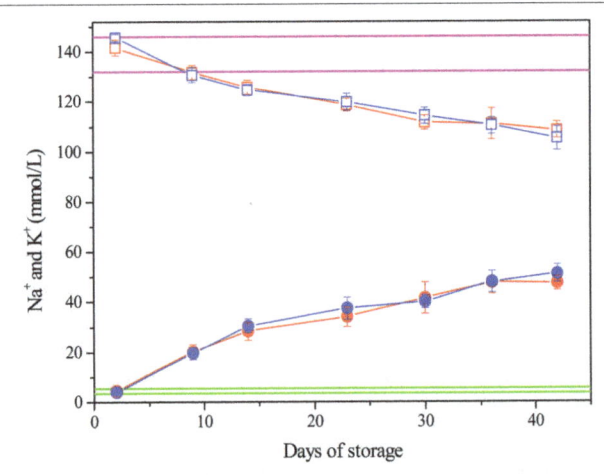

FIGURE 3 | Electrolytes concentration. K$^+$: LPEC (●), NLPEC (●). Na$^+$: LPEC (□) NLPEC (□). Normal ranges are marked in green for K$^+$, in magenta for Na$^+$. ANOVA showed a significant effect of storage time ($p < 0.0005$) for both ions. Interaction between variables is not significant. *Post hoc*-tests between storage times of the same preparation were always significant ($p < 0.05$) both for LPEC and NLPEC.

population. In particular, we hypothesize that leukodepleted RBCs are characterized by a higher deformability that would favor their passage through the filter and are more resistant to cell volume impairment during storage.

Changes in Free Hb and Lactate Dehydrogenase (LDH) Concentrations

The free hemoglobin content in the supernatant progressively increases over time for both preparation (**Figure 2**). This is expected to be the consequence of aging and death of the blood cells with consequent release of the intracellular content in the conservation medium. ANOVA reveals that free hemoglobin concentration is not significantly influenced by preparation (LPEC or NLPEC) but by a combined effect of storage time and preparation. Interestingly, as reported in **Figure 2A**, in the initial 20 days of storage, NLPEC (Hb = 2.8 μM, Day 2) exhibit a free hemoglobin content that is regularly lower than in LPEC (Hb = 7.9 μM, Day 2); a fact that we attribute to the consequence of leukodepletion, i.e., to the mechanical damage that some blood cells suffer passing through the sieve. However, after Day 20, the

trend is reversed and LPEC exhibit a reduced degree of hemolysis, proving that the leukodepleted RBCs are more resistant to the osmotic changes experienced during storage.

To remove any possible bias, data were normalized to the values measured at time 1, Day 2 (see Material and Methods). **Figure 2B** shows a higher free Hb content in the NLPEC units since the beginning of storage. More than that, in the case of LPEC units, the increment of free hemoglobin is clearly reduced throughout the storage time. The fact that the variation in free Hb is associated to the release of the intracellular enzyme LDH, a marker of cellular damage, assures that we are observing the result of cells' aging and death with subsequent lysis. The correlation between Hb and LDH increase is good in both preparation, LPEC and NLPEC ($r = 0.93$ and 0.97, respectively).

Electrolytes (K$^+$/Na$^+$) Concentration in the Supernantant during Storage

Figure 3 shows that for both LPEC and NLPEC the K$^+$ and Na$^+$ concentration in their supernatant changes across storage period with same trend and extension: K$^+$ concentration increases up to ~50 mM at Day 42, while the Na$^+$ concentration exhibits a 26% decrease. Our data are consistent with the values reported in the literature and, as generally accepted, result from the low storage temperature.

Proteins and Free Amino Acids

The supernatant of leukodepleted RBCs exhibits a significantly reduced total proteins content with respect to non-leukodepleted, since Day 2, ($p < 0.0001$), with a Δ-reduction of about 0.4 g/dL maintained over time (**Figure 4A**). Indeed, in the case of LPEC part of the residual plasma proteins are expected to be filtered out. During storage, an increase of the total protein content is observed in both preparations (≈0.2 g/dL, $p < 0.001$, Day 2 vs. Day 42).

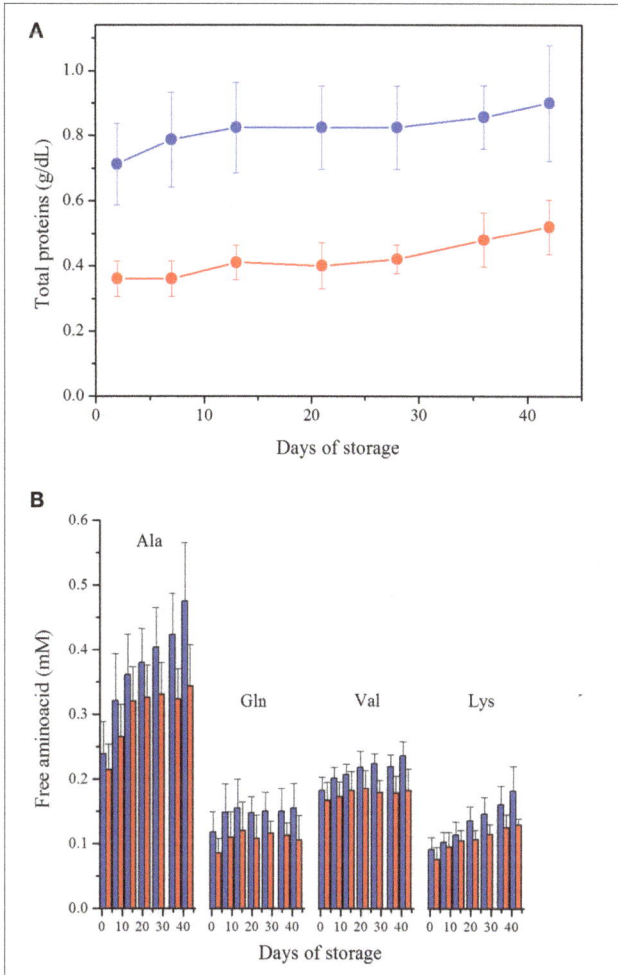

FIGURE 4 | (A). Protein content. ANOVA showed significance ($p < 0.0005$) for both storage time and preparation. Interaction between variables is not significant. LPEC vs. NLPEC comparison at each storage time is always significant ($p < 0.0001$). **(B)** Free amino acids quantification, in the supernatant, during storage. LPEC (red bar) vs. NLPEC (blue bar) comparisons at each storage time addressed all amino acids concentrations significantly higher ($p < 0.05$) in NLPEC units from Day 21.

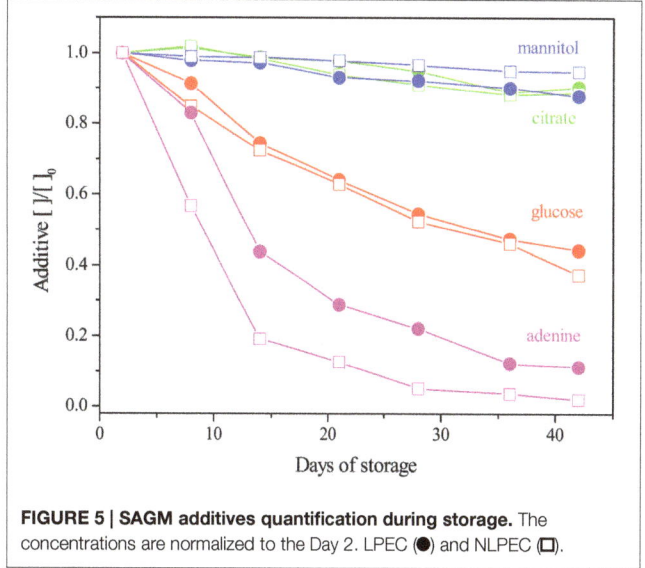

FIGURE 5 | SAGM additives quantification during storage. The concentrations are normalized to the Day 2. LPEC (●) and NLPEC (□).

Using ^{1}H-NMR spectroscopy, we have been able to identify 11 free amino acids (Gly, Ala, Gln, Phe, His, Tyr, Trp, Leu, Ile, Val, Lys) in the supernatants. Two-way ANOVA indicated a significant effect of both storage time and preparation ($p < 0.0005$) over amino acids concentrations. Ala, Gln, Val and Lys starting from Day 21 (**Figure 4B**) were always significantly higher in NLPEC ($p < 0.05$). As for the other amino acids, instead, we found a significant higher concentration in NLPEC ($p < 0.05$) starting either from Day 28 or Day 36 (data not shown). These results suggest that in the NLPEC units the extent of cell lysis throughout the storage time is constantly higher than in LPEC. This event leads to a concentration of proteases that is progressively higher in the NLPEC units than in LPEC and, therefore, to an enhanced protein degradation in the NLPEC units: note that during storage WBCs number, in NLPEC, decreases from 0.95×10^9/unit at Day 2 to 0.32×10^9/unit at Day 42.

Variation of Additives Concentration in the Storage Medium

Figure 5 reports the variation, during storage, of the concentrations of SAGM components measured by ^{1}H-NMR. Citrate concentration is stable and comparable, over time, in both NLPEC and LPEC samples as previously reported (Pertinhez et al., 2014). Gevi et al. (2012) reported that in RBCs, stored in SAGM, mannitol decreases overtime. Our results reveal a slight decrease of mannitol concentration in LPEC. Knowing that mannitol is not metabolized, this result points to a facilitated diffusion of this additive inside leukodepleted RBCs.

In NLPEC samples we measure a faster decrease of adenine concentration, this behavior being prominent in the initial 15 days. We interpret this trend as the result of the competition of WBCs and platelets with RBCs for that metabolite that is essential to produce ATP. After Day 15 when apparently most of those cells begins to lyse, as we inferred from the increase of the protein and free amino acid content in the supernatant (see above), the rate of adenine consumption equalizes between the two preparations.

As for glucose consumption, **Figure 5** shows it is comparable in both preparations. This result, that at first glance appears contradictory with the trend of adenine depletion, can be rationalized if we recall that in SAGM, the concentrations of glucose (45.4 mM) and adenine (1.25 mM) are quite different, and that the number of WBCs and platelets is considerably smaller than the number of RBCs. Therefore, considering the dynamic range of our measurements, the metabolic activity of WBCs and platelets results more evident when measuring the variation of adenine concentration with respect to the extent of glucose variation.

Overall, these results clearly indicate a negative effect of the presence of WBCs and platelets on RBCs viability during the storage time, as they significantly reduce adenine concentration that turns out to be a limiting factor. In fact, a significant difference is observed through the first 21 days ($p \leq 0.005$).

Metabolites Concentration in RBC Lysates

Table 1S reports a list of the 39 identified metabolites in RBC lysates and their concentrations on Days 2 and 42, for both preparations. A one-way ANOVA was performed to highlight which metabolites were influenced by preparation: 12 compounds presented $p < 0.005$ (**Table 1**). At Day 2, LPEC present a higher concentration of adenine, citrate, mannitol, and glucose. Alanine and urea, instead, are present

TABLE 1 | Metabolites identified on RBCs lysates of NLPEC and LPEC. Concentration as reported as mean ± SD at Day 2 and Day 42.

Metabolite	NLPEC (mM)		LPEC (mM)	
	Day 2	Day 42	Day 2	Day 42
Adenine*	0.119 ± 0.004	0.009 ± 0.006	0.220 ± 0.073	0.017 ± 0.024
Alanine*	0.388 ± 0.078	0.426 ± 0.073	0.309 ± 0.031	0.340 ± 0.045
AMP	0.140 ± 0.059	0.476 ± 0.160	0.137 ± 0.035	0.188 ± 0.048
ATP	1.228 ± 0.184	0.217 ± 0.093	1.425 ± 0.218	0.445 ± 0.089
Citrate*	0.017 ± 0.020	0.020 ± 0.011	0.056 ± 0.042	0.036 ± 0.022
Glucose*	6.319 ± 1.565	3.235 ± 1.105	11.960 ± 3.897	3.405 ± 1.069
Glutathione	2.378 ± 0.426	1.365 ± 0.165	2.692 ± 0.533	1.981 ± 0.529
Hypoxanthine	0.015 ± 0.001	0.393 ± 0.131	0.017 ± 0.025	0.280 ± 0.056
IMP	0.106 ± 0.038	0.145 ± 0.036	0.082 ± 0.020	0.104 ± 0.029
Mannitol*	0.324 ± 0.121	1.421 ± 0.486	0.483 ± 0.019	2.222 ± 0.261
5-oxoproline	0.086 ± 0.027	0.710 ± 0.064	0.076 ± 0.031	0.646 ± 0.031
Urea*	0.459 ± 0.120	1.005 ± 0.298	0.238 ± 0.113	0.490 ± 0.124

Metabolites were significantly affected by preparation (ANOVA, $p < 0.005$). Post hoc tests were performed. In red are highlighted the metabolites whose concentrations at Day 42 are significantly different ($p < 0.05$) between NLPEC and LPEC. *indicates the metabolites that significantly different ($p < 0.05$) at Day 2 between both preparations.

at lower concentration. Before any handling, the RBCs cytosolic concentration of those molecules, with respect to SAGM, is lower for adenine, citrate, mannitol and glucose and higher for alanine and urea; we interpret the data obtained for LPEC at Day 2 as the result of filtration. In fact during filtration, it is reasonable to expect that the perturbation of the membrane organization may enhance its osmotic permeability favoring the intake of the SAGM components, and the exit of the cytosolic metabolites. Similarly, we observed significant differences of the cytosolic concentration of some metabolite ($p < 0.05$), at Day 42. The ATP concentration in NLPEC samples is almost 50% lower with respect to LPEC, while purine's degradation catabolites, AMP, IMP and hypoxanthine show increased concentrations (2.5 fold AMP and 1.4 fold IMP and hypoxanthine). In addition, in NLPEC a decrease of glutathione (GSH) is accompanied by an increase of 5-oxoproline, thus pointing to the possibility that, during storage, non-leukodepleted RBCs are more keen to suffer a reduction of their antioxidant defenses, (Pertinhez et al., 2014).

Chemometric Analysis of RBC Lysates

As described in previous works (Pertinhez et al., 2014; Casali et al., 2015), using ^{1}H-NMR spectroscopy we identified a number of metabolites in the suspension medium of LPEC. However, due to the high concentration of the additives, as well as of the lactate produced by the cells during storage, that covered a wide region of the ^{1}H-NMR spectra, the Principal Component Analysis (PCA) of the supernatants composition could not be carried out. The PCA was then performed on LPEC and NLPEC lysates (**Figure 6**).

The first principal component (PC1) accounts for 63.3% of the total variation in the dataset and well separate LPEC (orange to violet symbols) from NLPEC (light to dark blue symbols).

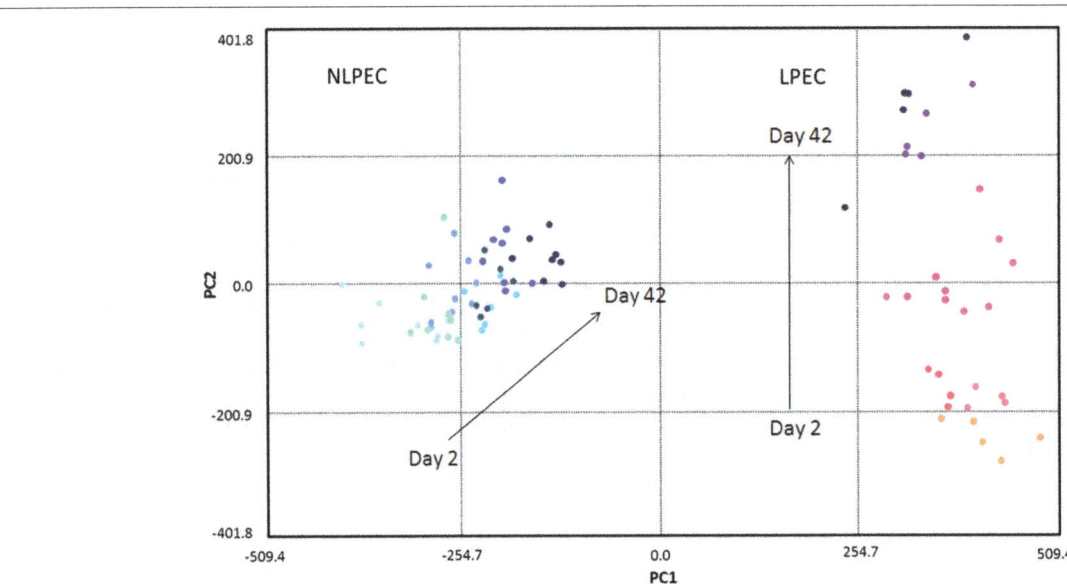

FIGURE 6 | Chemometric analysis of RBC lysates samples. The score plot reports NLPEC samples (light to dark blue symbols) and LPEC (orange to violet symbols). PC1 separates on the x axis the two preparations with the major contribution of glucose, lactate and alanine. Mannitol, ATP and GSH contribute to PC2, that on the y axis separate each preparations by day of storage.

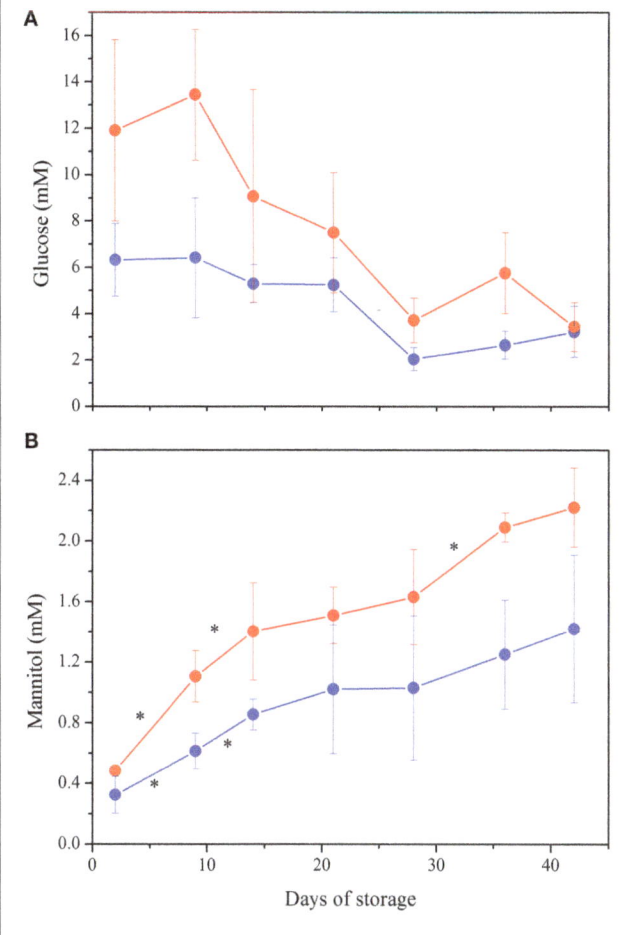

FIGURE 7 | NMR quantification of the major contributors to PCA plot: (A) Glucose. ANOVA indicated significant interaction between storage time and preparation ($p < 0.001$) as well as a significant effect of both variables alone ($p < 0.0005$). *Post hoc* tests showed higher concentration of glucose in LPEC from Day 2 to Day 36 ($p < 0.05$). **(B)** Mannitol. Storage time and preparation affect independently mannitol concentration ($p < 0.0005$ for both variables). *indicate $p < 0.05$ when comparing each storage time with the following one, in both preparations. Samples NLPEC (●), LPEC (●).

Glucose is the major contributor to PC1 together with alanine and lactate that however exert a minor influence (data not shown).

PC2 accounts for 11.6% of the discrimination among RBCs content and allows to evaluate the changes that occur in the metabolites present in the cytoplasm during the storage. The separation of RBCs as a function of their age is particularly evident for LPEC (**Figure 6**, right side). Mannitol turns out to be the major contributor to PC2, with ATP and GSH as additional factors. The ability of mannitol to discriminate RBCs during storage is confirmed by t-student's p-values obtained for both LPEC and NLPEC, when comparing each storage time with the following one of samples belonging to the same preparation. This is particularly evident between Days 2, 9, and 16 (**Figure 7B**).

The wide spread produced by PC2 only for LPEC supports the idea that filtration selects RBCs with similar morphological and biochemical features and that exhibit a comparable aging trend.

Figures 7A,B complement the PCA and allow to better understand the metabolic evolution of NLPEC and LPEC over time. **Figure 7A** shows that in LPEC, glucose concentration is significantly higher ($p < 0.01$) than in NLPEC, throughout storage, except for Day 42. Interestingly however, while in the initial 9 days of storage LPEC presents a concentration of glucose that is twice as much the one present inside NLPEC, from Day 14 throughout Day 36 the difference reduces to about 30% and equalize at Day 42. Overall these data support the hypothesis outlined above that filtration favors an initial burst of glucose uptake. **Figure 7B** shows that mannitol diffusion inside RBCs proceed more effectively in LPEC than in NLPEC throughout the storage time, consistently with the decrease of its concentration measured in the supernatant (**Figure 5**). All these data support the hypothesis that leukodepleted RBCs are characterized by a plasma membrane that preserve, for a longer time, its integrity, a fact compatible with their longer viability.

DISCUSSION

RBCs modifications over time are well-recognized to be one of the factors responsible for the negative outcomes associated to blood transfusion therapy (Sparrow, 2015). Thus, there is a stringent need to improve the protocols used to prepare RBCs concentrates for blood banks. Nowadays, around the world, both leukoreduction and leukodepletion are accepted as pre-storage protocols as they appears to prolong the viability/quality of RBCs units. A recent decree of the Italian Ministry of Health forces all Italian blood transfusion centers to include pre-storage leukodepletion in the preparation of RBCs concentrates. Because the scientific basis of the good clinical results obtained by reducing the number of leukocytes and platelets in the RBCs units to be stored is still matter of debate, and considering that leukodepletion is a more expensive protocol, we have been prompted to seek the biophysical/biochemical motivations that may justify that choice. Answering those questions will allow to devise a rationale improvement of the current storage protocols.

To tackle this issue we chose NMR spectroscopy in combination with standard biochemical (Gallo et al., 2015). Overall we have been able to follow the consumption of the additives necessary to RBCs survival, even in the presence of their high concentration we identified more than 30 metabolites in the supernatant, and 39 metabolites in the lysates RBCs. Interestingly, some of our data were confirmed by MS in the RBCs additive solution AS-3 (D'Alessandro et al., 2015).

Our results highlighted that leukodepletion, differently from leukoreduction: i) selects a homogeneous population of RBCs characterized by a healthier and more deformable plasma membrane that assures a prolonged viability. The stress due to filtration destroys the not perfectly fit RBCs, as revealed by the high hemolysis measured at Day 2 (**Figure 2**) and favors a burst of glucose uptake by the healthy RBCs (**Figure 7A**) able to pass through the filter. ii) by removing WBCs and platelets provides RBCs with an appropriate concentration of nutrients for a longer time (**Figure 5**). The removal of WBCs and platelets is particularly relevant with respect to the consumption of adenine that in NLPEC turns out to be a limiting factor for cells viability particularly after Day 28 (**Figure 5**). Indeed, recognizing that to preserve the RBCs energy requirement, the adenine nucleotide pool, under homeostatic conditions, it requires approximately 2 mM ATP, 0.1 mM ADP, and 0.04 mM AMP (Gibson and Harris, 2002), if we estimate the adenine nucleotide pool in NLPEC and LPEC toward the end of the allowed storage time, based on the data reported in **Table 1** and Table 1S, we find that LPEC better preserve the adenine nucleotides balance.

The PCA shows also that while the spread of LPEC determined by PC1 is limited, NLPEC are distributed over a wider range of values. Moreover, if we consider the PC2 component we observe it clearly separates LPEC according to their age, while it has practically no effect on NLPEC. This result further confirms the fact that filtration generates a homogeneous population of RBCs evolving uniformly in time.

We feel these results, *per se*, might already justify the advantage of pre-storage leukodepletion over leukoreduction.

In addition we found that the NLPEC suspension medium presents a higher concentration of free amino acids, especially on late storage times (**Figure 4B**) that, together with the associated higher proteins' content (**Figure 4A**), suggests the presence of proteases released by dead blood cells. It is worth noting that reduction of proteins and proteins derivatives in the RBCs units, as it occurs in LPEC, may also reduce unexpected immunoreaction in transfused patients.

In summary, our biochemical and biophysical data point to pre-storage leukodepletion as an appropriate and preferable, with respect to leukoreduction, protocol to preserve the quality of the RBCs over time.

We believe this more in depth vision of the behavior of leukoreduced and leukodepleted RBCs, during storage in blood bank conditions, will stimulate new ideas to further optimize storage protocols.

AUTHOR CONTRIBUTIONS

Study conception and design: TP, EC, RB, and AS; Acquisition of data: TP, EC, FB, and PB; Analysis and interpretation of data: TP, EC, FB, PB, RB, and AS; Drafting of manuscript: TP, EC, FB, PB, RB, and AS; Critical revision: RB and AS; Final approval of the version to be published: TP, EC, FB, PB, RB, AS.

FUNDING

This work was supported by Arcispedale Santa Maria Nuova - IRCCS Reggio Emilia, Italy.

ACKNOWLEDGMENTS

The authors thank the CIM Laboratory, Technopole Parma, University of Parma, Italy, for the use of NMR Spectrometer. Prof. G. Pedrazzi is acknowledge for help with the statistical analysis.

REFERENCES

Alexander, P. E., Barty, R., Fei, Y., Vandvik, P. O., Pai, M., Siemieniuk, R. A., et al. (2016). Transfusion of fresher vs older red blood cells in hospitalized patients: a systematic review and meta-analysis. *Blood* 127, 400–410. doi: 10.1182/blood-2015-09-670950

Bianchi, M., Vaglio, S., Pupella, S., Marano, G., Facco, G., Liumbruno, G. M., et al. (2015). Leucoreduction of blood components: an effective way to increase blood safety? *Blood Transfus.* 16, 1–14. doi: 10.2450/2015.0154-15

Bosman, G. J. (2013). Survival of red blood cells after transfusion: processes and consequences. *Front. Physiol.* 4:376. doi: 10.3389/fphys.2013.00376

Casali, E., Berni, P., Spisni, A., Baricchi, R., and Pertinhez, T. A. (2015). Hypoxanthine: a new paradigm to interpret the origin of transfusion toxicity. *Blood Transfus.* doi: 10.2450/2015.0177-15. [Epub ahead of print].

Chang, H., Hall, G. A., Geerts, W. H., Greenwood, C., McLeod, R. S., and Sher, G. D. (2000). Allogeneic red blood cell transfusion is an independent risk factor for the development of postoperative bacterial infection. *Vox Sang.* 78, 13–18. doi: 10.1046/j.1423-0410.2000.7810013.x

D'Alessandro, A., D'Amici, G. M., Vaglio, S., and Zolla, L. (2012). Time-course investigation of SAGM-stored leukocyte-filtered red bood cell concentrates: from metabolism to proteomics. *Haematologica* 97, 107–115. doi: 10.3324/haematol.2011.051789

D'Alessandro, A., Nemkov, T., Kelher, M., West, F. B., Schwindt, R. K., Banerjee, A., et al. (2015). Routine storage of red blood cell (RBC) units in additive solution-3: a comprehensive investigation of the RBC metabolome. *Transfusion* 55, 1155–1168. doi: 10.1111/trf.12975

Dzieciatkowska, M., Silliman, C. C., Moore, E. E., Kelher, M. R., Banerjee, A., Land, K. J., et al. (2013). Proteomic analysis of the supernatant of red blood cell units: the effects of storage and leukoreduction. *Vox Sang.* 105, 210–218. doi: 10.1111/vox.12042

Eisenfeld, L., Silver, H., McLaughlin, J., Klevjer-Anderson, P., Mayo, D., Anderson, J., et al. (1992). Prevention of transfusion-associated cytomegalovirus infection in neonatal patients by the removal of white cells from blood. *Transfusion* 32, 205–209. doi: 10.1046/j.1537-2995.1992.32392213801.x

Fischer, M., Chapman, J. R., Ting, A., and Morris, P. J. (1998). Alloimmunization to HLA antigens following transfusion with leukocyte-poor and purified platelet suspensions. *Vox Sang.* 49, 331–335. doi: 10.1111/j.1423-0410.1985.tb00807.x

Gallo, V., Intini, N., Mastrorilli, P., Latronico, M., Scapicchio, P., Triggiani, M., et al. (2015). Performance assessment in fingerprinting and multi component quantitative NMR analyses. *Anal. Chem.* 87, 6709–6717. doi: 10.1021/acs.analchem.5b00919

Gevi, F., D'Alessandro, A., Rinalducci, S., and Zolla, L. (2012). Alterations of red blood cell metabolome during cold liquid storage of erythrocyte concentrates in CPD-SAGM. *J. Proteomics.* 76 Spec No.:168–180. doi: 10.1016/j.jprot.2012.03.012

Gibson, D., and Harris, R. A. (2002). *Metabolic regulation in Mammals.* New York, NY: Taylor & Francis.

Gong, M. N., Thompson, B. T., Williams, P., Pothier, M. L., Boyce, P. D., and Christiani, D. C. (2005). Clinical predictors of mortality in acute respiratory distress syndrome: potential role of red cell transfusion. *Crit. Care Med.* 33, 1191–1198. doi: 10.1097/01.CCM.0000165566.82925.14

Hess, J. R., and Grazzini, G. (2010). Blood proteomics and transfusion safety. *J. Proteomics.* 73, 365–367. doi: 10.1016/j.jprot.2009.09.019

Lacroix, J., Hébert, P., Fergusson, D., Tinmouth, A., Blajchman, M. A., Callum, J., et al. (2011). The Age of Blood Evaluation (ABLE) randomized controlled trial: study design. *Transfus. Med. Rev.* 25, 197–205. doi: 10.1016/j.tmrv.2011.03.001

Leal-Noval, S. R., Jara-Lopez, I., García-Garmendia, J. L., Marín-Niebla, A., Herruzo-Avilés, A., Camacho-Laraña, P., et al. (2003). Influence of erythrocyte concentrate storage time on postsurgical morbidity in cardiac surgery patients. *Anesthesiology* 98, 815–822. doi: 10.1097/00000542-200304000-00005

Leal-Noval, S. R., Rincon-Ferrari, M. D., García-Curiel, A., Herruzo-Avilés, A., Camacho-Laraña, P., Garnacho-Montero, J., et al. (2001). Transfusion of blood components and postoperative infection in patients undergoing cardiac surgery. *Chest* 119, 1461–1468. doi: 10.1378/chest.119.5.1461

Moore, F. A., Moore, E. E., and Sauaia, A. (1997). Blood transfusion: an independent risk factor for post-injury multiple organ failure. *Arch Surg.* 132, 620–624. doi: 10.1001/archsurg.1997.01430300062013

Nemkov, T., Hansen, K. C., Dumont, L. J., and D'alessandro, A. (2016). Metabolomics in transfusion medicine. *Transfusion* 56, 980–993. doi: 10.1111/trf.13442

Novotny, V. M., van Doorn, R., Witvliet, M. D., Claas, F. H., and Brand, A. (1995). Occurrence of allogeneic HLA and non-HLA anti-bodies after transfusion of pre-storage filtered platelets and red blood cells: a prospective study. *Blood* 85, 1736–1741.

Pertinhez, T. A., Casali, E., Lindner, L., Spisni, A., Baricchi, R., and Berni, P. (2014). Biochemical assessment of red blood cells during storage by ^{1}H NMR spectroscopy. Identification of a biomarker of their oxidative stress protection level. *Blood Transfus.* 12, 548–556. doi: 10.2450/2014.0305-13

Prudent, M., Tissot, J. D., and Lion, N. (2015). *In vitro* assays and clinical trials in red blood cell aging: lost in translation. *Transfus. Apher. Sci.* 52, 270–276. doi: 10.1016/j.transci.2015.04.006. doi: 10.1016/j.transci.2015.04.006

Silliman, C. C., Moore, E. E., Kelher, M. R., Khan, S. Y., Gellar, L., and Elzi, D. J. (2011). Identification of lipids that accumulate during the routine storage of prestorage leukoreduced red blood cells and cause acute lung injury. *Transfusion* 51, 2549–2554. doi: 10.1111/j.1537-2995.2011.03186.x

Sparrow, R. L. (2015). Red blood cell storage duration and trauma. *Transfus. Med. Rev.* 29, 120–126. doi: 10.1016/j.tmrv.2014.09.007

Spitalnik, S. L., Triulzi, D., Devine, D. V., Dzik, W. H., Eder, A. F., Gernsheimer, T., et al. (2015). 2015 Proceedings of the national heart, lung, and blood institute's state of the science in transfusion medicine symposium. *Transfusion* 55, 2282–2290. doi: 10.1111/trf.13250

Steiner, M. E., Assmann, S. F., Levy, J. H., Marshall, J., Pulkrabek, S., Sloan, S. R., et al. (2010). Addressing the question of the effect of RBC storage on clinical outcomes: the red cell storage duration study (RECESS) (Section 7). *Transfus. Apher. Sci.* 43, 107–116. doi: 10.1016/j.transci.2010.05.014

Taylor, R. W., Manganaro, L., O'Brien, J., Trottier, S. J., Parkar, N., and Veremakis, C. (2002). Impact of allogenic packed red blood cell transfusion on nosocomial infection rates in the critically ill patient. *Crit. Care Med.* 30, 2249–2254. doi: 10.1097/01.CCM.0000030457.48434.17

Vamvakas, E. C., and Carven, J. H. (1999). Transfusion and postoperative pneumonia in coronary artery by pass graft surgery: effect of the length of storage of transfused red cells. *Transfusion* 39, 701–710. doi: 10.1046/j.1537-2995.1999.39070701.x

Veale, M. F., Healey, G., and Sparrow, R. L. (2011). Effect of additive solutions on RBC membrane properties of stored RBCs prepared from whole blood held for 24 hours at room temperature. *Transfusion* 51, 25S–33S. doi: 10.1111/j.1537-2995.2010.02960.x

Wang, D., Sun, J., Solomon, S. B., Klein, H. G., and Natanson, C. (2012). Transfusion of older stored blood and risk of death: a meta-analysis. *Transfusion* 52, 1184–1195. doi: 10.1111/j.1537-2995.2011.03466.x

Zolla, L., D'alessandro, A., Rinalducci, S., D'amici, G. M., Pupella, S., Vaglio, S., et al. (2015). Classic and alternative red blood cell storage strategies: seven years of "-omics" investigations. *Blood Transfus.* 13, 21–31. doi: 10.2450/2014.0053-14

13

Responses to Microbial Challenges by SLAMF Receptors

Boaz Job van Driel[1], Gongxian Liao[1], Pablo Engel[2] and Cox Terhorst[1]*

[1] *Division of Immunology, Beth Israel Deaconess Medical Center, Harvard Medical School, Boston, MA, USA,*
[2] *Immunology Unit, Department of Cell Biology, Immunology and Neurosciences, Medical School, University of Barcelona, Barcelona, Spain*

Edited by:
Abhay Satoskar,
The Ohio State University, USA

Reviewed by:
Christine Anne Biron,
Brown University, USA
Steve Oghumu,
The Ohio State University, USA

*Correspondence:
Cox Terhorst
cterhors@bidmc.harvard.edu

The SLAMF family (SLAMF) of cell surface glycoproteins is comprised of nine glycoproteins and while SLAMF1, 3, 5, 6, 7, 8, and 9 are self-ligand receptors, SLAMF2 and SLAMF4 interact with each other. Their interactions induce signal transduction networks *in trans*, thereby shaping immune cell–cell communications. Collectively, these receptors modulate a wide range of functions, such as myeloid cell and lymphocyte development, and T and B cell responses to microbes and parasites. In addition, several SLAMF receptors serve as microbial sensors, which either positively or negatively modulate the function of macrophages, dendritic cells, neutrophils, and NK cells in response to microbial challenges. The SLAMF receptor–microbe interactions contribute both to intracellular microbicidal activity as well as to migration of phagocytes to the site of inflammation. In this review, we describe the current knowledge on how the SLAMF receptors and their specific adapters SLAM-associated protein and EAT-2 regulate innate and adaptive immune responses to microbes.

Keywords: receptors, homophilic, SLAM, SAP, EAT-2, XLP, measles, *Escherichia coli*

SLAM FAMILY RECEPTORS AND THEIR ADAPTORS SAP AND EAT-2

The SLAMF Gene Family

Seven of the nine members of the signaling lymphocytic activation molecule (SLAM) gene Family (SLAMF1–7), a subfamily of the immunoglobulin superfamily, cluster on the long arm of human and mouse chromosome 1 (1). While SLAMF8 and SLAMF9, as well as the SLAM-associated adaptor EAT-2 (*SH2D1B*) are located in close proximity to the "core" SLAMF locus (shown in **Figure 1**), the SAP (SH2D1A) gene is on the X-chromosome [reviewed in Ref. (2, 3)]. The nine SLAMF genes encode cell surface receptors, whose expression is mostly confined to hematopoietic cells (**Table 1**). A wide range of these cells expresses at least one member. The activation state, presence of the adaptor molecules SAP and EAT-2, and the location of immune cells dictate SLAMF receptor expression and function (**Figure 2**). While SLAMF receptors share intracellular interaction partners and display overlapping features, the individual members of this family have a unique functional signature.

The consensus structure of SLAMF receptors consists of an extracellular membrane distal IgV domain linked to a proximal IgC2 domain, a transmembrane region, and an intracellular signaling domain that often contains several intracellular tyrosine-based switch motives (ITSM) (**Figure 1**). Notable exceptions to the consensus structure are SLAMF2, which lacks the intracellular and transmembrane region and instead harbors a glycosyl-phosphatidylinositol membrane anchor; SLAMF3,

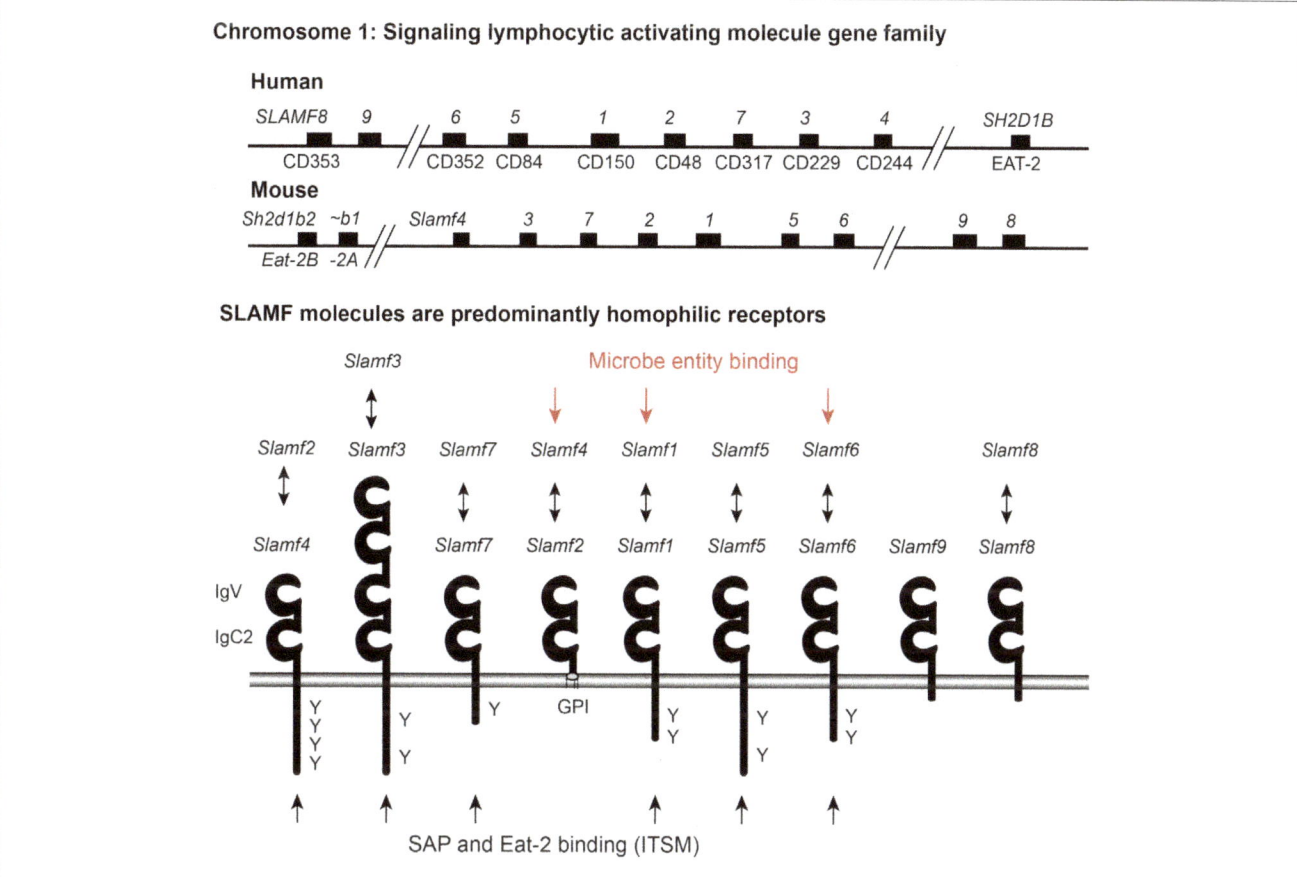

FIGURE 1 | Signaling lymphocytic activating molecule gene family (SLAMF receptors family) and proteins. Organizational overview of the SLAM family cluster on chromosome 1 in both human and mice. EAT-2 is also located proximal to this gene cluster and is duplicated in mice, encoding Eat-2a and Eat-2b. The SLAMF receptors are part of the Ig-superfamily and they have an IgV and an IgC2 domain. Seven of the SLAM receptors are homophilic ligands. SLAMF2 and SLAMF4 are co-ligands that bind each other. Three SLAM genes have been shown to possess bacterial binding capacity. Six of the SLAM receptors have docking domains for SAP (and EAT-2) represented by Y (tyrosine in ITSM). SLAMF2 is anchored to the plasma membrane by a GPI-anchor.

which has a duplication of the IgV–IgC2 domains; and SLAMF8 and SLAMF9, which only have ~30 intracellular amino acid residues and lack ITSMs.

Most SLAMF Receptors Are Homophilic

Most SLAMF receptors are self-ligands with signaling motifs, which function in cell–cell communication. Crystal structures of SLAMF1, SLAMF5, and SLAMF6 revealed an angled engagement of the IgV domains *in trans* (4, 5). Exceptions to this homotypic engagement are SLAMF2 and SLAMF4, which are counter-structures (6–8). Ligation of SLAMF receptors leads to inhibitory or activating signaling events through modulation of the cellular responses. Interestingly, SLAMF receptors can also engage microbial structures. For example, SLAMF1 partakes in a xenophilic interaction with the hemagglutinin MH-V of Measles virus, which facilitates viral entry as well as cell fusion (9, 10). As this interaction is thought to benefit the virus, it is *pathogen-centric*. Additional studies also revealed cognate interactions of SLAMF1, SLAMF2, and SLAMF6 with bacterial components (**Table 2**) (11–13). This class of xenophilic interactions appears to be beneficial for the host and is, therefore, *host-centric*.

The SLAMF-Specific Adaptor Proteins SAP and EAT-2

A little under two decades ago, three independent research groups discovered an association between mutations in *SH2D1A*, the gene that encodes the intracellular adaptor protein SLAM-associated protein (SAP) and X-linked lymphoproliferative syndrome (XLP) (14–16). At the same time, we showed that SAP is an intracellular binding partner of SLAMF1, which is required for proper functioning of SAP in response to Epstein–Barr virus (EBV) and other virus. In XLP patients, SAP is mutated or absent resulting in aberrant functioning of SLAMF1 (16).

SLAM-associated protein encodes a small adaptor protein (14 kDa) that consists almost entirely of a Src homology 2 (SH2) domain. SAP can interact with the ITSMs motif of six SLAMF receptors in phospho-tyrosine-dependent and independent modes (**Figure 1**) (16–19). Mice that are deficient for the gene that encodes SAP (*Sh2d1a$^{-/-}$*) have a range of specific immune malfunctions, which manifest the development and maturation of immune cells and during responses to microbial challenges (20–22). Although SAP expression by T-cells, NK cells, and NKT-cells is well established, B-cells express SAP only under certain

TABLE 1 | Slam receptor expression, associated effector molecules, and functions.

	Expression	Effectors	SAP-dependent	Eat2-dependent	Other/unknown
SLAMF1, SLAM, CD150	Act T, act B, mono, Mø, DC, plat, HSC	Fyn, Lck, SHIP-1, Src, Shp-1/2, PKCθ, Bcl-10, Beclin-1, PI3K, Nf-κB, Ras-GAP, Akt, JNK1/2, Dok-1/2	T: (+) IL-4, IL-13, proliferation, Th2/Th17 polarization, NKT: development (with Slamf6)	Unknown	T: (+) IFNγ, B: (+) proliferation and activation, (+) apoptosis, Mø: (+) ROS, IL-12, TNFα, NO, (−) IL-6, (+) myeloid cell migration, (+) platelet aggregation, (+) phagocytosis
SLAMF2, CD48	Pan-lymphocyte	Lck, Fyn, RhoA	N/A	N/A	T: (+) IL-2, proliferaton, B: (+) activation, (−) apoptosis Mast: (+) TNFα, eo: (+) activation, mobilization, Mø: (+) TNFα, IL-12, (+) phagocytosis, DC: (+) survival
SLAMF3, Ly-9, CD229	T, B, iCD8, NKT, mono, Mø, HSC	AP-2, Grb-2, ERK, PLZF, NFAT	Unknown	Unknown	T: (−) IFNγ, (+) proliferation, IL-2, IL-4, iCD8+ T-cells, iNKT (−) development
SLAMF4, 2B4, CD244	NK, NKT, T, γδ, CD8, DC, eo, mast, mono	LAT, PI3K, Vav-1, SHIP, c-Cbl, ERK, Shp-1/2, PLC-γ, 3BP2, Csk	T: (−) IFNγ, NK/CD8+: (+) cytotoxicity, proliferation	NK: (−) Cytotoxicity of Slamf2-neg target cells, (−) IFNγ	eo: (+) adhesion, chemotaxis, peroxidase, (+) IFNγ, IL-4
SLAMF5, CD84	Pan-lymphocyte plat, mast, eo	Dok-1, c-Cbl, ERK, JNK, Fes, Shp-1, Nf-κB	T-B: (+) GC response	NK: (+) Cytotoxicity Mast: (+) Degranulation	lat: (+) spreading
SLAMF6, NTB-A, Ly-108	NK, NKT, T, B, Mø, pDC	PLC-γ, SHIP, Shp-1/2, PI3K, PLZF, Lck, PKCθ, NFAT	T-B: (+) GC response, NK: (+) IFNγ, NKT: development (with Slamf1)	NK: (+) Cytotoxicity	T-B: (−) GC response, Neutro: (+) ROS, (+) IL-6, TNFα
SLAMF7, CRACC, CS1, CD319	T, B, mono, DC, NK	PLC-γ, c-Cbl, SHIP, Akt, Vav-1, Shp-1/2	Unknown/N/A	NK: (+) Cytotoxicity	NK: without Eat2 (−) Cytotoxicity, B: (+) proliferation
SLAMF8, BLAME	iCD8, mono, DC, Mø, Neu, endo, FRC	PKC, p40(phox)	N/A	N/A	(−) myeloid cell migration, (−) ROS, iCD8+ T-cells, iNKT (+) development
SLAMF9, SF2001	mono, DC	ND	N/A	N/A	Unknown

T, T cells; B, B cells; act, activated; Mø, macrophage; DC, dendritic cell; plat, platelet; HSC, hematopoietic stem cell; mono, monocyte; NKT, natural killer T cell; eo, eosinophil; γδ, γδ receptor-expressing T cell; mast, mast cell; endo, endothelial cell; FRC, fibroblastic reticular cell; ROS, reactive oxygen species.
Expression data are based on murine expression.

specific conditions (23, 24). Some EBV-transformed B-cells, Hodgkin's lymphomas, and germinal center (GC) B-cells appear to express SAP. The second SLAMF-associated adaptor, EAT-2, exhibits distinct functional features and is not associated with any primary human immune deficiency (25). EAT-2 binds different ITSMs in SLAMF receptors and is involved in the activation of antigen-presenting cells (APCs) and cytotoxicity of NK cells (25, 26). The expression profile of this adaptor also differs from SAP. NK cells express EAT-2 as do a range of APCs, including monocytes (25, 27).

Two SAP signaling modes exist: (1) blockade of the binding of SH2-domain-containing molecules, e.g., the tyrosine-phosphatases SHP-1 and SHP-2 to phosphorylated ITSMs and (2) recruitment of the Src kinase Fyn in its active ("open") configuration to SAP (3, 16, 28–30). The blocking function of SAP is due to its high affinity for ITSM motifs caused by an unusual three-pronged binding of the SH2 domain (31). In the absence of SAP, SLAMF1 and SLAMF6 bind the tyrosine phosphatases SHP-1 and/or SHP-2, which are negative regulators of T cell functions (16, 17, 32).

A set of functions of SAP in T-cells is dependent on the recruitment of the Src kinase Fyn, which is intricately involved

in T-cell receptor (TCR) signaling (**Figure 3**). SLAMF–SLAMF homophilic ligation leads to the recruitment of SAP to their ITSMs, which interacts with the SH3 domain of Fyn (28, 33). Binding of Fyn to SLAMF1-associated-SAP enhances IL-4 and IL-13 production (29). Structural analyses have shown that Arg78 of SAP is crucial to this interaction (28, 29). Indeed, SAP[R78A] mice showed a lack of IL-4 production, similar to that of Sh2d1a−/− mice (29). Lacking this arginine (28), EAT-2 does not interact with Fyn but associates with a variety of different Src kinases (27). Similar to Sh2d1a−/− T-cells, SLAMF1−/− CD4+ T-cells are also less prone to TCR-mediated IL-4 production (34). It was, therefore, concluded that SLAMF1 contributes to Th2 polarization. Subsequent studies showed that a signaling cascade involving SAP and Fyn as well as GATA-3 transcriptional promotion by NF-κB are responsible for this phenotype (22, 35, 36). This pathway in T-follicular helper cells effectively contributes to GC B-cell maintenance and optimal humoral responses (37).

Overall, these studies have demonstrated that SLAMF receptors and SAP have a complex involvement in mechanisms that fight intracellular infections, via their effect on cytokine production. Together, SAP and EAT-2 dictate the major part of the

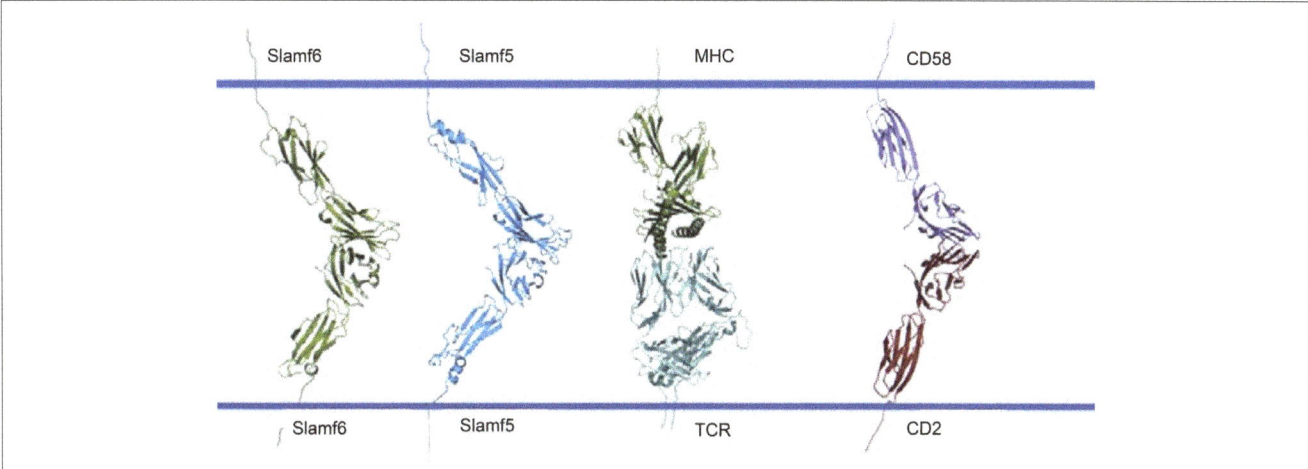

FIGURE 2 | Ribbon representation of Slamf6 and Slamf5 structures. Homophilic interactions of SLAMF6 and SLAMF5 as well as heterophilic interactions between two other Ig-superfamily receptors CD58 and CD2. MHC interacting with TCR functions as a reference for the molecular dimensions. Image adopted from Calpe et al. (2).

SLAMF signaling. However, other mediators dictate a distinct set of SLAMF receptor functions.

SEVERAL SLAMF RECEPTORS INTERACT WITH BACTERIA

SLAMF1 and SLAMF6 Interactions with Gram⁻ Bacteria

The importance of SLAMF receptors in phagocytes was high-lighted by our recent observations that SLAMF1 is involved in cognate interactions with bacterial entities. These interactions result in the defect in the clearance of *Salmonella typhimurium* SseB⁻ after peritoneal infection (11, 12, 38). Thus, direct cognate interactions with microbial components modulate SLAMF functions in phagocytes.

Evidence for direct interactions of SLAMF1 and SLAMF6 with *Escherichia coli* outer membrane porins C (OmpC) and OmpF was shown in a cell-based luciferase reporter assay (11). The specificity of these interactions extends to different Gram⁻ bacteria, but not Gram⁺ bacteria; SLAMF1 interacts with *S. typhimurium* (11); SLAMF6 interacts with *S. typhimurium* and to some degree with *Citrobacter rodentium* (38). Subsequent analyses demonstrated that this interaction depends on the IgV domain of SLAMF1 and SLAMF6. The structure of SLAMF1 has proven difficult to unravel due to the flexible (non-rigid) nature and high degree of glycosylation of SLAMF1. By a combination of techniques, several amino acid residues have been implicated in SLAMF1 homophilic engagement as well as SLAMF1 engagement with Measles virus protein MV-H (10). The FCC beta-sheet and the CC loop of SLAMF1 contain several conserved residues and substitution of Val63, Thr65, Ala67, Lys77, and Glu123 within these regions all resulted in a reduction in the binding of SLAMF1 to SLAMF1 as well as to MV-H. Single mutations of equivalent residues in mouse SLAMF1 resulted in little difference in the binding of OmpC/F containing *E. coli*. In line with this,

SLAMF6 engagement with *E. coli* structures does not require amino acid residues in the SLAMF6 IgV domain that are crucial for SLAMF6–SLAMF6 homophilic ligation (38). However, general masking of interaction domains by mAbs directed against epitopes in the IgV domains of SLAMF1 or SLAMF6 blocked their interactions with bacteria (11, 38). Thus, whereas there is overlap in the SLAMF1 residues that are essential for SLAMF1–SLAMF1 ligation with the residues involved in MV-H binding to SLAMF1, it is likely that OmpC/F binding involves a separate set of interacting SLAMF1 residues. This would suggest that the interaction of SLAMF1 with bacteria is of a separate origin, distinct from the SLAMF1–SLAMF1 interaction domain, and hence may represent a SLAMF1 function of separate evolutionary significance. Structural analyses of SLAMF1 or SLAMF6 and *E. coli* outer membrane porins should provide conclusive insights into the mode of these interactions.

SLAMF1 Enhances Phagocyte Effector Functions

The interaction of SLAMF1 with OmpC/F⁺ *E. coli* results in a more effective phagocytosis of these bacteria by macrophages (11). Clusters of SLAMF1 bound to OmpC/F remain proximal to the bacterium during phagocytosis, thus colocalizing to intracellular phagosomes. A signaling complex is recruited to the intracellular domain of SLAMF1 either directly upon bacterial ligation or shortly thereafter during internalization. The transient recruitment of the autophagy scaffold protein Beclin-1 is the initial event that leads to the formation of a functional complex that also contains Vps34, Vps15, and UVRAG (**Figure 4**) (13). This novel SLAMF1 signaling module is enhanced by, but not prerequisite of the presence of EAT-2 (13). Vps34 supported by its co-enzyme Vps15 is the sole Class III phosphatidylinositol kinase and produces the docking lipid phosphatidylinositol-3′-phosphate (PI₃P) (39). This SLAMF1-enhanced production of PI₃P affects two important phagosomal processes. First, formation and activation

TABLE 2 | Slamf receptors and their adaptor SAP modulate susceptibility to microbes.

	Deficiency: resistant	Deficiency: susceptible	SLAMF ligand	Microbial ligand
SLAMF1	*T. cruzi*	Gram⁻ bacteria, *L. major*	Slamf1	Measles virus, *E. coli* (OmpC/F⁺) *S. typhimurium*
SLAMF2	*S. aureus*	FimH⁺ enterobacterae	Slamf4, CD2	*E. coli (FimH⁺)*
SLAMF3		MCMV	Slamf3	
SLAMF4		LCMV, γHV-68	Slamf2	
SLAMF5			Slamf5	
SLAMF6	*L. mexicana, C. rodentium*	*S. typhimurium*	Slamf6	*E. coli, C. rodentium*
SLAMF7			Slamf7	
SLAMF8			Slamf8	
SLAMF9			?	
SAP		Mouse: γHV-68, LCMV, influenza, human: EBV, some other viruses	Slamf1, 3, 4, 5, 6 human: Slamf7	N/A

SAP (Sh2d1a), SLAM-associated protein; LCMV, lymphocytic choriomeningitis virus; Omp, outer membrane porin; EBV, Epstein–Barr virus; FimH, bacterial lectin; MCMV, murine cytomegalovirus; γHV-68, murine gamma-herpes virus 68.
Deficiency: resistant and deficiency: susceptible refer to observations made in Slamf-deficient mice; resistant indicates that knock out animals have milder disease, susceptible indicates that knock out animals have stronger disease manifestations.
? Unknown.

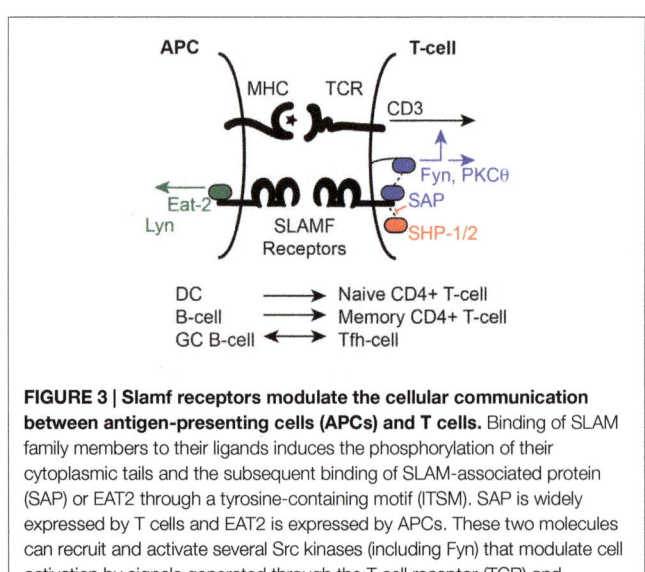

FIGURE 3 | Slamf receptors modulate the cellular communication between antigen-presenting cells (APCs) and T cells. Binding of SLAM family members to their ligands induces the phosphorylation of their cytoplasmic tails and the subsequent binding of SLAM-associated protein (SAP) or EAT2 through a tyrosine-containing motif (ITSM). SAP is widely expressed by T cells and EAT2 is expressed by APCs. These two molecules can recruit and activate several Src kinases (including Fyn) that modulate cell activation by signals generated through the T cell receptor (TCR) and costimulatory proteins, such as CD28. Signals mediated by the SLAM receptors can also affect the function of APCs. SLAM receptors recruit various SH2-domain-containing proteins giving rise to different signals that determine distinct and, in some cases, divergent biological outcomes.

of the classical phagocytic NADPH oxidase (Nox2) complex is a tightly regulated process that involves assembly of the membrane bound catalytic gp91^{phox} and p22^{phox} with at least four cytosolic subunits p40^{phox}, p47^{phox}, p67^{phox}, Rac1/2 (40). By recruiting the p40^{phox} subunit to the maturing phagosome, PI₃P initiates the formation of this superoxide-producing complex (39). Second, PI₃P enables the recruitment of the tethering molecule EEA1, which is critically involved in phagolysosomal fusion. Thus, in the absence of SLAMF1 from phagocytes, the phagocytic process of specific Gram⁻ bacteria is compromised.

SLAMF2 Interactions with Gram⁻ Bacteria

SLAMF2 is implicated in the recognition of non-opsonized *E. coli* via surface type-1 fimbriae, which contain the lectin FimH (12). Microscopy and genetic analysis suggest that SLAMF2 binds to FimH, which is dependent on the presence of mannose on SLAMF2 (41). Uptake of FimH⁻ *E. coli* is not mediated by SLAMF2 (42).

SLAMF2 internalizes with FimH upon phagocytosis of FimH⁺ *E. coli* by mast cells and macrophages, which can be inhibited by mAb directed against SLAMF2. The "force catch" interactions between SLAMF2 and FimH are strengthened by the motility that is implicit to fimbriae and, therefore, represents a unique mode of interaction between phagocytes and *E. coli* (43). Studies utilizing mast cells show that the SLAMF2-FimH-mediated phagocytosis, which results in cholesterol-dense *E. coli*⁺ caveolae (44), has a distinct outcome compared to phagocytosis of

opsonized *E. coli* (**Figure 5**). SLAMF2-aided uptake results in the expulsion of the bacterium rather than its intracellular killing (42). Thus, SLAMF2 mediates uptake of FimH⁺ *E. coli* via the formation of caveolin⁺ phagocytes that represent recycling vesicles that release their content to the extracellular milieu within several hours.

SLAMF Receptors Alter Cytokine Production by Phagocytes

Beside the delayed phagocytosis of *E. coli*, SLAMF1⁻/⁻ macrophages display impaired responses to crude LPS (bacterial homogenate) (11, 13, 34). Stimulation with IFNγ and LPS, but not GpC or PGN, induced an ameliorated production of IL-12, TNF-α, and nitric oxide in SLAMF1⁻/⁻ macrophages (34). Conversely, human DCs that were stimulated with CD40-L expressing cells produced less IL-12 and TNF-α when SLAMF1 costimulation was induced, even in the presence of IFNγ and LPS (45). This discrepancy could suggest that SLAMF1 plays distinct roles on cytokine production in phagocytes, depending on whether SLAMF1 engages in homophilic interactions and/ or bacterial interactions (i.e., OmpC/F). Although SLAMF2 has no intracellular signaling domain, SLAMF2 induces signaling events in human brain microvascular endothelial cells that involve an influx of intracellular Ca²⁺ and the phosphorylation of RhoA (46). In mast cells, SLAMF2 engagement results in an increase in their TNF-α production and histamine release (41, 47, 48). Stimulation of SLAMF2⁻/⁻ macrophages with LPS results in reduced induction of TNF-α and IL-12 production (49). No specific interactions of SLAMF5 with bacterial entities have currently been reported, yet SLAMF5 also affects phagocyte functions. Transfection studies in mast cells and macrophages have shown that SLAMF5 signaling enhances phagocyte activation. SLAMF5

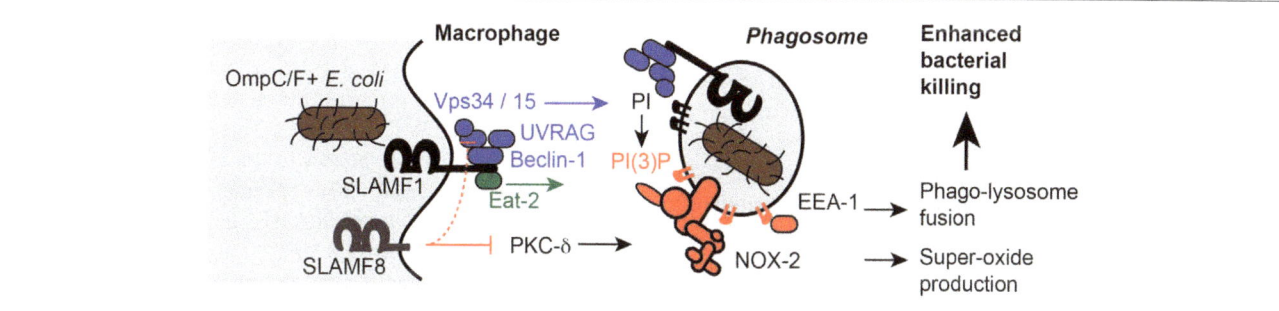

FIGURE 4 | Slamf1 affects phagosome functions in two ways, after binding to *E. coli*. OmpC/F⁺ *E. coli* can be bound by SLAMF1. Subsequently, SLAMF1 is internalized into the progressing phagosome. The Vps34/15 > UVRAG > Beclin-1 complex is formed. PI is converted to PI3P, which is the docking lipid for subunits of the Nox2 complex as well as the tethering molecule EEA-1. The result of the docking of these proteins is the progression of phagosomes toward bactericidal phagolysosomes that are able to kill the internalized bacteria. The positive modulation of Nox2 complex formation by PKC-delta is inhibited by SLAMF8. There is preliminary evidence for an inhibition by SLAMF8 of Vps34/15 > UVRAG > Beclin-1 complex recruitment to SLAMF1.

engagement induces FcεRI-mediated mast cell degranulation, which depends on Dok1 phosphorylation (50). Interestingly, LPS stimulation of macrophages results in phosphorylation of SLAMF5 at the second ITSM domain (Y300), which enhances the production of MCP-1 and TNF-α in an NF-κB dependent fashion (51). These observations indicate that SLAMF receptors initiate the signaling through the phosphorylated ITSM motif in phagocytic cells.

EAT-2 may modulate cytokine production. Indeed, recent reports suggest that EAT-2 mediates the production of TNF-α through several SLAMF receptors in human DCs (52). Although specific mechanisms need to be further identified, it is clear that SLAMF receptors modulate inflammatory effector functions of phagocytes in the presence of bacteria or LPS.

SLAMF8 INHIBITS NOX2 ACTIVITY IN BACTERIAL PHAGOSOMES

SLAMF8 is a member of the SLAMF receptor family that exhibits unique characteristics, as SLAMF8⁻/⁻ macrophages appear over-activated. The presence of SLAMF8 in phagocytes inhibits the maturation of phagosomes, irrespective whether the cargoes are Gram⁺ or Gram⁻ bacteria (53). We have recently reported that SLAMF8 negatively regulates the activity of PKC-δ, which phosphorylates the p40^phox subunit of the NOX2 complex (53). The presence of SLAMF8, therefore, negatively regulates the production of superoxide. However, the molecular intermediates that facilitate this SLAMF8 function have yet to be determined. Because SLAMF8 does not contain an intracellular domain with known signaling motives, it is unlikely that SLAMF8 recruits adaptor molecules that in turn inhibit PKC. Speculatively, competitive inhibition of SLAMF1 by SLAMF8 represents a possible mechanism. Although interactions *in trans* between SLAMF1 and SLAMF8 did not occur (54), the SLAMF1-Beclin1-Vps34/15-UVRAG complex is more readily formed in the absence of SLAMF8. This preliminary finding alludes to a functional interplay between these two SLAMF receptors.

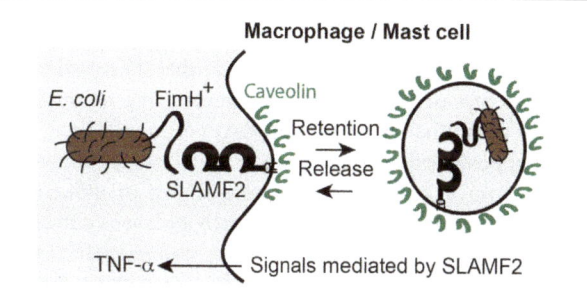

FIGURE 5 | SLAMF2 mediated the temporary retention of FimH⁺ *E. coli* in phagocytes. SLAMF2 can associate with the bacterial lectin FimH on the flagella of *E. coli*. The bacteria are internalized into caveolin⁺ vesicles to subsequently be released. The presence of SLAMF2 on macrophages and mast cells induced an LPS- or bacteria-mediated enhanced burst of TNF-α production.

SLAMF1 AND SLAMF8 REGULATE MIGRATION OF MYELOID CELLS TO SITES OF INFLAMMATION

Differential Expression of SLAMF1 and SLAMF8 by Phagocytes

Several SLAMF receptors are highly expressed by phagocytes after activation by inflammatory signals, suggesting a time-sensitive functional significance of SLAMF receptor surface expression in these cells. SLAMF1 expression is induced by stimulation with either LPS or IL-1β and in phagocytes during active colitis (34, 55, 56). Resting blood leukocytes are virtually devoid of SLAMF8 transcripts and protein (57). LPS only marginally induces SLAMF8 expression, rather its expression in phagocytes is mainly dependent on IFNγ signals, which result in a strong upregulation of SLAMF8 (53, 54, 57). Thus, during an ongoing infectious inflammation, phagocytes initially increase SLAMF1 surface expression and subsequently induce SLAMF8 expression.

SLAMF1 and SLAMF8 Modulate Myeloid Cell Motility

Phagocyte-expressed SLAMF1 positively affects cell migration to sites of ongoing inflammation. Our study that focused on cell motility during inflammation revealed that phagocyte-intrinsic functions of SLAMF1 enhance the capacity to migrate into sites of inflammation (54). Inflammatory phagocytes are required to infiltrate the lamina propria of the colon to establish persisting colitis after transfer of CD45RB^{hi} CD4$^+$ T-cell into $Rag1^{-/-}$ mice. The impairment of inflammatory phagocytes in $SLAMF1^{-/-}$ $Rag1^{-/-}$ mice to migrate to the lamina propria, therefore, resulted in ameliorated colitis (55). The poor outcome in SLAMF1-deficient mice of experimental infections with *Leishmania major*, which rely on macrophages for effective clearance, may also be partly explained by impaired migration of macrophage-forming monocytes (34). Opposed to the positive effect that SLAMF1 has on myeloid migration, SLAMF8 has a phagocyte-intrinsic negative effect on cell motility (54). Given the timing of the surface expression of SLAMF1 and SLAMF8 and their opposite effect on phagocyte activation, we hypothesize that these two SLAMF molecules represent a rheostat mechanism that modulates the extent of inflammation at different stages of an infection.

The opposite effects on reactive oxygen production displayed by these two SLAMF receptors were shown to influence cell motility. Specific inhibition of NOX2 activity canceled the *in vitro* migration phenotypes of both $SLAMF1^{-/-}$ and $SLAMF8^{-/-}$ phagocytes (54). These two phenomena can be linked by the mounting evidence that hydrogen peroxide, which is the more stable intermediate of superoxide, can act as a "second messenger" by oxidizing phosphatases and – as such – modulate cell motility (40, 58, 59).

SLAMF1, 2, 4, AND 6 REGULATE ENTEROCOLITIS

In line with the observations that SLAMF members modulate the function of phagocytes, three SLAMF receptors (SLAMF1, SLAMF2, and SLAMF6) also affect the pathogenesis of murine models of colitis, which are complex, multifaceted immune events, including activation of the mucosal immune system by microbes. Accumulating evidence by our group and by others shows a role of SLAMF receptors in cognate interactions with bacteria. The infiltration of pro-inflammatory phagocyte into the lamina propria of the colon is also prerequisite of the pathogenesis of colitis and some SLAMF receptors affect the extent of the colitis by influencing this process. Additionally, modulation of cytokine production may also contribute to these colitis phenotypes. No strong intestinal inflammation phenotype has been ascribed to XLP (60), thus SAP-independent functions of SLAMF receptors likely modulate mucosal immune processes.

SLAMF6 Enhances *C. rodentium* Colitis

Citrobacter rodentium are attaching bacteria that harbor a pathogenicity island, which renders them capable of colonizing the colonic epithelia of mice. Colonized *C. rodentium* causes lesions that result in a compromised mucosal barrier. Colitis induced by oral infection with *C. rodentium* is remarkably reduced in mice lacking both the *Rag1* and the *SLAMF6* genes compared to their *Rag-1*-deficient controls, but not in mice that only lack the *SLAMF6* gene (single knock out) as compared to their WT littermates. This shows an involvement of SLAMF6 in innate responses to the mucosal infections with specific enterobacteriae (38). Specific interactions between *E. coli* or *C. rodentium* and SLAMF6 have also been reported. Lacking this interaction in $SLAMF6^{-/-}$ mice manifests in impaired functions of phagocytes that first detect the effacing *C. rodentium* bacteria, hence driving the phenotype of reduced pathology (38).

Phagocyte Functions of SLAMF1 Contribute to Colitis

SLAMF1 in phagocytes also contributes to the development of colitis. By adoptive transfer of CD45RB^{hi} CD4$^+$ T-cells into $Rag^{-/-}$ or $SLAMF1^{-/-}Rag^{-/-}$ mice, we found that only SLAMF1 expression by innate cells, and not T-cells, is required for the full induction of experimental colitis (55). Activation of macrophages and DCs via CD40-stimulation alone was not sufficient to overcome the reduced inflammation in $SLAMF1^{-/-}Rag^{-/-}$ mice, further establishing a phagocyte-intrinsic cause of this phenotype. The hampered migratory capacity of SLAMF1-deficient inflammatory phagocytes was shown to be the primary cause of this phenotype (55). The enhanced phagosomal maturation and ROS production that results from the interaction of SLAMF1 with *E. coli* could represent an additional mechanism if these SLAMF1-mediated functions lead to a higher activation state of the lamina propria phagocytes. The production of pro-inflammatory cytokines that are implicated in colitis development are also impaired by SLAMF1-deficiency (55).

SLAMF2 Enhances Colitis while SLAMF4 Negatively Regulates Inflammation of the Small Intestine by the Control of Cytotoxic IELs

SLAMF2 is abundantly expressed in all myeloid cells (61). $SLAMF2^{-/-}$ T-cells induced colitis in $Rag^{-/-}$ mice, but not in $SLAMF2^{-/-}Rag^{-/-}$ mice, indicating that SLAMF2 expression by both innate cells and transferred T-cells contributes to the development of colitis (49). Indeed, SLAMF2-deficient mice were shown to have severely impaired CD4$^+$ T-cell activation and SLAMF2 expression is required on both T-cells and APCs for proper activation (62). Beside T-cell activation, which is a prerequisite for the development of colitis in this model, macrophage-expressed SLAMF2 could contribute to colitis by inducing TNF-α production, as suggested by *in vitro* experiments (41, 49). Whether both SLAMF2 interactions with SLAMF4 and bacteria drive this *in vivo* remains to be determined.

SLAMF4 also affects gut-mucosal immune responses. CD8$^+$ T-cell transfer experiments showed that SLAMF4 expression specifically correlated with localization to the intestinal lamina propria, where SLAMF4 modulates homeostasis by negative regulation of the expansion of cytotoxic CD8$^+$ IELs (61). SLAMF2 expression in myeloid cells, especially the CX3CR1$^+$ and CX3CR1$^-$ phagocytes in the lamina propria of the small intestine, facilitates this negative regulation (61). Vice versa, under specific conditions these cytotoxic IELs are capable of controlling the phagocyte population (61).

SAP AND SLAMF RECEPTORS MEDIATE PROTECTION FROM EBV AND OTHER VIRUSES

Whereas SLAMF receptor-mediated immune responses to bacteria are mostly mediated by SLAMF–bacteria interactions, the involvement of SLAMF receptors in antiviral immunity relies mostly on SLAMF–SLAMF homophilic interactions.

XLP and Epstein–Barr Virus

X-linked lymphoproliferative disease finds its primary cause in dysfunctional SAP (14–16). Often, but not always (63), patients develop fulminant infectious mononucleosis with a fatal outcome upon the first encounter with EBV. Although SAP-deficient patients who survive EBV infections or never encounter EBV will develop aberrant B-cell response such as dysgammaglobulinemia and B-cell lymphomas as well as a lack of innate type lymphocytes such as NKT-cells, the most prominent manifestations of this genetic defect arise in the context of EBV infections. Excellent reviews about EBV-independent immunologic manifestations of the aberrant response in SAP-deficient patients are published elsewhere (3, 64–66). In sum, in the absence of functional SAP, EBV-infected B-cells are not cleared and massive B- and T-lymphocytic expansion is found in most organs. CD4+ T-cells, CD8+ CTLs, NKT cells, and NK cells are implemented in the defective immune mechanisms that result in uncontrolled or ineffective immune responses to EBV infections in XLP patients. The phenotypic manifestations of non-EBV viral infections in XLP patients are sometimes also more severe than those in SAP-proficient individuals, although the disease manifestations are usually less increased.

SAP and CD8+ T-cell Expansion and Cytotoxic Responses

T-cell receptor signals in naïve T-cells induce a proliferative burst. SAP and SLAMF receptors control both the extent of the CD8+ T-cell expansion as well as the cytotoxicity of these cells, thereby influencing the effectiveness of the immune response to viruses as well as potential immunopathology.

In an effort to delineate the complex phenotypes of EBV infections of XLP patients, $Sh2d1a^{-/-}$ mice were generated and infected with γHV-68 (67) or LCMV (22, 68). The murine virus γHV-68 is, like EBV and Kaposi's sarcoma-associated herpes virus, a gamma-herpes virus but has coevolved with rodents and, therefore, does not infect humans. In addition to B-cells, γHV-68 also infects macrophages and DCs, which should be noted when comparing EBV infections of XLP patients with γHV-68 in $Sh2d1a^{-/-}$ mice. After infection with γHV-68, $Sh2d1a^{-/-}$ mice have an expanded population of CD8+ T-cells (69, 70), which produce higher levels of IFNγ as compared to CD8+ T-cells from infected WT mice (70). This higher amount of IFNγ controls

γHV-68 in macrophages in the peritoneum, but not in the B-cell reservoir (71). In accordance with reports on γHV-68 infected $Sh2d1a^{-/-}$ mice, LCMV-Armstrong infections induce a stronger expansion of CD4+ and CD8+ IFNγ-producing T-cells (22, 68). However, exacerbated immune pathology caused by the over-expansion of CD8+ T cells in this infection results in a higher mortality (22, 68).

One of the mechanisms that drive the massive expansion of T-cells is the deregulation of reactivation-induced cell death (RICD). A second TCR activation leads to proapoptotic signals in some expanding T-cells, thereby controlling the extent of the expansion of the collective T-cell pool. XLP patients that suffer fulminant mononucleosis typically lack this T-cell restricting phase of the response to EBV, which is also not observed in virus-infected $Sh2d1a^{-/-}$ mice. SAP expression was shown to correlate with the extent of RICD in several cell lines and a lack of cell cycle arrest was found in irradiated lymphocytes from XLP patients (72). The observation that SAP immuno-precipitates with the proapoptotic valosin-containing protein (VCP) alludes to a potential mechanism. A later study showed that SLAMF6 recruitment of SAP and Lck rather than Fyn in these restimulated T-cells results in a proapoptotic signal, which was not observed in T-cells obtained from XLP patients (73).

The expanded population of γHV-68-specific CD8+ CTLs in $Sh2d1a^{-/-}$ mice does reduce the amount of infected B-cells (69, 70). However, cytotoxicity per cell appears not to be affected by SAP (69). In contrast to these murine T-cells, CD8+ T-cells from XLP patients are selectively impaired in their cytotoxic response to B-cells (74). These human CTLs showed similar cytokine production and proliferation when they are stimulated in vitro with anti-CD3 and anti-CD28 or anti-SLAMF1 mAbs (75, 76). However, incubation with anti-SLAMF4 mAb markedly reduces cytotoxicity of the EBV-specific CD8+ CTLs and lowered IFNγ production (76). Because this defect is associated with aberrant lipid rafts, perforin release, and SAP recruitment to the cytolytic synapse, it can be concluded that SLAMF4–SAP pathway plays a critical role in the cytotoxic response of CD8+ T-cells to EBV-infected autologous B-cells (75). Indeed, whereas virtually all EBV-specific CD8+ T-cells in SAP-proficient individuals are SAP+, other viruses induce a mixed pool of SAP+ and SAP− virus-specific CTLs (77). The dependence of EBV-specific CD8+ T-cells on the SLAMF4–SAP pathway to target infected B-cells together with the narrow B-cells tropism of EBV may represent two of the underlining principles for the strong susceptibility of XLP patients to this virus.

SAP and CD4+ T-Cell Responses and Germinal Centers

Like XLP patients, γHV-68 infected $Sh2d1a^{-/-}$ mice had a strong reduction in the amount of GC B-cells (69). These mice also displayed the typical hypo-gammaglobulinemia (67, 69). Whereas SAP-deficient mice develop normal acute IgG responses upon infection with LCMV, they lack a humoral memory response (78). When the (chronic-infectious) LCMV$_{cl13}$ strain was used, GCs were grossly absent from $Sh2d1a^{-/-}$ mice (68). Lacking adequate help from CD4$^+$ T cells, humoral response and cytotoxicity of CD8$^+$ T cells are impaired, which renders the immune system not sufficient to clear the virus (68). Protection against secondary influenza infections is best established by CD4$^+$ T-cell-mediated humoral responses through the generation of memory B-cells and long-lived plasma cells. Experimental exposure of *Sh2d1a$^{-/-}$* mice to a second influenza challenge established the observation that these mice have a severely impaired IgG antibody response and, therefore, succumb to this infection (20). Thus, in the late stages of infections with LCMV, γHV-68, and influenza virus, profound defects in humoral immunity become apparent in *Sh2d1a$^{-/-}$* mice.

SLAM-associated protein is critical for the development of GCs, the anatomical site for B/T-cell cooperation. The observation that T-cell-independent humoral responses are unaffected by SAP deficiency, showed that this phenotype depends on T-cell interactions with B-cells (79). Whereas a B-cell intrinsic SAP component in IgG antibody production was reported in some transfer experiments but not in others, SAP expression by helper T-cells is indispensible for early GC responses (21, 80–82). The contact time of T-B-cell interactions is reduced in SAP-deficient mice, which is the likely underlining mechanism of the impaired GC response (83). Sustained adhesion of T-cells to B-cells is dependent on SLAMF5 (84). An additional study showed that SLAMF6, in the absence of SAP, conveys a negative signal resulting in an insufficient contact time between B-cells and T-cells (32). This negative signal is mediated by SLAMF6 as *SLAMF6$^{-/-}$ Sh2d1a$^{-/-}$* mice (lacking both SLAMF6 and SAP) have normal developing GCs. Recruitment of SHP-1 to SLAMF6 is the signaling event that is responsible for the impaired cognate B/T-cell interaction (32). Although SLAMF1 signaling contributes to GC IL-4 production (37), SLAMF1 and Fyn are not involved in proper GC formation (85). SLAMF3-deficiency does not notably affect GC formation either (86).

NKT Cell Development Depends on SAP, SLAMF1, and SLAMF6

NKT-cells are implicated in responses to a wide range of microbes and are reactive to lipid antigens. Positive selection of NKT cells is mediated by semi-invariant TCR interactions with lipid antigens in the MHC-I-like CD1d molecule from one double-positive (DP) thymocyte to a neighboring DP thymocyte. Thus, commitment of NKT cells, which takes place in the thymus, is dependent on CD1d stimulation from proximal lymphocytes instead of stromal cells. A secondary signal is required to induce differentiation and expansion. Either SLAMF1 or SLAMF6 homophilic ligation is required for this second signal that induces SAP recruitment to their ITSM (87). SAP-mediated signals are crucial for the development of NKT cells as *Sh2d1a$^{-/-}$* mice completely lack these cells (88). Upon SAP recruitment to either SLAMF1 or SLAMF6, Fyn binds to the SLAMF–SAP complex to induce signals that facilitate the requirements for differentiation and expansion. In contrast to SLAMF1 and SLAMF6, SLAMF3-deficient mice present elevated numbers of thymic NKT cells, indicating that SLAMF3 plays a unique role as an inhibitory receptor regulating the development of NKT cells (89). An in-depth review of SLAMF receptors in NKT-cells and other innate lymphocyte populations has recently been published (90).

Role for SAP, SLAMF4, and Other SLAMF Receptors in NK Cells

The capacity of chronic infections with lymphotropic viruses to transform their host cells makes targeted killing of infected cells an important requirement in the immunity to such viruses. SLAMF4 is the major SLAMF receptor to mediate cytotoxicity in both NK cells as well as CD8$^+$ CTLs. Initial studies have shown that SLAMF4 interactions with SLAMF2 on target cells induced perforin-mediated killing, which is dependent on SAP (91–95). SLAMF4 phosphorylation is dependent on its sublocation in lipid rafts (96). Within these rafts, association with linker for activation of T-cells (LAT) is prerequisite for SLAMF4 phosphorylation and, hence, SLAMF4-mediated killing of target cells (97). SLAMF4 has four ITSM domains and the membrane proximal ITSM recruits SAP to the cytotoxic immune synapse upon phosphorylation (98). This SLAMF4–SAP complex inhibits the recruitment of inhibitory phosphatases and, hence, is required for a sustained interaction between the NK cell and the target cells (99). However, SLAMF4 can also mediate inhibitory signals in cytotoxic cells (100, 101). The levels of SLAMF4 surface expression on NK cells as well as the abundance of SAP appear to dictate whether signals induce or inhibit targeted killing (95, 102, 103). Naïve human NK cells do not express SAP, but IL-2 or IL-12 stimulation results in the upregulation of SAP expression. Only NK cells that express SAP had the potential to kill target cells by SLAMF4 ligation (104). A recent review describes the intricacies of the dual function of SLAMF4 on cytotoxicity of NK cells in more detail (103).

Whereas SLAMF4 appears to be dominated by SAP, other SLAMF receptors have a stronger dependence of EAT-2. Analysis of EAT-2-mediated signals revealed that EAT-2 induces calcium fluxes and ERK phosphorylation, which results in exocytosis of cytotoxic granules (105). SLAMF6 ligation was shown to induce a cytotoxicity signal by recruiting EAT-2 to its second phosphorylated ITSM, which does not bind to SAP (106). In addition, EAT-2-deficient mice were incapable of SLAMF5- or SLAMF6-mediated targeted killing of SLAMF2$^+$ tumors (107). Thus, SLAMF6 signaling through EAT-2 in addition to SAP enhances the cytotoxicity of NK cells. SLAMF7 expression on target cells enhanced NK cell cytotoxicity, which was solely dependent on EAT-2, as *EAT-2$^{-/-}$* NK cells conveyed a signal that inhibits cytotoxicity through SLAMF7 (26).

VIRAL USE OF SLAMF RECEPTORS

Thus far, we have discussed how SLAMF receptors perform functions by interactions with bacterial entities and by interaction with SLAMF receptors. SLAMF receptors are also actively targeted by pathogens that seek to use or to alter functions of SLAMF receptors for their benefit. Three such modes of interaction have been postulated to date. First, Morbilliviruses (most prominently Measles virus) utilize SLAMF1 as entry receptors. Second, certain cytomegaloviruses (CMVs) express SLAMF receptors or molecules that closely resemble the structure of SLAMF receptor, potentially representing (negative) competitors of endogenous SLAMF receptors to modulate their functions. Third, several other viruses encode molecules that interfere with cell surface expression of SLAMF receptors and inhibit their functions.

SLAMF1 on the Surface of Myeloid Cells Binds to the Measles Virus H Protein and Is Involved in Virus Entry

The human pathogenic Measles virus belongs to the lymphotropic Morbillivirus genus. Measles virus and other Morbilliviruses utilize SLAMF1 as one of two entry receptors (9, 108). Crystal structures of SLAMF1 and Measles virus protein MV-H reveal four binding domains that are conserved between marmoset and human but not between mice and human, which determines the tropism of Measles virus (10). Mechanistically, the interaction between SLAMF1 and MV-H reduces the distance between the membranes of the target cell and the virus. The subsequent release of the viral protein MV-F enables fusion of the membranes and, hence, facilitates infection.

Measles virus has evolved a mechanism to induce SLAMF1 surface expression, thereby gaining access to its entry receptor (109, 110). Acidic Sphingomyelinase (ASMase)-containing vesicles, which are also SLAMF1⁺, play an interesting role in this process (**Figure 6**). ASMases convert sphingolipids into ceramide, creating a lipid environment that favors endocytosis or internalization of small membrane fractures. Thus, under non-infectious conditions, the recruitment of these vesicles to the surface of cells provides a membrane repair mechanism. Activation of the lectin receptor DC-SIGN by Measles virus induces a signaling cascade that involves Raf-1 and ERK (109). This signal relies on the expression of ASM and results in the relocation of ASM⁺ vesicles to the surface of DCs (109). Thus, by activating DC-SIGN,

Measles virus induces surface expression of its entry receptor (110). This observation, thus, provides evidence of a coupling between SLAMF1 localization and membrane dynamics and shows that SLAMF1 resides in intracellular membranes, suggesting that SLAMF1 has distinct intracellular location with putative intracellular functions. These functions may represent events that are similar to the functions that were described for SLAMF1 in *E. coli*⁺ phagosomes.

Viral Expression of SLAMF Receptor Homologs

SLAMF3 has stronger sequence homology with the human CMV protein UL-7 than with other human SLAMF receptors (111). Only one other CMV, which infects chimpanzees, bears a similar gene, suggesting that this gene was hijacked relatively late during the evolutionary arms race between mammals and β-herpes viruses. While no binding of UL7 to SLAMF3 could be detected, this viral protein has been shown to be secreted from infected cells and to reduce the production of TNFα, IL-8, and IL-6 by DCs (111).

Recently, seven SLAMF gene-homologs encoded by the genomes of two CMVs that infect New World monkeys have been identified. Several of these viral SLAMFs exhibit exceptional preservation of their N-terminal immunoglobulin domains, which results in maintenance of their ligand-binding capacities. The observation that large DNA viruses have captured SLAMF family homologs further underscores the importance of these molecules as critical immune regulators and as convenient scaffolds for viral evolution (112).

HIV-1 Protein Vpu and CMV m154 Modulate SLAMF Expression

Assessment of SLAMF expression in HIV-1 infected cells showed a negative correlation between SLAMF4 expression by NK cells and viral load, suggesting a positive role for SLAMF4 in the killing of HIV-1 infected cells (113). Indeed, NK cell treatment with specific antibodies for SLAMF4 or SLAMF6 decreased their *in vitro* killing potential of infected T-cells (114). Surface expression of both of these SLAMF receptors is actively down-modulated by HIV-1. CD8⁺ CTLs of patients required both SLAMF2-to-SLAMF4 signaling and TCR stimulation for the downmodulation of SLAMF4 surface expression (115). HIV-1 infection also down-modulates the expression of SLAMF2 and SLAMF6 in infected CD4⁺ T-cells, suggesting active modulation of cytotoxicity by the virus. The HIV-1 protein Vpu associates with SLAMF6 by interacting at the transmembrane regions. This interaction interferes with the glycosylation of SLAMF6 and results in retention in the Golgi-complex (116, 117). SLAMF6 downmodulation leads to insufficient degranulation, and hence impaired targeted killing of HIV-1 infected cells (116).

Murine CMV encodes a different viral protein that interferes with NK cell cytotoxicity. During CMV infection, m154 expression leads to proteolytic degradation of SLAMF2 that reduces the capacity of NK cells to kill infected cells (118).

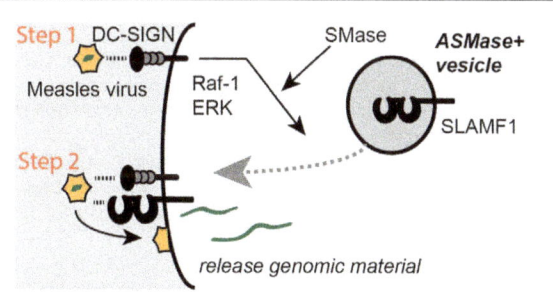

FIGURE 6 | Measles virus actively recruits its entry receptors Slamf1 to the cell surface. Binding of the lectin receptor DC-SIGN to a Measles virus particle induces a signaling cascade that involves Raf-1 and ERK and requires the activation of acidic SMase to induce a membrane trafficking event. Slamf1⁺ intracellular vesicles are recruited to the plasma membrane and fuse. This releases Slamf1 to the plasma membrane where Measles viral MV-H protein can bind to it to induce a fusion event between the viral membrane and the plasma membrane, consequently resulting in the delivery of the viral genomic material to the cytosol.

Detrimental Effects of SLAMF4 During Chronic Hepatitis Infection

Lysis of non-MHC HCV-infected cells by activated CD8[+] T-cells is mediated by SLAMF4 (119). However, during chronic HCV infections, SLAMF4 predominates as an inhibitor of cytotoxic functions in CD8[+] T-cells (95). In line with this notion, recombinant IFN-α therapy of HCV-infected patients induces NK cell-mediated enhanced immunity but reduces SLAMF4 expression of these cells (120). SLAMF4 expression by CD8[+] T-cells also correlated with poor clinical outcomes in HBV-infected patients (121). Blockade of SLAMF4 signaling effectively enhanced IFNγ production and virus-specific CD8[+] T-cell proliferation in approximately one-third of HCV[+] patients (122). Overall, SLAMF4 expression correlates with the T-cell exhaustion that is typically observed during HCV infections. However, functionally exhausted T-cells are not universally revived by blockade of SLAMF4 alone, but other CTL inhibitory receptors are involved (122). Thus, these β-herpes virus infections cause the expression and function of specific SLAMF receptors to be detrimental to the immune outcome.

CONCLUDING REMARKS

SLAMF receptors and their adaptors are intricately involved in the responses to microbial challenges. Modulation of immune responses as a result of SLAMF receptor homophilic interactions represents an important category of functions for these receptors. We can also observe an emerging theme that places SLAMF receptors in a possibly underappreciated category of functions; they can engage microbial ligands. SLAMF receptors are direct microbial sensors and are part of functional anti-microbial mechanisms. Thus, SLAMF receptors fulfill a unique role within the immune system, as they are both microbial sensors and cell–cell communicators of immunologic conditions. Additionally, we can distinguish a category of microbe-encoded genes that directly interfere with SLAMF functions. Interestingly, some of these genes have strong homology with endogenous SLAMF receptors.

AUTHOR CONTRIBUTIONS

BvD, CT: initial writing and collection of literature. GL: writing and editing. PE: expertise on virus – SLAMF interactions, editing.

FUNDING

This work is supported by a grant from the National Institutes of Health RO1 AI-15066, PO1-AI-076210, and PO1-AI-065687.

REFERENCES

1. Wu C, Sayos J, Wang N, Howie D, Coyle A, Terhorst C. Genomic organization and characterization of mouse SAP, the gene that is altered in X-linked lymphoproliferative disease. *Immunogenetics* (2000) **51**:805–15. doi:10.1007/s002510000215

2. Calpe S, Wang N, Romero X, Berger SB, Lanyi A, Engel P, et al. The SLAM and SAP gene families control innate and adaptive immune responses. *Adv Immunol* (2008) **97**:177–250. doi:10.1016/S0065-2776(08)00004-7

3. Cannons JL, Tangye SG, Schwartzberg PL. SLAM family receptors and SAP adaptors in immunity. *Annu Rev Immunol* (2011) **29**:665–705. doi:10.1146/annurev-immunol-030409-101302

4. Yan Q, Malashkevich VN, Fedorov A, Fedorov E, Cao E, Lary JW, et al. Structure of CD84 provides insight into SLAM family function. *Proc Natl Acad Sci U S A* (2007) **104**:10583–8. doi:10.1073/pnas.0703893104

5. Cao E, Ramagopal UA, Fedorov A, Fedorov E, Yan Q, Lary JW, et al. NTB-A receptor crystal structure: insights into homophilic interactions in the signaling lymphocytic activation molecule receptor family. *Immunity* (2006) **25**:559–70. doi:10.1016/j.immuni.2006.06.020

6. Velikovsky CA, Deng L, Chlewicki LK, Fernández MM, Kumar V, Mariuzza RA. Structure of natural killer receptor 2B4 bound to CD48 reveals basis for heterophilic recognition in signaling lymphocyte activation molecule family. *Immunity* (2007) **27**:572–84. doi:10.1016/j.immuni.2007.08.019

7. Brown MH, Boles K, van der Merwe PA, Kumar V, Mathew PA, Barclay AN. 2B4, the natural killer and T cell immunoglobulin superfamily surface protein, is a ligand for CD48. *J Exp Med* (1998) **188**:2083–90. doi:10.1084/jem.188.11.2083

8. Latchman Y, McKay PF, Reiser H. Identification of the 2B4 molecule as a counter-receptor for CD48. *J Immunol* (1998) **161**:5809–12.

9. Tatsuo H, Ono N, Tanaka K, Yanagi Y. SLAM (CDw150) is a cellular receptor for measles virus. *Nature* (2000) **406**:893–7. doi:10.1038/35022579

10. Hashiguchi T, Ose T, Kubota M, Maita N, Kamishikiryo J, Maenaka K, et al. Structure of the measles virus hemagglutinin bound to its cellular receptor SLAM. *Nat Struct Mol Biol* (2011) **18**:135–41. doi:10.1038/nsmb.1969

11. Berger SB, Romero X, Ma C, Wang G, Faubion WA, Liao G, et al. SLAM is a microbial sensor that regulates bacterial phagosome functions in macrophages. *Nat Immunol* (2010) **11**:920–7. doi:10.1038/ni.1931

12. Baorto DM, Gao Z, Malaviya R, Dustin ML, van der Merwe A, Lublin DM, et al. Survival of FimH-expressing enterobacteria in macrophages relies on glycolipid traffic. *Nature* (1997) **389**:636–9. doi:10.1038/39376

13. Ma C, Wang N, Detre C, Wang G, O'Keeffe M, Terhorst C. Receptor signaling lymphocyte-activation molecule family 1 (Slamf1) regulates membrane fusion and NADPH oxidase 2 (NOX2) activity by recruiting a Beclin-1/Vps34/ultraviolet radiation resistance-associated gene (UVRAG) complex. *J Biol Chem* (2012) **287**:18359–65. doi:10.1074/jbc.M112.367060

14. Coffey AJ, Brooksbank RA, Brandau O, Oohashi T, Howell GR, Bye JM, et al. Host response to EBV infection in X-linked lymphoproliferative disease results from mutations in an SH2-domain encoding gene. *Nat Genet* (1998) **20**:129–35. doi:10.1038/2424

15. Nichols KE, Harkin DP, Levitz S, Krainer M, Kolquist KA, Genovese C, et al. Inactivating mutations in an SH2 domain-encoding gene in X-linked lymphoproliferative syndrome. *Proc Natl Acad Sci U S A* (1998) **95**:13765–70. doi:10.1073/pnas.95.23.13765

16. Sayos J, Wu C, Morra M, Wang N, Zhang X, Allen D, et al. The X-linked lymphoproliferative-disease gene product SAP regulates signals induced through the co-receptor SLAM. *Nature* (1998) **395**:462–9. doi:10.1038/26683

17. Sayós J, Nguyen KB, Wu C, Stepp SE, Howie D, Schatzle JD, et al. Potential pathways for regulation of NK and T cell responses: differential X-linked lymphoproliferative syndrome gene product SAP interactions with SLAM and 2B4. *Int Immunol* (2000) **12**:1749–57. doi:10.1093/intimm/12.12.1749

18. Sayós J, Martín M, Chen A, Simarro M, Howie D, Morra M, et al. Cell surface receptors *Ly-9* and CD84 recruit the X-linked lymphoproliferative disease gene product SAP. *Blood* (2001) **97**:3867–74. doi:10.1182/blood.V97.12.3867

19. Li SC, Gish G, Yang D, Coffey AJ, Forman-Kay JD, Ernberg I, et al. Novel mode of ligand binding by the SH2 domain of the human XLP disease gene product SAP/SH2D1A. *Curr Biol* (1999) **9**:1355–62. doi:10.1016/S0960-9822(00)80080-9

20. Kamperschroer C, Dibble JP, Meents DL, Schwartzberg PL, Swain SL. SAP is required for Th cell function and for immunity to influenza. *J Immunol* (2006) 177:5317–27. doi:10.4049/jimmunol.177.8.5317

21. Chen Q, Cannons JL, Paton JC, Akiba H, Schwartzberg PL, Snapper CM. A novel ICOS-independent, but CD28- and SAP-dependent, pathway of T cell-dependent, polysaccharide-specific humoral immunity in response to intact Streptococcus pneumoniae versus pneumococcal conjugate vaccine. *J Immunol* (2008) 181:8258–66. doi:10.4049/jimmunol.181.12.8258

22. Wu C, Nguyen KB, Pien GC, Wang N, Gullo C, Howie D, et al. SAP controls T cell responses to virus and terminal differentiation of TH2 cells. *Nat Immunol* (2001) 2:410–4. doi:10.1038/ni0901-823

23. Al-Alem U, Li C, Forey N, Relouzat F, Fondanèche MC, Tavtigian SV, et al. Impaired Ig class switch in mice deficient for the X-linked lymphoproliferative disease gene Sap. *Blood* (2005) 106:2069–75. doi:10.1182/blood-2004-07-2731

24. Kis LL, Nagy N, Klein G, Klein E. Expression of SH2D1A in five classical Hodgkin's disease-derived cell lines. *Int J Cancer* (2003) 104:658–61. doi:10.1002/ijc.10986

25. Morra M, Lu J, Poy F, Martin M, Sayos J, Calpe S, et al. Structural basis for the interaction of the free SH2 domain EAT-2 with SLAM receptors in hematopoietic cells. *EMBO J* (2001) 20:5840–52. doi:10.1093/emboj/20.21.5840

26. Cruz-Munoz ME, Dong Z, Shi X, Zhang S, Veillette A. Influence of CRACC, a SLAM family receptor coupled to the adaptor EAT-2, on natural killer cell function. *Nat Immunol* (2009) 10:297–305. doi:10.1038/ni.1693

27. Calpe S, Erdos E, Liao G, Wang N, Rietdijk S, Simarro M, et al. Identification and characterization of two related murine genes, Eat2a and Eat2b, encoding single SH2-domain adapters. *Immunogenetics* (2006) 58:15–25. doi:10.1007/s00251-005-0056-3

28. Chan B, Lanyi A, Song HK, Griesbach J, Simarro-Grande M, Poy F, et al. SAP couples Fyn to SLAM immune receptors. *Nat Cell Biol* (2003) 5:155–60. doi:10.1038/ncb920

29. Davidson D, Shi X, Zhang S, Wang H, Nemer M, Ono N, et al. Genetic evidence linking SAP, the X-linked lymphoproliferative gene product, to Src-related kinase FynT in T(H)2 cytokine regulation. *Immunity* (2004) 21:707–17. doi:10.1016/j.immuni.2004.10.005

30. Simarro M, Lanyi A, Howie D, Poy F, Bruggeman J, Choi M, et al. SAP increases FynT kinase activity and is required for phosphorylation of SLAM and Ly9. *Int Immunol* (2004) 16:727–36. doi:10.1093/intimm/dxh074

31. Poy F, Yaffe MB, Sayos J, Saxena K, Morra M, Sumegi J, et al. Crystal structures of the XLP protein SAP reveal a class of SH2 domains with extended, phosphotyrosine-independent sequence recognition. *Mol Cell* (1999) 4:555–61. doi:10.1016/S1097-2765(00)80206-3

32. Kageyama R, Cannons JL, Zhao F, Yusuf I, Lao C, Locci M, et al. The receptor Ly108 functions as a SAP adaptor-dependent on-off switch for T cell help to B cells and NKT cell development. *Immunity* (2012) 36:986–1002. doi:10.1016/j.immuni.2012.05.016

33. Latour S, Roncagalli R, Chen R, Bakinowski M, Shi X, Schwartzberg PL, et al. Binding of SAP SH2 domain to FynT SH3 domain reveals a novel mechanism of receptor signalling in immune regulation. *Nat Cell Biol* (2003) 5:149–54. doi:10.1038/ncb919

34. Wang N, Satoskar A, Faubion W, Howie D, Okamoto S, Feske S, et al. The cell surface receptor SLAM controls T cell and macrophage functions. *J Exp Med* (2004) 199:1255–64. doi:10.1084/jem.20031835

35. Cannons JL, Yu LJ, Hill B, Mijares LA, Dombroski D, Nichols KE, et al. SAP regulates T(H)2 differentiation and PKC-theta-mediated activation of NF-kappaB1. *Immunity* (2004) 21:693–706. doi:10.1016/j.immuni.2004.09.012

36. Cannons JL, Wu JZ, Gomez-Rodriguez J, Zhang J, Dong B, Liu Y, et al. Biochemical and genetic evidence for a SAP-PKC-theta interaction contributing to IL-4 regulation. *J Immunol* (2010) 185:2819–27. doi:10.4049/jimmunol.0902182

37. Yusuf I, Kageyama R, Monticelli L, Johnston RJ, Ditoro D, Hansen K, et al. Germinal center T follicular helper cell IL-4 production is dependent on signaling lymphocytic activation molecule receptor (CD150). *J Immunol* (2010) 185:190–202. doi:10.4049/jimmunol.0903505

38. van Driel B, Wang G, Liao G, Halibozek PJ, Keszei M, O'Keeffe MS, et al. The cell surface receptor Slamf6 modulates innate immune responses during *Citrobacter rodentium* induced colitis. *Int Immunol* (2015) 27:447–57. doi:10.1093/intimm/dxv029

39. Anderson KE, Boyle KB, Davidson K, Chessa TA, Kulkarni S, Jarvis GE, et al. CD18-dependent activation of the neutrophil NADPH oxidase during phagocytosis of *Escherichia coli* or *Staphylococcus aureus* is regulated by class III but not class I or II PI3Ks. *Blood* (2008) 112:5202–11. doi:10.1182/blood-2008-04-149450

40. Bedard K, Krause KH. The NOX family of ROS-generating NADPH oxidases: physiology and pathophysiology. *Physiol Rev* (2007) 87:245–313. doi:10.1152/physrev.00044.2005

41. Malaviya R, Gao Z, Thankavel K, van der Merwe PA, Abraham SN. The mast cell tumor necrosis factor alpha response to FimH-expressing *Escherichia coli* is mediated by the glycosylphosphatidylinositol-anchored molecule CD48. *Proc Natl Acad Sci U S A* (1999) 96:8110–5. doi:10.1073/pnas.96.14.8110

42. Shin JS, Gao Z, Abraham SN. Bacteria-host cell interaction mediated by cellular cholesterol/glycolipid-enriched microdomains. *Biosci Rep* (1999) 19:421–32. doi:10.1023/A:1020216323271

43. Moller J, Luhmann T, Chabria M, Hall H, Vogel V. Macrophages lift off surface-bound bacteria using a filopodium-lamellipodium hook-and-shovel mechanism. *Sci Rep* (2013) 3:2884. doi:10.1038/srep02884

44. Shin JS, Gao Z, Abraham SN. Involvement of cellular caveolae in bacterial entry into mast cells. *Science* (2000) 289:785–8. doi:10.1126/science.289.5480.785

45. Réthi B, Gogolák P, Szatmari I, Veres A, Erdõs E, Nagy L, et al. SLAM/SLAM interactions inhibit CD40-induced production of inflammatory cytokines in monocyte-derived dendritic cells. *Blood* (2006) 107:2821–9. doi:10.1182/blood-2005-06-2265

46. Khan NA, Kim Y, Shin S, Kim KS. FimH-mediated *Escherichia coli* K1 invasion of human brain microvascular endothelial cells. *Cell Microbiol* (2007) 9:169–78. doi:10.1111/j.1462-5822.2006.00779.x

47. Rocha-de-Souza CM, Berent-Maoz B, Mankuta D, Moses AE, Levi-Schaffer F. Human mast cell activation by *Staphylococcus aureus*: interleukin-8 and tumor necrosis factor alpha release and the role of Toll-like receptor 2 and CD48 molecules. *Infect Immun* (2008) 76:4489–97. doi:10.1128/IAI.00270-08

48. Munoz S, Hernandez-Pando R, Abraham SN, Enciso JA. Mast cell activation by *Mycobacterium tuberculosis*: mediator release and role of CD48. *J Immunol* (2003) 170:5590–6. doi:10.4049/jimmunol.170.11.5590

49. Abadía-Molina AC, Ji H, Faubion WA, Julien A, Latchman Y, Yagita H, et al. CD48 controls T-cell and antigen-presenting cell functions in experimental colitis. *Gastroenterology* (2006) 130:424–34. doi:10.1053/j.gastro.2005.12.009

50. Oliver-Vila I, Saborit-Villarroya I, Engel P, Martin M. The leukocyte receptor CD84 inhibits Fc epsilon RI-mediated signaling through homophilic interaction in transfected RBL-2H3 cells. *Mol Immunol* (2008) 45:2138–49. doi:10.1016/j.molimm.2007.12.006

51. Sintes J, Romero X, de Salort J, Terhorst C, Engel P. Mouse CD84 is a pan-leukocyte cell-surface molecule that modulates LPS-induced cytokine secretion by macrophages. *J Leukoc Biol* (2010) 88:687–97. doi:10.1189/jlb.1109756

52. Aldhamen YA, Seregin SS, Aylsworth CF, Godbehere S, Amalfitano A. Manipulation of EAT-2 expression promotes induction of multiple beneficial regulatory and effector functions of the human innate immune system as a novel immunomodulatory strategy. *Int Immunol* (2014) 26:291–303. doi:10.1093/intimm/dxt061

53. Wang G, Abadía-Molina AC, Berger SB, Romero X, O'Keeffe MS, Rojas-Barros DI, et al. Cutting edge: Slamf8 is a negative regulator of Nox2 activity in macrophages. *J Immunol* (2012) 188:5829–32. doi:10.4049/jimmunol.1102620

54. Wang G, van Driel BJ, Liao G, O'Keeffe MS, Halibozek PJ, Flipse J, et al. Migration of myeloid cells during inflammation is differentially regulated by the cell surface receptors Slamf1 and Slamf8. *PLoS One* (2015) 10:e0121968. doi:10.1371/journal.pone.0121968

55. van Driel B, Liao G, Romero X, O'Keeffe MS, Wang G, Faubion WA, et al. Signaling lymphocyte activation molecule regulates development of colitis in mice. *Gastroenterology* (2012) 143:1544–1554e1547. doi:10.1053/j.gastro.2012.08.042

56. Kruse M, Meinl E, Henning G, Kuhnt C, Berchtold S, Berger T, et al. Signaling lymphocytic activation molecule is expressed on mature CD83+ dendritic cells and is up-regulated by IL-1 beta. *J Immunol* (2001) 167:1989–95. doi:10.4049/jimmunol.167.4.1989

57. Kingsbury GA, Feeney LA, Nong Y, Calandra SA, Murphy CJ, Corcoran JM, et al. Cloning, expression, and function of BLAME, a novel member of the CD2 family. *J Immunol* (2001) 166:5675–80. doi:10.4049/jimmunol.166.9.5675

58. Lam GY, Huang J, Brumell JH. The many roles of NOX2 NADPH oxidase-derived ROS in immunity. *Semin Immunopathol* (2010) 32:415–30. doi:10.1007/s00281-010-0221-0

59. Schroder K. NADPH oxidases in redox regulation of cell adhesion and migration. *Antioxid Redox Signal* (2014) 20:2043–58. doi:10.1089/ars.2013.5633

60. Pachlopnik Schmid J, Canioni D, Moshous D, Touzot F, Mahlaoui N, Hauck F, et al. Clinical similarities and differences of patients with X-linked lymphoproliferative syndrome type 1 (XLP-1/SAP deficiency) versus type 2 (XLP-2/XIAP deficiency). *Blood* (2011) 117:1522–9. doi:10.1182/blood-2010-07-298372

61. O'Keeffe MS, Song JH, Liao G, De Calisto J, Halibozek PJ, Mora JR, et al. SLAMF4 is a negative regulator of expansion of cytotoxic intraepithelial CD8(+) T cells that maintains homeostasis in the small intestine. *Gastroenterology* (2015) 148:991–1001e1004. doi:10.1053/j.gastro.2015.02.003

62. González-Cabrero J, Wise CJ, Latchman Y, Freeman GJ, Sharpe AH, Reiser H. CD48-deficient mice have a pronounced defect in CD4(+) T cell activation. *Proc Natl Acad Sci U S A* (1999) 96:1019–23. doi:10.1073/pnas.96.3.1019

63. Sumegi J, Huang D, Lanyi A, Davis JD, Seemayer TA, Maeda A, et al. Correlation of mutations of the SH2D1A gene and Epstein-Barr virus infection with clinical phenotype and outcome in X-linked lymphoproliferative disease. *Blood* (2000) 96:3118–25.

64. Engel P, Eck MJ, Terhorst C. The SAP and SLAM families in immune responses and X-linked lymphoproliferative disease. *Nat Rev Immunol* (2003) 3:813–21. doi:10.1038/nri1202

65. Howie D, Sayos J, Terhorst C, Morra M. The gene defective in X-linked lymphoproliferative disease controls T cell dependent immune surveillance against Epstein-Barr virus. *Curr Opin Immunol* (2000) 12:474–8. doi:10.1016/S0952-7915(00)00123-0

66. Morra M, Howie D, Grande MS, Sayos J, Wang N, Wu C, et al. X-linked lymphoproliferative disease: a progressive immunodeficiency. *Annu Rev Immunol* (2001) 19:657–82. doi:10.1146/annurev.immunol.19.1.657

67. Yin L, Al-Alem U, Liang J, Tong WM, Li C, Badiali M, et al. Mice deficient in the X-linked lymphoproliferative disease gene sap exhibit increased susceptibility to murine gammaherpesvirus-68 and hypo-gammaglobulinemia. *J Med Virol* (2003) 71:446–55. doi:10.1002/jmv.10504

68. Crotty S, McCausland MM, Aubert RD, Wherry EJ, Ahmed R. Hypogammaglobulinemia and exacerbated CD8 T-cell-mediated immunopathology in SAP-deficient mice with chronic LCMV infection mimics human XLP disease. *Blood* (2006) 108:3085–93. doi:10.1182/blood-2006-04-018929

69. Kim IJ, Burkum CE, Cookenham T, Schwartzberg PL, Woodland DL, Blackman MA. Perturbation of B cell activation in SLAM-associated protein-deficient mice is associated with changes in gammaherpesvirus latency reservoirs. *J Immunol* (2007) 178:1692–701. doi:10.4049/jimmunol.178.3.1692

70. Chen G, Tai AK, Lin M, Chang F, Terhorst C, Huber BT. Signaling lymphocyte activation molecule-associated protein is a negative regulator of the CD8 T cell response in mice. *J Immunol* (2005) 175:2212–8. doi:10.4049/jimmunol.175.4.2212

71. Steed A, Buch T, Waisman A, Virgin HW. Gamma interferon blocks gammaherpesvirus reactivation from latency in a cell type-specific manner. *J Virol* (2007) 81:6134–40. doi:10.1128/JVI.00108-07

72. Nagy N, Matskova L, Kis LL, Hellman U, Klein G, Klein E. The proapoptotic function of SAP provides a clue to the clinical picture of X-linked lymphoproliferative disease. *Proc Natl Acad Sci U S A* (2009) 106:11966–71. doi:10.1073/pnas.0905691106

73. Katz G, Krummey SM, Larsen SE, Stinson JR, Snow AL. SAP facilitates recruitment and activation of LCK at NTB-A receptors during restimulation-induced cell death. *J Immunol* (2014) 192:4202–9. doi:10.4049/jimmunol.1303070

74. Tangye SG. XLP: clinical features and molecular etiology due to mutations in SH2D1A encoding SAP. *J Clin Immunol* (2014) 34:772–9. doi:10.1007/s10875-014-0083-7

75. Dupré L, Andolfi G, Tangye SG, Clementi R, Locatelli F, Aricò M, et al. SAP controls the cytolytic activity of CD8+ T cells against EBV-infected cells. *Blood* (2005) 105:4383–9. doi:10.1182/blood-2004-08-3269

76. Sharifi R, Sinclair JC, Gilmour KC, Arkwright PD, Kinnon C, Thrasher AJ, et al. SAP mediates specific cytotoxic T-cell functions in X-linked lymphoproliferative disease. *Blood* (2004) 103:3821–7. doi:10.1182/blood-2003-09-3359

77. Palendira U, Low C, Chan A, Hislop AD, Ho E, Phan TG, et al. Molecular pathogenesis of EBV susceptibility in XLP as revealed by analysis of female

carriers with heterozygous expression of SAP. *PLoS Biol* (2011) 9:e1001187. doi:10.1371/journal.pbio.1001187

78. Crotty S, Kersh EN, Cannons J, Schwartzberg PL, Ahmed R. SAP is required for generating long-term humoral immunity. *Nature* (2003) 421:282–7. doi:10.1038/nature01318

79. Hron JD, Caplan L, Gerth AJ, Schwartzberg PL, Peng SL. SH2D1A regulates T-dependent humoral autoimmunity. *J Exp Med* (2004) 200:261–6. doi:10.1084/jem.20040526

80. Morra M, Barrington RA, Abadia-Molina AC, Okamoto S, Julien A, Gullo C, et al. Defective B cell responses in the absence of SH2D1A. *Proc Natl Acad Sci U S A* (2005) 102:4819–23. doi:10.1073/pnas.0408681102

81. Cannons JL, Yu LJ, Jankovic D, Crotty S, Horai R, Kirby M, et al. SAP regulates T cell-mediated help for humoral immunity by a mechanism distinct from cytokine regulation. *J Exp Med* (2006) 203:1551–65. doi:10.1084/jem.20052097

82. Veillette A, Zhang S, Shi X, Dong Z, Davidson D, Zhong MC. SAP expression in T cells, not in B cells, is required for humoral immunity. *Proc Natl Acad Sci U S A* (2008) 105:1273–8. doi:10.1073/pnas.0710698105

83. Qi H, Cannons JL, Klauschen F, Schwartzberg PL, Germain RN. SAP-controlled T-B cell interactions underlie germinal centre formation. *Nature* (2008) 455:764–9. doi:10.1038/nature07345

84. Cannons JL, Qi H, Lu KT, Dutta M, Gomez-Rodriguez J, Cheng J, et al. Optimal germinal center responses require a multistage T cell: B cell adhesion process involving integrins, SLAM-associated protein, and CD84. *Immunity* (2010) 32:253–65. doi:10.1016/j.immuni.2010.01.010

85. McCausland MM, Yusuf I, Tran H, Ono N, Yanagi Y, Crotty S. SAP regulation of follicular helper CD4 T cell development and humoral immunity is independent of SLAM and Fyn kinase. *J Immunol* (2007) 178:817–28. doi:10.4049/jimmunol.178.2.817

86. Graham DB, Bell MP, McCausland MM, Huntoon CJ, van Deursen J, Faubion WA, et al. Ly9 (CD229)-deficient mice exhibit T cell defects yet do not share several phenotypic characteristics associated with SLAM- and SAP-deficient mice. *J Immunol* (2006) 176:291–300. doi:10.4049/jimmunol.176.1.291

87. Griewank K, Borowski C, Rietdijk S, Wang N, Julien A, Wei DG, et al. Homotypic interactions mediated by Slamf1 and Slamf6 receptors control NKT cell lineage development. *Immunity* (2007) 27:751–62. doi:10.1016/j.immuni.2007.08.020

88. Nichols KE, Hom J, Gong SY, Ganguly A, Ma CS, Cannons JL, et al. Regulation of NKT cell development by SAP, the protein defective in XLP. *Nat Med* (2005) 11:340–5. doi:10.1038/nm1189

89. Sintes J, Cuenca M, Romero X, Bastos R, Terhorst C, Angulo A, et al. Cutting edge: Ly9 (CD229), a SLAM family receptor, negatively regulates the development of thymic innate memory-like CD8+ T and invariant NKT cells. *J Immunol* (2013) 190:21–6. doi:10.4049/jimmunol.1202435

90. Romero X, Sintes J, Engel P. Role of SLAM family receptors and specific adapter SAP in innate-like lymphocytes. *Crit Rev Immunol* (2014) 34:263–99. doi:10.1615/CritRevImmunol.2014010538

91. Benoit L, Wang X, Pabst HF, Dutz J, Tan R. Defective NK cell activation in X-linked lymphoproliferative disease. *J Immunol* (2000) 165:3549–53. doi:10.4049/jimmunol.165.7.3549

92. Tangye SG, Cherwinski H, Lanier LL, Phillips JH. 2B4-mediated activation of human natural killer cells. *Mol Immunol* (2000) 37:493–501. doi:10.1016/S0161-5890(00)00076-6

93. Tangye SG, Phillips JH, Lanier LL, Nichols KE. Functional requirement for SAP in 2B4-mediated activation of human natural killer cells as revealed by the X-linked lymphoproliferative syndrome. *J Immunol* (2000) 165:2932–6. doi:10.4049/jimmunol.165.6.2932

94. Lee KM, Bhawan S, Majima T, Wei H, Nishimura MI, Yagita H, et al. Cutting edge: the NK cell receptor 2B4 augments antigen-specific T cell cytotoxicity through CD48 ligation on neighboring T cells. *J Immunol* (2003) 170:4881–5. doi:10.4049/jimmunol.170.10.4881

95. Schlaphoff V, Lunemann S, Suneetha PV, Jaroszewicz J, Grabowski J, Dietz J, et al. Dual function of the NK cell receptor 2B4 (CD244) in the regulation of HCV-specific CD8+ T cells. *PLoS Pathog* (2011) 7:e1002045. doi:10.1371/journal.ppat.1002045

96. Watzl C, Long EO. Natural killer cell inhibitory receptors block actin cytoskeleton-dependent recruitment of 2B4 (CD244) to lipid rafts. *J Exp Med* (2003) 197:77–85. doi:10.1084/jem.20020427

97. Klem J, Verrett PC, Kumar V, Schatzle JD. 2B4 is constitutively associated with linker for the activation of T cells in glycolipid-enriched microdomains: properties required for 2B4 lytic function. *J Immunol* (2002) **169**:55–62. doi:10.4049/jimmunol.169.1.55

98. Roda-Navarro P, Mittelbrunn M, Ortega M, Howie D, Terhorst C, Sánchez-Madrid F, et al. Dynamic redistribution of the activating 2B4/SAP complex at the cytotoxic NK cell immune synapse. *J Immunol* (2004) **173**:3640–6. doi:10.4049/jimmunol.173.6.3640

99. Eissmann P, Beauchamp L, Wooters J, Tilton JC, Long EO, Watzl C. Molecular basis for positive and negative signaling by the natural killer cell receptor 2B4 (CD244). *Blood* (2005) **105**:4722–9. doi:10.1182/blood-2004-09-3796

100. Mooney JM, Klem J, Wülfing C, Mijares LA, Schwartzberg PL, Bennett M, et al. The murine NK receptor 2B4 (CD244) exhibits inhibitory function independent of signaling lymphocytic activation molecule-associated protein expression. *J Immunol* (2004) **173**:3953–61. doi:10.4049/jimmunol.173.6.3953

101. Lee KM, McNerney ME, Stepp SE, Mathew PA, Schatzle JD, Bennett M, et al. 2B4 acts as a non-major histocompatibility complex binding inhibitory receptor on mouse natural killer cells. *J Exp Med* (2004) **199**:1245–54. doi:10.1084/jem.20031989

102. Chlewicki LK, Velikovsky CA, Balakrishnan V, Mariuzza RA, Kumar V. Molecular basis of the dual functions of 2B4 (CD244). *J Immunol* (2008) **180**:8159–67. doi:10.4049/jimmunol.180.12.8159

103. Waggoner SN, Kumar V. Evolving role of 2B4/CD244 in T and NK cell responses during virus infection. *Front Immunol* (2012) **3**:377. doi:10.3389/fimmu.2012.00377

104. Endt J, Eissmann P, Hoffmann SC, Meinke S, Giese T, Watzl C. Modulation of 2B4 (CD244) activity and regulated SAP expression in human NK cells. *Eur J Immunol* (2007) **37**:193–8. doi:10.1002/eji.200636341

105. Perez-Quintero LA, Roncagalli R, Guo H, Latour S, Davidson D, Veillette A. EAT-2, a SAP-like adaptor, controls NK cell activation through phospholipase Cgamma, Ca++, and Erk, leading to granule polarization. *J Exp Med* (2014) **211**:727–42. doi:10.1084/jem.20132038

106. Eissmann P, Watzl C. Molecular analysis of NTB-A signaling: a role for EAT-2 in NTB-A-mediated activation of human NK cells. *J Immunol* (2006) **177**:3170–7. doi:10.4049/jimmunol.177.5.3170

107. Wang N, Calpe S, Westcott J, Castro W, Ma C, Engel P, et al. Cutting edge: the adapters EAT-2A and -2B are positive regulators of CD244- and CD84-dependent NK cell functions in the C57BL/6 mouse. *J Immunol* (2010) **185**:5683–7. doi:10.4049/jimmunol.1001974

108. Tatsuo H, Ono N, Yanagi Y. Morbilliviruses use signaling lymphocyte activation molecules (CD150) as cellular receptors. *J Virol* (2001) **75**:5842–50. doi:10.1128/JVI.75.13.5842-5850.2001

109. Avota E, Gulbins E, Schneider-Schaulies S. DC-SIGN mediated sphingomyelinase-activation and ceramide generation is essential for enhancement of viral uptake in dendritic cells. *PLoS Pathog* (2011) **7**:e1001290. doi:10.1371/journal.ppat.1001290

110. Schneider-Schaulies J, Schneider-Schaulies S. Sphingolipids in viral infection.

111. Engel P, Pérez-Carmona N, Albà MM, Robertson K, Ghazal P, Angulo A. Human cytomegalovirus UL7, a homologue of the SLAM-family receptor CD229, impairs cytokine production. *Immunol Cell Biol* (2011) **89**:753–66. doi:10.1038/icb.2011.55

112. Pérez-Carmona N, Farré D, Martínez-Vicente P, Terhorst C, Engel P, Angulo A. Signaling lymphocytic activation molecule family receptor homologs in new world monkey cytomegaloviruses. *J Virol* (2015) **89**:11323–36. doi:10.1128/JVI.01296-15

113. Ostrowski SR, Ullum H, Pedersen BK, Gerstoft J, Katzenstein TL. 2B4 expression on natural killer cells increases in HIV-1 infected patients followed prospectively during highly active antiretroviral therapy. *Clin Exp Immunol* (2005) **141**:526–33. doi:10.1111/j.1365-2249.2005.02869.x

114. Ward J, Bonaparte M, Sacks J, Guterman J, Fogli M, Mavilio D, et al. HIV modulates the expression of ligands important in triggering natural killer cell cytotoxic responses on infected primary T-cell blasts. *Blood* (2007) **110**:1207–14. doi:10.1182/blood-2006-06-028175

115. Pacheco Y, McLean AP, Rohrbach J, Porichis F, Kaufmann DE, Kavanagh DG. Simultaneous TCR and CD244 signals induce dynamic downmodulation of CD244 on human antiviral T cells. *J Immunol* (2013) **191**:2072–81. doi:10.4049/jimmunol.1300435

116. Shah AH, Sowrirajan B, Davis ZB, Ward JP, Campbell EM, Planelles V, et al. Degranulation of natural killer cells following interaction with HIV-1-infected cells is hindered by downmodulation of NTB-A by Vpu. *Cell Host Microbe* (2010) **8**:397–409. doi:10.1016/j.chom.2010.10.008

117. Bolduan S, Hubel P, Reif T, Lodermeyer V, Höhne K, Fritz JV, et al. HIV-1 Vpu affects the anterograde transport and the glycosylation pattern of NTB-A. *Virology* (2013) **440**:190–203. doi:10.1016/j.virol.2013.02.021

118. Zarama A, Pérez-Carmona N, Farré D, Tomic A, Borst EM, Messerle M, et al. Cytomegalovirus m154 hinders CD48 cell-surface expression and promotes viral escape from host natural killer cell control. *PLoS Pathog* (2014) **10**:e1004000. doi:10.1371/journal.ppat.1004000

119. Garland RJ, El-Shanti N, West SE, Hancock JP, Goulden NJ, Steward CG, et al. Human CD8+ CTL recognition and in vitro lysis of herpes simplex virus-infected cells by a non-MHC restricted mechanism. *Scand J Immunol* (2002) **55**:61–9. doi:10.1046/j.1365-3083.2002.01021.x

120. Ahlstiel G, Edlich B, Hogdal LJ, Rotman Y, Noureddin M, Feld JJ, et al. Early changes in natural killer cell function indicate virologic response to interferon therapy for hepatitis C. *Gastroenterology* (2011) **141**:e1231–2. doi:10.1053/j.gastro.2011.06.069

121. Raziorrouh B, Schraut W, Gerlach T, Nowack D, Grüner NH, Ulsenheimer A, et al. The immunoregulatory role of CD244 in chronic hepatitis B infection and its inhibitory potential on virus-specific CD8+ T-cell function. *Hepatology* (2010) **52**:1934–47. doi:10.1002/hep.23936

122. Owusu Sekyere S, Suneetha PV, Kraft AR, Zhang S, Dietz J, Sarrazin C, et al. A heterogeneous hierarchy of co-regulatory receptors regulates exhaustion of HCV-specific CD8 T cells in patients with chronic hepatitis C. *J Hepatol* (2015) **62**:31–40. doi:10.1016/j.jhep.2014.08.008

Biol Chem (2015) **396**:585–95. doi:10.1515/hsz-2014-0273

14

Immune Responses of Ducks Infected with Duck Tembusu Virus

Ning Li [1,2,3†], Yao Wang [1,2†], Rong Li [1,2], Jiyuan Liu [1,2], Jinzhou Zhang [1,2], Yumei Cai [1,2], Sidang Liu [1,2], Tongjie Chai [1,2,3] and Liangmeng Wei [1,2,3*]*

[1] College of Animal Science and Veterinary Medicine, Shandong Agricultural University, Tai'an, China, [2] Sino-German Cooperative Research Centre for Zoonosis of Animal Origin Shandong Province, Tai'an, China, [3] Collaborative Innovation Centre for the Origin and Control of Emerging Infectious Diseases of Taishan Medical College, Tai'an, China

Edited by:
Diana Bahia,
Universidade Federal de Minas
Gerais, Brazil

Reviewed by:
Heinrich Korner,
Menzies Research Institute Tasmania,
Australia
Adam C. Silver,
University of Hartford, USA

*Correspondence:
Tongjie Chai and Liangmeng Wei,
College of Animal Science
and Veterinary Medicine,
Shandong Agricultural University,
61 Daizong Road, Tai'an 271000,
Shandong Province, China
chaitj117@163.com;
lmwei@sdau.edu.cn

[†] Ning Li and Yao Wang are co-first
authors.

Duck Tembusu virus (DTMUV) can cause serious disease in ducks, characterized by reduced egg production. Although the virus has been isolated and detection methods developed, the host immune responses to DTMUV infection are unclear. Therefore, we systematically examined the expression of immune-related genes and the viral distribution in DTMUV-infected ducks, using quantitative real-time PCR. Our results show that DTMUV replicates quickly in many tissues early in infection, with the highest viral titers in the spleen 1 day after infection. *Rig-1*, *Mda5*, and *Tlr3* are involved in the host immune response to DTMUV, and the expression of proinflammatory cytokines (*Il-1β*, *−2*, *−6*, *Cxcl8*) and antiviral proteins (*Mx*, *Oas*, etc.) *are also upregulated* early in infection. The expression of *Il-6* increased most significantly in the tissues tested. The upregulation of *Mhc-I* was observed in the brain and spleen, but the expression of *Mhc-II* was upregulated in the brain and downregulated in the spleen. The expression of the interferons was also upregulated to different degrees in the spleen but that of the brain was various. Our study suggests that DTMUV replicates rapidly in various tissues and that the host immune responses are activated early in infection. However, the overexpression of cytokines may damage the host. These results extend our understanding of the immune responses of ducks to DTMUV infection, and provide insight into the pathogenesis of DTMUV attributable to host factors.

Keywords: duck, DTMUV, host innate immune response, proinflammatory cytokines, antiviral proteins

Introduction

Duck Tembusu virus (DTMUV) is an enveloped, positive-sense, single-stranded RNA virus, classified in the genus *Flavivirus*, which includes West Nile virus (WNV), dengue virus (DENV) and other zoonotic viruses (Tang et al., 2012). DTMUV was isolated in the major duck-producing regions of China in 2010, and can cause an acute contagious infection characterized by heavy egg drop in egg-laying and breeder ducks. It is the first flavivirus reported to cause a serious epidemic disease in ducks (Cao et al., 2011; Su et al., 2011; Yan et al., 2011). Almost all species of duck can be infected with DTMUV, including Cherry Valley ducks, Pekin ducks, and shelducks (Tang et al., 2015), as can chickens, geese, and sparrows (Liu et al., 2012; Tang et al., 2013a). Most importantly, a recent study has shown that DTMUV can infect humans (Tang et al., 2013b). The serious threat DTMUV poses to the development of the duck industry and the concerns it raises for public health mean that this virus must be taken seriously.

The innate immune response is the first line of defense protecting the host from pathogenic organisms. It is well known that pattern recognition receptors (PRRs), such as Toll-like receptors

(*Tlr*) 3, 7, and 8, retinoic acid inducible gene I (*Rig-1*), and melanoma differentiation factor 5 (*Mda5*), *can identify viral molecular patterns* and trigger the activation of specific signaling pathways, leading to the transcription of proinflammatory cytokines, apoptotic responses, and the expression of type I interferons (*Ifns*; Akira et al., 2006). *Tlr3*, *Rig-1*, and *Mda5* are involved in the host response to DENV and induce the production of interleukin 8 (*Cxcl8*) and *Ifn-α/β in vitro* (Loo et al., 2008; Green et al., 2014). Both *Rig-1* and *Mda5* recognize WNV, upregulating the expression of type I *Ifn*, *Il-1β* and antiviral effector proteins (Fredericksen et al., 2008; Quicke and Suthar, 2013).

Because DTMUV is a newly emerging virus, most studies have focused on its isolation, genetic analysis, and the establishment of diagnostic methods (Jiang et al., 2012; Li et al., 2012a,b; Zhu et al., 2012), although preliminary investigations of the pathogenicity of DTMUV have also been reported (Li et al., 2013). However, the immune responses of ducks infected with DTMUV have not been fully explored. Therefore, to clarify the innate immune responses to DTMUV in infected ducks and the tropism of the virus, we systematically investigated the expression of immune-related genes in the duck spleen and brain, and the viral titers in various tissues of infected ducks. Our study extends our understanding of the immune responses of ducks to DTMUV infection.

Materials and Methods

Virus Preparation

DTMUV strain FX2010, used in this study, was a gift from Zejun Li, a researcher at the Shanghai Veterinary Research Institute, Chinese Academy of Agricultural Sciences. The virus was propagated in specific-pathogen-free embryonated chicken eggs and the titers were shown to be $10^{5.2}$ median tissue culture infective doses ($TCID_{50}$)/mL in infected duck embryonic fibroblasts, calculated with the Reed and Muench method (Reed and Muench, 1938).

Animal Experiment

One-day-old Cherry Valley ducks were purchased from a duck farm (Tai'an, Shandong) and housed in isolators until use. The ducks were confirmed to be serologically negative for DTMUV using a blocking enzyme-linked immunosorbent assay (Li et al., 2012b). All animal experiments were performed according to the guidelines of the Committee on the Ethics of Animals of Shandong and the appropriate biosecurity guidelines. At 5 days old, the ducks were randomly divided into two groups, each containing 25 animals. The ducks of one group were infected intramuscularly with 0.4 mL of $10^{5.2}$ $TCID_{50}$ virus. The control group was inoculated in the same manner with 0.4 mL of sterile phosphate-buffered saline (PBS). Three live ducks, except the dead ducks, from each group were euthanized at 1, 2, 3, 4, and 5 days post infection (dpi) and their parenchymatous organs (heart, liver, spleen, lung, kidney, brain, and pancreas) were collected and stored at −70°C until viral titration and the analysis of immune-related gene expression. The remaining ducks were observed for clinical symptoms for 9 days and were euthanized with an intravenous injection of sodium pentobarbital (100 mg/kg

bodyweight) at the end of the study (Pantin-Jackwood et al., 2012).

RNA and cDNA Preparation

The collected tissues (0.1 g) were ground in liquid nitrogen and the total RNAs were extracted from the tissues with TRIzol Reagent (Takara, Dalian, China), according to the manufacturer's instructions. The concentrations of the total RNAs were measured with an ultraviolet spectrophotometer (Shimadzu, Shimazu, Japan). A sample of each RNA (1 μg) was treated with DNase I (Thermo Scientific, Lithuania) and reverse transcribed with M-MLV reverse transcriptase (Promega, Madison, WI, USA). The synthesized cDNA was stored at −20°C until analysis.

Quantitative Real-time PCR

The relative expression of immune-related genes was quantified after infection using previously described primers (Wei et al., 2013a, 2014). The primers for the *Ifn-β* gene were designed using the Primer 3 software, based on the published GenBank sequence (GenBank: KM035791.1). The primers for the *E* gene of DTMUV were as previously reported (Yu et al., 2012). To confirm the copy numbers of DTMUV in the affected ducks, the viral titers (log_{10}) were normalized to 1 μg of total RNA. Quantitative real-time PCR was performed with the 7500 Fast Real-Time PCR System (Applied Biosystems, Carlsbad, CA, USA) using the SYBR Green PCR kit (Takara, Dalian, China). All primer pairs (**Table 1**) were selected according to their specificity, determined with dissociation curves. Quantitative real-time PCR was performed in a reaction volume of 20 μL, according to the manufacturer's instructions. The PCR cycling conditions were: one cycle at 95°C for 30 s, 40 cycles of denaturation at 95°C for 5 s and extension at 60°C for 34 s, followed by a dissociation curve analysis step. To validate the assay, the purified PCR products were cloned into the pMD18-T plasmid and sequenced to confirm the proper amplification. Each sample was analyzed in triplicate.

Statistical Analysis

The relative expression of the target genes in the infected and control groups was calculated with the $2^{-\Delta\Delta Ct}$ method and expressed as the fold changes in gene expression. The housekeeping gene encoding β-actin (*Actb*) was used as the endogenous control against which to normalize the expression levels of the target genes. The fold changes were logarithmically transformed. All data were analyzed with Student's *t*-test using GraphPad Prism 5 (GraphPad Software Inc., San Diego, CA, USA). Statistical significance was set at $P < 0.05$.

Results

Clinical Symptoms and Viral Titers in DTMUV-infected Ducks

The clinical symptoms of the affected ducks were observed at 3 dpi and the ducks showed loss of appetite and depression and were reluctant to move. At 4–6 dpi, some ducks appeared neurological signs, such as dystaxia and paralysis. In present study, four infected ducks died at 4 dpi and three died at 5 dpi. The

TABLE 1 | Primers used in this study.

Primer name	Sequence (5′–3′)	Product size (bp)	GenBank no.
Rig-1 F	GCTACCGCCGCTACATCGAG	224	EU363349
Rig-1 R	TGCCAGTCCTGTGTAACCTG		
Mda5 F	GCTACAGAAGATAGAAGTGTCA	120	KJ451070.1
Mda5 R	CAGGATCAGATCTGGTTCAG		
Tlr3 F	GAGTTTCACACAGGATGTTTAC	200	JQ910167
Tlr3 R	GTGAGATTTGTTCCTTGCAG		
IL-1β F	TCATCTTCTACCGCCTGGAC	149	DQ393268
Il-1β R	GTAGGTGGCGATGTTGACCT		
Il-2 F	GCCAAGAGCTGACCAACTTC	137	AF294323
Il-2 R	ATCGCCCACACTAAGAGCAT		
Il-6 F	TTCGACGAGGAGAAATGCTT	150	AB191038
Il-6 R	CCTTATCGTCGTTGCCAGAT		
Cxcl8 F	AAGTTCATCCACCCTAAATC	182	DQ393274
Cxcl8 R	GCATCAGAATTGAGCTGAGC		
Ifn-α F	TCCTCCAACACCTCTTCGAC	232	EF053034
Ifn-α R	GGGCTGTAGGTGTGGTTCTG		
Ifn-β F	AGATGGCTCCCAGCTCTACA	210	KM035791.1
Ifn-β R	AGTGGTTGAGCTGGTTGAGG		
Ifn-γ F	GCTGATGGCAATCCTGTTTT	247	AJ012254
Ifn-γ R	GGATTTTCAAGCCAGTCAGC		
Mx F	TGCTGTCCTTCATGACTTCG	153	GU202170.1
Mx R	GCTTTGCTGAGCCGATTAAC		
Oas F	TCTTCCTCAGCTGCTTCTCC	187	KJ126991.1
Oas R	ACTTCGATGGACTCGCTGTT		
Pkr F	AATTCCTTGCCTTTTCATTCAA	109	Unpublished
Pkr R	TTTGTTTTGTGCCATATCTTGG		
Mhc-I F	GAAGGAAGAGACTTCATTGCCTTGG	196	AB115246
Mhc-I R	CTCTCCTCTCCAGTACGTCCTTCC		
Mhc-II F	CCACCTTTACCAGCTTCGAG	229	AY905539
Mhc-II R	CCGTTCTTCATCCAGGTGAT		
DTMUV-E F	CGCTGAGATGGAGGATTATGG	225	KC990541.1
DTMUV-E R	ACTGATTGTTTGGTGGCGTG		
β-actin F	GGTATCGGCAGCAGTCTTA	160	EF667345.1
β-actin R	TTCACAGAGGCGAGTAACTT		

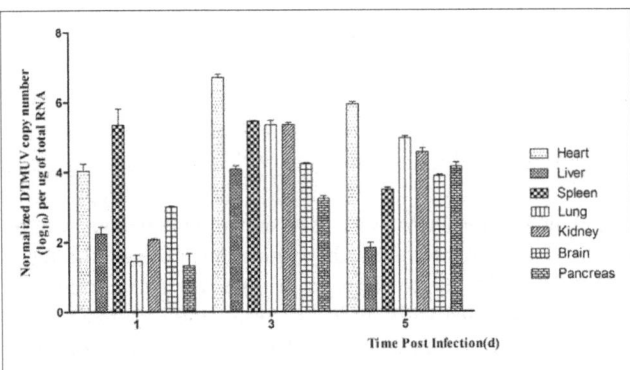

FIGURE 1 | Viral titers in DTMUV-infected ducks at 1, 2, and 3 dpi. Data are expressed as means ± standard deviations ($n = 3$), and each sample was analyzed in triplicate.

brain, to a high titer. The viral titers in all tissues peaked at 3 dpi, except in the spleen and pancreas. The viral titer in the heart reached $10^{6.7}$ copies, and those in the spleen, lung and kidney were basically identical. The viral titers in most of the tested tissues began to decline at 5 dpi, but were still highest in the heart. The viral titer in the spleen decreased dramatically at 5 dpi compared with that at 3 dpi, whereas the viral titer in the brain reached $10^{3.8}$ copies. No virus was detected in the control group. In summary, DTMUV replicated quickly in many organs, leading to systemic impairment.

Expression of PRR mRNAs in DTMUV-infected Ducks

We detected the expression of PRRs (*Rig-1*, *Mda5*, and *Tlr3*) in the brains and spleens of ducks infected with DTMUV during the early post infection period. In the brain, the expression of *Rig-1* and *Mda5* was upregulated during the first 3 days of infection, and peaked at 2 dpi and 3 dpi, respectively (4.13-fold and 20.60-fold, respectively, $P < 0.05$; **Figures 2A,B**). *Tlr3* was expressed at 1 dpi (1.35-fold), peaked at 2 dpi (28.54-fold, $P < 0.05$), and remained high at 3 dpi (13.49-fold, $P < 0.05$; **Figure 2C**).

In the spleen, the expression of *Rig-1* was upregulated at 1 dpi (2.89-fold) and peaked at 2 dpi (13.62-fold, $P < 0.05$), and then decreased slightly at 3 dpi (9.71-fold, $P < 0.05$; **Figure 3A**). *Mda5* transcripts were detected at 1 dpi (10.29-fold, $P < 0.05$), peaked at 3 dpi (18.77-fold, $P < 0.05$; **Figure 3B**). There was an 18.34-fold increase in *Tlr3* mRNA at 1 dpi, which then decreased significantly at 2 dpi (1.57-fold) and decreased further at 3 dpi (0.81-fold; **Figure 3C**). These data indicate that *Rig-1*, *Mda5*, and *Tlr3* are involved in the host immune response to DTMUV, and that the roles they play might differ with time.

Cytokine Expression in DTMUV-infected Ducks

To determine the induction of proinflammatory cytokines in ducks infected with DTMUV, we determined the expression levels of *Il-1β*, *Il-2*, *Il-6*, *Cxcl8*, and the type I and II *Ifn* genes. In the brain, the expression of *Il-1β* was downregulated at 1 dpi and 2 dpi, but upregulated at 3 dpi (5.30-fold, $P < 0.05$; **Figure 2D**). *Il-2* expression was downregulated 0.53-fold at 1 dpi but upregulated 1.73-fold at 2 dpi, after which it decreased slightly at 3 dpi (1.08-fold; **Figure 2E**). The expression of *Il-6* and *Cxcl8* showed similar tendencies during the 3 days tested, with reduced expression (0.72-fold and 0.18-fold, respectively) at 1 dpi, which gradually increased to a peak at 3 dpi (47.78-fold and 16.05-fold, respectively, $P < 0.05$; **Figures 2F,G**). The patterns of type I and II *Ifn* expression differed in the brain. The expression of *Ifn-α* was downregulated at 1 dpi (0.55-fold) and gradually upregulated at 2 dpi, reaching its highest level at 3 dpi (1.96-fold; **Figure 2K**). The expression of *Ifn-β* was downregulated at 1 dpi, slightly upregulated at 2 dpi (1.29-fold), and downregulated again at 3 dpi (**Figure 2L**). The expression of *Ifn-γ* was downregulated at all time points (**Figure 2M**).

In the spleen, the expression of *Il-1β* and *Cxcl8* was highest at 1 dpi (3.94-fold and 10.36-fold, respectively, $P < 0.05$), and then decreased slightly in the following 2 days (**Figures 3D and G**). The expression of *Il-2* was upregulated at 1 dpi (4.17-fold, $P < 0.05$), decreased gradually by 2 dpi, and was downregulated

symptoms of ducks affected with DTMUV gradually lessened and disappeared at 9 dpi.

In this study, we detected DTMUV replication in the parenchymatous organs of infected ducks in the first 5 days after infection. As shown in **Figure 1**, the viral titers could be detected in all the tissues tested at 1 dpi and the highest titer was observed in the spleen. In the same period, DTMUV replicated rapidly in the

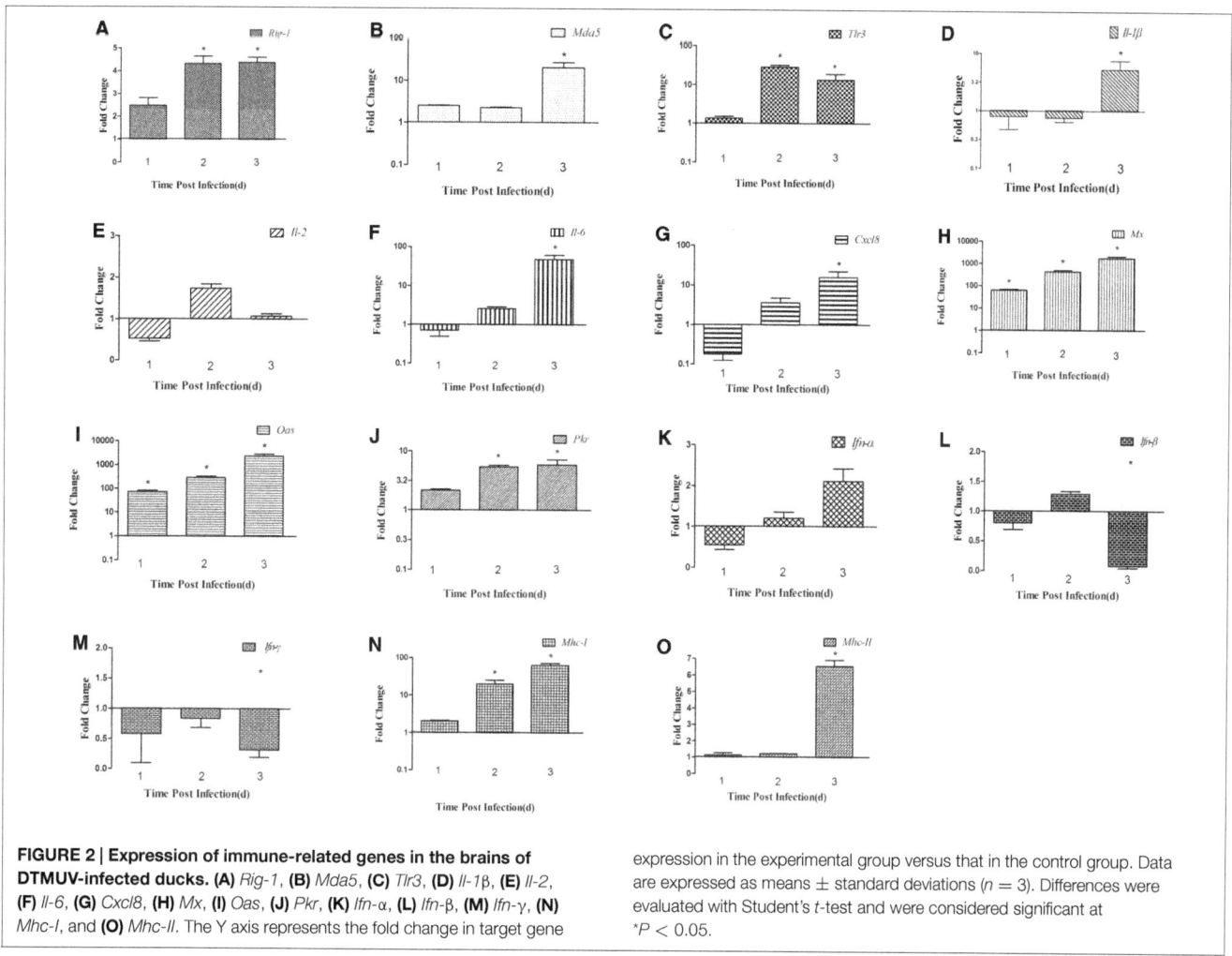

FIGURE 2 | Expression of immune-related genes in the brains of DTMUV-infected ducks. (A) *Rig-1*, **(B)** *Mda5*, **(C)** *Tlr3*, **(D)** *Il-1β*, **(E)** *Il-2*, **(F)** *Il-6*, **(G)** *Cxcl8*, **(H)** *Mx*, **(I)** *Oas*, **(J)** *Pkr*, **(K)** *Ifn-α*, **(L)** *Ifn-β*, **(M)** *Ifn-γ*, **(N)** *Mhc-I*, and **(O)** *Mhc-II*. The Y axis represents the fold change in target gene expression in the experimental group versus that in the control group. Data are expressed as means ± standard deviations ($n = 3$). Differences were evaluated with Student's *t*-test and were considered significant at *$P < 0.05$.

at 3 dpi (0.82-fold; **Figure 3E**). *Il-6* mRNA expression increased constantly, peaking at 3 dpi (160.10-fold, $P < 0.05$; **Figure 3F**). The high expression of *Ifn-α* was maintained for 2 days (11.84-fold at 1 dpi and 11.02-fold at 2 dpi, $P < 0.05$), but decreased to 3.10-fold at 3 dpi (**Figure 3K**). The expression of the *Ifn-β* gene was markedly upregulated at 1 dpi and 3 dpi (38.38-fold and 46.63-fold, respectively, $P < 0.05$), but less so at 2 dpi (9.78-fold; **Figure 3L**). Unlike its expression in the brain, the expression of the *Ifn-γ* gene was higher than in the control at 1 dpi and 2 dpi (3.05-fold and 2.18-fold, respectively), but was downregulated at 3 dpi (0.09-fold, $P < 0.05$; **Figure 3M**). These results show that the expression of various cytokines is induced in DTMUV-infected ducks, and that the expression patterns of some cytokines are variable. In summary, the expression of *Il-6* was most significantly increased during the early period of DTMUV infection and the type I *Ifns* played a key role in the duck's response to DTMUV in the same period.

Expression of Antiviral Proteins in DTMUV-infected Ducks

Antiviral proteins are effective components of the response to viral infections, so we investigated the expression of several antiviral

proteins, including MX, OAS, and PKR. The three antiviral proteins showed similar trends in the brain: the expression of all of them increased during the 3 days tested. In the brain, the expression of the *Mx* and *Oas* genes was upregulated 65.01-fold and 72.84-fold, respectively, at 1 dpi ($P < 0.05$), increased 431.61-fold and 298.52-fold, respectively, at 2 dpi ($P < 0.05$), and showed dramatic increases of 1733.20-fold and 2375.16-fold, respectively, at 3 dpi ($P < 0.05$; **Figures 2H,I**). Pkr expression was gradually increasing during the tested days and peaked at 3 dpi (5.91-fold, $P < 0.05$; **Figure 2J**). However, in the spleen, the expression of these antiviral proteins was variable. *Mx* and *Oas* expression was significantly increased at 1 dpi (67.26-fold and 60.97-fold, respectively; $P < 0.05$) and at 2 dpi (172.67-fold and 144.55-fold, respectively; $P < 0.05$), and then decreased at 3 dpi (**Figures 3H,I**). *Pkr* mRNA expression was significantly upregulated at 1 dpi (21.07-fold, $P < 0.05$), and gradually declined in the following 2 days (9.81-fold and 5.06-fold, respectively, $P < 0.05$; **Figure 3J**).

These data demonstrate that the expression of some antiviral proteins, especially MX and OAS, increased significantly in the brains and spleens of DTMUV-infected ducks, indicating that they play important roles in resisting DTMUV infection.

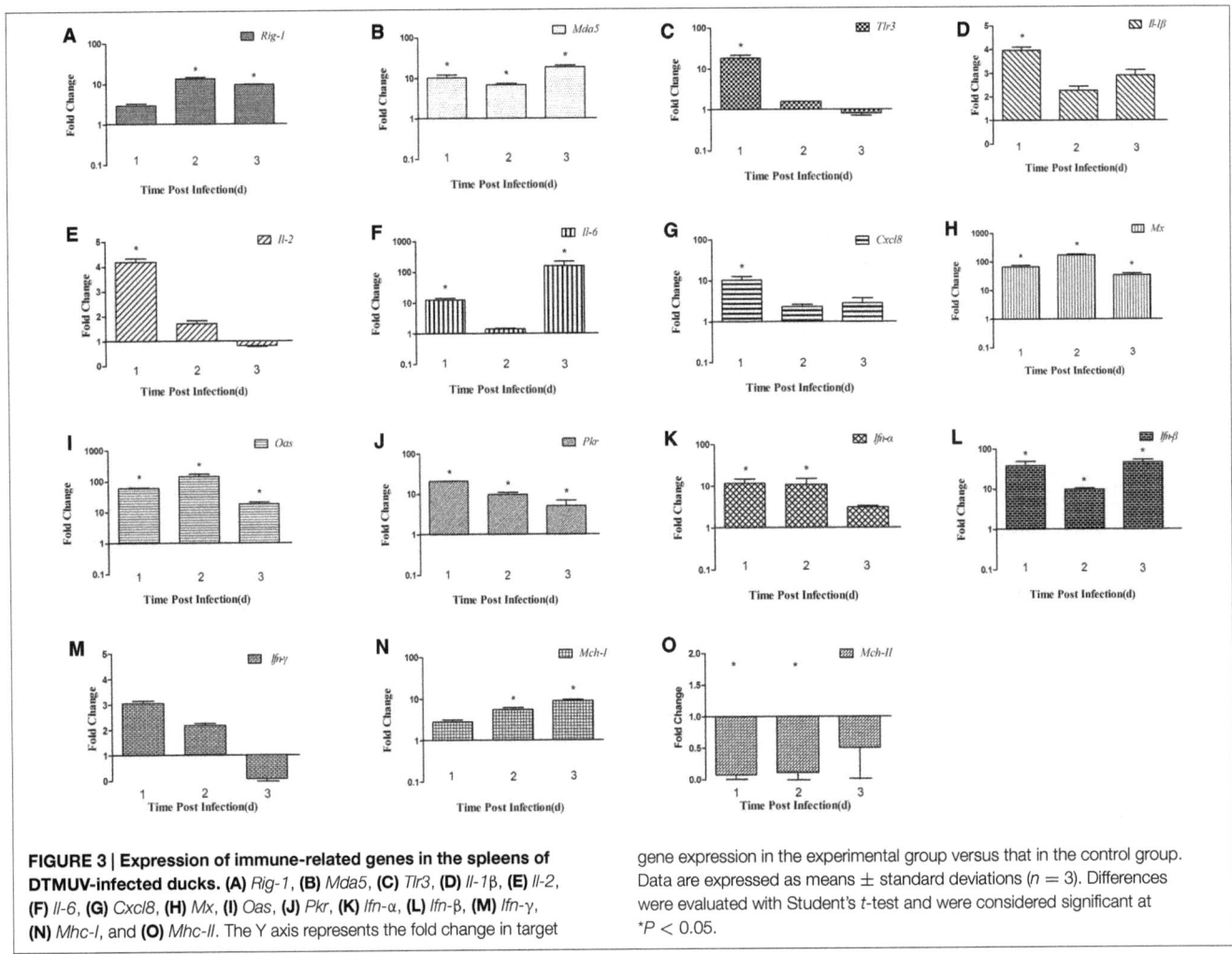

FIGURE 3 | Expression of immune-related genes in the spleens of DTMUV-infected ducks. (A) *Rig-1*, **(B)** *Mda5*, **(C)** *Tlr3*, **(D)** *Il-1*β, **(E)** *Il-2*, **(F)** *Il-6*, **(G)** *Cxcl8*, **(H)** *Mx*, **(I)** *Oas*, **(J)** *Pkr*, **(K)** *Ifn-*α, **(L)** *Ifn-*β, **(M)** *Ifn-*γ, **(N)** *Mhc-I*, and **(O)** *Mhc-II*. The Y axis represents the fold change in target gene expression in the experimental group versus that in the control group. Data are expressed as means ± standard deviations ($n = 3$). Differences were evaluated with Student's *t*-test and were considered significant at *$P < 0.05$.

Expression of MHC Class I and II Molecules in DTMUV-infected Ducks

To confirm whether MHC-I and -II molecules are involved in the host immune responses to DTMUV, we examined their expression in the first 3 days after infection. The expression of both *Mhc-I* and *-II* molecules was upregulated in the brains of the infected ducks, peaking at 3 dpi (64.08-fold and 6.26-fold, respectively, $P < 0.05$; **Figures 2N,O**). In the spleen, *Mhc-I* gene expression was upregulated 2.84-fold at 1 dpi and 5.55-fold at 2 dpi, peaking at 3 dpi (8.99-fold, $P < 0.05$; **Figure 3N**). However, the expression of *Mhc-II* molecules was downregulated in the spleen on all 3 days examined (**Figure 3O**). These results indicate that both MHC-I and -II molecules are involved in the duck immune responses to DTMUV.

Discussion

It has been reported that Tembusu virus causes a disease in chickens that is characterized by encephalitis and growth retardation (Kono et al., 2000), but does not do so in ducks. However, DTMUV caused serious outbreaks of disease in ducks in 2010, involving severe economic losses (Su et al., 2011; Yan et al., 2011). This difference in the pathogenicity of Tembusu virus is

determined by many factors, especially the host immune response. Here, we systematically examined the expression of immune-related genes at the mRNA level and the distribution of the virus in DTMUV-infected ducks.

The pathogenicity of DTMUV in ducks correlates directly with the level of virus in the tissues. In this study, the viral titer at 1 dpi was highest in the spleen, indicating that the spleen is the target organ of DTMUV (Jiang et al., 2012). The viral titers in all the tissues tested peaked at 3 dpi, except in the spleen and pancreas, and gradually decreased to 5 dpi (**Figure 1**). These results show that DTMUV replicates rapidly in the parenchymal organs, including the brain, and that DTMUV causes viremia and disrupts the blood–brain barrier in a short period of time. High levels of virus in their tissues may have been the main cause of death in some infected ducks.

The innate immune response of the host is the primary mechanism for resisting and clearing viruses during the early stage of infection. Viral genomes and replication products are sensed by key PRRs, such as *Rig-1*, *Mda5*, and *Tlr3/7/8* (Pichlmair and Reis e Sousa, 2007), and both WNV and DENV-2 trigger RIG-1 and MDA5 signaling (Fredericksen et al., 2008; Green et al., 2014). However, The role of TLR3 involved in Flavivirus infection is controversial. It has recently been demonstrated that DENV activates TLR3 signaling cascades, leading to the transcription of IFN-α/β in mononuclear cells (Tsai et al., 2009). WNV inhibited the TLR3-mediated production of IL-6 and an antiviral state (Scholle and Mason, 2005; Wilson et al., 2008), and Wang et al. (2004) had proved that viral titers and neuropathology were reduced in the brain of WNV-infected TLR3-deficient mice comparing to the control (Wang et al., 2004), which suggesting TLR3-mediated inflammatory response may disrupt the blood-brain barrier and accelerate the WNV into the CNS (Fredericksen and Gale, 2006; Matsumoto et al., 2011). In our study, the expression of *Rig-1*, *Mda5*, and *Tlr3* was upregulated in the brain and spleen during the period of infection tested, although the expression of *Tlr3* was not upregulated in the spleen at 3 dpi (**Figures 2C and 3C**). We also found that *Tlr3* expression was significantly upregulated in the brain at 2 dpi (28.54-fold, $P < 0.05$), but decreased to 13.49-fold at 3 dpi, whereas *Mda5* was markedly increased (20.60-fold) at that time (**Figures 2B,C**). In the spleen, the expression of *Tlr3* increased at 1 dpi (18.34-fold, $P < 0.05$), but decreased to 1.57-fold at 2 dpi and was further downregulated at 3 dpi, whereas *Rig-1* and *Mda5* were significantly upregulated at 2 dpi and 3 dpi (13.62-fold and 18.77-fold, respectively, $P < 0.05$; **Figures 3A,B**). These results suggest that the different PRRs may play key roles at different times. It was recently reported that RIG-1 and MDA5 are required for the recognition of WNV: RIG-1 is considered to trigger the expression of immune-related genes early in infection, whereas MDA5 signaling occurs later (Errett et al., 2013).

The activation of PRRs induces the expression of cytokines and antiviral proteins, including IL-1β, IL-2, TNF-α, MX, and OAS, which alert the immune system to viral infection. Here, we examined several cytokines and found that the expression of *Il-1β*, *Il-2*, *Il-6*, and *Cxcl8* increased in the spleen on the days examined, with *Il-6* expression particularly elevated (160.12-fold at 3 dpi, $P < 0.05$). In the brain, all the cytokines tested were downregulated at 1 dpi, but upregulated at 3 dpi, and *Il-6* was again most strongly upregulated (47.78-fold, $P < 0.05$). A previous study suggested that *Il-6* is more robustly induced in chickens than in ducks, which may be responsible for the different symptoms observed in the two species after influenza virus infection (Liang et al., 2011). In mammals, "cytokine storms" are believed to contribute to more severe pathological lesions and higher rates of death (Chan et al., 2005). Similar results were also observed in ducks infected with the highly pathogenic avian influenza virus H5N1 (Wei et al., 2013a). In the present study, 28% of the DTMUV-infected ducks died, suggesting that the excessive expression of cytokines, such as *Il-6* and *Cxcl8*, and the rapid replication of DTMUV in various tissues may have caused the deaths of the infected ducks.

Type I IFN production is a typical innate defense against viral infection and the expression of antiviral proteins contributes to viral clearance. The expression of the most genes including the Mx, Oas have increased in brain and spleen from the mice infected with WNV, which suggesting that the gene products may be involved in the protection against WNV (Venter et al., 2005). In this study, *Ifn-α/β* expression was significantly induced in the spleen early in infection. The expression of *Mx* and *Oas* increased significantly in the brain and spleen, but failed to prevent massive viral replication and was insufficient to protect the ducks from DTMUV. A similar phenomenon has been observed in geese and chickens infected with highly pathogenic avian influenza virus H5N1 (Daviet et al., 2009; Wei et al., 2013b).

MHC molecules can activate the acquired immune response to eliminate a viral infection, and some viruses inhibit MHC-I expression. In our study, the upregulation of *Mhc-I* was observed in the brains and spleens of infected ducks (**Figures 2N and 3N**), which is not surprising because the upregulation of *Mhc-I* has been observed during infection with DENV and WNV. This phenomenon has only been observed in the genus *Flavivirus*, and not in the other two genera, hepatitis C virus and the pestiviruses. However, the definitive role of *Mhc-I* upregulation during *Flavivirus* infection is unclear (Lobigs et al., 2003). In the present study, the production of *Mhc-II* increased slightly in the brain, but was downregulated in the spleen throughout the experimental period (**Figures 2O and 3O**). *Mhc-II* is also reportedly downregulated in response to avian influenza virus infection *in vivo* and *in vitro* (Adams et al., 2009; Liang et al., 2011; Cagle et al., 2012).

In summary, DTMUV induces the upregulation of *Rig-1*, *Mda5*, and *Tlr3* expression in ducks, resulting in the activation of *Ifns* and several interferon-stimulated genes, including proinflammatory cytokines and antiviral proteins. Although various antiviral proteins and *Ifns* were induced, they did not provide adequate protection against DTMUV infection in ducks, and the excessive host immune responses, including massive *Il-6* expression, and the rapid replication of DTMUV damaged the host, leading to serious disease and even death. As far as we know, this is the first report of the immune-related gene expression in response to DTMUV infection in ducks. We have attempted to provide a comprehensive picture of the duck immune responses to DTMUV infection. Our results provide useful information concerning the relationship between DTMUV and the host immune response, and insight into the pathogenesis of DTMUV attributable to host factors.

Acknowledgments

This work was supported by the National Science and Technology Support Project (2012BAD39B02); National Natural Science Foundation of China (31270172 and 31470258); The State Key Laboratory of Pathogen and Biosecurity (Academy of Military Medical Science, SKLPBS1449); The Project of Natural Science Foundation of Shandong Province (ZR2014CQ050) and China Postdoctoral Science Foundation (2014M560569).

References

Adams, S. C., Xing, Z., Li, J., and Cardona, C. J. (2009). Immune-related gene expression in response to H11N9 low pathogenic avian influenza virus infection in chicken and Pekin duck peripheral blood mononuclear cells. *Mol. Immunol.* 46, 1744–1749. doi: 10.1016/j.molimm.2009.01.025

Akira, S., Uematsu, S., and Takechi, O. (2006). Pathogen recognition and innate immunity. *Cell* 124, 783–801. doi: 10.1016/j.cell.2006.02.015

Cagle, C., Wasilenko, J., Adams, S. C., Cardona, C. J., To, T. L., Nguyen, T., et al. (2012). Differences in pathogenicity, response to vaccination, and innate immune responses in different types of ducks infected with a virulent H5N1 highly pathogenic avian influenza virus from Vietnam. *Avian Dis.* 56, 479–487. doi: 10.1637/10030-120511-Reg.1

Cao, Z., Zhang, C., Liu, Y., Liu, Y., Ye, W., Han, J., et al. (2011). Tembusu virus in ducks, china. *Emerg. Infect. Dis.* 17, 1873–1875. doi: 10.3201/eid1710.101890

Chan, M. C., Cheung, C. Y., Chui, W. H., Tsao, S. W., Nicholls, J. M., Chan, Y. O., et al. (2005). Proinflammatory cytokine responses induced by influenza A (H5N1) viruses in primary human alveolar and bronchial epithelial cells. *Respir. Res.* 6:135. doi: 10.1186/1465-9921-6-135

Daviet, S., Van Borm, S., Habyarimana, A., Ahanda, M. L., Morin, V., Oudin, A., et al. (2009). Induction of Mx and PKR failed to protect chickens from H5N1 infection. *Viral Immunol.* 22, 467–472. doi: 10.1089/vim.2009.0053

Errett, J. S., Suthar, M. S., McMillan, A., Diamond, M. S., and Gale, M. J. (2013). The essential, nonredundant roles of RIG-I and MDA5 in detecting and controlling West Nile virus infection. *J. Virol.* 87, 11416–11425. doi: 10.1128/JVI.01488-13

Fredericksen, B. L., Keller, B. C., Fornek, J., Katze, M. G., and Gale, M. J. (2008). Establishment and maintenance of the innate antiviral response to West Nile Virus involves both RIG-I and MDA5 signaling through IPS-1. *J. Virol.* 82, 609–616. doi: 10.1128/JVI.01305-07

Fredericksen, B. L., and Gale, M. J. (2006). West Nile Virus evades activation of interferon regulatory factor 3 through RIG-I-dependent and -independent pathways without antagonizing host defense signaling. *J. Virol.* 80, 2193–2923. doi: 10.1128/JVI.80.6.2913-2923.2006

Green, A. M., Beatty, P. R., Hadjilaou, A., and Harris, E. (2014). Innate immunity to dengue virus infection and subversion of antiviral responses. *J. Mol. Biol.* 426, 1148–1160. doi: 10.1016/j.jmb.2013.11.023

Jiang, T., Liu, J., Deng, Y. Q., Su, J. L., Xu, L. J., Liu, Z. H., et al. (2012). Development of RT-LAMP and real-time RT-PCR assays for the rapid detection of the new duck Tembusu-like BYD virus. *Arch. Virol.* 157, 2273–2280. doi: 10.1007/s00705-012-1431-7

Kono, Y., Tsukamoto, K., Abd, H. M., Darus, A., Lian, T. C., Sam, L. S., et al. (2000). Encephalitis and retarded growth of chicks caused by Sitiawan virus, a new isolate belonging to the genus Flavivirus. *Am. J. Trop. Med. Hyg.* 63, 94–101.

Li, L., An, H., Sun, M., Dong, J., Yuan, J., and Hu, Q. (2012a). Identification and genomic analysis of two duck-origin Tembusu virus strains in southern China. *Virus Genes* 45, 105–112. doi: 10.1007/s11262-012-0753-6

Li, X., Li, G., Teng, Q., Yu, L., Wu, X., and Li, Z. (2012b). Development of a blocking ELISA for detection of serum neutralizing antibodies against newly emerged duck Tembusu virus. *PLoS ONE* 7:e53026. doi: 10.1371/journal.pone.0053026

Li, S., Zhang, L., Wang, Y., Wang, S., Sun, H., Su, W., et al. (2013). An infectious full-length cDNA clone of duck Tembusu virus, a newly emerging flavivirus causing duck egg drop syndrome in China. *Virus Res.* 171, 238–241. doi: 10.1016/j.virusres.2012.10.019

Liang, Q. L., Luo, J., Zhou, K., Dong, J. X., and He, H. X. (2011). Immune-related gene expression in response to H5N1 avian influenza virus infection in chicken and duck embryonic fibroblasts. *Mol. Immunol.* 48, 924–930. doi: 10.1016/j.molimm.2010.12.011

Liu, M., Chen, S., Chen, Y., Liu, C., Chen, S., Yin, X., et al. (2012). Adapted Tembusu-like virus in chickens and geese in China. *J. Clin. Microbiol.* 50, 2807–2809. doi: 10.1128/JCM.00655-12

Lobigs, M., Müllbacher, A., and Regner, M. (2003). MHC class I up-regulation by flaviviruses: immune interaction with unkonwn advantage to host or pathogen. *Immunol. Cell Biol.* 81, 217–223. doi: 10.1046/j.1440-1711.2003.01161.x

Loo, Y. M., Fornek, J., Crochet, N., Bajwa, G., Perwitasari, O., Martinez-Sobrido, L., et al. (2008). Distinct RIG-I and MDA5 signaling by RNA viruses in innate immunity. *J. Virol.* 82, 335–345. doi: 10.1128/JVI.01080-07

Matsumoto, M., Oshiumi, H., and Seya, T. (2011). Antiviral responses induced by the TLR3 pathway. *Rev. Med. Virol.* 21, 67–77. doi: 10.1002/rmv.680

Pantin-Jackwood, M. J., Smith, D. M., Wasilenko, J. L., Cagle, C., Shepherd, E., Sarmento, L., et al. (2012). Effect of age on the pathogenesis and innate immune responses in Pekin ducks infected with different H5N1 highly pathogenic avian influenza viruses. *Virus Res.* 167, 196–206. doi: 10.1016/j.virusres.2012.04.015

Pichlmair, A., and Reis e Sousa, C. (2007). Innate recognition of viruses. *Immunity*

27, 370–383. doi: 10.1016/j.immuni.2007.08.012

Quicke, K. M., and Suthar, M. S. (2013). The innate immune playbook for restricting West Nile virus infection. *Viruses* 5, 2643–2658. doi: 10.3390/v5112643

Reed, L. J., and Muench, H. (1938). A simple method of estimating fifty percent endpoints. *Am. J. Epidemiol.* 27, 493–497.

Scholle, F., and Mason, P. W. (2005). West Nile virus replication interferes with both poly(I:C)-induced interferon gene transcription and response to interferon treatment. *Virology* 342, 77–87. doi: 10.1016/j.virol.2005.07.021

Su, J., Li, S., Hu, X., Yu, X., Wang, Y., Liu, P., et al. (2011). Duck egg-drop syndrome caused by BYD virus, a new Tembusu-related flavivirus. *PLoS ONE* 6:e18106. doi: 10.1371/journal.pone.0018106

Tang, Y., Diao, Y., Yu, C., Gao, X., Ju, X., Xue, C., et al. (2013a). Characterization of a Tembusu virus isolated from naturally infected house sparrows (*Passer domesticus*) in Northern China. *Transbound. Emerg. Dis.* 60, 152–158. doi: 10.1111/j.1865-1682.2012.01328.x

Tang, Y., Gao, X., Diao, Y., Feng, Q., Chen, H., Liu, X., et al. (2013b). Tembusu virus in human, china. *Transbound. Emerg. Dis.* 60, 193–196. doi: 10.1111/tbed.12085

Tang, Y., Diao, Y., Chen, H., Ou, Q., Liu, X., Gao, X., Yu, C., Wang, L., (2015). Isolation and genetic characterization of a tembusu virus strain isolated from mosquitoes in Shandong, China. *Transbound. Emerg. Dis.* 62, 209–216. doi: 10.1111/tbed.12111

Tang, Y., Diao, Y., Gao, X., Yu, C., Chen, L., and Zhang, D. (2012). Analysis of the complete genome of Tembusu virus, a flavivirus isolated from ducks in china. *Transbound. Emerg. Dis.* 59, 336–343. doi: 10.1111/j.1865-1682.2011.01275.x

Tsai, Y. T., Chang, S. Y., Lee, C. N., and Kao, C. L. (2009). Human TLR3 recognizes dengue virus and modulates viral replication *in vitro*. *Cell Microbiol.* 11, 604–615. doi: 10.1111/j.1462-5822.2008.01277.x

Venter, M., Myers, T. G., Wilson, M. A., Kindt, T. J., Paweska, J. T., Burt, F. J., et al. (2005). Gene expression in mice infected with West Nile virus strains of different neurovirulence. *Virology* 342, 119–140. doi: 10.1016/j.virol.2005.07.013

Wang, T., Town, T., Alexopoulou, L., Anderson, J. F., Fikrig, E., and Flavell, R. A., (2004). Toll-like receptor 3 mediates West Nile virus entry into the brain causing lethal encephalitis. *Nat. Med.* 10, 1366–1373. doi: 10.1038/nm1140

Wei, L., Cui, J., Song, Y., Zhang, S., Han, F., Yuan, R., et al. (2014). Duck MDA5 functions in innate immunity against H5N1 highly pathogenic avian influenza virus infections. *Vet. Res.* 45:66. doi: 10.1186/1297-9716-45-66

Wei, L., Jiao, P., Song, Y., Cao, L., Yuan, R., Gong, L., et al. (2013a). Host immune responses of ducks infected with H5N1 highly pathogenic avian influenza viruses of different pathogenicities. *Vet. Microbiol.* 166, 386–393. doi: 10.1016/j.vetmic.2013.06.019

Wei, L., Jiao, P., Yuan, R., Song, Y., Cui, P., Guo, X., et al. (2013b). Goose Toll-like receptor 7 (TLR7), myeloid differentiation factor 88 (MyD88) and antiviral molecules involved in anti-H5N1 highly pathogenic avian influenza virus response. *Vet. Immunol. Immunopathol.* 153, 99–106. doi: 10.1016/j.vetimm.2013.02.012

Wilson, J. R., de Sessions, P. F., Leon, M. A., and Scholle, F. (2008). West Nile virus nonstructural protein 1 inhibits TLR3 signal transduction. *J. Virol.* 82, 8262–8271. doi: 10.1128/JVI.00226-08

Yan, P., Zhao, Y., Zhang, X., Xu, D., Dai, X., Teng, Q., et al. (2011). An infectious disease of ducks caused by a newly emerged Tembusu virus strain in mainland China. *Virology* 417, 1–8. doi: 10.1016/j.virol.2011.06.003

Yu, C., Diao, Y., Tang, Y., Cui, J., Gao, X., Zhang, Y., et al. (2012). Fluorescence quantitative RT-PCR assay for detection of Tembusu virus. *China Agric. Sci.* 45, 4492–4500. doi: 10.3864/j.issn.0578-1752.2012.21.018

Zhu, W., Chen, J., Wei, C., Wang, H., Huang, Z., Zhang, M., et al. (2012). Complete genome sequence of duck Tembusu virus, isolated from Muscovy ducks in southern China. *J. Virol.* 86, 13119. doi: 10.1128/JVI.02361-12

15

Suilysin Stimulates the Release of Heparin Binding Protein from Neutrophils and Increases Vascular Permeability in Mice

Shaolong Chen, Wenlong Xie, Kai Wu, Ping Li, Zhiqiang Ren, Lin Li, Yuan Yuan, Chunmao Zhang, Yuling Zheng, Qingyu Lv, Hua Jiang*† and Yongqiang Jiang*†

State Key Laboratory of Pathogen and Biosecurity, Beijing Institute of Microbiology and Epidemiology, Beijing, China

Edited by:
Olivier Dussurget,
Université Paris Diderot/Institut
Pasteur/INSERM/INRA, France

Reviewed by:
Yukihiro Akeda,
Osaka University, Japan
Hridayesh Prakash,
University of Hyderabad, India
Peter Valentin-Weigand,
University of Veterinary Medicine
Hannover, Germany

*Correspondence:
Yongqiang Jiang
jiangyq@bmi.ac.cn
Hua Jiang
jhua76@126.com

†These authors have contributed
equally to this work and are senior
authors.

Most of the deaths that occurred during two large outbreaks of *Streptococcus suis* infections in 1998 and 2005 in China were caused by streptococcal toxic shock syndrome (STSS), which is characterized by increased vascular permeability. Heparin-binding protein (HBP) is thought to mediate the vascular leakage. The purpose of this study was to investigate the detailed mechanism underlying the release of HBP and the vascular leakage induced by *S. suis*. Significantly higher serum levels of HBP were detected in Chinese patients with STSS than in patients with meningitis or healthy controls. Suilysin (SLY) is an exotoxin secreted by the highly virulent strain 05ZYH33, and it stimulated the release of HBP from the polymorphonuclear neutrophils and mediated vascular leakage in mice. The release of HBP induced by SLY was caused by a calcium influx-dependent degranulation. Analyses using a pharmacological approach revealed that the release of HBP induced by SLY was related to Toll-like receptor 4, p38 mitogen-activated protein kinase, and the 1-phosphatidylinositol 3-kinase pathway. It was also dependent on a G protein-coupled seven-membrane spanning receptor. The results of this study provide new insights into the vascular leakage in STSS associated with non-Group A streptococci, which could lead to the discovery of potential therapeutic targets for STSS associated with *S. suis*.

Keywords: heparin binding protein, *Streptococcus suis*-associated streptococcal toxic shock syndrome, suilysin, vascular permeability

INTRODUCTION

Streptococcus suis has been recognized as an emerging zoonotic pathogen (Staats et al., 1997; Lun et al., 2007; Segura, 2009; Wertheim et al., 2009; Gottschalk et al., 2010). Since the first human infection was identified in 1968 in Denmark (Perch et al., 1968), more than 400 cases of *S. suis* infection have been reported worldwide during the subsequent four decades (Lun et al., 2007). Most of these cases presented as meningitis, septicemia, endocarditis, arthritis, or pneumonia (Staats et al., 1997; Lun et al., 2007). Two large outbreaks of *S. suis* serotype 2 (*S. suis* 2) infection in 1998 and 2005 in China affected more than 200 people and led to 52 deaths. Most of these deaths were caused by streptococcal toxic shock syndrome (STSS; Hu et al., 2000; Tang et al., 2006; Yu et al., 2006).

Streptococcal toxic shock syndrome associated with group A streptococci (GAS) is well recognized. A massive over-stimulation of T-cells is believed to be associated with STSS (Brown, 2004; Low, 2013). Superantigens (SAgs), including the streptococcal pyrogenic exotoxin serotypes A, C, and G–M as well as the streptococcal mitogenic exotoxin Z are involved in the molecular and pathological mechanism underlying STSS associated with GAS (Brosnahan and Schlievert, 2011). In addition to the SAgs, M protein, which is a highly conserved cell-surface protein of GAS that can induce strong inflammation, might also contribute to the development of STSS (Pahlman et al., 2006). However, S. suis is a non-GAS pathogen, and does not contain DNA sequences that are homologous to the genes encoding the SAgs or M protein, indicating that molecules other than the SAgs or M protein might be involved in the mechanism underlying the STSS outbreaks associated with S. suis in China (Tang et al., 2006).

Streptococcal toxic shock syndrome caused by S. suis 2, is characterized by acute high fever, vascular collapse, hypotension, shock, petechia, disseminated intravascular coagulation, and multiple organ failure (Tang et al., 2006; Yu et al., 2006). Vascular leakage is a fundamental mechanism of shock. Active polymorphonuclear neutrophils (PMNs) have been shown to release a broad spectrum of cytokines and other molecules that induce increased vascular permeability (Gautam et al., 1998, 2000, 2001). One of these molecules, heparin-binding protein (HBP), is thought to be a key mediator that induces vascular leakage (Gautam et al., 2001; Edens and Parkos, 2003). HBP is also known as azurocidin or CAP37, and has diverse functions. It is usually stored in the azurophilic granules and secretory vesicles in the PMNs (Tapper et al., 2002). The molecular mechanism underlying the release of HBP from the PMNs has been extensively investigated. Bacterial-derived M protein has been shown to bind to fibrinogen and interact with the β_2-integrins on the surface of the PMNs, stimulating the release of HBP (Herwald et al., 2004). HBP can also be released from the PMNs by other mechanisms, including PMN degranulation mediated by streptolysin O (Nilsson et al., 2006) and the lipid leukotriene B4 (LTB4)-mediated stimulation of the BLT1 receptor and phosphatidylinositol 3-kinase (PI3K) intracellular pathway (Di Gennaro et al., 2009). The molecular and pathological mechanisms underlying STSS that are not associated with GAS remain poorly understood (Hashikawa et al., 2004; Ekelund et al., 2005). In this study, we focused on the highly virulent S. suis 2 strain 05ZYH33 to investigate the molecular mechanism underlying the release of HBP from the PMNs and the induced vascular leakage. This strain was originally isolated from a patient who died from STSS during the S. suis outbreak in 2005 in Sichuan, China.

MATERIALS AND METHODS

Reagents

Recombinant human HBP (rHBP, Cat.No.2200-SE-050/CF), polyclonal goat anti-human HBP antibody (Cat.No.AF2200), and monoclonal mouse anti-human HBP antibody (Cat.No.MAB2200) were purchased from R&D (Minneapolis, MN, USA). CLI-095 [a specific inhibitor of Toll-like receptor 4 (TLR4), Cat.No.tlrl-cli95] and OxPAPC (a TLR4 non-specific inhibitor, Cat.No.tlrl-oxp1) were obtained from Invivogen (Hong Kong, China). Escherichia coli O26:B6 lipopolysaccharide (LPS, Cat.No.L2654), EGTA (a calcium chelator, Cat.No.E3889), U73122 (a phospholipase C inhibitor, Cat.No.U6756), PD98059 (a mitogen-activated protein kinase pathway inhibitor, Cat.No.P215), SB202474 (a negative control for SB203580), SB203580 (a p38 MAPK inhibitor), Wortmannin (a PI3K inhibitor, Cat.No.F9128) and Genistein (a receptor tyrosine kinase inhibitor, Cat.No.G6649) were purchased from Sigma–Aldrich (St. Louis, MO, USA). Pertussis toxin (PTX; a receptor inhibitor coupled to G protein) was kindly provided by the National Vaccine and Serum Institute (Beijing, China). The PMN Elastase Human ELISA Kit (Cat.No.ab119553) and the Lactoferrin Human ELISA kit (Cat.No.ab108882) were from Cayman Chemical (Ann Arbor, MI, USA).

Blood Specimens from Patients

All the blood specimens were kindly provided by the Chinese Center for Disease Control and Prevention and kept anonymous. The Chinese Center for Disease Control Human Research Protection Office approved the retrospective testing using anonymous samples. Blood specimens from 13 healthy individuals and 14 patients, including eight with meningitis and six with STSS, were used in this study.

Isolation of Polymorphonuclear Neutrophils (PMNs)

Human PMNs were isolated from freshly heparinized blood that was collected by gradient centrifugation. Red blood cells were separated by 6% dextran (Sigma–Aldrich), and then the PMNs were isolated using density gradient media containing 70% Percoll (Pharmacia, New York, NY, USA) and Ficoll (GE Healthcare, Little Chalfont, UK). The purified PMNs were suspended in Hanks' balanced salt solution (HBSS, Invitrogen, Carlsbad, CA, USA) at a concentration of 2×10^6 cells/mL. The cellular purity was greater than 97% as indicated by Wright-Giemsa staining. The survival rate was greater than 97% as assessed by Trypan blue staining. The purified PMNs were used in the following experiments.

PMN Stimulation

One hundred microliters of human whole blood or 1.0×10^6 isolated PMNs were diluted in HBSS to a final volume of 1.0 mL and incubated with various putative stimulants at 37°C for 30 min. The HBP levels in the blood samples or the release of HBP from PMNs were subsequently determined. Cells were centrifuged at $300 \times g$ for 15 min and HBP levels in the supernatant were analyzed by a sandwich ELISA or a Western blot. The amount of HBP in the cell lysate of the whole blood samples or the PMNs obtained by treatment with 1% Triton X-100 was determined and considered as the total HBP. The percentage of the HBP level in the supernatant relative to the total HBP level was calculated.

Determination of HBP Levels by ELISA

The HBP levels in the supernatant of the whole blood sample or the PMN suspension were determined by a sandwich-based ELISA as previously described (Tapper et al., 2002). Briefly, microtiter plates (Costar, Corning Inc., Corning, NY, USA) were coated with a polyclonal goat anti-human HBP antibody (100 μL of 0.5 μg/mL antibody solution). After washing and blocking, 100 μL of sample was added to duplicate wells and incubated at 37°C for 1 h or 4°C overnight. The standard for quantifying the HBP levels was rHBP. After a thorough washing, the plates were incubated with a monoclonal mouse anti-human HBP antibody (100 μL of 1 μg/mL antibody solution) at 37°C for 1 h or 4°C overnight. A horseradish peroxidase (HRP)-conjugated antibody against mouse IgG (1:8,000, ZSGB-Bio, Beijing, China) was added to detect the primary antibody. The plate was read in a microtiter plate reader (Thermo Fisher Scientific, Waltham, MA, USA) at 450 nm.

Detection of HBP by Western Blotting

Thirty microliters of the supernatant of a whole blood sample or a PMN suspension were separated by 12% (w/v) polyacrylamide gel electrophoresis in the presence of 1% (w/v) sodium dodecyl sulfate (SDS-PAGE) and transferred to a polyscreen polyvinylidene difluoride membrane (Merck Millipore, Billerica, MA, USA). The membrane was first blocked in 5% (w/v) non-fat milk in phosphate-buffered saline (PBS) supplemented with 0.05% (v/v) Tween 20 at room temperature for 1 h, and incubated with the polyclonal goat anti-human HBP antibody (1 μg/mL) at 4°C overnight. After intensive washing, the membrane was incubated with HRP-labeled mouse anti-goat IgG (1:8,000, ZSGB-Bio) for 1 h. The immunoblots were analyzed using a Gel Imaging System and the Quantity One Software version 4.0 (Bio-Rad, Hercules, CA, USA).

Bacterial Strains and Culture Conditions

The highly virulent *S. suis* serotype type 2 strain 05ZYH33 (GenBank accession number NC_009442), which was originally isolated from a patient who died from STSS during the *S. suis* outbreak in 2005 in Sichuan, China, was used in this study. A suilysin gene (*sly*) deletion mutant of 05ZYH33 (Δ*sly*) was constructed via the in-frame replacement of the SSU05_1403 gene with a chloramphenicol (Cm) resistance cassette. The complement strain of the Δ*sly* mutant (CΔ*sly*) was established by transforming the mutant with a pAT18 vector that expressed the *sly* gene. These strains were previously constructed in our laboratory (He et al., 2014). All the bacterial strains were grown on Colombia agar plates (BD Biosciences, San Jose, CA, USA) containing 5% sheep's blood at 37°C under 5% CO_2. The bacterial suspensions were grown in Todd-Hewitt broth (THB, BD) for 8 h without agitation, and the supernatants were harvested for future experiments. Five microgram per mL of Cm or 8 μg/mL Em (Sigma–Aldrich) were added to the media for selection.

Purification of Native and Recombinant Suilysin (SLY)

Native SLY was purified as previously described (Jacobs et al., 1994; Lv et al., 2014). Briefly, the supernatant from a large-scale 05ZYH33 culture was collected by continuous-flow centrifugation and filtered through a ceramic filer with a 0.8 μm pore size at 4°C, and concentrated to 150 ml using 10,000-nominal-molecular-weight-limit filters (PTCG; Minitan; Merck Millipore). The concentrated supernatants were sterilized by filtration through a 0.2 μm pore size filer (Falcon, Thermo Fisher Scientific), loaded onto a Superose-12 gel filtration column (FPLC, Pharmacia), and eluted into 40 mM PBS (pH 7.2) supplemented with 0.5 M NaCl. The elution fractions were collected and analyzed by SDS-PAGE. The recombinant SLY and the non-hemolytic *sly* mutant SLY (P353V) used in this study were previously constructed and purified in our laboratory (Ren et al., 2012). The endotoxin level in the native suilysin (SLY) was less than 0.03 EU/mL. The endotoxins that remained in the purified recombinant SLY were removed with Triton X-114 (Liu et al., 1997) and were less than 0.5 EU/mL before the purified recombinant SLY was used in experiments.

Measurement of Superoxide Anion Production

Superoxide anion production was determined as previously described (Nilsson et al., 2006). PMNs were stimulated with 0.50, 0.75, or 1.0 μg/mL native SLY or with the supernatants from 05ZYH33 cultures at 37°C for 20 min. The PMNs were kept on ice for 10 min to terminate the stimulation, collected by centrifugation at 200 × g for 15 min and re-suspended in 0.25 ml Krebs-Ringer phosphate (KRG, 120 mM NaCl, 4.9 mM KCl, 1.7 mM KH_2PO_4, 1.2 mM $MgSO_4 \cdot 7H_2O$, 8.3 mM $Na_2HPO_4 \cdot 2H_2O$, 10 mM glucose, pH 7.3) buffer containing 0.1 mM cytochrome C. Superoxide production from the PMNs was induced by incubation with 200 ng/mL phorbol myristate acetate (PMA) at 37°C for 20 min. The PMNs were pelleted, and the absorbance of the supernatants was determined at 550 nm. To determine baseline levels of superoxide anion production, the PMNs were incubated with KRG buffer without cytochrome C. KRG buffer without PMNs was used as the blank reference. The HBP in the supernatant was also determined by Western blotting.

Electron Microscopy

The integrity of the PMNs was assessed by electron microscopy. After incubation with 1.0 μg/mL native SLY, the PMNs were spread onto poly-L-lysine coated coverslips, incubated at 37°C for 1 h, and fixed in 2.5% (v/v) glutaraldehyde at room temperature for 30 min. The PMNs were dehydrated in an ascending ethanol series from 50% (v/v) to absolute ethanol (10 min per step). The specimens were subjected to critical point drying in CO_2 with absolute ethanol as an intermediate solvent, mounted on copper holders, sputtered with 30 nm palladium and gold, and examined in an S-3400N scanning electron microscope.

Flow Cytometry

A total of 2×10^6 PMNs were incubated with 1.0 µg/mL native SLY at 37°C for 30 min with rotation and subsequently kept on ice. The expression of markers for PMN degranulation was assessed with PE conjugated mouse anti-human CD63 (1:100), APC conjugated mouse anti-human CD11b (1:100), and FITC conjugated mouse anti-human CD66b (1:100, BD Biosciences). Isotype-matched antibodies were used as negative controls. The fluorescence intensity of each sample was determined on an Accuri C6 flow cytometer (BD Biosciences). The data were analyzed using the FlowJo software (FlowJo LLC, Ashland, OR, USA).

Determination of Calcium Mobilization

Polymorphonuclear neutrophils were loaded with 5 µM Fura-3 AM (Invitrogen) in HBSS at 37°C for 1 h and plated into a confocal dish (Thermo Fisher Scientific). Putative PMN stimulants were added to the confocal dish. The PMNs were monitored with a FV1000 confocal laser-scanning microscope (Olympus, Tokyo, Japan) for 200 s. The mean fluorescence intensity of at least eight cells was recorded.

In vivo Miles Assay for the Assessment of Vascular Permeability

The Miles assay was performed on 8-week-old female C57BL/6J mice to evaluate the effects of SLY on vascular permeability. C57BL/6J mice were purchased from the Animal Care Center of the Academy of Military Medical Science (AMMS; Beijing, China) and housed in a clean room with unlimited access to food and water. TLR4 knockout mice C57BL/10ScNJNju and control mice were obtained from the Model Animal Research Center of Nanjing University. The mice were randomly divided into four groups and intradermally injected in the abdomen with 100 µL THB, 05ZYH33-supernatant, Δ*sly*-supernatant, or CΔ*sly*-supernatant. Four hours after the injection, 100 µL Evans blue dye solution (2.5%, Ourchem, Sinopharm) was injected via the tail vein. The mice were sacrificed by cervical dislocation 30 min after the Evans blue injection, and equal areas (20 mm × 20 mm) of skin surrounding the intradermal injection site were removed from each mouse and completely dried in an oven. The Evans blue dye was eluted from the oven-dried skin into 1 mL formamide (Sinopharm) at 55°C for 2 days, and quantified by spectrophotometry at 630 nm.

Statistics

Statistical analyses were performed using the Prism 5 software (GraphPad Software, La Jolla, CA, USA). For normally distributed data, a comparison between two groups was made using Student's *t*-test, and comparisons among multiple groups were done with a one-way analysis of variance. The *P*-value was for a 2-sided test, and $P < 0.05$ was considered as statistically significant. All data were expressed as the mean ± standard deviation. A rank-sum test was performed to compare the HBP levels in the patients versus the controls.

Study Approval

All the protocols for handling the patient blood specimens or experiments using blood from healthy donors were approved by the Institutional Medical Ethics Committee of AMMS. A signed informed consent form was obtained from each patient or a guardian and healthy volunteer. The entire study was conducted in accordance with the Declaration of Helsinki.

All experimental procedures involving mice were conducted in strict accordance with the recommendations in the Guide for the Care and Use of Laboratory Animals of the National Institutes of Health and State Key Laboratory of Pathogens and Biosecurity of the Beijing Institute of Microbiology and Epidemiology and approved by the Institutional Animal Care and Use Committee of the AMMS under Permit No. IACUC of AMMS-2014-031. Animal welfare was considered and suffering was minimized.

RESULTS

Suilysin from the Strain 05ZYH33 Induced the Release of HBP from PMNs

Heparin-binding protein has been reported as a potential biomarker of severe sepsis and septic shock (Linder et al., 2009, 2012; Chew et al., 2012) and as an indicator of vascular leakage (Johansson et al., 2009). A comparison of serum HBP levels between patients with *S. suis*-associated STSS versus patients without STSS or healthy individuals revealed that the serum HBP levels in patients with STSS were significantly higher than those in patients with meningitis or healthy controls (46.61 ± 9.49 ng/mL vs. 11.52 ± 5.20 ng/mL or 1.84 ± 0.96 ng/mL, respectively; $P = 0.0007$, **Figure 1**), which suggested that HBP is released during the progression of STSS in patients infected with *S. suis*. Both the ELISA and Western blotting analyses demonstrated that the bacterial culture of the highly virulent *S. suis* strain 05ZYH33 significantly increased the release of HBP in the whole blood of healthy individuals (**Figure 2A**) and in purified PMN suspension (**Figures 2B,C**), whereas THB media without 05ZYH33 failed to induce the release of HBP in a whole blood sample or a PMN suspension. The positive control for the Western blot was rHBP. The molecular weight of native HBP is approximately 31 kDa, while rHBP appears to be approximately 39 kDa on a Western blot owning to its carboxyl-terminal polyhistidine tag (**Figure 2C**). These results suggest that strain 05ZYH33 stimulates the release of HBP from PMNs.

The molecule responsible for inducing the release of HBP was identified. Supernatants of the 05ZYH33 culture, but not the 05ZYH33 bacterial cells, significantly induced the release of HBP in whole blood samples ($P < 0.001$, **Figure 3A**). Heat treatment of the supernatants significantly reduced the release of HBP in the whole blood samples ($P < 0.001$, **Figure 3B**), indicating that the specific factor(s) could be proteins. Therefore, the proteins in the supernatants were isolated by anion-exchange and hydrophobic interaction chromatography. The proteins in the elution fractions that markedly stimulated the release of HBP were separated by SDS-PAGE, and two major protein bands at approximately 58 and 35 kDa appeared on the gel

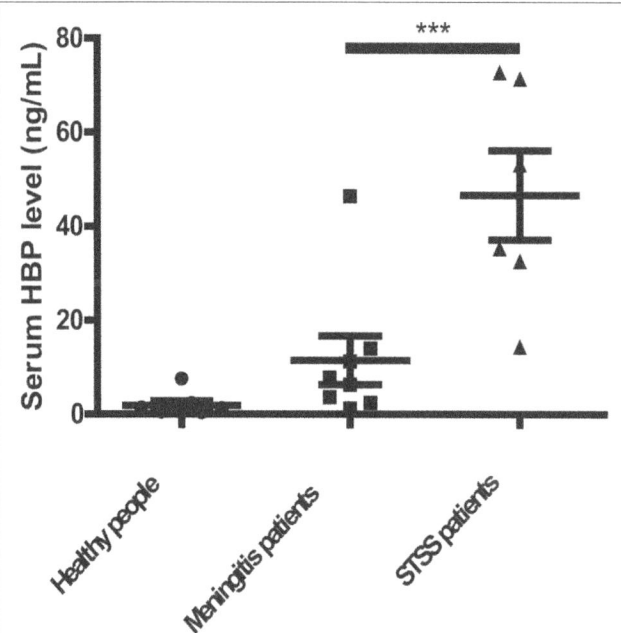

FIGURE 1 | The serum HBP levels were significantly increased in patients with STSS compared with patients without STSS or healthy individuals. The serum HBP levels of healthy individuals ($n = 13$), patients infected with *S. suis* presenting meningitis ($n = 8$), and patients infected with *S. suis* presenting STSS ($n = 6$) were determined by ELISA. Data were analyzed by a rank-sum test. *** indicates a significant difference between patients with STSS vs. controls, $P = 0.0007$.

(**Figure 3C**). Mass spectrometry revealed that the 58 kDa protein was SLY, and the 35 KDa protein was L-lactate dehydrogenase (LDH; Supplementary Table S1). Recombinant LDH failed to stimulate the release of HBP from PMNs (**Figure 3D**), which suggests that LDH was not the molecule that induced the release of HBP. Therefore, SLY is a likely candidate for the substance that releases HBP. Supernatants of the *sly* gene deletion mutant strain of 05ZYH33 (Δsly) did not stimulate the release of HBP, while the supernatants of the complement strain for Δsly (CΔsly) restored the stimulation of the release of HBP ($P < 0.01$, **Figure 3E**). Furthermore, Western blotting showed that purified native SLY (nSLY) stimulated the release of HBP from the PMNs, but a bacterial culture of Δsly strain failed to do so (**Figure 3F**). Taken together, these findings suggest that SLY secreted by 05ZYH33 induces the release of HBP from the PMNs.

Suilysin Stimulated the Release of HBP by Promoting PMN Degranulation

Heparin-binding protein is usually stored in azurophilic granules and secretory vesicles in the PMNs and released by cell lysis or via degranulation after the activation of a series of signal transduction pathways (Borregaard and Cowland, 1997; Tapper et al., 2002). We can infer that a high concentration of SLY could lyse the PMNs and release the stored HBP, while a sub-cytolytic concentration would interact with the cell membrane

receptor and could possibly trigger PMN degranulation. Different concentrations of nSLY (0.50–1.0 µg/mL) did not affect the production of superoxide anions from the PMNs (**Figure 4A**), while the release of HBP from the PMNs was SLY dose-dependent (**Figure 4B**), suggesting that the PMNs remained viable when treated with nSLY. These findings indicate that cell lysis is not related to the release of HBP mediated by SLY. Scanning electron microscopy demonstrated that the PMNs exhibited membrane blebs when treated with 1.0 µg/mL nSLY (**Figure 4D**) and the control did not exhibit such blebs (**Figure 4C**), suggesting that the degranulation might be induced by nSLY. A flow cytometry analysis revealed the significant up-regulation of markers for azurophilic granules (CD63), secondary/tertiary granules and secretory vesicles (CD11b), and specific granules (CD66b, also named CD67) at the plasma membrane after the PMNs were incubated with recombinant SLY (rSLY), but the recombinant factor H-binding protein (rFhb – a *S. suis* cell wall protein that was purified using the same methods) was not up-regulated. This implies that the granular vesicles in the PMNs were mobilized for degranulation (**Figures 4E–G**). Other proteins stored in the neutrophil granules, including elastase from the azurophilic granules and lactoferrin from the secondary granules, were released when the neutrophils were incubated with nSLY or the 05ZYH33-supernatant (**Figures 4H,I**). All of these results suggest that SLY induces the release of HBP via PMN degranulation.

The Release of HBP Induced by Suilysin Was Dependent on the Ca^{2+} Influx

An increased cytosolic Ca^{2+} concentration has been shown to promote PMN degranulation via the mobilization of intracellular granules and cytoskeleton rearrangements (O'Flaherty et al., 1991; Lacy, 2006; Lacy and Eitzen, 2008). Thus, the SLY-induced release of HBP might also be Ca^{2+}-dependent. Indeed, the release of HBP induced by the bacterial culture supernatants was completely abolished in the presence of EGTA ($P < 0.0001$, **Figure 5A**), suggesting that the release of HBP is dependent on extracellular Ca^{2+}. The measurement of intracellular Ca^{2+} revealed that nSLY (1 µg/mL) induced a rapid increase in the intracellular Ca^{2+} concentration in standard HBSS, and that the Ca^{2+} increase was completely abolished by EGTA or in Ca^{2+}-free HBSS (**Figure 5B**). N-formylmethionyl-leucylphenylalanine (fMLP) served as a positive control and LPS was used as a negative control (**Figures 5B,C**). These results indicate that SLY induced the Ca^{2+} influx. A constructed non-hemolytic *sly* mutant SLY (P353V; Xu et al., 2010; Du et al., 2013) could neither induce calcium mobilization (**Figure 5B**), nor evoke the secretion of HBP (Supplementary Figure S1), suggesting that the SLY-induced release of HBP was dependent on the Ca^{2+} influx. Supernatants of 05ZYH33 consistently induced a Ca^{2+} influx, whereas THB media and supernatants of the Δsly mutant did not induce a Ca^{2+} influx (**Figure 5C**). The Ca^{2+} influx induced by the 05ZYH33 supernatant was also completely abolished by EGTA (**Figure 5C**). All of these results indicated that the Ca^{2+} influx plays a critical role in the release of HBP induced by SLY.

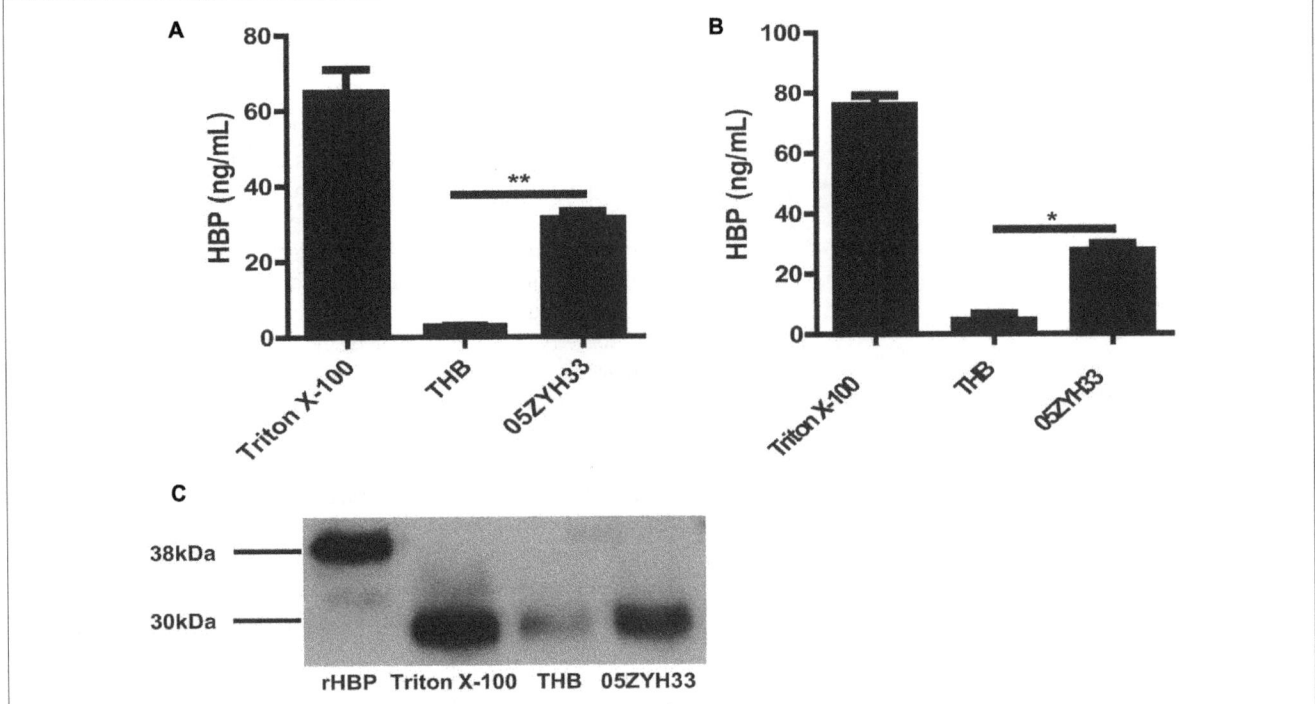

FIGURE 2 | 05ZYH33 cultures induced the release of HBP in whole blood samples and purified PMN suspension. (A,B) 05ZYH33 cultures significantly induced the release of HBP in human whole blood samples **(A)** and purified PMN suspensions **(B)**. Whole blood samples or purified PMN suspensions were incubated with 05ZYH33 cultures or THB (control) for 30 min at 37°C. The HBP level in the supernatants was measured by ELISA. The total cell lysate following Triton X-100 treatment was used as the positive control. Data are expressed as the mean ± SD. An unpaired Student's *t*-test was used. ***P* < 0.01, **P* < 0.05. **(C)** Western blotting results of release of HBP from human purified PMN suspensions. rHBP and Triton X-100 were used as positive controls.

The Release of HBP Induced by Suilysin Was Dependent on the TLR4 Receptor, p38 MAPK, and the PI3K Pathways, as well as GPCR

To investigate possible mechanisms and the key molecules involved in release of HBP, inhibitors were commonly used in this study because of the lack of suitable cell lines for investigating the release of HBP induced by SLY, and direct siRNA transfection into PMNs could cause activation of the PMNs and the release of HBP (data not shown).

Some cholesterol-dependent cytolysin (CDC) family members, such as anthrolysin O, perfringolysin O, listeriolysin O, streptolysin O, and pneumolysin, have been reported to interact with TLR4 (Malley et al., 2003; Park et al., 2004; Srivastava et al., 2005). SLY also belongs to the CDC family, and results from our previous study have shown that SLY interacts with TLR4 to induce an inflammatory response (Bi et al., 2015). In this study, our results demonstrated that both the TLR4 specific inhibitor CLI-095 (**Figure 6A**), which blocks the intracellular domain of TLR4, and the TLR4 non-specific inhibitor OxPAPC (**Figure 6B**), which competes with other ligands including CD14, LBP, and MD2 for receptor binding, significantly reduced the release of HBP induced by the 05ZYH33 supernatant. The results for both inhibitors were significant at $P < 0.0001$. This suggests that the release of HBP induced by SLY could be dependent on TLR4. We then used a pharmacological approach to further investigate

the signaling pathway underlying the SLY-induced release of HBP. The phospholipase C inhibitor U73122 (**Figure 6C**) and the MAPK pathway inhibitor PD98059 (**Figure 6D**) did not affect the release of HBP, whereas the p38 MAPK inhibitor SB203580 (**Figure 6E**) and the PI3K inhibitor Wortmannin (**Figure 6F**) markedly reduced HBP secretion (both $P < 0.0001$). We confirmed that a time-dependent increase in the level of phosphorylated p38 MAPK occurred in the presence of nSLY (1.0 μg/ml; Supplementary Figure S2). Other receptor inhibitors such as the GPCR inhibitor PTX significantly inhibited the release of HBP ($P < 0.0001$, **Figure 6G**), but Genistein, an inhibitor of the receptor tyrosine kinases (RTK), had no effect (**Figure 6H**). Therefore, a GPCR might also be involved in the release of HBP induced by SLY in addition to TLR4. The mechanism could be complex, and further studies are necessary to explain why two receptors are involved in the release of HBP induced by SLY.

Suilysin Increased Vascular Leakage in Mice

The effects of SLY on vascular leakage were tested in 8-week-old female C57BL mice using the Miles assay. Supernatants of the wild-type 05ZYH33 strain and the CΔ*sly* mutant significantly increased the leakage of Evans blue dye in mice compared with a THB control or the supernatants from the Δ*sly* mutant (05ZYH33-sup: 18.96 ± 1.35 μg/mL vs. Δ*sly*-sup:

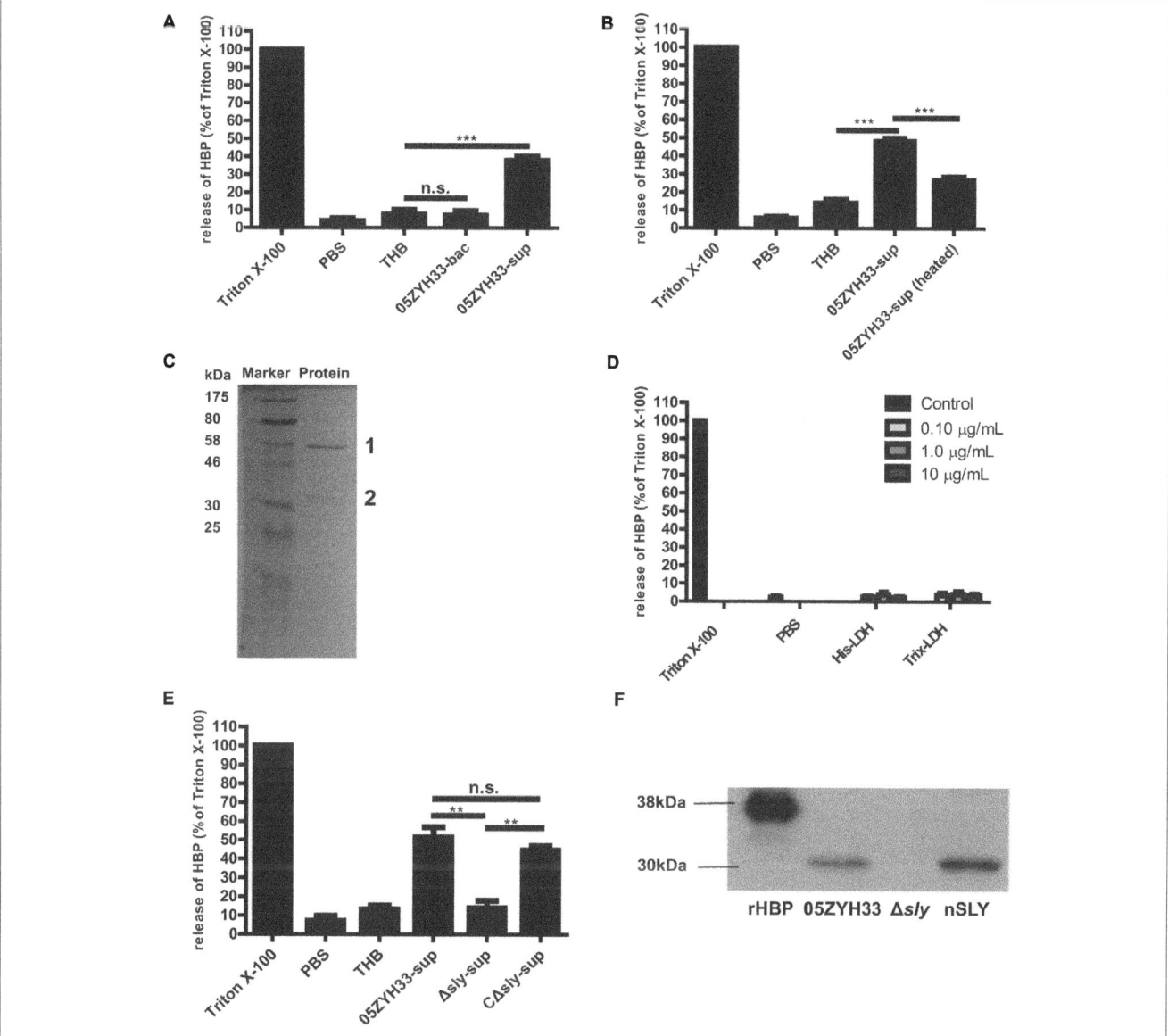

FIGURE 3 | Suilysin secreted from 05ZYH33 induced the release of HBP from PMNs. (A) The release of HBP from whole blood samples was induced by supernatants of 05ZYH33 but not by the bacterial cells. Human whole blood samples were incubated with PBS, THB, 05ZYH33 bacterial cells, or 05ZYH33 supernatants for 30 min at 37°C. The HBP level in the supernatants was measured by ELISA. PBS was considered as the background and THB was used as a negative control. An unpaired Student's t-test was used for the statistical analysis. **(B)** Heat treatment of the supernatants significantly reduced the release of HBP from whole blood samples. The supernatants were treated at 100°C for 5 min before being added to whole blood samples. **(C)** A SDS-PAGE analysis of the elution fraction from anion-exchange and hydrophobic interaction chromatography, which induced the maximal release of HBP. **(D)** LDH failed to induce the release of HBP from PMNs. The PMNs were incubated with His-LDH or Trix-LDH at 0.1, 1.0, or 10 μg/mL for 30 min at 37°C. **(E)** Suilysin in the 05ZYH33-supernatant stimulated the PMNs to secret HBP. Human whole blood samples were incubated with PBS, THB, the 05ZYH33-supernatant, the Δ*sly*-supernatant or the CΔ*sly*-supernatant for 30 min at 37°C. **(F)** Western-blot analysis of the HBP that was released from PMNs incubated with cultures of wild-type 05ZYH33, the Δ*sly* mutant, or purified nSLY. A representative image is presented. ***$P < 0.001$, **$P < 0.01$, n.s., not significant. ELISA data are expressed as the mean ± SD of at least four independent experiments.

12.38 ± 0.57 μg/mL, $P = 0.0008$, **Figures 7A,B**), implying that SLY contributes to vascular leakage. In addition, the 05ZYH33 supernatant failed to induce vascular leakage in *Tlr4* knockout mice (TLR4$^{+/+}$: 26.80 ± 1.96 μg/mL vs. TLR4$^{-/-}$: 11.35 ± 0.98 μg/mL, $P < 0.0001$, **Figures 7C,D**). These *in vivo* data support the essential role of TLR4 in vascular leakage mediated by SLY.

DISCUSSION

Since the two large outbreaks of *S. suis* human infections occurred in China, sporadic cases of *S. suis*-associated STSS have been reported worldwide (Tramontana et al., 2008; Gomez et al., 2014). This indicates that *S. suis* has already become a persistent problem, and the potential for a large-scale outbreak exists.

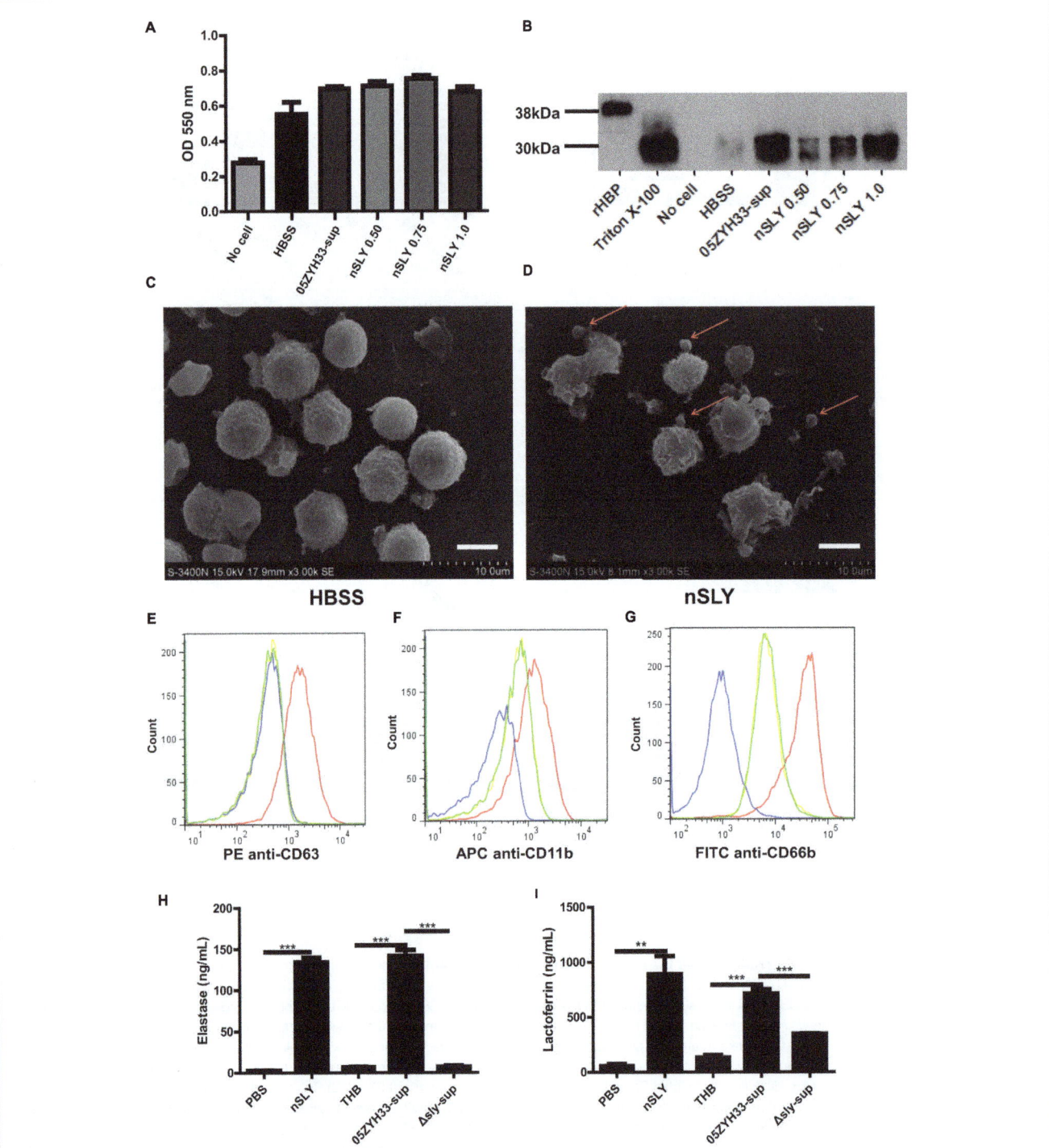

FIGURE 4 | Suilysin induced PMN degranulation. (A) Superoxide anion production was not affected by treatment with nSLY or 05ZYH33 supernatants. Purified PMNs were incubated with 0.50, 0.75, or 1.0 μg/mL nSLY, the 05ZYH33-supernatant, or HBSS for 30 min at 37°C. Following treatment, the cells were stimulated with 200 ng/ml PMA in the presence of 0.1 mM cytochrome C. The superoxide anion production was determined by measuring the absorbance at 550 nm. Data are expressed as the mean ± SD of three independent experiments. **(B)** Western blot analysis of release of HBP from PMNs that were treated as described for the measurement of superoxide anion production. Thirty microliters of each sample was loaded in each lane. A representative image is presented. **(C,D)** Scanning electron microscopy images of purified PMNs that were pre-incubated with HBSS **(C)** or 1 μg/mL nSLY **(D)**. Red arrows indicate excessive membrane ruffles. Representative images are presented. The scale bar represents 5 μm. **(E–G)**. Levels of membrane-associated marker proteins for PMN degranulation, CD63 **(E)**, CD11b **(F)**, and CD66b **(G)** were increased by 1 μg/mL rSLY. PMNs were stimulated with HBSS as a negative control (−), SLY-isotype (−), rSLY (−), or rFhb (−) and analyzed by flow cytometry. **(H,I)** Levels of granule marker proteins for PMN degranulation, elastase **(H)** and lactoferrin **(I)** were increased by 1 μg/mL nSLY or the 05ZYH33 supernatant. ***$P < 0.001$, **$P < 0.01$. ELISA data are presented as the mean ± SD of at least three independent experiments.

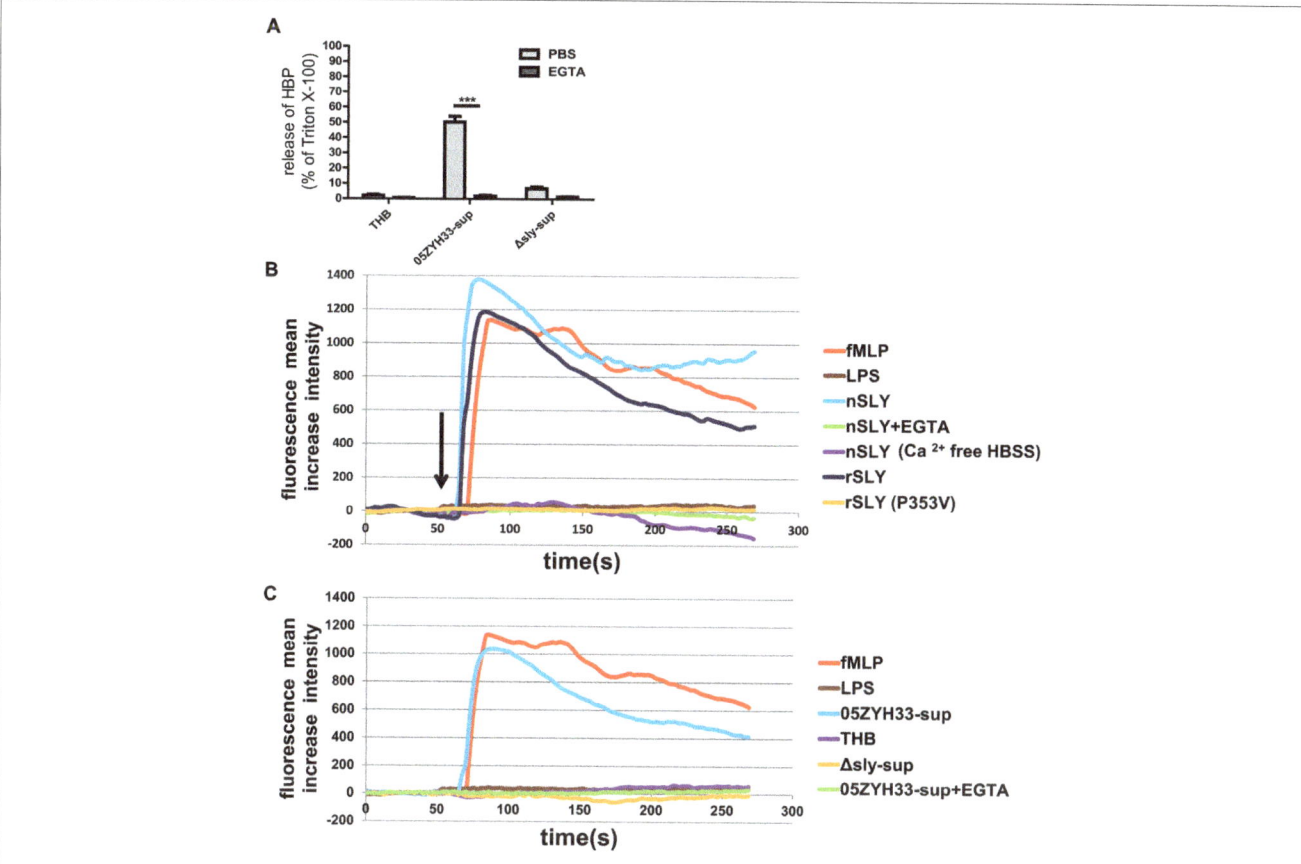

FIGURE 5 | SLY induced a Ca^{2+} influx and the release of HBP was dependent on the Ca^{2+} influx. (A) The SLY-induced release of HBP from human whole blood samples was abolished by EGTA. PMNs were pre-incubated with EGTA (10 mM) for 1 h at 37°C. ***P < 0.001. **(B)** SLY induced a Ca^{2+} influx. Positive control fMLP, negative control LPS, nSLY, nSLY with pre-incubation of EGTA or with Ca^{2+}-free HBSS, rSLY, non-hemolytic SLY mutant P353V was added to a confocal small dish containing purified PMNs loaded with the fluorescent Ca^{2+} indicator Fluo-3/AM at the indicated time (60 s, arrow), and Ca^{2+} mobilization was monitored by real-time fluorescence microscopy for 250 s. **(C)** SLY in the 05ZYH33-supernatant stimulated a Ca^{2+} influx. The positive control was fMLP, the negative control was LPS, and the 05ZYH33-supernatant, THB, the Δsly-supernatant, the 05ZYH33-supernatant with pre-incubation of EGTA were used as stimulants. The fluorescence intensity of at least eight cells in one vision was measured and the mean increase intensity is presented.

Although a great deal of effort has been spent on *S. suis*, the pathogenic mechanism of STSS associated with *S. suis* remains poorly understood. The results of this study demonstrated that SLY secreted by the highly virulent *S. suis* strain 05ZYH33 stimulated the release of HBP from the PMNs and induced vascular leakage in mice, indicating a role for SLY in the development of STSS. An in-depth investigation of the molecular mechanism underlying the release of HBP from PMNs mediated by SLY revealed that the release of the HBP occurred via Ca^{2+}-influx dependent degranulation and required both the p38 MAPK and PI3K signaling pathways.

Suilysin is a member of the CDC toxin family. It comprises 497 amino acids and exhibits hemolytic activity. SLY was considered as a putative virulence factor when it was first purified (Jacobs et al., 1994). However, studies investigating the association of SLY with the virulence of *S. suis* yielded conflicting results. A *sly* deletion mutation did not appear to affect the virulence of *S. suis* in piglets although it did impact the virulence in mice (Allen et al., 2001; Lun et al., 2003). Immunization with purified SLY failed to induce significant protection against *S. suis*

infection in piglets (Jacobs et al., 1996). Moreover, virulent *S. suis* strains from North America do not have the *sly* gene in their genome (Gottschalk et al., 1998). These findings seem to support the view that SLY is not an essential virulence factor of *S. suis*. In contrast, we conducted a study in which SLY levels in different *S. suis* strains were compared. That study showed that the highly virulent strain 05ZYH33, which caused higher mortality and greater damage to PMNs and human umbilical vein endothelial cells than non-epidemic strains, expressed a higher level of SLY than less virulent strains (He et al., 2014). In addition, a *sly* deletion markedly reduced the virulence of 05ZYH33 in mice (He et al., 2014). However, Gottschalk et al. thought that the higher toxicity of the Chinese ST7 strain for human peripheral blood mononuclear cells could simply be the result of its higher capacity to release SLY (Ye et al., 2006, 2009; Gottschalk et al., 2010). These results suggest that SLY is associated with the high virulence of the Chinese strain.

The role of SLY in the mediation of the inflammatory response is also controversial. Segura et al. (2006) suggested that SLY

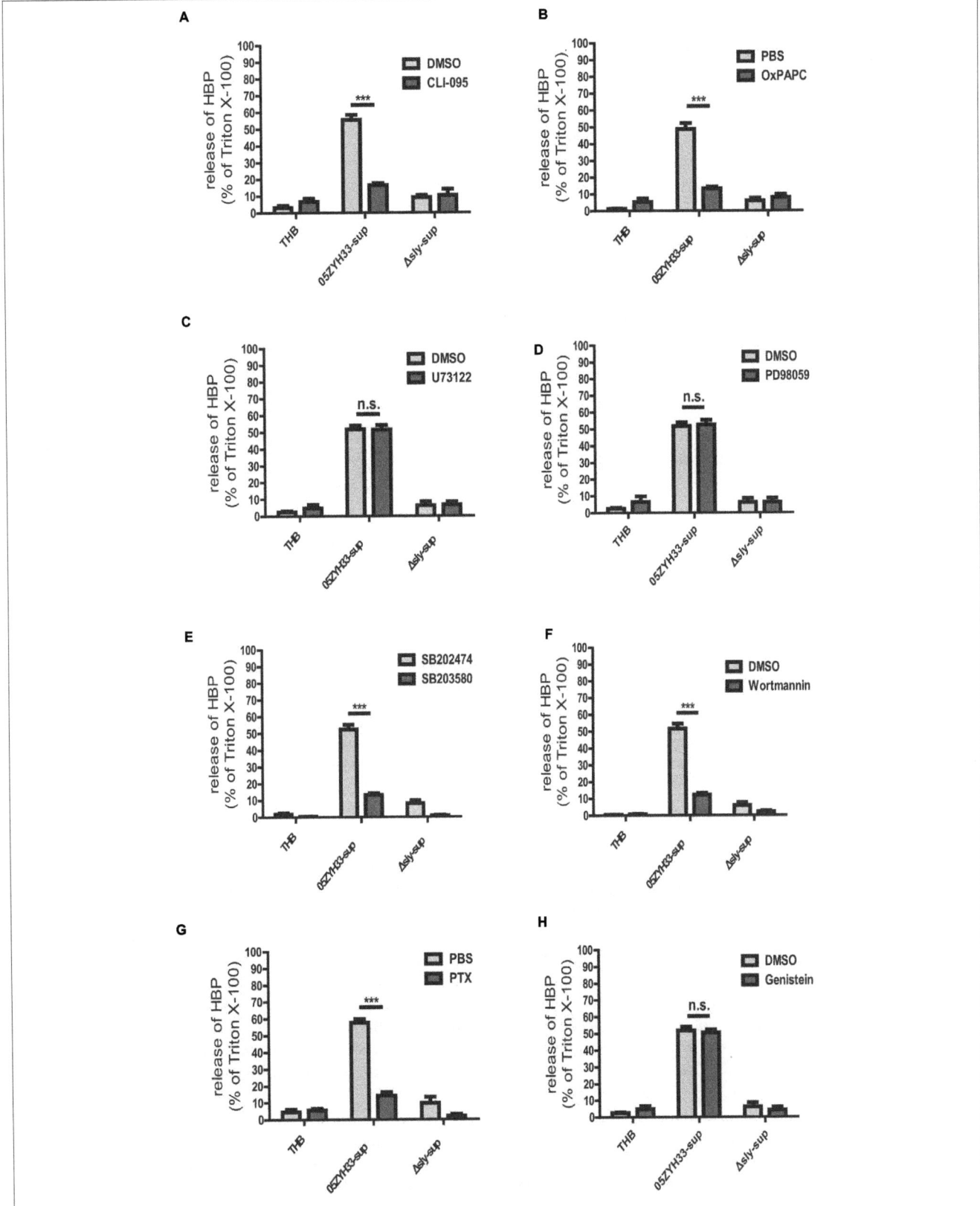

FIGURE 6 | Receptor and signal molecule were involved in the Suilysin induced release of HBP. (A,B) The release of HBP was blocked by the TLR4-specific inhibitor CLI-095 (3 μM; **A**) and the TLR4-non-specific inhibitor OxPAPC (30 μg/mL; **B**). **(C,D)** The release of HBP was not affected by the PLC and

(Continued)

FIGURE 6 | Continued

PLA2 inhibitor U73122 (10 µM; **C**) and the ERK1/2 inhibitor PD98095 (20 µM; **D**). (**E–H**) The release of HBP was blocked by the p38 MAPK inhibitor SB203580 (10 µM; **E**), the PI3K inhibitor Wortmannin (1.0 µM; **F**), the GPRP inhibitor PTX (0.15 µg/mL; **G**) but not by the RTK inhibitor Genistein (100 µM; **H**). The PMNs were pre-treated with or without the inhibitors at 37°C for 1 h (PTX for 3h and CLI-095 for 6 h) before a 30 min incubation with the stimulants at 37°C. The HBP level was measured by ELISA. The HBP level relative to the total amount of HBP in the cell lysate from treatment with Triton X-100 was calculated. The data are expressed as the mean ± SD of at least four independent experiments. ***$P < 0.001$. n.s., not significant.

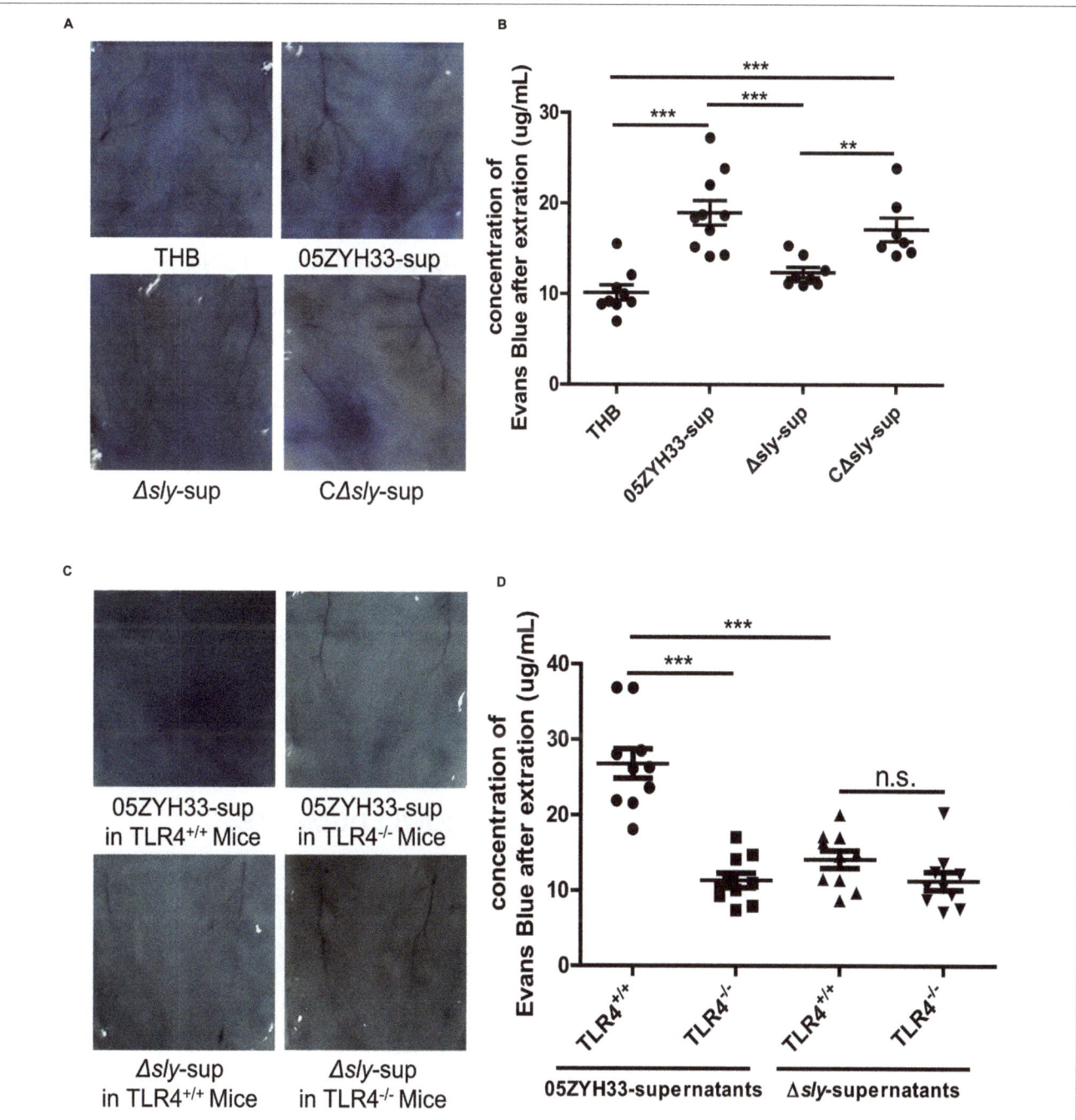

FIGURE 7 | Suilysin increased the vascular leakage in C57BL mice. (A,B) SLY increased the vascular leakage in mice. C57BL/6J mice were intradermally injected with THB, the 05ZYH33-supernatant, the Δ*sly*-supernatant, or the CΔ*sly*-supernatant. Four hours later, Evans blue dye (2.5%, 100 µL) was injected via the tail vein. After 30 min, photographs of the skin area containing the extravasated protein-bound dye were taken, and the dye was extracted from the skin using formamide. Dye concentrations were measured at 630 nm using a spectrophotometer. **(C,D)** The 05ZYH33 supernatant did not induce vascular leakage in *Tlr4* knockout mice. TLR4$^{+/+}$ and TLR4$^{-/-}$ were treated in the similar ways as in **(A,B)**. ***$P < 0.001$, **$P < 0.01$, n.s., not significant.

might only play a limited role in stimulating the release of pro-inflammatory cytokines, and the consequent inflammatory response. They found an increase in the mRNA and protein levels of pro-inflammatory cytokines that did not appear to be affected in a SLY-negative mutant strain compared with a wild-type *S. suis* strain (Segura et al., 2006). They also showed that purified SLY did not stimulate the production of tumor necrosis factor alpha (TNF-α) or interleukin-6 (IL-6) in murine macrophages (Segura et al., 1999). In contrast, Lun et al. found that rSLY triggered the release of TNF-α and IL-6 from human and pig monocytes (Lun et al., 2003). The intraperitoneal injection of a recombinant SLY resulted in a significant increase in the serum IL-6 levels in C57BL/6 mice (Du et al., 2009). This discrepancy might be attributable to the various SLY activities obtained with different purification methods or because different *S. suis* strains were used in these studies. HBP has been proven to be an important inflammatory mediator that can recruit and activate monocytes (Pereira et al., 1990; Lee et al., 2003; Soehnlein et al., 2005). To our knowledge, SLY from the highly virulent strain 05ZYH33 was found to induce the release of HBP from PMNs for the first time in this study.

In this study, we also found that, in addition to activating intracellular signaling pathways including the p38 MAPK and PI3K pathways, SLY also induced a Ca^{2+} influx in the PMNs. Mobilization of cytosolic free Ca^{2+} and cytoskeletal rearrangement have been demonstrated as requirements for degranulation in PMNs (O'Flaherty et al., 1991; Lacy, 2006; Lacy and Eitzen, 2008). In this study, removal of the extracellular Ca^{2+} by chelation with EGTA completely abolished the release of HBP, further supporting a critical role for a Ca^{2+} influx in PMN degranulation. Furthermore, we found that a non-hemolytic SLY mutant (P353V) failed to induce a Ca^{2+} influx (**Figure 5B**) and the release of HBP (Supplementary Figure S1), also indicating that the release of HBP is Ca^{2+} dependent. By directly monitoring the intracellular Ca^{2+} concentration, we provided direct evidence for the first time to support that SLY induced a Ca^{2+} influx. The underlying mechanism of the Ca^{2+} influx induced by SLY will be determined in future studies.

Until now, the molecular and pathological mechanisms underlying STSS that is not associated with GAS have been poorly understood. In this study, we found that mitilysin secreted from *Streptococcus mitis* and vaginolysin released from *Streptococcus viridans*, also belong to the CDC toxin family and may share a similar crystal structure with SLY, because they share 49 and 51% sequence identities, respectively, with the primary protein sequence of SLY. This indicates that these two toxins may have a similar function in the host cell. These two opportunistic pathogens may induce the release of HBP and cause vascular leakage when they enter the human bloodstream, and lead to STSS with severe vascular leakage. To some extent, these phenomena may be involved in the STSS cases caused by *S. mitis* in 1991 in the Yangtze River

Delta of China (Lu et al., 2003). On the other hand, other CDC toxins such as anthrolysin O, perfringolysin O, listeriolysin O, and streptolysin O, may also lead to the release of HBP in humans. Because HBP has already been considered as a potential biomarker for septic shock (Linder et al., 2009, 2012; Chew et al., 2012), these toxins could be a possible reason for the release of HBP in septic shock caused by related strains.

In summary, we reported for the first time in this study that SLY from the highly virulent *S. suis* strain 05ZYH33 induced the release of HBP from PMNs in a Ca^{2+} influx-dependent manner and increased vascular leakage in mice. We provided new insights into vascular leakage in STSS not associated with the GAS, which might lead to the discovery of potential therapeutic targets for *S. suis*-associated STSS.

AUTHOR CONTRIBUTIONS

All authors significantly contributed to this study. SC, WX, KW, PL, HJ, and YJ conceived and designed the entire study. SC, WX, KW, PL, ZR, LL, and YY performed experiments. SC, WX, KW, PL, and CZ analyzed data and prepared Figures. ZR, YZ, and QL provided important reagents. SC wrote the first draft of the paper and YJ revised the manuscript. All authors read and approved the manuscript.

FUNDING

This work was supported by the National Natural Science Foundation of China (81171528, 81441062, and 81371766), the National Basic Research Program (973) of China (2012CB518804), AWS15J006 and the State Key Laboratory of Pathogen and Biosecurity (Academy of Military Medical Science; SKLPBS 1421).

ACKNOWLEDGMENTS

We are grateful to Bing Xu from the National Center of Biomedical Analysis in Beijing, who performed the Mass analyses, and Yuchuan Li from the Beijing Institute of Microbiology and Epidemiology, who did the sample preparation for Electron microscopy.

REFERENCES

Allen, A. G., Bolitho, S., Lindsay, H., Khan, S., Bryant, C., Norton, P., et al. (2001). Generation and characterization of a defined mutant of *Streptococcus suis* lacking suilysin. *Infect. Immun.* 69, 2732–2735. doi: 10.1128/IAI.69.4.2732-2735.2001

Bi, L., Pian, Y., Chen, S., Ren, Z., Liu, P., Lv, Q., et al. (2015). Toll-like receptor 4 confers inflammatory response to suilysin. *Front. Microbiol.* 6:644. doi: 10.3389/fmicb.2015.00644

Borregaard, N., and Cowland, J. B. (1997). Granules of the human neutrophilic polymorphonuclear leukocyte. *Blood* 89, 3503–3521.

Brosnahan, A. J., and Schlievert, P. M. (2011). Gram-positive bacterial superantigen outside-in signaling causes toxic shock syndrome. *FEBS J.* 278, 4649–4667. doi: 10.1111/j.1742-4658.2011.08151.x

Brown, E. J. (2004). The molecular basis of streptococcal toxic shock syndrome. *N. Engl. J. Med.* 350, 2093–2094. doi: 10.1056/NEJMcibr040657

Chew, M. S., Linder, A., Santen, S., Ersson, A., Herwald, H., and Thorlacius, H. (2012). Increased plasma levels of heparin-binding protein in patients with shock: a prospective, cohort study. *Inflamm. Res.* 61, 375–379. doi: 10.1007/s00011-011-0422-6

Di Gennaro, A., Kenne, E., Wan, M., Soehnlein, O., Lindbom, L., and Haeggstrom, J. Z. (2009). Leukotriene B4-induced changes in vascular permeability are mediated by neutrophil release of heparin-binding protein (HBP/CAP37/azurocidin). *FASEB J.* 23, 1750–1757. doi: 10.1096/fj.08-121277

Du, H., Huang, W., Xie, H., Ye, C., Jing, H., Ren, Z., et al. (2013). The genetically modified suilysin, rSLY(P353L), provides a candidate vaccine that suppresses proinflammatory response and reduces fatality following infection with *Streptococcus suis*. *Vaccine* 31, 4209–4215. doi: 10.1016/j.vaccine.2013.07.004

Du, H., Xu, L., Wang, X., Li, X., Ye, C., and Xu, J. (2009). [Biological profiles of recombinant suilysin]. *Wei Sheng Wu Xue Bao* 49, 792–798.

Edens, H. A., and Parkos, C. A. (2003). Neutrophil transendothelial migration and alteration in vascular permeability: focus on neutrophil-derived azurocidin. *Curr. Opin. Hematol.* 10, 25–30. doi: 10.1097/00062752-200301000-00005

Ekelund, K., Skinhoj, P., Madsen, J., and Konradsen, H. B. (2005). Invasive group A, B, C and G streptococcal infections in Denmark 1999-2002: epidemiological and clinical aspects. *Clin. Microbiol. Infect.* 11, 569–576. doi: 10.1111/j.1469-0691.2005.01169.x

Gautam, N., Hedqvist, P., and Lindbom, L. (1998). Kinetics of leukocyte-induced changes in endothelial barrier function. *Br. J. Pharmacol.* 125, 1109–1114. doi: 10.1038/sj.bjp.0702186

Gautam, N., Herwald, H., Hedqvist, P., and Lindbom, L. (2000). Signaling via beta(2) integrins triggers neutrophil-dependent alteration in endothelial barrier function. *J. Exp. Med.* 191, 1829–1839. doi: 10.1084/jem.191.11.1829

Gautam, N., Olofsson, A. M., Herwald, H., Iversen, L. F., Lundgren-Akerlund, E., Hedqvist, P., et al. (2001). Heparin-binding protein (HBP/CAP37): a missing link in neutrophil-evoked alteration of vascular permeability. *Nat. Med.* 7, 1123–1127. doi: 10.1038/nm1001-1123

Gomez, E., Kennedy, C. C., Gottschalk, M., Cunningham, S. A., Patel, R., and Virk, A. (2014). *Streptococcus suis*-related prosthetic joint infection and streptococcal toxic shock-like syndrome in a pig farmer in the United States. *J. Clin. Microbiol* 52, 2254–2258. doi: 10.1128/JCM.02934-13

Gottschalk, M., Lebrun, A., Wisselink, H., Dubreuil, J. D., Smith, H., and Vecht, U. (1998). Production of virulence-related proteins by Canadian strains of *Streptococcus suis* capsular type 2. *Can. J. Vet. Res.* 62, 75–79.

Gottschalk, M., Xu, J., Calzas, C., and Segura, M. (2010). *Streptococcus suis*: a new emerging or an old neglected zoonotic pathogen? *Future Microbiol.* 5, 371–391. doi: 10.2217/fmb.10.2

Hashikawa, S., Iinuma, Y., Furushita, M., Ohkura, T., Nada, T., Torii, K., et al. (2004). Characterization of group C and G streptococcal strains that cause streptococcal toxic shock syndrome. *J. Clin. Microbiol.* 42, 186–192. doi: 10.1128/JCM.42.1.186-192.2004

He, Z., Pian, Y., Ren, Z., Bi, L., Yuan, Y., Zheng, Y., et al. (2014). Increased production of suilysin contributes to invasive infection of the *Streptococcus suis* strain 05ZYH33. *Mol. Med. Rep.* 10, 2819–2826. doi: 10.3892/mmr.2014.2586

Herwald, H., Cramer, H., Morgelin, M., Russell, W., Sollenberg, U., Norrby-Teglund, A., et al. (2004). M protein, a classical bacterial virulence determinant, forms complexes with fibrinogen that induce vascular leakage. *Cell* 116, 367–379. doi: 10.1016/S0092-8674(04)00057-1

Hu, X., Zhu, F., Wang, H., Chen, S., Wang, G., Sun, J., et al. (2000). [Studies on human streptococcal infectious syndrome caused by infected pigs]. *Zhonghua Yu Fang Yi Xue Za Zhi* 34, 150–152.

Jacobs, A. A., Loeffen, P. L., van den Berg, A. J., and Storm, P. K. (1994). Identification, purification, and characterization of a thiol-activated hemolysin (suilysin) of *Streptococcus suis*. *Infect. Immun.* 62, 1742–1748.

Jacobs, A. A., van den Berg, A. J., and Loeffen, P. L. (1996). Protection of experimentally infected pigs by suilysin, the thiol-activated haemolysin of *Streptococcus suis*. *Vet. Rec.* 139, 225–228. doi: 10.1136/vr.139.10.225

Johansson, J., Lindbom, L., Herwald, H., and Sjoberg, F. (2009). Neutrophil-derived heparin binding protein-a mediator of increased vascular permeability after burns? *Burns* 35, 1185–1187. doi: 10.1016/j.burns.2009.02.021

Lacy, P. (2006). Mechanisms of degranulation in neutrophils. *Allergy Asthma Clin. Immunol.* 2, 98–108. doi: 10.1186/1710-1492-2-3-98

Lacy, P., and Eitzen, G. (2008). Control of granule exocytosis in neutrophils. *Front. Biosci.* 13, 5559–5570. doi: 10.2741/3099

Lee, T. D., Gonzalez, M. L., Kumar, P., Grammas, P., and Pereira, H. A. (2003). CAP37, a neutrophil-derived inflammatory mediator, augments leukocyte adhesion to endothelial monolayers. *Microvasc. Res.* 66, 38–48. doi: 10.1016/S0026-2862(03)00010-4

Linder, A., Akesson, P., Inghammar, M., Treutiger, C. J., Linner, A., and Sunden-Cullberg, J. (2012). Elevated plasma levels of heparin-binding protein in intensive care unit patients with severe sepsis and septic shock. *Crit. Care* 16, R90. doi: 10.1186/cc11353

Linder, A., Christensson, B., Herwald, H., Bjorck, L., and Akesson, P. (2009). Heparin-binding protein: an early marker of circulatory failure in sepsis. *Clin. Infect. Dis.* 49, 1044–1050. doi: 10.1086/605563

Liu, S., Tobias, R., McClure, S., Styba, G., Shi, Q., and Jackowski, G. (1997). Removal of endotoxin from recombinant protein preparations. *Clin. Biochem.* 30, 455–463. doi: 10.1016/S0009-9120(97)00049-0

Low, D. E. (2013). Toxic shock syndrome: major advances in pathogenesis, but not treatment. *Crit. Care Clin.* 29, 651–675. doi: 10.1016/j.ccc.2013.03.012

Lu, H. Z., Weng, X. H., Zhu, B., Li, H., Yin, Y. K., Zhang, Y. X., et al. (2003). Major outbreak of toxic shock-like syndrome caused by *Streptococcus mitis*. *J. Clin. Microbiol.* 41, 3051–3055. doi: 10.1128/JCM.41.7.3051-3055.2003

Lun, S., Perez-Casal, J., Connor, W., and Willson, P. J. (2003). Role of suilysin in pathogenesis of *Streptococcus suis* capsular serotype 2. *Microb. Pathog.* 34, 27–37. doi: 10.1016/S0882-4010(02)00192-4

Lun, Z. R., Wang, Q. P., Chen, X. G., Li, A. X., and Zhu, X. Q. (2007). *Streptococcus suis*: an emerging zoonotic pathogen. *Lancet Infect. Dis.* 7, 201–209. doi: 10.1016/S1473-3099(07)70001-4

Lv, Q., Hao, H., Bi, L., Zheng, Y., Zhou, X., and Jiang, Y. (2014). Suilysin remodels the cytoskeletons of human brain microvascular endothelial cells by activating RhoA and Rac1 GTPase. *Protein Cell* 5, 261–264. doi: 10.1007/s13238-014-0037-0

Malley, R., Henneke, P., Morse, S. C., Cieslewicz, M. J., Lipsitch, M., Thompson, C. M., et al. (2003). Recognition of pneumolysin by Toll-like receptor 4 confers resistance to pneumococcal infection. *Proc. Natl. Acad. Sci. U.S.A.* 100, 1966–1971. doi: 10.1073/pnas.0435928100

Nilsson, M., Sorensen, O. E., Morgelin, M., Weineisen, M., Sjobring, U., and Herwald, H. (2006). Activation of human polymorphonuclear neutrophils by streptolysin O from *Streptococcus pyogenes* leads to the release of proinflammatory mediators. *Thromb. Haemost.* 95, 982–990.

O'Flaherty, J. T., Rossi, A. G., Jacobson, D. P., and Redman, J. F. (1991). Roles of Ca2+ in human neutrophil responses to receptor agonists. *Biochem. J.* 277, 705–711. doi: 10.1042/bj2770705

Pahlman, L. I., Morgelin, M., Eckert, J., Johansson, L., Russell, W., Riesbeck, K., et al. (2006). Streptococcal M protein: a multipotent and powerful inducer of inflammation. *J. Immunol.* 177, 1221–1228. doi: 10.4049/jimmunol.177.2.1221

Park, J. M., Ng, V. H., Maeda, S., Rest, R. F., and Karin, M. (2004). Anthrolysin O and other gram-positive cytolysins are toll-like receptor 4 agonists. *J. Exp. Med.* 200, 1647–1655. doi: 10.1084/jem.20041215

Perch, B., Kristjansen, P., and Skadhauge, K. (1968). Group R streptococci pathogenic for man. Two cases of meningitis and one fatal case of sepsis. *Acta Pathol. Microbiol. Scand.* 74, 69–76. doi: 10.1111/j.1699-0463.1968.tb03456.x

Pereira, H. A., Shafer, W. M., Pohl, J., Martin, L. E., and Spitznagel, J. K. (1990). CAP37, a human neutrophil-derived chemotactic factor with monocyte specific activity. *J. Clin. Invest.* 85, 1468–1476. doi: 10.1172/JCI114593

Ren, Z. Q., Zheng, Y. L., Gan, S. Z., Lv, Q. Y., Hao, H. J., Jiang, Y. Q., et al. (2012). [Construction and activities of suilysin mutants]. *Xi Bao Yu Fen Zi Mian Yi Xue Za Zhi* 28, 580–582.

Segura, M. (2009). *Streptococcus suis*: an emerging human threat. *J. Infect. Dis.* 199, 4–6. doi: 10.1086/594371

Segura, M., Stankova, J., and Gottschalk, M. (1999). Heat-killed *Streptococcus suis* capsular type 2 strains stimulate tumor necrosis factor alpha and interleukin-6 production by murine macrophages. *Infect. Immun.* 67, 4646–4654.

Segura, M., Vanier, G., Al-Numani, D., Lacouture, S., Olivier, M., and Gottschalk, M. (2006). Proinflammatory cytokine and chemokine modulation by *Streptococcus suis* in a whole-blood culture system. *FEMS Immunol. Med. Microbiol.* 47, 92–106. doi: 10.1111/j.1574-695X.2006.00067.x

Soehnlein, O., Xie, X., Ulbrich, H., Kenne, E., Rotzius, P., Flodgaard, H., et al. (2005). Neutrophil-derived heparin-binding protein (HBP/CAP37) deposited on endothelium enhances monocyte arrest under flow conditions. *J. Immunol.* 174, 6399–6405. doi: 10.4049/jimmunol.174.10.6399

Srivastava, A., Henneke, P., Visintin, A., Morse, S. C., Martin, V., Watkins, C., et al. (2005). The apoptotic response to pneumolysin is toll-like receptor 4 dependent and protects against pneumococcal disease. *Infect. Immun.* 73, 6479–6487. doi: 10.1128/IAI.73.10.6479-6487.2005

Staats, J. J., Feder, I., Okwumabua, O., and Chengappa, M. M. (1997). *Streptococcus suis*: past and present. *Vet. Res. Commun.* 21, 381–407. doi: 10.1023/A:1005870317757

Tang, J., Wang, C., Feng, Y., Yang, W., Song, H., Chen, Z., et al. (2006). Streptococcal toxic shock syndrome caused by *Streptococcus suis* serotype 2. *PLoS Med.* 3:e151. doi: 10.1371/journal.pmed.0030151

Tapper, H., Karlsson, A., Morgelin, M., Flodgaard, H., and Herwald, H. (2002). Secretion of heparin-binding protein from human neutrophils is determined by its localization in azurophilic granules and secretory vesicles. *Blood* 99, 1785–1793. doi: 10.1182/blood.V99.5.1785

Tramontana, A. R., Graham, M., Sinickas, V., and Bak, N. (2008). An Australian case of *Streptococcus suis* toxic shock syndrome associated with occupational exposure to animal carcasses. *Med. J. Aust.* 188, 538–539.

Wertheim, H. F., Nghia, H. D., Taylor, W., and Schultsz, C. (2009). *Streptococcus suis*: an emerging human pathogen. *Clin. Infect. Dis.* 48, 617–625. doi: 10.1086/596763

Xu, L., Huang, B., Du, H., Zhang, X. C., Xu, J., Li, X., et al. (2010). Crystal structure of cytotoxin protein suilysin from *Streptococcus suis*. *Protein Cell* 1, 96–105. doi: 10.1007/s13238-010-0012-3

Ye, C., Zheng, H., Zhang, J., Jing, H., Wang, L., Xiong, Y., et al. (2009). Clinical, experimental, and genomic differences between intermediately pathogenic, highly pathogenic, and epidemic *Streptococcus suis*. *J. Infect. Dis.* 199, 97–107. doi: 10.1086/594370

Ye, C., Zhu, X., Jing, H., Du, H., Segura, M., Zheng, H., et al. (2006). *Streptococcus suis* sequence type 7 outbreak, Sichuan, China. *Emerg. Infect. Dis.* 12, 1203–1208. doi: 10.3201/eid1708.060232

Yu, H., Jing, H., Chen, Z., Zheng, H., Zhu, X., Wang, H., et al. (2006). Human *Streptococcus suis* outbreak, Sichuan, China. *Emerg. Infect. Dis.* 12, 914–920. doi: 10.3201/eid1206.051194

16

Live *Staphylococcus aureus* Induces Expression and Release of Vascular Endothelial Growth Factor in Terminally Differentiated Mouse Mast Cells

Carl-Fredrik Johnzon[1], Elin Rönnberg[2], Bengt Guss[3] and Gunnar Pejler[1,2]*

[1] Department of Anatomy, Physiology and Biochemistry, Swedish University of Agricultural Sciences, Uppsala, Sweden,
[2] Department of Medical Biochemistry and Microbiology, Uppsala University, Uppsala, Sweden, [3] Department of Biomedical Science and Veterinary Public Health, Swedish University of Agricultural Sciences, Uppsala, Sweden

Edited by:
Diana Bahia,
Federal University of
Minas Gerais, Brazil

Reviewed by:
Elizabeth Hong-Geller,
Los Alamos National
Laboratory, USA
Raquel Bartz,
Duke University
School of Medicine, USA
Luciola Silva Barcelos,
Federal University of
Minas Gerais, Brazil

*Correspondence:
Gunnar Pejler
gunnar.pejler@imbim.uu.se

Mast cells have been shown to express vascular endothelial growth factor (VEGF), thereby implicating mast cells in pro-angiogenic processes. However, the mechanism of VEGF induction in mast cells and the possible expression of VEGF in fully mature mast cells have not been extensively studied. Here, we report that terminally differentiated peritoneal cell-derived mast cells can be induced to express VEGF in response to challenge with *Staphylococcus aureus*, thus identifying a mast cell–bacteria axis as a novel mechanism leading to VEGF release. Whereas live bacteria produced a robust upregulation of VEGF in mast cells, heat-inactivated bacteria failed to do so, and bacteria-conditioned media did not induce VEGF expression. The induction of VEGF was not critically dependent on direct cell–cell contact between bacteria and mast cells. Hence, these findings suggest that VEGF can be induced by soluble factors released during the co-culture conditions. Neither of a panel of bacterial cell-wall products known to activate toll-like receptor (TLR) signaling promoted VEGF expression in mast cells. In agreement with the latter, VEGF induction occurred independently of Myd88, an adaptor molecule that mediates the downstream events following TLR engagement. The VEGF induction was insensitive to nuclear factor of activated T-cells inhibition but was partly dependent on the nuclear factor kappa light-chain enhancer of activated B cells signaling pathway. Together, these findings identify bacterial challenge as a novel mechanism by which VEGF is induced in mast cells.

Keywords: mast cells, VEGF family, *Staphylococcus aureus*, NF-κB, peritoneal cavity

INTRODUCTION

Mast cells are tissue-resident cells located at the host–environment interface. They express numerous immune receptors and host a multitude of immunological mediators (1–3). While generally known for their involvement in allergy (4), mast cells have also implicated in numerous additional pathological settings, ranging from defense against bacterial infections (5) to an involvement in malignant processes (6). In the latter context, mast cells have, in particular, been implicated to support angiogenesis, thereby promoting tumor growth and metastasis (7, 8).

In the angiogenic process, vascular endothelial growth factor (VEGF) has a key role by promoting endothelial cell survival, proliferation, and migration (9). Moreover, VEGF has been found to exert a chemoattractant effect on immune cells (10). In support for the notion that mast cells are involved in angiogenesis, human, rat, and murine mast cells have been shown to synthesize and secrete VEGF. Previous work has shown that mast cells can release VEGF in response to IgE receptor cross-linking, through stimulation of c-kit, by challenge with a protein kinase C activator (phorbol myristate acetate) or calcium ionophore (11, 12). Additionally, mast cells have been shown to release VEGF in response to PGE_2 activation through the EP(2) receptor (13), and the adenosine analog [5'-N-ethylcarboxamido adenosine (NECA)] has been reported to increase VEGF expression in human lung mast cells (14).

In a previous study, we studied the impact of *Staphylococcus aureus* (*S. aureus*) on gene expression patterns in mast cells (15). As judged by gene array data, we found that live *S. aureus* induced the expression of numerous pro-inflammatory genes such as various cytokines and chemokines. Somewhat unexpectedly, we also found that VEGF was markedly upregulated. In fact, the gene array experiment indicated that the VEGF gene was induced to a higher extent than most other genes. Since no previous study has suggested a link between bacterial infection and induction of VEGF in mast cells, we, here, investigated the relevance of this finding. Moreover, since most of the previous studies in which mast cells have been shown to express VEGF were performed using relatively immature mast cells, we also sought to investigate whether fully mature mast cells can be induced to express VEGF. Indeed, we here provide evidence that live *S. aureus* induces high levels of VEGF expression in terminally differentiated mast cells. Hence, our findings provide a hitherto unrecognized link between mast cells and VEGF expression in the context of bacterial infection.

MATERIALS AND METHODS

Bacteria and Conditioned Media

Staphylococcus aureus (strain 8325-4) was streaked on horse blood agar plates (5%; National Veterinary Institute, Uppsala, Sweden) and incubated at 37°C for 24 h. Liquid cultures were started by inoculating 20 ml of Tryptone Soy Broth (TSB; BD) followed by incubation at 37°C and 150 rpm for 16 h. Two hundred microliters of this overnight culture were used to inoculate 20 ml of fresh TSB followed by incubation at 37°C and 150 rpm to an OD_{600} of 1.0. Conditioned media were produced by culturing *S. aureus* in 20 ml of TSB or antibiotic-free peritoneal cell-derived mast cells (PCMCs) media for 24 h at 37°C. The bacteria were removed by centrifugation (6000 × *g* for 5 min) followed by sterile filtration through 0.2 μm filters. Sterility was checked by plating a 100-μl aliquot onto horse blood agar followed by incubation at 37°C for 24 h.

Peritoneal Cell-Derived Mast Cells

Peritoneal cell-derived mast cells (PCMCs) were established according to a published protocol (16). Briefly, peritoneal lavage of mice was performed, followed by culture of the peritoneal cells in DMEM plus GlutaMAX (Gibco, Invitrogen, Paisley, UK) supplemented with 10% supernatant of stem cell factor-transfected Chinese hamster ovary cells (a gift from Dr. M. Daeron, Pasteur Institute, France), 10% fetal bovine serum, 50 μg/ml streptomycin, 60 μg/ml penicillin, 10 mM MEM non-essential amino acids, and 50 μM 2-mercaptoethanol. The medium was changed every 3–4 days. The inclusion of stem cell factor in the medium promotes the expansion of mast cells at the expense of other peritoneal cell populations. After ~1 month, pure mast cell cultures were obtained, as judged by toluidine blue staining.

Mice

Female mice of the C57BL/6 background were used for the experiments. All animal experiments were approved by the local ethical committee (Uppsala djurförsöksetiska nämnd; C31/14).

In Vitro Exposure of PCMCs to *S. aureus*, Bacterial Cell-Wall Components, and Conditioned Media

PCMCs were washed twice in PBS and re-suspended in antibiotic-free PCMC medium and plated in 24-well tissue plates at a density of 0.5×10^6 cells per replicate. Alternatively, PCMCs were plated in Transwell plates (0.4 μm pores; Costar). For inhibition experiments, PCMCs were pretreated with 10 μM nuclear factor of activated T-cells (NFAT) inhibitor (11R-VIVIT; Calbiochem, Darmstadt, Germany), 200 nM nuclear factor kappa light-chain enhancer of activated B cells (NF-κβ) inhibitor [6-amino-4-(4-phenoxyphenylethylamino)quinazoline] for 1 h, or with 45 μM Myd88 inhibitor (Pepinh-MYD and control peptide Pepinh-Control) for 6 h prior to infection. The bacteria were washed twice in PBS and added to the PCMC cultures at a final concentration of ~1.25×10^7 CFU/ml; multiplicity of infection (MOI) 25. For inactivation experiments, the bacteria were heat inactivated (HIA) at 60°C for 1 h. Conditioned media were added at a volume corresponding to that of the added bacteria. Purified bacterial cell-wall components: lipopolysaccharide (LPS; 1 μg/ml), lipoteichoic acid (LTA; 1 μg/ml), peptidoglycan (PGN; 10 μg/ml), or Pam3CSK4 (PAM3; 0.5 μg/ml) were added in some experiments. At various time points after infection, cells were collected by centrifugation (6000 × *g*; 5 min). Media and cell fractions were frozen and stored at −20 and −80°C, respectively. All experiments were performed in quadruplicates.

RNA Preparation and Quantitative Real-Time PCR

Quantitative real-time PCR (qPCR) was performed as described (17). Briefly, total RNA from the co-culture pellets was isolated using the NucleoSpin RNA II kit (Macherey-Nagel, Düren, Germany). qPCR was performed using the SYBR GreenER qPCR Supermix Universal Mastermix (Invitrogen, Waltham, MA, USA) on an ABI 7900HT Fast Real-Time PCR System (Applied Biosystems). The following primers were used: VEGF forward, 5'-GGAGTCTGTGCTCTGGGATT and VEGF reverse, 5'-AACCAACCTCCTCAAACCGT. HPRT forward, 5'-GATTAGCGATGATGAACCAGGTTA and HPRT reverse,

5′-GACATCTCGAGCAAGTCTTTCAGTC. Melt curve analyses of all qPCR products were performed. Relative expression of the VEGF gene in comparison with the house keeping gene (HPRT) was calculated, as previously described (17).

ELISAs

ELISAs for murine VEGF (PeproTech) were performed according to the manufacturer's instructions.

Statistical Analysis

Statistical analyses were performed using one-way ANOVA without matching and Fisher's LSD *post hoc* test. The analyses were carried out with GraphPad Prism 6 (GraphPad Software). The results shown are from individual experiments, representative for at least two experiments.

RESULTS

S. aureus Induces VEGF Expression and Release in Cultured Peritoneal Cell-Derived Mast Cells

To investigate whether *S. aureus* can affect VEGF expression in mast cells, terminally differentiated peritoneal cell-derived mast cells (PCMCs; 0.5×10^6) were co-cultured with live *S. aureus* (MOI = 25). Cell pellets and supernatants were collected after 2, 6, and 24 h. Total RNA from cell pellets was used for reverse transcription and qPCR. Supernatants were analyzed by ELISA for content of VEGF protein. As shown in **Figure 1A**, VEGF gene expression was highly induced in the PCMCs after co-culture with the live *S. aureus*. Notably, VEGF expression was modest after 2 h of co-culture and reached a maximum after 6 h.

Increased VEGF gene expression was also accompanied by release of VEGF protein as determined by ELISA. As depicted in **Figure 1B**, VEGF release was seen from 2 h and onward, with gradually increasing accumulation of VEGF in the medium up to 24 h.

Live *S. aureus* Induces VEGF Expression in Mast Cells Independent of Bacterial Cell-Wall Components

To approach the mechanism by with the bacteria induce VEGF expression in mast cells, we investigated the possibility that VEGF is induced by various toll-like receptor (TLR) ligands that are expressed by bacteria. To this end, PCMCs were stimulated with typical cell-wall components of Gram-positive bacteria: lipoteichoic acid (LTA), PGN, or Pam3CSK4 (PAM3). In addition, we assessed the effect of LPS, i.e., the prototype cell-wall component of Gram-negative bacteria. The bacterial products were added to the mast cells, either alone or in combination, followed by measurement of VEGF gene expression. However, neither of these TLR ligands induced VEGF expression in PCMCs (**Figure 2A**), suggesting that bacteria induce VEGF by mechanisms in mature mast cells independent of bacterial cell-wall compounds and TLR signaling. In agreement with this notion, inhibition of Myd88, an adaptor molecule common for

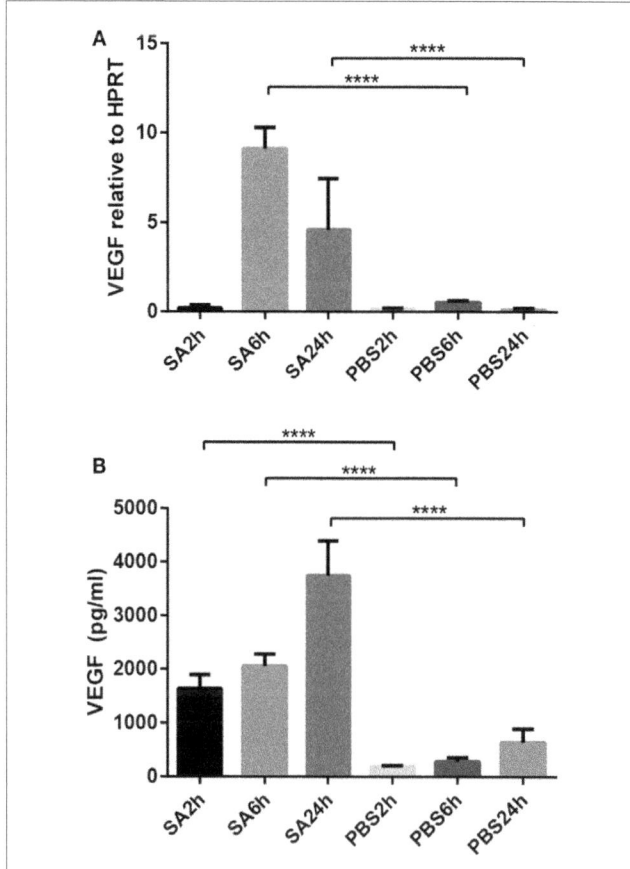

FIGURE 1 | Co-culture of PCMCs and *S. aureus* induces the expression and release of vascular endothelial growth factor (VEGF). Mast cells (PCMCs) were co-cultured with *S. aureus* (SA) with PBS as negative control. Cell fractions were taken at indicated time points, and VEGF expression was measured by qPCR **(A)**. Release of VEGF protein was measured by ELISA **(B)**. Results are given as mean ± SD, ****$p < 0.0001$ ($n = 4$).

most TLR signaling pathways, did not reduce the expression of VEGF in response to stimulation of mast cells by live *S. aureus* (**Figure 2B**).

Next, we assessed whether VEGF induction requires that the bacteria are alive, by investigating the effect of heat-inactivated bacteria on VEGF expression in mast cells. As seen in **Figure 2C**, HIA *S. aureus* did not induce VEGF expression above baseline levels, indicating that it is essential that the bacteria are alive to be able to induce VEGF expression in mature mast cells. Moreover, since most bacterial cell-wall components are not affected by heat inactivation, this finding further supports that the VEGF induction in mast cells is not mediated by cell-wall components of *S. aureus*.

As VEGF expression was not induced by any of the tested cell-wall compounds, we assessed the possibility that *S. aureus* secrete soluble compounds that might drive VEGF expression. To test this, we collected conditioned media from *S. aureus*, having non-conditioned medium as control. For this experiment, *S. aureus* was either cultured in bacterial growth medium (TSB) or in the medium used for culture of the PCMCs. However,

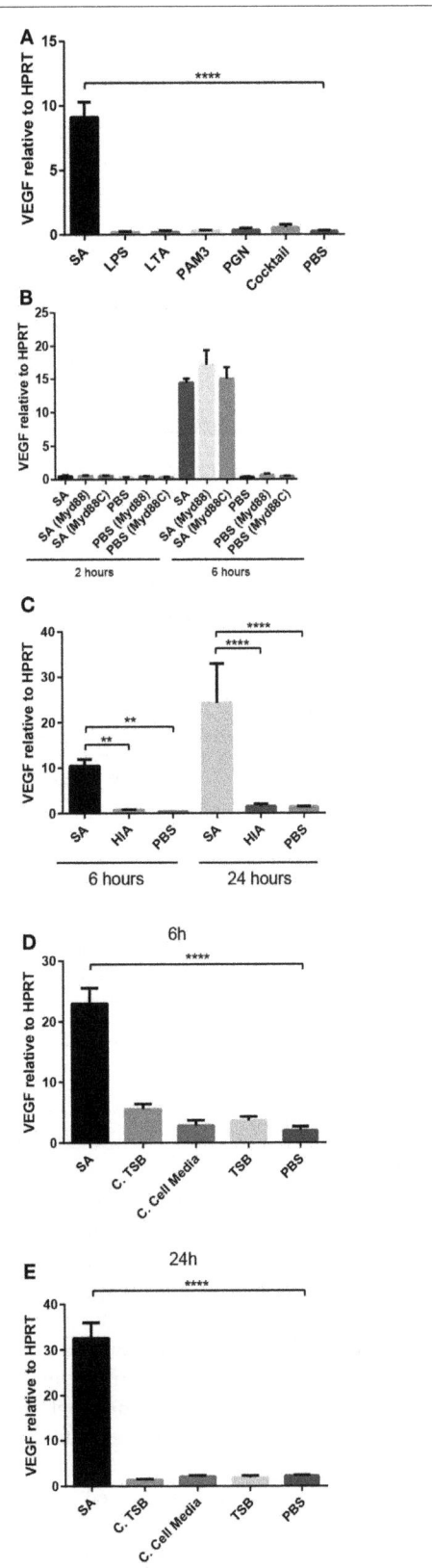

FIGURE 2 | The induction of VEGF expression in PCMCs requires live bacteria.

(Continued)

neither of these variants of conditioned media induced the expression of VEGF in PCMCs (**Figures 2D,E**).

Optimal Induction of VEGF Expression in Mast Cell Is Dependent of Direct Contact between Mast Cells and Bacteria

To determine whether the induction of VEGF expression in mast cells by *S. aureus* is dependent on physical contact between the bacteria and mast cells, PCMCs were co-cultured with *S. aureus* in a Transwell system in which the mast cells and bacteria were separated by a 0.4-µm membrane. PCMCs were collected after 6 and 24 h and were assessed for VEGF expression. As shown in **Figure 3**, the separation of bacteria and mast cells did not obviate the upregulation of VEGF in mast cells at 6 h, suggesting that soluble factors released during co-culture of mast cells and *S. aureus* can account for the induction of VEGF in mast cells. However, it is notable that increased VEGF induction between 6 and 24 h was seen in the direct contact situation, whereas the induction was reduced over time when mast cells and bacteria were separated. This indicates that direct cell–cell contact may be required for sustained VEGF induction at high levels.

VEGF Upregulation in *S. aureus*-Stimulated Mast Cells Is Independent on NFAT but Partly Dependent on NF-κB

The inhibition of NFAT, a signaling molecule that previously has been shown to have role in the induction of pro-inflammatory genes in mast cells (18, 19), did not significantly affect the induction of VEGF by *S. aureus* (**Figure 4**). In contrast, NF-κB inhibition produced a modest, yet significant, reduction in VEGF expression in mast cells (**Figure 4**). Hence, the upregulated VEGF expression in mast cells stimulated with live *S. aureus* is partly dependent on the NF-κB signaling pathway.

DISCUSSION

Mast cells are emerging as major detrimental effector cells in numerous pathophysiological conditions, not only in allergy but also in diverse processes such as autoimmune disease, atherosclerosis, cancer, obesity, and contact dermatitis (20, 21). On the other hand, mast cells can also be beneficial to their

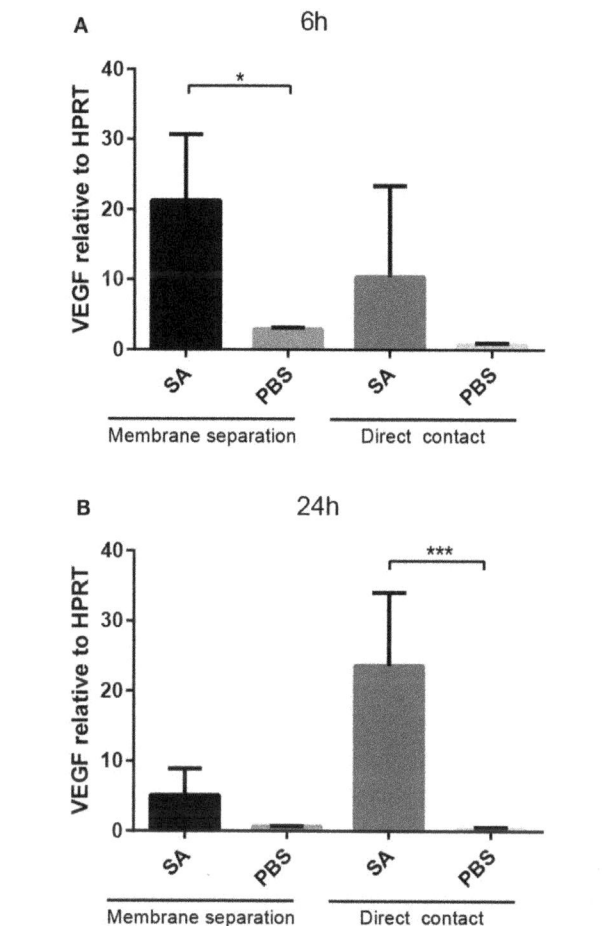

FIGURE 3 | Upregulation of VEGF does not require direct contact between *S. aureus* and PCMCs. *Staphylococcus aureus* (SA) was co-cultured for 6 h **(A)** or 24 h **(B)**, either in direct contact or in Transwell conditions as indicated. Cells were recovered and were analyzed for VEGF expression by qPCR. Results are given as mean ± SD, $*p < 0.05$, and $***p < 0.001$ ($n = 3$).

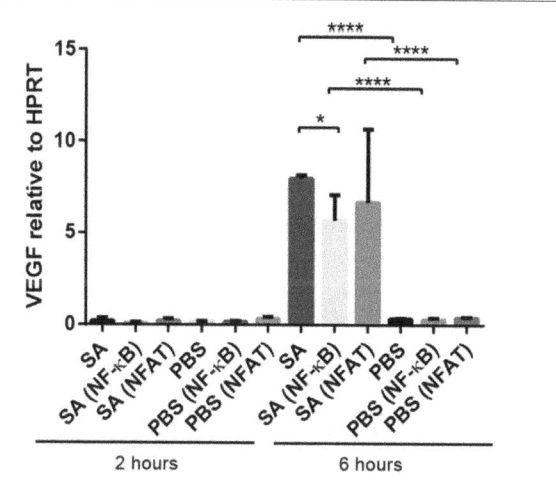

FIGURE 4 | Effect of NF-κB, and NFAT inhibition on VEGF expression in *S. aureus*-stimulated mast cells. Mast cells (PCMCs) were preincubated with inhibitors of NFAT (11R-VIVIT) or NF-κB [6-amino-4-(4-phenoxyphenylethylamino)quinazoline] followed by co-culture with *S. aureus* (SA) for 2 or 6 h. VEGF expression was assessed by qPCR. Results are given as mean ± SD, $*p < 0.05$, and $****p < 0.0001$ ($n = 3$–4).

host, as exemplified by the contribution of mast cells to the host response against bacterial insult (1–3). Although the mechanism by which mast cells influence these processes can vary, there is a widespread notion that mast cells are a source of numerous growth factors, such as basic fibroblast growth factor, nerve growth factor, platelet-derived growth factor, and VEGF (21). Among these, the expression of VEGF by mast cells has attracted particular attention because of the implication of mast cells in malignant processes, where mast cells are thought to promote tumor angiogenesis by secreting growth factors including VEGF (22–25). However, although mast cells are emerging as major VEGF-producing cells, there is still limited knowledge of the mechanisms of VEGF induction in mast cells.

Here, we report that mast cells can be induced to express high levels of VEGF in response to bacterial insult, thus introducing a bacteria–mast cell axis as a mechanism for production of this growth factor. It is also important to stress that most previous studies in which mast cells were shown to express VEGF were

focused on relatively immature mast cells, whereas we here report VEGF expression by fully mature mast cells. Interestingly, we noted a robust release of VEGF already after 2 h, whereas the onset of VEGF gene expression occurred at a later stage. This indicates that the early VEGF secretion is due to release of preformed VEGF from stores in granules, whereas the induction of VEGF gene expression may serve to maintain high levels of VEGF release after the preformed stores have been emptied.

The expression of VEGF in response to bacterial challenge could potentially have various pathophysiological consequences. One obvious scenario could be that mast cell-derived VEGF could have a role in the angiogenesis that accompanies the wound healing process following a bacterial infection. Alternatively, mast cell-expressed VEGF could promote vascular permeability and leukocyte attraction, thereby contributing to the primary host response following a bacterial insult. On a different angle, it is possible that VEGF expression induced in mast cells by bacteria, in fact, could be of relevance for the progress of malignant processes. Several species of bacteria and bacterial strains are known to populate tumors. These bacteria may either be the cause of the tumor or may represent an opportunistic infection occurring as a consequence of the immunosuppressed status of the tumor tissue (26). It is also well known that mast cells populate a wide range of tumors, often being located at the tumor periphery but also within the actual tumor (7, 8). Possibly, bacteria populating the tumor may thus cause mast cells to upregulate their expression of VEGF, and the mast cell-derived VEGF could then have a pathogenic impact by promoting tumor angiogenesis.

Our findings suggest that the induction of VEGF is critically dependent on the interaction of mast cells with live bacteria, whereas various isolated bacterial cell-wall components and heat-inactivated bacteria were without effect. These findings are somewhat surprising considering that mast cells express several

TLRs, and that stimulation of these by various PAMPs have previously been shown to induce the expression of pro-inflammatory cytokines in mast cells (27–32). This suggests that the induction of VEGF occurs independently of TLR stimulation and, in support of this, we did not see any effect of an Myd88 inhibitor on VEGF expression in response to bacterial challenge. It was also noted that VEGF induction was not critically dependent on direct cell–cell contact between the bacteria and mast cells, suggesting that VEGF can to some extent be induced by soluble factors released by the bacteria. Intriguingly though, such factors were not found in conditioned medium obtained by culturing *S. aureus* alone, either in bacterial growth medium or in the medium used for culture of the mast cells. Hence, the release of factors promoting VEGF expression in mast cells appears to require communication between live bacteria and mast cells, leading to the induction and release of VEGF-driving soluble factors. Although we are at present not able to specify the nature of such factors, we noted that the induction of VEGF expression in mast cells was partly dependent on the NF-κB pathway. Altogether, the present findings suggest that soluble, VEGF-driving factors are released by live *S. aureus* as a consequence of crosstalk between live *S. aureus* and mast cells.

AUTHOR CONTRIBUTIONS

C-FJ planned and performed most of the experimental work, interpreted data, and wrote the manuscript; ER performed experimental work, interpreted data, contributed to the planning of the study, and to the writing of the paper; BG contributed to the experiments and to the design of the study; GP planned the study, interpreted data, and wrote the manuscript.

FUNDING

This work was supported by grants from Formas, The Swedish Research Council, The Swedish Cancer Society, The Torsten Söderberg Foundation, and The Swedish Heart and Lung Foundation.

REFERENCES

1. Abraham SN, St John AL. Mast cell-orchestrated immunity to pathogens. *Nat Rev* (2010) 10:440–52. doi:10.1038/nri2782
2. Gri G, Frossi B, D'Inca F, Danelli L, Betto E, Mion F, et al. Mast cell: an emerging partner in immune interaction. *Front Immunol* (2012) 3:120. doi:10.3389/fimmu.2012.00120
3. Wernersson S, Pejler G. Mast cell granules: armed for battle. *Nat Rev Immunol* (2014) 14:478–94. doi:10.1038/nri3690
4. Forsythe P. Microbes taming mast cells: implications for allergic inflammation and beyond. *Eur J Pharmacol* (2016) 778:169–75. doi:10.1016/j.ejphar.2015.06.034
5. Johnzon CF, Rönnberg E, Pejler G. The role of mast cells in bacterial infection. *Am J Pathol* (2016) 186:4–14. doi:10.1016/j.ajpath.2015.06.024
6. Tlsty TD, Coussens LM. Tumor stroma and regulation of cancer development. *Annu Rev Pathol* (2006) 1:119–50. doi:10.1146/annurev.pathol.1.110304.100224
7. Marichal T, Tsai M, Galli SJ. Mast cells: potential positive and negative roles in tumor biology. *Cancer Immunol Res* (2013) 1:269–79. doi:10.1158/2326-6066.CIR-13-0119
8. Rigoni A, Colombo MP, Pucillo C. The role of mast cells in molding the tumor microenvironment. *Cancer Microenviron* (2014) 8:167–76. doi:10.1007/s12307-014-0152-8
9. Byrne AM, Bouchier-Hayes DJ, Harmey JH. Angiogenic and cell survival functions of vascular endothelial growth factor (VEGF). *J Cell Mol Med* (2005) 9:777–94. doi:10.1111/j.1582-4934.2005.tb00379.x
10. Hoeben A, Landuyt B, Highley MS, Wildiers H, Van Oosterom AT, De Bruijn EA. Vascular endothelial growth factor and angiogenesis. *Pharmacol Rev* (2004) 56:549–80. doi:10.1124/pr.56.4.3
11. Boesiger J, Tsai M, Maurer M, Yamaguchi M, Brown LF, Claffey KP, et al. Mast cells can secrete vascular permeability factor/vascular endothelial cell growth factor and exhibit enhanced release after immunoglobulin E-dependent upregulation of fc epsilon receptor I expression. *J Exp Med* (1998) 188:1135–45. doi:10.1084/jem.188.6.1135
12. Grutzkau A, Kruger-Krasagakes S, Baumeister H, Schwarz C, Kogel H, Welker P, et al. Synthesis, storage, and release of vascular endothelial growth factor/vascular permeability factor (VEGF/VPF) by human mast cells: implications for the biological significance of VEGF206. *Mol Biol Cell* (1998) 9:875–84. doi:10.1091/mbc.9.4.875
13. Abdel-Majid RM, Marshall JS. Prostaglandin E2 induces degranulation-independent production of vascular endothelial growth factor by human mast cells. *J Immunol* (2004) 172:1227–36. doi:10.4049/jimmunol.172.2.1227
14. Detoraki A, Staiano RI, Granata F, Giannattasio G, Prevete N, de Paulis A, et al. Vascular endothelial growth factors synthesized by human lung mast cells exert angiogenic effects. *J Allergy Clin Immunol* (2009) 123:1142–9. doi:10.1016/j.jaci.2009.01.044
15. Rönnberg E, Johnzon CF, Calounova G, Garcia Faroldi G, Grujic M, Hartmann K, et al. Mast cells are activated by *Staphylococcus aureus* in vitro but do not influence the outcome of intraperitoneal *S. aureus* infection in vivo. *Immunology* (2014) 143:155–63. doi:10.1111/imm.12297
16. Malbec O, Roget K, Schiffer C, Iannascoli B, Dumas AR, Arock M, et al. Peritoneal cell-derived mast cells: an in vitro model of mature serosal-type mouse mast cells. *J Immunol* (2007) 178:6465–75. doi:10.4049/jimmunol.178.10.6465
17. Rönnberg E, Pejler G. Serglycin: the master of the mast cell. *Methods Mol Biol* (2012) 836:201–17. doi:10.1007/978-1-61779-498-8_14
18. Rönnberg E, Calounova G, Guss B, Lundequist A, Pejler G. Granzyme D is a novel murine mast cell protease that is highly induced by multiple pathways of mast cell activation. *Infect Immun* (2013) 81:2085–94. doi:10.1128/IAI.00290-13
19. Klein M, Klein-Hessling S, Palmetshofer A, Serfling E, Tertilt C, Bopp T, et al. Specific and redundant roles for NFAT transcription factors in the expression of mast cell-derived cytokines. *J Immunol* (2006) 177:6667–74. doi:10.4049/jimmunol.177.10.6667
20. Voehringer D. Protective and pathological roles of mast cells and basophils. *Nat Rev Immunol* (2013) 13:362–75. doi:10.1038/nri3427
21. Galli SJ, Nakae S, Tsai M. Mast cells in the development of adaptive immune responses. *Nat Immunol* (2005) 6:135–42. doi:10.1038/ni1158
22. Sawatsubashi M, Yamada T, Fukushima N, Mizokami H, Tokunaga O, Shin T. Association of vascular endothelial growth factor and mast cells with angiogenesis in laryngeal squamous cell carcinoma. *Virchows Arch* (2000) 436:243–8. doi:10.1007/s004280050037
23. Toth-Jakatics R, Jimi S, Takebayashi S, Kawamoto N. Cutaneous malignant melanoma: correlation between neovascularization and peritumor accumulation of mast cells overexpressing vascular endothelial growth factor. *Hum Pathol* (2000) 31:955–60. doi:10.1053/hupa.2000.16658
24. Imada A, Shijubo N, Kojima H, Abe S. Mast cells correlate with angiogenesis and poor outcome in stage I lung adenocarcinoma. *Eur Respir J* (2000) 15:1087–93. doi:10.1034/j.1399-3003.2000.01517.x
25. Aoki M, Pawankar R, Niimi Y, Kawana S. Mast cells in basal cell carcinoma express VEGF, IL-8 and RANTES. *Int Arch Allergy Immunol* (2003) 130:216–23. doi:10.1159/000069515
26. Cummins J, Tangney M. Bacteria and tumours: causative agents or opportunistic inhabitants? *Infect Agent Cancer* (2013) 8:11. doi:10.1186/1750-9378-8-11

27. Marshall JS. Mast-cell responses to pathogens. *Nat Rev Immunol* (2004) 4:787 99. doi:10.1038/nri1460

28. Supajatura V, Ushio H, Nakao A, Akira S, Okumura K, Ra C, et al. Differential responses of mast cell toll-like receptors 2 and 4 in allergy and innate immunity. *J Clin Invest* (2002) 109:1351–9. doi:10.1172/JCI0214704

29. Ikeda T, Funaba M. Altered function of murine mast cells in response to lipopolysaccharide and peptidoglycan. *Immunol Lett* (2003) 88:21–6. doi:10.1016/S0165-2478(03)00031-2

30. Mrabet-Dahbi S, Metz M, Dudeck A, Zuberbier T, Maurer M. Murine mast cells secrete a unique profile of cytokines and prostaglandins in response to distinct TLR2 ligands. *Exp Dermatol* (2009) 18:437–44. doi:10.1111/j.1600-0625.2009.00878.x

31. Qiao H, Andrade MV, Lisboa FA, Morgan K, Beaven MA. FcepsilonR1 and toll-like receptors mediate synergistic signals to markedly augment production of inflammatory cytokines in murine mast cells. *Blood* (2006) 107:610–8. doi:10.1182/blood-2005-06-2271

32. Sandig H, Bulfone-Paus S. TLR signaling in mast cells: common and unique features. *Front Immunol* (2012) 3:185. doi:10.3389/fimmu.2012.00185

17

A Quantitative Modular Modeling Approach Reveals the Effects of Different A20 Feedback Implementations for the NF-κB Signaling Dynamics

Janina Mothes[1], Inbal Ipenberg[2†], Seda Çöl Arslan[2], Uwe Benary[1], Claus Scheidereit[2] and Jana Wolf[1*]

[1] Mathematical Modelling of Cellular Processes, Max Delbrück Center for Molecular Medicine, Berlin, Germany, [2] Signal Transduction in Tumor Cells, Max Delbrück Center for Molecular Medicine, Berlin, Germany

Edited by:
Zhike Zi,
Max Planck Institute for Molecular
Genetics, Germany

Reviewed by:
Mariko Okada-Hatakeyama,
Osaka University, Japan
Didier Gonze,
Université Libre de Bruxelles, Belgium

*Correspondence:
Jana Wolf
jana.wolf@mdc-berlin.de

† Present address:
Inbal Ipenberg,
Institut für Neuroimmunologie und
Multiple Sklerose, Zentrum für
Molekulare Neurobiologie Hamburg,
Universitätsklinikum
Hamburg-Eppendorf, Hamburg,
Germany

Signaling pathways involve complex molecular interactions and are controled by non-linear regulatory mechanisms. If details of regulatory mechanisms are not fully elucidated, they can be implemented by different, equally reasonable mathematical representations in computational models. The study presented here focusses on NF-κB signaling, which is regulated by negative feedbacks via IκBα and A20. A20 inhibits NF-κB activation indirectly through interference with proteins that transduce the signal from the TNF receptor complex to activate the IκB kinase (IKK) complex. A number of pathway models has been developed implementing the A20 effect in different ways. We here focus on the question how different A20 feedback implementations impact the dynamics of NF-κB. To this end, we develop a modular modeling approach that allows combining previously published A20 modules with a common pathway core module. The resulting models are fitted to a published comprehensive experimental data set and therefore show quantitatively comparable NF-κB dynamics. Based on defined measures for the initial and long-term behavior we analyze the effects of a wide range of changes in the A20 feedback strength, the IκBα feedback strength and the TNFα stimulation strength on NF-κB dynamics. This shows similarities between the models but also model-specific differences. In particular, the A20 feedback strength and the TNFα stimulation strength affect initial and long-term NF-κB concentrations differently in the analyzed models. We validated our model predictions experimentally by varying TNFα concentrations applied to HeLa cells. These time course data indicate that only one of the A20 feedback models appropriately describes the impact of A20 on the NF-κB dynamics in this cell type.

AUTHOR SUMMARY

Models are abstractions of reality and simplify a complex biological process to its essential components and regulations while preserving its particular spatial-temporal characteristics. Modeling of biological processes is based on assumptions, in part to implement the necessary simplifications but also to cope with missing knowledge and experimental information. In consequence, biological processes have been implemented by different, equally reasonable mathematical representations in computational models. We here focus on the NF-κB signaling pathway and develop a modular modeling approach to investigate how different implementations of a negative feedback regulation impact the dynamical behavior of a computational model. Our analysis shows similarities of the models with different implementations but also reveals implementation-specific differences. The identified differences are used to design and perform informative experiments that elucidate unknown details of the regulatory feedback mechanism.

Keywords: quantitative modeling, interlocked feedback loops, regulation, NF-κB signaling, A20, IKK regulation, response time, signaling dynamics

INTRODUCTION

Transcription factor NF-κB regulates cell differentiation, proliferation, and survival. In line with its broad range of normal physiological functions, aberrant activation of NF-κB can lead to severe diseases, e.g., autoimmune, neurodegenerative, and cardiovascular diseases as well as cancer and diabetes (Hayden and Ghosh, 2012; Perkins, 2012). In resting cells, the transcription factor NF-κB is located in the cytoplasm bound to IκBα, which prevents the translocation of NF-κB into the nucleus. Upon stimulation, e.g., with TNFα, the IκB kinase (IKK) complex is activated. The IKK complex phosphorylates IκBα, marking it for proteasomal degradation. Released NF-κB translocates into the nucleus and activates the transcription of a number of target genes (Hinz and Scheidereit, 2014). Two of these are NFKBIA, encoding IκBα, and TNFAIP3, encoding A20. Both proteins exhibit negative feedbacks on NF-κB activation. IκBα binds to NF-κB retrieving it from the DNA and thus exhibiting a direct negative feedback (Huxford et al., 1998). A20 inhibits NF-κB activity indirectly through interference with proteins mediating the signal from the TNF receptor complex to the IKK complex (Lork et al., 2017). The exact molecular mechanism of the inhibitory effect of A20 on the IKK complex is still under discussion (Skaug et al., 2011; De et al., 2014; Wertz et al., 2015).

In the last decades, several mathematical models describing the NF-κB signaling in different cell lines have been published (Hoffmann et al., 2002; Lipniacki et al., 2004; Longo et al., 2013; Zambrano et al., 2014; Fagerlund et al., 2015; Mothes et al., 2015; Murakawa et al., 2015; Benary and Wolf, 2019), and reviewed (Lipniacki and Kimmel, 2007; Cheong et al., 2008; Basak et al., 2012; Williams et al., 2014). These models describe the transient NF-κB activation or the oscillatory dynamics observed experimentally. It was also studied which factors can lead to a switch between oscillatory and non-oscillatory NF-κB dynamics (Mothes et al., 2015). All models comprise the core processes of the canonical NF-κB signaling, e.g., the interaction of NF-κB and IκBα and the transcription and translation of IκBα as well as the IKK-induced degradation of IκBα. The majority of those models include only the negative feedback via IκBα, which has been well-studied and characterized (Fagerlund et al., 2015).

Until today, only a small number of mathematical models has been developed that include the A20-dependent negative feedback mechanism (Lipniacki et al., 2004; Werner et al., 2008; Ashall et al., 2009; Murakawa et al., 2015). These models utilize similar implementations of the core signaling processes but differ in their implementation of the A20 feedback. Since the exact inhibitory mechanism of A20 on IKK has not yet been fully elucidated and may also vary between cell lines, the models implement different mechanisms. While the model of Lipniacki et al. (2004) and the derived model by Ashall et al. (2009) implement the inhibitory action of A20 on the level of IKK, the models of Werner et al. (2008) and Murakawa et al. (2015) basically implement the hypothesis that A20 blocks the signaling upstream of IKK by binding to TNF receptor associated proteins. In particular, the models by Lipniacki et al. (2004) and Ashall et al. (2009) comprise three different states of IKK: neutral, active and inactive. In the model proposed by Lipniacki et al. (2004), A20 promotes the inactivation of activated IKK, whereas, in the model by Ashall et al. (2009) A20 inhibits the "recycling" of inactive IKK to neutral IKK and consequently the activation of IKK. In the models by Werner et al. (2008) and Murakawa et al. (2015), A20 inhibits basal and TNFα-induced IKK activation, although Werner et al. (2008) consider the signaling mechanisms upstream of IKK with substantially more molecular detail than Murakawa et al. (2015). In short, all four models share a feedback inhibition of IKK activity by A20 but differ in the specifics of their A20 feedback implementations.

Here, we ask whether these different A20 implementations have effects for the NF-κB dynamics. This knowledge is required when choosing an available published model for the description of a new data set. For our comparison we selected the different A20 feedback structures implemented in the models of Lipniacki et al. (2004), Ashall et al. (2009), and Murakawa et al. (2015), because these capture three different hypotheses and the models are comparable at their level of detailedness. In contrast, the model by Werner et al. (2008) is very detailed, including 38 parameters for the upstream part. We addressed the question whether the different feedback implementations affect NF-κB dynamics in similar or distinct ways. To this end, we used a computational approach in which we established three ordinary differential equation (ODE) models. Each model is composed of a core module and an upstream module (**Figure 1A**). The core module is identical

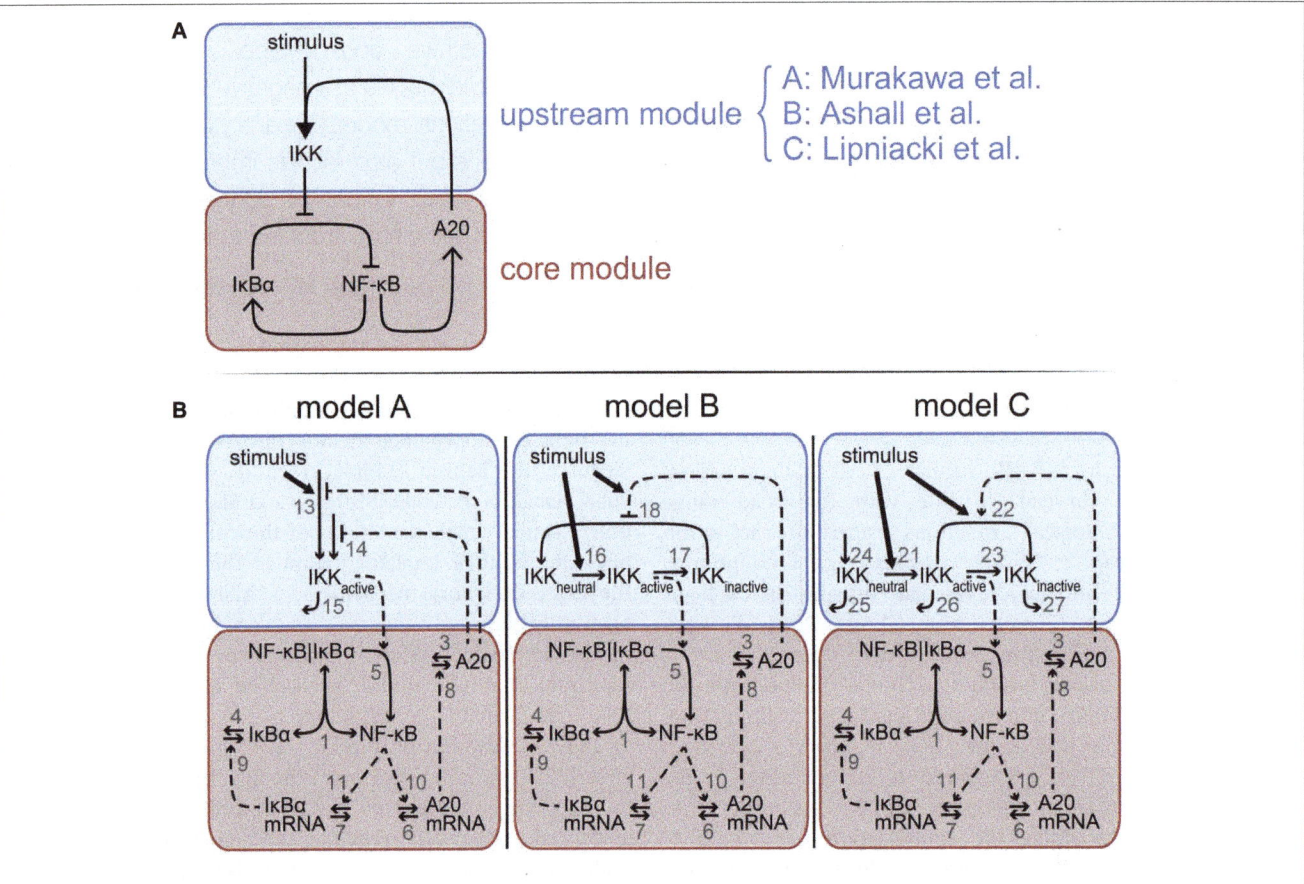

FIGURE 1 | Model schemes comprising the common core module and distinct upstream modules. **(A)** Each model is composed of a core module (red) and an upstream module (blue). The core module is identical in each model but the upstream module differs between model A, B, and C, implementing the A20 feedback mechanisms proposed by Lipniacki et al. (2004), Ashall et al. (2009), and Murakawa et al. (2015), respectively. **(B)** Schematic representations of the three models A–C. Vertical bars separate components in a complex. One-headed arrows indicate the direction of the reaction; double-headed arrows illustrate reversible binding reactions. Dashed arrows represent activation processes; the dashed lines ending in T-shape denote inhibition. The number next to an arrow specifies the number of the reaction. Model equations and the reference parameters are provided in the **Supplementary Information**.

in all three models and describes the interaction of NF-κB and IκBα, transcription and translation of IκBα, and IKK-induced degradation of IκBα. The three upstream modules comprise the three distinct mechanisms of IKK inhibition by A20 that Lipniacki et al. (2004), Ashall et al. (2009), and Murakawa et al. (2015) have proposed. In this way, we applied a modular concept to derive three models that share an identical core module but differ in their implementations of the A20 feedback in the upstream module. By fitting these models to a set of published experimental data, we derive three models showing quantitatively similar NF-κB dynamics. We use this computational approach to directly compare the influences of the structural difference in the upstream modules on the response of the NF-κB dynamics. In particular, we focused on the impact of the A20 and IκBα feedback strength. Moreover, we analyze in each model how the A20 feedback modulates the effect of varied TNFα stimulations on the NF-κB dynamics. We find that the different A20 feedback implementations exert similar but also model-specific effects. To demonstrate how the predicted distinct dynamic responses can be employed for model selection we compare our simulations results for incremental

alterations of TNFα stimulation strength to corresponding experiments in Hela cells.

MATERIALS AND METHODS

Model Structures

In order to compare the three distinct implementations of the inhibitory mechanism of A20, we modularly designed three models. These models comprise an identical core module to which different upstream modules are attached (**Figures 1A,B**). The upstream modules are those proposed by Lipniacki et al. (2004), Ashall et al. (2009), and Murakawa et al. (2015) capturing different A20 feedback implementations. The overall models are hereafter referred to as model A–C.

The common core module of models A–C (**Figure 1B**) describes the reversible binding of free NF-κB and IκBα (reaction 1). Activated IKK (IKKactive) induces the IκBα degradation releasing NF-κB from the complex (reaction 5). Unbound NF-κB induces the transcription of IκBα mRNA (reaction 11), which is translated to IκBα (reactions 9). IκBα mRNA and IκBα protein degrade via reactions 7 and 4, respectively. In addition to IκBα mRNA, NF-κB induces the transcription of A20 mRNA (reaction 10). A20 mRNA is translated to A20 (reaction 8). A20 mRNA and protein are degraded via reactions 6 and 3, respectively. Taken together, the core module consists of five ordinary differential equations (ODEs) and one conservation relation for NF-κB. A detailed description of the corresponding rates and a list of the parameters are provided in the **Supplementary Information**.

The upstream module of model A (**Figure 1B**, left) comprises a very condensed representation of the activation of the IKK complex. The abundance of IKKactive increases in a TNFα-dependent and independent manner (reactions 13 and 14, respectively), both of which are inhibited by A20. IKKactive is inactivated via reaction 15.

In the upstream module of model B (**Figure 1B**, middle), IKK cycles between three distinct states: IKKneutral, IKKactive, and IKKinactive. TNFα stimulation converts IKKneutral into IKKactive (reaction 16), IKKactive is converted to IKKinactive (reaction 17) and IKKinactive is finally turned over to IKKneutral again (reaction 18). A20 inhibits this last reaction in a stimulus-sensitive manner.

The upstream module of model C (**Figure 1B**, right) includes the same states of IKK as described in model B, but IKKneutral, IKKactive, and IKKinactive do not interconvert in a cycle, i.e., obey a conservation relation. Instead, IKKneutral is continuously produced (reaction 24) and all three forms of IKK are subject to degradation (reactions 25–27). Similar to model B, TNFα stimulation in model C also converts IKKneutral into IKKactive (reaction 21), which in turn forms IKKinactive (reaction 23). In contrast to model B, model C includes an additional mechanism to convert IKKactive into IKKinactive (reaction 22). TNFα stimulation as well as A20 enhance this conversion.

Taken together, model A consist of one ODE in its upstream module in addition to the five ODEs and one conservation relation of NF-κB in the core module; model B incorporates two additional ODEs and an additional conservation relation of IKK in the upstream module; and model C includes three additional ODEs in its upstream module. Detailed descriptions of all three models are given in the **Supplementary Information**.

Model Parameterizations

To parameterize the ODEs of the core module, we decided to use the parameters from our previously published model (Murakawa et al., 2015). This approach was based on two arguments. First, this model is based on a comprehensive data set characterizing the modulation of A20 feedback strength and its impact on NF-κB dynamics. Secondly, the core processes of this model perfectly match the reactions of the core module of our models A–C.

To parameterize the three different upstream modules of models A–C, we initially used the parameters published for the corresponding models (Lipniacki et al., 2004; Ashall et al., 2009; Murakawa et al., 2015). However, simulations of models A–C showed very diverse dynamics of unbound NF-κB in response to identical TNFα stimulation conditions (**Figure 2A**). For instance, the concentration of free NF-κB transiently increases in models A and B, but on a slower time scale in model

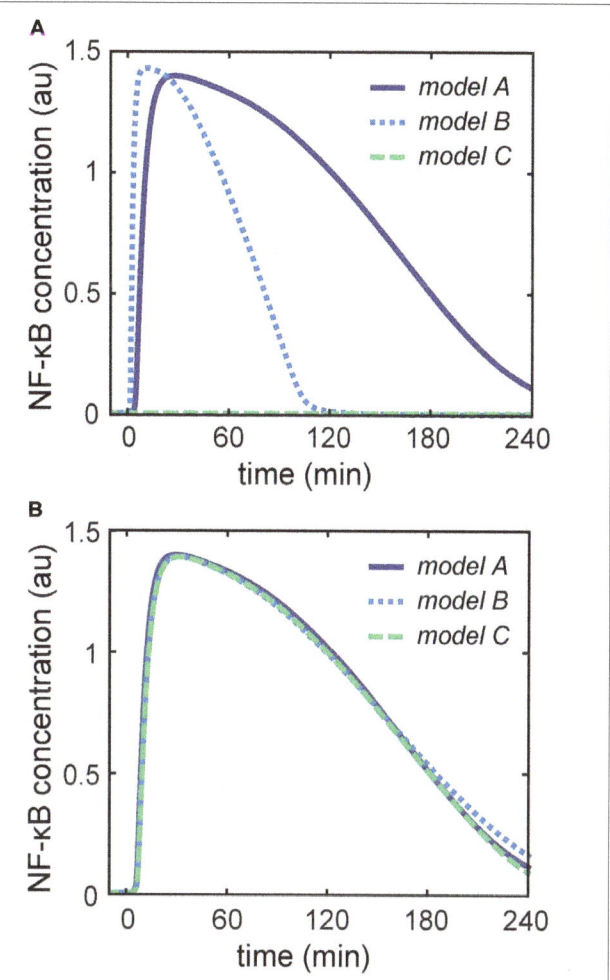

FIGURE 2 | NF-κB dynamics of the three models comprising the core module and the indicated upstream module. **(A)** Differences in NF-κB dynamics can be observed for the three models using the originally published parameters. **(B)** Nearly identical NF-κB dynamics can be observed for the three models with newly estimated parameters for the upstream modules.

A. In contrast, unbound NF-κB hardly increases upon TNFα stimulation in model C.

In order to compare models A–C directly, it is necessary that NF-κB exhibits the same dynamics upon TNFα stimulation in all three models. Thus, we estimated new parameters of the reactions in the upstream modules such that all components of the core module show the same dynamics in all three models. We used the D2D Toolbox (Raue et al., 2013) to estimate these parameters while keeping the parameters of the core module fixed. With this restriction on the parameters of the core module, we were able to reasonably minimize the parameter search space and obtain identical dynamics of the components of the core module. The details of the parameter estimation are explained in the **Supplementary Information**. Simulations of models A–C with these estimated parameters showed nearly identical dynamics of NF-κB activation upon TNFα stimulation (**Figure 2B**) and all remaining components of the core module (**Supplementary Figures S1, S2**).

Next, we checked whether the new parameterization changed the inhibitory effect of A20 on the activation of IKK. To do so, we simulated A20 knockout conditions by setting the A20 transcription rate k10 to zero and compared the resulting dynamics to those of wild-type conditions, i.e., using the reference value of k10 (**Supplementary Table S1**). The simulations show that the A20 knockout causes a prolonged increase in NF-κB, IKK and IκBα mRNA upon TNFα stimulation compared to wild-type (Lee et al., 2000) in all three models (**Supplementary Figures S3–S5**). The simulations furthermore show that the absence of A20 leads to a decrease in IκBα concentration in all three models. These results demonstrate that the parameterizations of the models A–C do represent the inhibitory effect of A20 on the activation of IKK.

Taken together, models A, B, and C were derived by modular design from an identical core module and different upstream modules specifying distinct implementations of the A20 feedback and TNFα stimulation. The models exhibit almost identical dynamics of their common model components, and show similar dynamical behavior in A20 knockout simulations.

Quantitative Characterization of the NF-κB Dynamics

To quantitatively compare the dynamics of unbound NF-κB between the models A–C, we used three established quantitative measures for signaling characteristics, in particular: (i) the maximal NF-κB concentration (x_{max}), (ii) the time of the maximal NF-κB concentration (t_{max}), and (iii) the response time (t_r) (**Figure 3**). The response time has been defined in Llorens et al. (1999), and quantifies the time required for a complete NF-κB response after stimulation. The function f is transformed to the gray line by taking the absolute gradient of f. The area above the transformed function is calculated and normalized by the steady state f^* of the transformed function. While x_{max} and t_{max} describe the initial response of NF-κB to TNFα stimulation, t_r represents a normalized duration of NF-κB signaling and can therefore be used as a measure for the long-term dynamics.

Numerical Simulations

The model equations are listed in the **Supplementary Information**. Calculations were done with MathWorks Matlab R2013b. Steady state solutions were numerically obtained. Starting from those steady state solutions, the models are always simulated for 57,600 min in order to definitely reach a steady state and thus ensure convergence of the response time.

Experimental Methods

HeLa cells were stimulated with 10, 25, or 100 ng/ml TNFα (human recombinant TNFα, Alexis Corporation) for the time periods indicated (120, 100, 80, 60, 40, 20, and 10 min) or were left untreated. Following stimulation, cells were lysed in 20 mM Hepes pH = 7.9, 450 mM NaCl, 1 mM MgCl2, 0.5 mM EDTA pH = 8.0, 0.1 mM EGTA, 1% NP-40, 20% glycerol, supplemented with complete protease inhibitor mixture and Phosphostop

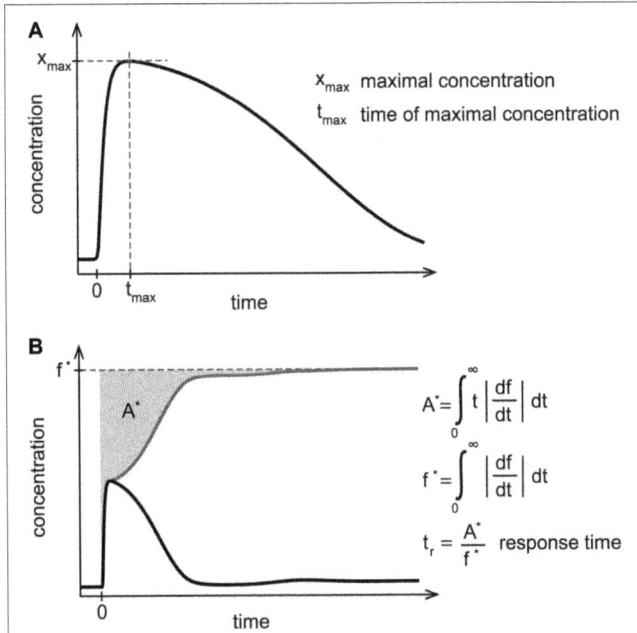

FIGURE 3 | Measures to quantify NF-κB dynamics. **(A)** The maximal concentration of NF-κB (x_{max}) and the time of the maximal concentration of NF-κB (t_{max}) characterize the initial NF-κB response. **(B)** The response time (t_r) defined in Llorens et al. (1999) is determined by the gray area (A*) normalized to the steady state (f*) of the absolute gradient of the dynamics of NF-κB. The response time quantifies the time required for the activation and deactivation of NF-κB upon stimulation and can be interpreted as a characterization of the NF-κB long-term behavior.

(Roche Applied Science), 50 nM Calyculin A, 10 mM NaF, 10 mM β-glycerophosphate, 0.3 mM Na3VO4, and 1 mM Dithiothreitol. Lysates were centrifuged at 14,000 rpm for 10 min.

NF-κB DNA-binding activity was assayed by Electrophoretic Mobility Shift Assay (EMSA) as previously described (Stilmann et al., 2009).

EMSA quantification was made using the phosphor-imager Typhoon FLA 9500, GE Healthcare. Data were quantified using ImageQuant software. After background subtraction, the NF-κB band was normalized to a respective constant non-specific band.

RESULTS

Effects of Different A20 Feedback Strength on NF-κB Dynamics

As a starting point, we studied the impact of the A20 feedback on the NF-κB dynamics upon a constant TNFα stimulation. To do so, we varied the A20 feedback strength and studied its effects on the temporal change of the concentration of unbound NF-κB (hereafter denoted NF-κB) in all models. The strength of the A20 feedback is varied by multiplying the transcription rate constant of the A20 mRNA (k10) with a factor, i.e., feedback strength. A low value of the feedback strength corresponds to a weak negative feedback, whereas a high feedback strength results in a strong negative feedback. Local sensitivity analyses showed that a variation of the translation rate constants of A20 (k8) and of the transcription rate constant have a comparable effect on the three measures of the NF-κB dynamics (**Supplementary Figures S6–S8**). Thus, our choice to vary the transcription rate constant by a factor, i.e., the feedback strength, rather than the translation rate constant does not affect our conclusions.

The NF-κB dynamics of the models A–C for the A20 feedback strength 0.1 and 10 are shown in **Figure 4A**. In case of a high A20 feedback strength of factor 10, models B and C show a fast and transient increase of NF-κB concentration upon a constant TNFα stimulation (**Figure 4A** – top). In model A, NF-κB increases later and to a lesser extent compared to model B and C, yet it decreases to a similar final concentration. In the case of a low A20 feedback strength of factor 0.1 (**Figure 4A** – bottom), all three models show an almost identical increase in the NF-κB concentration. However, NF-κB decreases faster and to a lower final concentration in model C compared to model A and B. Comparing the simulations of the high with the low A20 feedback strength, all three models show a faster decrease in NF-κB in the case of high compared with low A20 feedback strength.

These results reflect the strong influence of the A20 feedback on the deactivation of NF-κB. A high A20 feedback strength causes a stronger and faster deactivation in all three models. Moreover, in model A a strong A20 feedback strength notably reduces and also delays NF-κB activation.

The IκBα Feedback Modulates the Effect of the A20 Feedback on NF-κB

Besides A20, IκBα is an important negative regulator of NF-κB dynamics. We next analyzed whether the interplay of these two feedbacks in the regulation of NF-κB dynamics is similar in the three models. To address this question, we varied the IκBα feedback strength in addition to that of A20. Similar to the A20 feedback strength, we multiplied the transcription rate constant of the IκBα mRNA (k11) by a factor to change the IκBα feedback strength.

The NF-κB dynamics of the three models for four exemplary combinations of different A20 and IκBα feedback strength are shown in **Figure 4B** (cases I–IV). The simulations show a rapid increase of NF-κB concentration upon TNFα stimulation for all models and in all four cases (I–IV), with one exception (model A, case I). The subsequent decrease of NF-κB concentration differs in strength and pace. For a combination of a high A20 feedback strength and a low IκBα feedback strength (case I), NF-κB concentrations in models B and C decrease to the half-maximum level at around 250 min whereas model A shows no NF-κB response to TNFα stimulation. When A20 and IκBα feedback strength are both low (case II), NF-κB concentration decreases at a much slower pace and to lesser extent than in case I for models B and C; here (case II) model A also shows a transient NF-κB activation. If the feedback strength of A20 and IκBα are high (case III), a fast increase can be observed that is followed by a nearly complete decrease of NF-κB concentration at 100 min for all models. For combinations of a high IκBα feedback strength with a low A20 feedback strength (case IV), the decrease in NF-κB concentration is slightly prolonged compared to case

III, depending also on the model. These results are in agreement with our earlier finding that higher A20 feedback strength cause a faster and stronger decrease in NF-κB than lower A20 feedback strength (**Figure 4A**).

In the comparison of case I and case III, which both comprise the same A20 feedback strength but differ in their IκBα feedback strength, a stronger as well as faster decrease in the NF-κB concentration can be observed for high IκBα feedback strength. The comparison of case II and case IV yields a similar result, showing that a higher IκBα feedback strength leads to a faster and stronger decrease in NF-κB concentrations and therefore influencing its short-term and long-term dynamics.

In summary, both feedbacks lead to the deactivation of NF-κB after a transient increase. Thus, if only one of the two feedbacks is strong, it can compensate for the other. If A20 and IκBα feedback strength are both strong, the effect on the deactivation of NF-κB is enhanced resulting in an even faster and stronger NF-κB deactivation.

Beside these general observations, we find model-specific effects of the feedbacks. Most obviously, the maximal NF-κB activation and the deactivation pace seem to vary between the models. An interesting combination is a strong A20 with a low IκBα feedback strength (case I) for model A, which prevents an NF-κB response to TNFα stimulation.

Quantification of the Influences of the A20 and the IκBα Feedback on NF-κB Dynamics

To determine to what extent the models A–C differ in their NF-κB response under the various feedback strength, we quantified the dynamics of NF-κB by three measures: the maximal concentration of NF-κB, the time of the maximal concentration, and the response time (**Figure 3**). The first two measures characterize the initial NF-κB dynamics whereas the last measure characterizes the long-term NF-κB dynamics. For each model we then continuously varied the A20 and the IκBα feedback strength over a broad range of four orders of magnitude, covering very low (e.g., 0.01) as well as very high (e.g., 100) feedback strength (**Figure 4C**).

In model A, the maximal NF-κB concentration barely changes at A20 feedback strength below 1 (**Figure 4C** – first column, first row). In those cases, only an increase in the IκBα feedback strength leads to a decrease in the maximal concentration of NF-κB. For strong A20 feedback strength above 1, the A20 feedback can prevent the NF-κB response almost completely for a wide range of different IκBα feedback strength (**Figure 4C** – first row, black area). This is in agreement with case I in **Figure 4B** showing no NF-κB response for high A20 and low IκBα feedback strength. For A20 feedback strength below 1 in combination with a wide range of different IκBα feedback strength, the maximal concentration of NF-κB is reached in the first 80 min (**Figure 4C** – first column, second row – blue area). For A20 feedback strength above 1, an increase in the A20 feedback strength can lead to a delay in the time of the maximal concentration of NF-κB. Very high A20 feedback

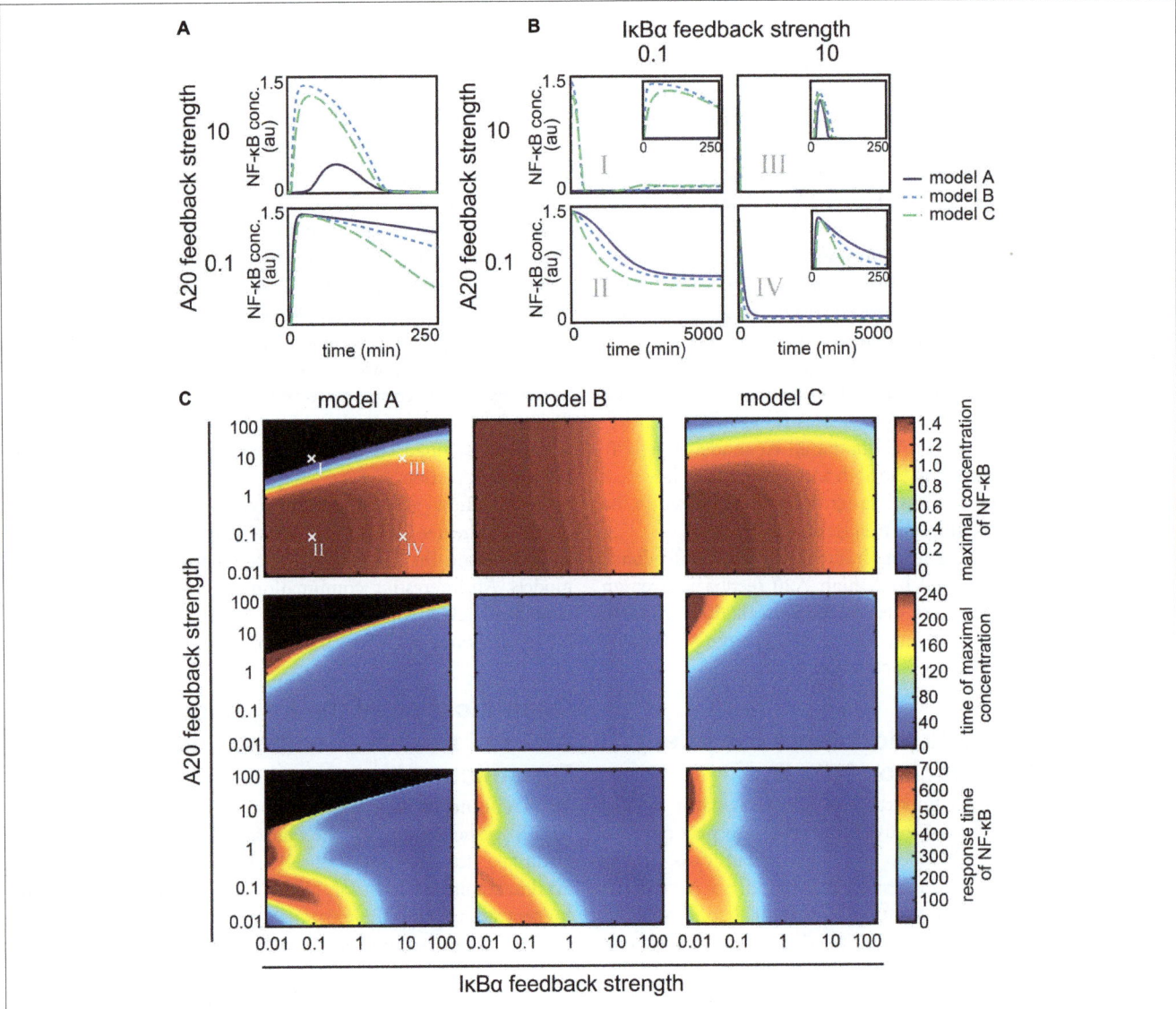

FIGURE 4 | Influence of the A20 feedback strength and the IκBα feedback strength on NF-κB dynamics. **(A)** NF-κB dynamics of the three models for two different A20 feedback strength. **(B)** NF-κB dynamics of the three models for four exemplary combinations of A20 and IκBα feedback strength. Insets zoom into the early time points of the dynamics. **(C)** The effect of the different combinations of feedback strength on the maximal concentration of NF-κB (first row), the time of the maximal concentration (second row), and the response time of NF-κB (third row) in the case of model A (first column), model B (second column) and model C (third column). The four exemplary combinations of feedback strength shown in panel B (I, II, III, and IV) are indicated. Black areas mark the combinations of feedback strength where hardly any NF-κB response is observed, i.e., the difference between maximal concentration of NF-κB and initial concentration of NF-κB is less than the threshold value of 0.001 μM.

strength completely diminish the NF-κB response. The effect of the A20 feedback on the response time of NF-κB is also modulated by the IκBα feedback (**Figure 4C** – first column, third row). The increase in the response time of NF-κB for confined combinations of low A20 and IκBα feedback strength is due to a prolonged higher concentration of NF-κB at later time points. The response time of NF-κB remains low for a wide range of different A20 feedback strength for IκBα feedback strength above 1. To summarize, the effects of the two feedbacks, A20 and IκBα, in model A can be subdivided into three main areas. The first area comprises combinations of A20 and IκBα

feedback strength below 1. Those combinations result in a rapid but prolonged first peak of NF-κB and a higher NF-κB concentration at later time points similar to case II in **Figure 4B**. The second area is determined by high A20 feedback strength, where the NF-κB response is completely inhibited for low IκBα feedback strength similar to case I in **Figure 4B**. However, if the IκBα feedback strength is high, NF-κB remains responsive. The third area comprises high IκBα feedback strength resulting in a slightly decreased first peak of NF-κB and no response at later time points similar to case III and IV in **Figure 4B**.

In model B, the A20 feedback strength hardly influences the height and time of the maximal concentration of NF-κB. Both measures are mainly determined by the IκBα feedback strength (**Figure 4C** – second column, first and second row). However, the A20 feedback strength influences the response time of NF-κB (**Figure 4C** – second column, third row). Especially, if the A20 and IκBα feedback strength are both low, the NF-κB response time is higher. Thus, in model B the initial NF-κB response is mainly determined by the IκBα feedback, whereas the combination of both feedbacks influences the NF-κB dynamics at later time points.

In model C, an increase in the A20 feedback strength reduces the maximal concentration of NF-κB for A20 feedback strength above 1 (**Figure 4C** – third column, first row). For feedback strength below 1, the A20 feedback barely influences the maximal concentration of NF-κB. In those cases, an increase in the IκBα feedback strength can gradually decrease the maximal concentration of NF-κB. The time of the maximal concentration of NF-κB appears to be mainly robust toward changes in the two feedback strength (**Figure 4C** – third column, second row). Only combinations of A20 feedback strength above 1 and IκBα feedback strength below 0.1 delay the time of the maximal concentration of NF-κB. Considering the response time of NF-κB, the influence of the A20 feedback can be strongly modulated by the IκBα feedback (**Figure 4C** – third column, third row). The NF-κB response time remains low for IκBα feedback strength above 1 independent of the A20 feedback strength. For an IκBα feedback strength below 1, the A20 feedback strength can increase the NF-κB response time for A20 feedback strength either above 10 or for feedback strength between 1 and 0.1. To summarize, the effects of the two feedbacks in model C can be subdivided into three areas. The first area comprises combinations of A20 and IκBα feedback strength below 1. Those combinations result in a rapid, but prolonged first peak of NF-κB and a higher NF-κB concentration at later time points similar to case II in **Figure 4B**. The second area is confined by A20 feedback strength above 10 and IκBα feedback strength below 0.1 resulting in a reduced as well as a delayed maximal NF-κB concentration similar to case I in **Figure 4B**. The third area comprises IκBα feedback strength above 1 leading to a fast but decreased first peak of maximal NF-κB and no response at later time points similar to case III and IV in **Figure 4B**.

Altogether, the models show similar, but also different influences of the feedbacks on the NF-κB dynamics. For model A and C, the two negative feedbacks, IκBα and A20, have an impact on the initial dynamics. Both can independently reduce the maximal NF-κB concentration. However, in both models the two feedbacks are not completely redundant but have distinct functions in modulating the NF-κB response. If both feedback strength are below 1, the inhibitory effect of A20 and IκBα is weak. In that case, the initial NF-κB response is slightly delayed and a prolonged activation of NF-κB can be observed at later time points. If A20 feedback strength are high, the NF-κB response is completely inhibited in model A. In model C, a reduced as well as delayed NF-κB response can be observed. If the IκBα feedback strength is high, both models show a reduced but fast initial NF-κB increase and no response at later time points.

To summarize, in models A and C both feedbacks inhibit the maximal concentration of NF-κB, but the A20 feedback delays the initial response and prolongs the response at later time points, whereas the IκBα feedback results in a faster initial activation and rapid deactivation of NF-κB. In contrast, in model B the initial NF-κB response is hardly influenced by the A20 feedback but mainly regulated by the IκBα feedback. Also in model B both feedbacks have an effect on the later phase of the NF-κB dynamics.

Characterization of the Interplay of TNFα Stimulation and A20 Feedback Strength

In all three considered mechanisms, the A20 feedback modulates the signal transduction of the TNFα stimulus toward the activation of IKK. We are therefore interested in the influence of the A20 feedback strength on the NF-κB response upon different strength of TNFα stimulation. To address this question, we simultaneously varied the stimulation strength of TNFα and the strength of the A20 feedback and quantified their influence on the maximal concentration of NF-κB, time of the maximal concentration and the response time of NF-κB (**Figure 5**). Here, the IκBα feedback strength is fixed to the value of 1.

In model A, variations in TNFα stimulation change the initial and long term dynamics of NF-κB (**Figure 5** – first column). In particular, an increase in TNFα stimulation strength leads to a faster and stronger increase in the maximal NF-κB value (**Figure 5** – first column, first and second row). This effect can be strongly modulated by the A20 feedback: for feedback strength above 1 a reduction and delay of the maximal NF-κB concentration can be observed. High A20 feedback strength above 10 result in a complete prevention of the NF-κB response for various TNFα stimulation strength (**Figure 5** – first column, black area). The response time of NF-κB is influenced by TNFα stimulation and A20 feedback strength in a complex way (**Figure 5** – first column, third row). For instance, for the combination of A20 feedback strength below 1 and TNFα stimulation strength above 1 the response time of NF-κB increases, indicating a prolonged NF-κB activation. In contrast, the combination of A20 feedback strength around 0.01 and TNFα stimulation strength above 10 leads to a decrease in the response time of NF-κB. The underlying reason is the change in the deactivation of NF-κB. For A20 feedback strength of 0.01 and TNFα stimulation strength of 100, NF-κB is not deactivated. Thus, NF-κB concentration does not decrease after its initial increase, resulting in a low response time (**Supplementary Figure S9**). However, for A20 feedback strength of 0.1 and TNFα stimulation strength of 100, NF-κB concentration slowly decreases after its initial increase, resulting in a high response time (**Supplementary Figure S9**).

In model B, the amount and time of the maximal concentration of NF-κB depend on the TNFα stimulation strength, but are mostly robust toward changes in A20 feedback strength (**Figure 5** – second column, first and second row). However, both TNFα stimulation strength and A20 feedback strength affect the response time of NF-κB (**Figure 5** – second column, third row). The effect is non-monotonous: low TNFα

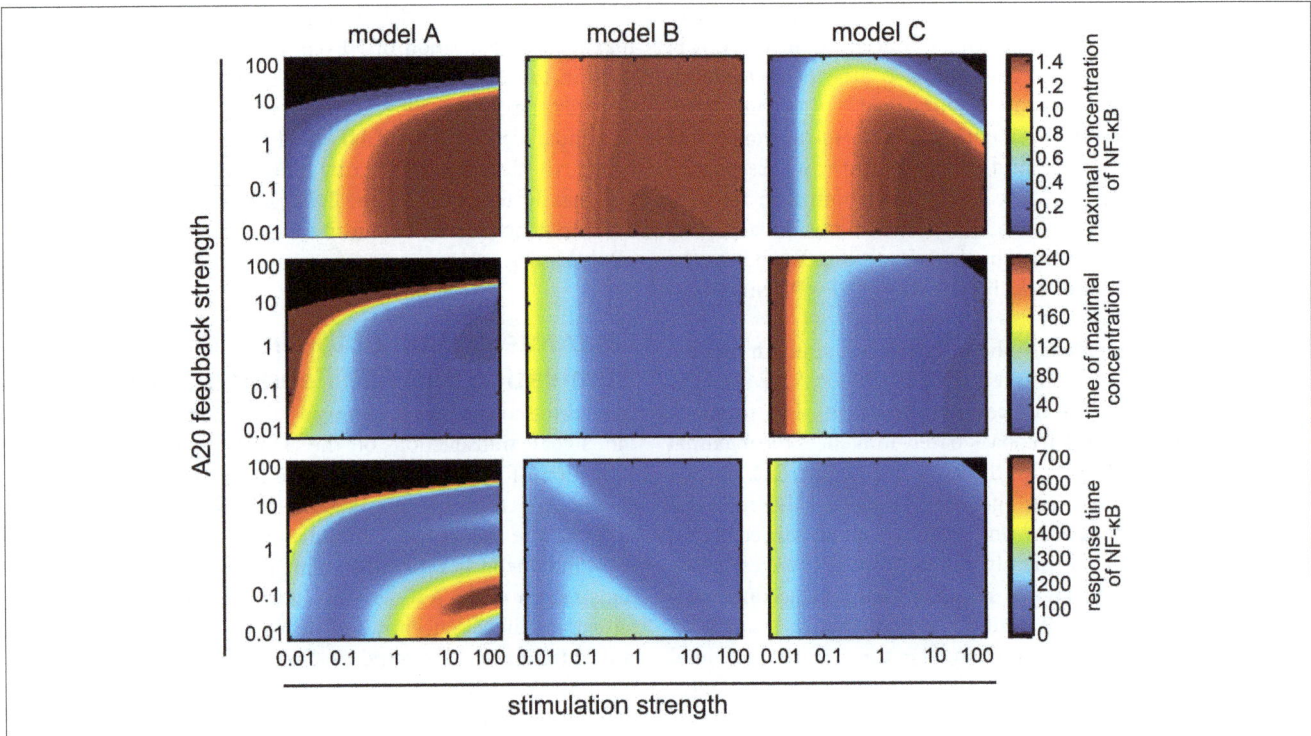

FIGURE 5 | Influence of A20 feedback strength and TNFα stimulation strength on NF-κB dynamics. NF-κB dynamics of model A (first column), model B (second column) and model C (third column) are characterized by the maximal concentration of NF-κB (first row), the time of the maximal concentration of NF-κB (second row) and the response time of NF-κB (third row). Black areas mark combinations of A20 feedback strength and TNFα stimulation strength with hardly any observable NF-κB response; the difference between maximal and initial NF-κB concentrations is less than 0.001 μM.

stimulation strength between 0.1 and 1 and very low A20 feedback strength below 0.1 show an increase in the response time of NF-κB, indicating a prolonged activation of NF-κB. However, in the case of TNFα stimulation strength between 10 and 100, a decrease in the response time is observed.

In model C, the maximal concentration of NF-κB and the timing of its peak mostly depend on TNFα stimulation strength (**Figure 5** – third column, first and second row). A20 feedback strength can lead to a reduction and a slight delay of the maximal NF-κB concentration for high TNFα stimulation strength. In particular, if A20 feedback strength as well as TNFα stimulation strength are high, the maximal concentration of NF-κB decreases and can result in a complete prevention of the NF-κB response (**Figure 5** – third column, black area). The response time of NF-κB mainly depends on TNFα stimulation strength and hardly on A20 feedback strength (**Figure 5** – third column, third row).

In conclusion, the maximal NF-κB concentration and its timing, are strongly determined by the TNFα stimulation strength in all models. In models A and C the A20 feedback can strongly modify that impact. However, in model B, we see no significant effect of the A20 feedback on the amount and time of maximal NF-κB. The effect of the TNFα stimulation strength and the A20 feedback on the long-term dynamics is more complex. However, if we consider the effect of TNFα stimulation (for factors > 1) and a given A20 feedback strength (factor = 1), we observe opposite effects in the models: while a higher TNFα stimulation strength leads to an increase of the response time in

model A, such a stimulus increase would cause a decrease in the response time in models B and C.

Comparison of Simulations With Experimental Data for the Effect of Varied TNFα Stimulation Strength

The qualitative differences between the models suggest an experimental setup to scrutinize the A20 feedback implementations. To predict the outcome of such an experiment, we simulated the NF-κB dynamics of the models A–C in response to three different TNFα concentrations (**Figure 6A**). We selected TNFα stimulation because changes in TNFα concentration are easier to perform experimentally than changes in A20 feedback strength. Our simulations predict for model A that NF-κB levels remain high for stimulation with 100 ng/ml TNFα compared with 10 ng/ml TNFα at later time points (**Figure 6A**). In contrast, in models B and C, NF-κB levels decrease faster at later time points upon stimulation with 100 ng/ml TNFα compared to 10 ng/ml TNFα. These predictions are independent of the assumed A20 feedback strength (**Supplementary Figure S10**) and are furthermore verified by simulations of the models published by Lipniacki et al. (2004), Ashall et al. (2009), and Murakawa et al. (2015) (**Supplementary Figure S11**).

We compared our model predictions to experimental data applying 10, 25, and 100 ng/ml TNFα to HeLa cells. The time course measurements of NF-κB's DNA-binding activity by EMSA

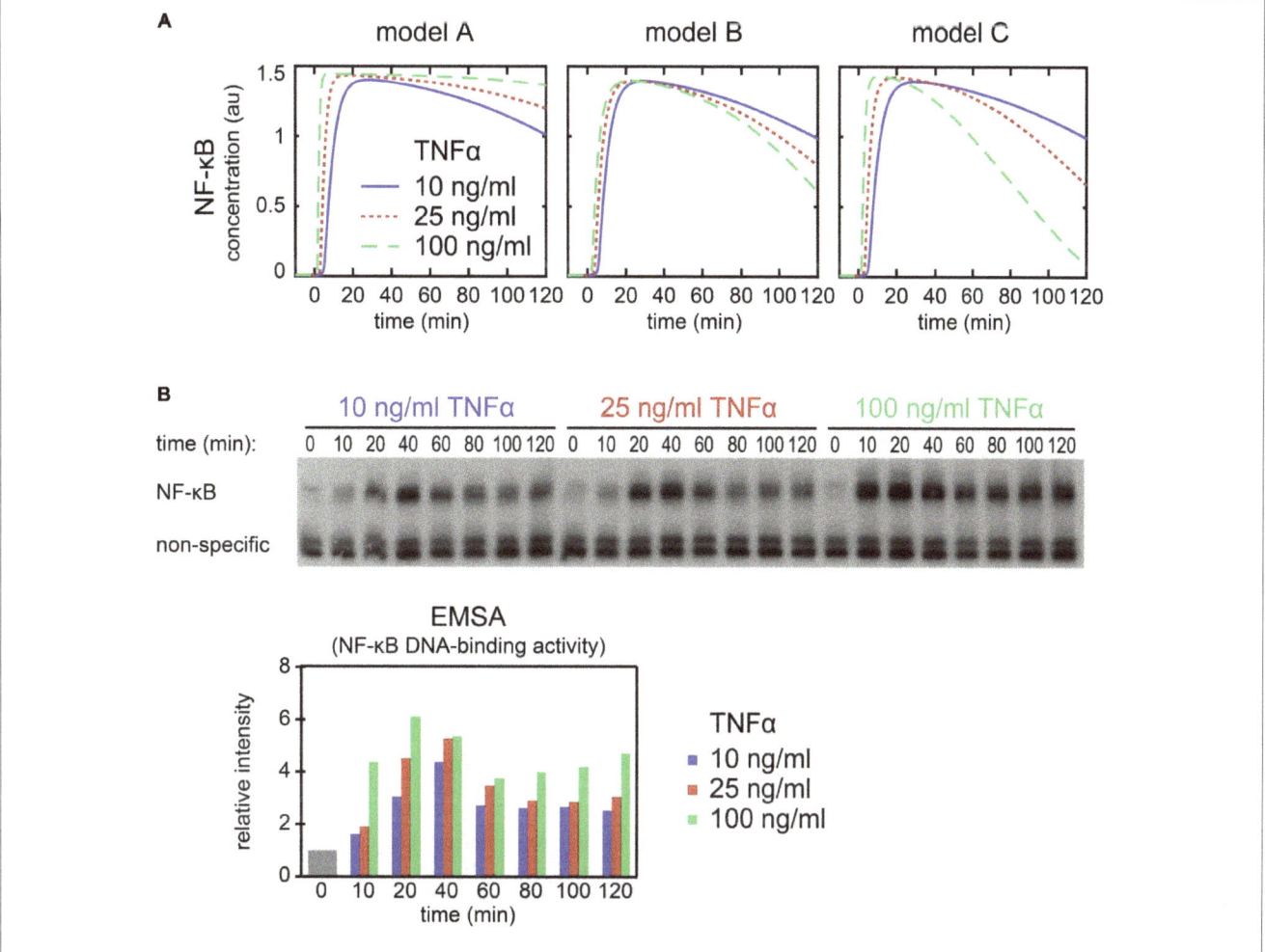

FIGURE 6 | Dynamics of NF-κB upon stimulation with different TNFα concentrations. **(A)** Simulation of NF-κB assuming a stimulation with 10 ng/ml (solid line), 25 ng/ml (dotted line), and 100 ng/ml TNFα (dashed line) in model A (left), model B (middle) and model C (right). **(B)** Exemplary EMSA experiment measuring NF-κB DNA-binding activity over a time course of 120 min in HeLa cells upon stimulation with 10, 25, and 100 ng/ml TNFα. The histogram shows the quantification of the EMSA experiment. The mean value of the relative intensities at $t = 0$ is set to 1 and used as a normalization for all other values. Two replicate experiments are shown as **Supplementary Figure S12**.

showed NF-κB dynamics as predicted for model A but not model B or C (**Figure 6B**). The comparison of model results and experiments thus suggests that in HeLa cells the implementation of the A20 feedback structure of model A is appropriate to describe the effect of A20 on the dynamics of NF-κB.

DISCUSSION

In this study, we developed a modular modeling approach to analyze the impact of different A20 inhibition mechanisms on the dynamics of NF-κB. In particular, we compared three distinct implementations of the A20 feedback by combining upstream modules of available models with a common core pathway module. By fitting the resulting models to a published comprehensive experimental data set, we derive models with quantitatively comparable NF-κB dynamics. When analysing the effect of variations of the strength of the A20 and IκBα feedbacks, as well as of TNFα stimulation in these models, we observe

similarities, but also model-specific differences. Increasing IκBα feedback strength attenuate the initial as well as the long-term NF-κB response in all three models, that is, reduce the maximum and response time, respectively. Increasing A20 feedback strength reduce the maximum and duration of the NF-κB response in models A and C. In model A, the NF-κB response is even completely diminished for very high A20 feedback strength. However, in model B the A20 feedback has no impact on the initial dynamics. Moreover, our simulations predicted that changes in the TNFα stimulation strength influence initial and long-term dynamics of NF-κB. Here, we observed qualitative differences in the long-term NF-κB response between the different models. We used these predictions for an experimental validation in HeLa cells. The experimental observations support model A, but not model B or C in this cell type.

Models A–C differ in the implementation of the A20 feedback. We compared the effect of this feedback implementation for a carefully derived parametrization of the models. While the detailed NF-κB dynamics might change for other model parametrizations, we expect the effect of the model structure to more generally valid. In all three models, A20 acts conjointly with the stimulus in order to inhibit IKK activation. Model A includes in addition a basal IKK activation rate that is inhibited by A20 (reaction 14). Such a composite, non-linear description of the inhibitory influence of A20 seems necessary to reproduce the NF-κB dynamics of HeLa cells. This indicates that the regulation of IKK activity by A20 in this cell type may result from a combination of several mechanisms and is thus more complex than anticipated. Indeed, A20 seems to fulfill multiple functions *in vivo*, such as a deubiquitinating activity mediated by its N-terminal ovarian tumor (OTU) domain and an E3 ubiquitin ligase activity mediated by its C-terminal zinc finger domain (Lork et al., 2017). These distinct functions of A20 may regulate the activity of upstream signal mediators and constitute potential mechanisms that may explain the complex non-linearity in the signal transduction from TNFα stimulation to IKK activation (Hymowitz and Wertz, 2010). In a stochastic pathway model the different A20 effects have been combined to better explain experimental data (Lipniacki et al., 2007). A recent analysis of temperature effects on the NF-κB pathway also highlights the importance of the A20 feedback and the necessity to extend and modify its implementation in model B (Ashall et al., 2009; Harper et al., 2018). Moreover, it will be interesting to explore the role of additional negative regulators on the pathway, e.g., the deubiquitinating enzymes CYLD and OTULIN (Lork et al., 2017) as well as the effect of the cross-talk with the non-canonical pathway (Ashall et al., 2009; Yilmaz et al., 2014; Mukherjee et al., 2017).

Our analyses of the three models revealed redundant but also distinct functions of the two negative feedbacks, A20 and IκBα. This confirms and extends earlier findings by Werner et al. (2008), demonstrating distinct roles of the two feedbacks in a very detailed pathway model. In that publication, IκBα has been reported to modulate mostly the initial NF-κB response while A20 mainly shapes the late response. In our current study, we characterize the output based on quantitative measures for a wide range of different feedback strength. We find that the IκBα feedback fine-tunes the initial NF-κB response in all models. However, it can also influence the response-time and therefore the long-term dynamics. The A20 feedback has different effects in models A, B, and C. In models A and C, it modulates the initial as well as long-term dynamics. Moreover, in model A it has a bimodal on-off effect on the NF-κB response, i.e., preventing the NF-κB response at high A20 feedback strength. While our analysis revealed a lower sensitivity of model B to changes in the A20 feedback, a comprehensive analysis (**Supplementary Figures S6–S8**) showed comparable sensitivities of all three models to parameter changes in general, only the distribution of the sensitivities between processes differs in the models.

The non-redundant functions of the two negative feedbacks could be due to their structural properties: the two feedbacks are interlocked, with the IκBα feedback serving as an inner feedback loop and the A20 feedback as an outer feedback loop. Previous studies indicted distinct functions of interlocked feedback loops with respect to the oscillatory behavior of a system (Nguyen, 2012; Baum et al., 2016). Here, a weak or strong outer feedback loop may cause an on or off response, respectively, independent of the strength of the inner feedback loop. However, the inner feedback loop can fine-tune the response in the case of a weak outer feedback loop. Such interlocked feedback loops are very common regulatory motifs in signaling pathways in general (Batchelor et al., 2011; Benary et al., 2015; Kochanczyk et al., 2017; Zhang et al., 2017).

Taken together, our quantitative modular modeling approach employs the regulation of NF-κB signaling by the A20 feedback as an example case to study the impact of different implementations of an inhibition mechanism on the model's response to perturbations. Comparing the simulations of the three models A–C to experimental data suggests that model A is an appropriate choice to describe TNFα stimulation in HeLa cells. Our results emphasize the need to further explore the molecular details of processes upstream of IKK regulation.

AUTHOR CONTRIBUTIONS

JM and JW contributed to the conceptualization and design of study. JM contributed to the development, simulation and analysis of ODE models, design and implementation of computer code. JM, UB, and JW contributed to the data interpretation and curation. II and SÇ contributed to the experimental work. CS contributed to the supervision of experimental work. JW contributed to the supervision of project. JM, UB, CS, and JW contributed to the preparation of manuscript. All authors contributed to the article and approved the submitted version.

FUNDING

The project was supported by a grant from the German Federal Ministry of Education and Research BMBF (Project ProSiTu, 0316047A) awarded to JW and CS and by the Personalized Medicine Initiative "iMed" of the Helmholtz Association to JW. The funders had no role in study design, data collection and analysis, decision to publish, or preparation of the manuscript.

ACKNOWLEDGMENTS

This manuscript has been released as a pre-print at https://www. biorxiv.org (Mothes et al., 2019).

REFERENCES

Ashall, L., Horton, C. A., Nelson, D. E., Paszek, P., Harper, C. V., Sillitoe, K., et al. (2009). Pulsatile stimulation determines timing and specificity of NF-kappaB-dependent transcription. *Science* 324, 242–246. doi: 10.1126/science.1164860

Basak, S., Behar, M., and Hoffmann, A. (2012). Lessons from mathematically modeling the NF-kappaB pathway. *Immunol. Rev.* 246, 221–238. doi: 10.1111/j.1600-065x.2011.01092.x

Batchelor, E., Loewer, A., Mock, C., and Lahav, G. (2011). Stimulus-dependent dynamics of p53 in single cells. *Mol. Syst. Biol.* 7:488. doi: 10.1038/msb.2011.20

Baum, K., Politi, A. Z., Kofahl, B., Steuer, R., and Wolf, J. (2016). Feedback, mass conservation and reaction kinetics impact the robustness of cellular oscillations. *PLoS Comput Biol.* 12:e1005298. doi: 10.1371/journal.pcbi.1005298

Benary, U., and Wolf, J. (2019). Controlling nuclear NF-kappaB dynamics by beta-TrCP-insights from a computational model. *Biomedicines* 7:40. doi: 10.3390/biomedicines7020040

Benary, U., Kofahl, B., Hecht, A., and Wolf, J. (2015). Mathematical modelling suggests a differential impact of beta-transducin repeat-containing protein paralogues on Wnt/beta-catenin signalling dynamics. *FEBS J.* 282, 1080–1096. doi: 10.1111/febs.13204

Cheong, R., Hoffmann, A., and Levchenko, A. (2008). Understanding NF-kappaB signaling via mathematical modeling. *Mol. Syst. Biol.* 4:192. doi: 10.1038/msb.2008.30

De, A., Dainichi, T., Rathinam, C. V., and Ghosh, S. (2014). The deubiquitinase activity of A20 is dispensable for NF-kappaB signaling. *EMBO Rep.* 15, 775–783. doi: 10.15252/embr.201338305

Fagerlund, R., Behar, M., Fortmann, K. T., Lin, Y. E., Vargas, J. D., and Hoffmann, A. (2015). Anatomy of a negative feedback loop: the case of IkappaBalpha. *J. R. Soc. Interf.* 12:0262.

Harper, C. V., Woodcock, D. J., Lam, C., Garcia-Albornoz, M., Adamson, A., Ashall, L., et al. (2018). Temperature regulates NF-kappaB dynamics and function through timing of A20 transcription. *Proc. Natl. Acad. Sci. U.S.A.* 115, E5243–E5249.

Hayden, M. S., and Ghosh, S. (2012). NF-kappaB, the first quarter-century: remarkable progress and outstanding questions. *Genes Dev.* 26, 203–234. doi: 10.1101/gad.183434.111

Hinz, M., and Scheidereit, C. (2014). The IκB kinase complex in NF−κB regulation and beyond. *EMBO Rep.* 15, 46–61. doi: 10.1002/embr.201337983

Hoffmann, A., Levchenko, A., Scott, M. L., and Baltimore, D. (2002). The IkappaB-NF-kappaB signaling module: temporal control and selective gene activation. *Science* 298, 1241–1245. doi: 10.1126/science.1071914

Huxford, T., Huang, D. B., Malek, S., and Ghosh, G. (1998). The crystal structure of the IkappaBalpha/NF-kappaB complex reveals mechanisms of NF-kappaB inactivation. *Cell* 95, 759–770.

Hymowitz, S. G., and Wertz, I. E. (2010). A20: from ubiquitin editing to tumour suppression. *Nat. Rev. Cancer* 10, 332–341. doi: 10.1038/nrc2775

Kochanczyk, M., Kocieniewski, P., Kozlowska, E., Jaruszewicz-Blonska, J., Sparta, B., Pargett, M., et al. (2017). Relaxation oscillations and hierarchy of feedbacks in MAPK signaling. *Sci. Rep.* 7:38244.

Lee, E. G., Boone, D. L., Chai, S., Libby, S. L., Chien, M., Lodolce, J. P., et al. (2000). Failure to regulate TNF-induced NF-kappaB and cell death responses in A20-deficient mice. *Science* 289, 2350–2354. doi: 10.1126/science.289.5488.2350

Lipniacki, T., and Kimmel, M. (2007). Deterministic and stochastic models of NFkappaB pathway. *Cardiovasc. Toxicol.* 7, 215–234. doi: 10.1007/s12012-007-9003-x

Lipniacki, T., Paszek, P., Brasier, A. R., Luxon, B., and Kimmel, M. (2004). Mathematical model of NF-kappaB regulatory module. *J. Theor. Biol.* 228, 195–215. doi: 10.1016/j.jtbi.2004.01.001

Lipniacki, T., Puszynski, K., Paszek, P., Brasier, A. R., and Kimmel, M. (2007). Single TNFalpha trimers mediating NF-kappaB activation: stochastic robustness of NF-kappaB signaling. *BMC Bioinform.* 8:376. doi: 10.1186/1471-2105-8-376

Llorens, M., Nuno, J. C., Rodriguez, Y., Melendez-Hevia, E., and Montero, F. (1999). Generalization of the theory of transition times in metabolic pathways: a geometrical approach. *Biophys. J.* 77, 23–36. doi: 10.1016/s0006-3495(99)76869-4

Longo, D. M., Selimkhanov, J., Kearns, J. D., Hasty, J., Hoffmann, A., and Tsimring, L. S. (2013). Dual delayed feedback provides sensitivity and robustness to the NF-kappaB signaling module. *PLoS Comput. Biol.* 9:e1003112. doi: 10.1371/journal.pcbi.1003112

Lork, M., Verhelst, K., and Beyaert, R. (2017). CYLD, A20 and OTULIN deubiquitinases in NF-kappaB signaling and cell death: so similar, yet so different. *Cell Death Differ.* 24, 1172–1183. doi: 10.1038/cdd.2017.46

Mothes, J., Busse, D., Kofahl, B., and Wolf, J. (2015). Sources of dynamic variability in NF-kappaB signal transduction: a mechanistic model. *Bioessays* 37, 452–462. doi: 10.1002/bies.201400113

Mothes, J., Ipenberg, I., Arslan, S. Ç, Benary, U., Scheidereit, C., and Wolf, J. (2019). A quantitative modular modeling approach reveals the consequences of different A20 feedback implementations for the NF-kB signaling dynamics. *bioRxiv* [Preprint].

Mukherjee, T., Chatterjee, B., Dhar, A., Bais, S. S., Chawla, M., Roy, P., et al. (2017). A TNF-p100 pathway subverts noncanonical NF-kappaB signaling in inflamed secondary lymphoid organs. *EMBO J.* 36, 3501–3516. doi: 10.15252/embj.201796919

Murakawa, Y., Hinz, M., Mothes, J., Schuetz, A., Uhl, M., Wyler, E., et al. (2015). RC3H1 post-transcriptionally regulates A20 mRNA and modulates the activity of the IKK/NF-kappaB pathway. *Nat. Commun.* 6:7367.

Nguyen, L. K. (2012). Regulation of oscillation dynamics in biochemical systems with dual negative feedback loops. *J. R. Soc. Interf.* 9, 1998–2010. doi: 10.1098/rsif.2012.0028

Perkins, N. D. (2012). The diverse and complex roles of NF-kappaB subunits in cancer. *Nat. Rev. Cancer* 12, 121–132. doi: 10.1038/nrc3204

Raue, A., Schilling, M., Bachmann, J., Matteson, A., Schelker, M., Kaschek, D., et al. (2013). Lessons learned from quantitative dynamical modeling in systems biology. *PLoS One* 8:e74335. doi: 10.1371/journal.pone.0074335

Skaug, B., Chen, J., Du, F., He, J., Ma, A., and Chen, Z. J. (2011). Direct, noncatalytic mechanism of IKK inhibition by A20. *Mol. Cell* 44, 559–571. doi: 10.1016/j.molcel.2011.09.015

Stilmann, M., Hinz, M., Arslan, S. C., Zimmer, A., Schreiber, V., and Scheidereit, C. (2009). A nuclear poly(ADP-ribose)-dependent signalosome confers DNA damage-induced IkappaB kinase activation. *Mol. Cell* 36, 365–378. doi: 10.1016/j.molcel.2009.09.032

Werner, S. L., Kearns, J. D., Zadorozhnaya, V., Lynch, C., O'Dea, E., Boldin, M. P., et al. (2008). Encoding NF-kappaB temporal control in response to TNF: distinct roles for the negative regulators IkappaBalpha and A20. *Genes Dev.* 22, 2093–2101. doi: 10.1101/gad.1680708

Wertz, I. E., Newton, K., Seshasayee, D., Kusam, S., Lam, C., Zhang, J., et al. (2015). Phosphorylation and linear ubiquitin direct A20 inhibition of inflammation. *Nature* 528, 370–375. doi: 10.1038/nature16165

Williams, R., Timmis, J., and Qwarnstrom, E. (2014). Computational models of the NF-κB signalling pathway. *Computation* 2:131. doi: 10.3390/computation2040131

Yilmaz, Z. B., Kofahl, B., Beaudette, P., Baum, K., Ipenberg, I., Weih, F., et al. (2014). Quantitative dissection and modeling of the NF-kappaB p100-p105 module reveals interdependent precursor proteolysis. *Cell Rep.* 9, 1756–1769. doi: 10.1016/j.celrep.2014.11.014

Zambrano, S., Bianchi, M. E., and Agresti, A. (2014). A simple model of NF-kappaB dynamics reproduces experimental observations. *J. Theor. Biol.* 347, 44–53.

Zhang, Z. B., Wang, Q. Y., Ke, Y. X., Liu, S. Y., Ju, J. Q., Lim, W. A., et al. (2017). Design of tunable oscillatory dynamics in a synthetic NF-kappaB signaling circuit. *Cell Syst.* 5, 460-70.e5.

18

New Data on Human Macrophages Polarization by *Hymenolepis diminuta* Tapeworm—An *In Vitro* Study

Anna Zawistowska-Deniziak[1]*, Katarzyna Basałaj[1], Barbara Strojny[2] and Daniel Młocicki[1,3]

[1] Witold Stefański Institute of Parasitology, Polish Academy of Sciences, Warsaw, Poland, [2] Division of Nanobiotechnology, Faculty of Animal Sciences, Department of Animal Feeding and Biotechnology, Warsaw University of Life Sciences, Warsaw, Poland, [3] Department of General Biology and Parasitology, Medical University of Warsaw, Warsaw, Poland

Edited by:
Diana Bahia,
Universidade Federal de Minas
Gerais, Brazil

Reviewed by:
David Voehringer,
University of Erlangen-Nuremberg,
Germany
Patricia Talamás-Rohana,
CINVESTAV, Mexico

*Correspondence:
Anna Zawistowska-Deniziak
anna.zawistowska@twarda.pan.pl

Helminths and their products can suppress the host immune response to escape host defense mechanisms and establish chronic infections. Current studies indicate that macrophages play a key role in the immune response to pathogen invasion. They can be polarized into two distinct phenotypes: M1 and M2. The present paper examines the impact of the adult *Hymenolepis diminuta* (HD) tapeworm and its excretory/secretory products (ESP) on THP-1 macrophages. Monocytes were differentiated into macrophages and cultured with a living parasite or its ESP. Our findings indicate that HD and ESP have a considerable impact on human THP-1 macrophages. Macrophages treated with parasite ESP (with or without LPS) demonstrated reduced expression of cytokines (i.e., IL-1α, TNFα, TGFβ, IL-10) and chemokines (i.e., IL-8, MIP-1α, RANTES, and IL-1ra), while s-ICAM and CxCL10 expression rose after ESP stimulation. In addition, inflammatory factor expression rose significantly when macrophages were exposed to living parasites. Regarding induced and repressed pathways, significant differences were found between HD and ESP concerning their influence on the phosphorylation of ERK1/2, STAT2, STAT3, AMPKα1, Akt 1/2/3 S473, Hsp60, and Hck. The superior immunosuppressive properties of ESP compared to HD were demonstrated with lower levels of IL-1β, TNF-α, IL-6, IL-23, and IL-12p70 following stimulation. The presence of HD and its ESP were found to stimulate mixed M1/M2 macrophage phenotypes. Our findings indicate new molecular mechanisms involved in the response of human macrophages to tapeworm infection, this could be a valuable tool in understanding the mechanisms underlying the processes of immune regulation during cestodiasis.

Keywords: human, macrophages, Cestoda, *Hymenolepis diminuta*, immunomodulation, immunology, host–parasite interactions

INTRODUCTION

Macrophages are versatile cells that play crucial roles in the complex process of the immune response to pathogen invasion. As macrophages are key modulator and effector cells in the immune response, their activation influences and responds to other arms of the immune system. It is generally considered that macrophages represent a spectrum of activated phenotypes rather than stable subpopulations (1–3). Typically, macrophages can be polarized into two distinct phenotypes: M1—classically activated macrophages induced by T helper 1 (Th1) cytokines, and M2—alternatively activated macrophages classified as M2a, M2b, M2c, and M2d induced by Th2 cytokines (1–3). Regulation of macrophage function and activity is essential to balance tissue homeostasis, forcing or solving inflammation in most disease processes. The inflammatory or anti-inflammatory activities of macrophages are shaped in a tissue- and signal-specific manner, enabling macrophages to induce various activation patterns and develop specific functional programs (4, 5).

Helminths are known to have coevolved with their hosts for millennia, and the principal goal of the adult parasite is, arguably, not to kill the host but to survive as long as possible by generating a state of tolerance. This state of affairs is beneficial for the parasitic organism as the host provides nutrition, protection, and stable conditions for growth. Therefore, dendritic cells and macrophages are among the first cells to be encountered by the parasite, which, by expressing certain molecules, has developed complex mechanisms to escape and modulate host immunity. One of these mechanisms exploits the impact of parasite surface proteins or their excretory/secretory products (ESP) on macrophage polarization type.

Both the M1 and M2 phenotypes are involved in the parasite invasion to various extents depending on parasite type and life cycle. In general, macrophages undergo a dynamic switch toward the M2 phenotype. In the case of *Taenia crassiceps*, while the M1 phenotype was observed during the early stage of infestation, the M2 phenotype later become dominant as the infection progressed, with a decreased parasite burden (6).

The influence of helminth-derived products on immune systems has been extensively studied (7–14), particularly with regard to the value of helminth products as antigens displaying immunomodulatory properties. The immunomodulatory properties of helminth-derived molecules have been screened for *Hymenolepis diminuta* (HD) (9, 15–18); these data show that HD may represent a source of anti-inflammatory and immunomodulatory molecules.

Experiments performed on animal models of human autoimmune diseases have shown that parasites can be beneficial and may have therapeutic potential in treatment of autoimmune disorders (10–12, 19, 20). Despite increasing knowledge of the influence of parasites on the host immune system, numerous mechanisms involved in this process seem to be unknown. Therefore, the ultimate goal of our study was to find new molecular pathways present in macrophages exposed to adult tapeworms. To achieve our goal, we used adult HD, commonly known as rat tapeworm, which is able to establish a chronic infection in the small intestine of the host with minimal influence on the intestinal tissue. *Hymenolepis* does not cause serious damage and influences the rat host immune system at the molecular level, producing proteins with antigenic properties (9, 15–18, 21–24). In addition, the ability to infect both animals and man makes this parasite a valuable model to study the influence of the parasite on its host, and since the regulatory mechanisms of rats and humans are comparable, host–parasite interactions such as immunomodulation can also be examined.

In light of the immunomodulatory properties of parasites and the importance of macrophages in numerous serious diseases, there is a need for more comprehensive research regarding the interactions and role of macrophages during parasite infections. Therefore, the aim of the present study was to characterize the polarization type of human THP-1 macrophages following stimulation with living HD and its ESP. We chose the THP-1 human leukemia monocytic cell line as it has been extensively used to study monocyte/macrophage functions, mechanisms, and signaling pathways. As our analysis was aimed at screening for changes and looking for new possible pathways induced by the parasite, we decided to use a cell line to select the most interesting factors, which will be carefully studied in the future using primary cells. The results obtained using this model are comparable to primary human PBMC–monocytes as indicated by a number of publications that have compared responses of both cell types and in most cases showed relatively similar response patterns (25–27). Certainly, THP-1 macrophages represent an alternative to PBMC–macrophages for screening purposes, when looking for new mechanisms and a homogeneous genetic background is wanted (28). This is especially the case when the availability of PBMC-derived macrophages is often limited, and insufficient quantities are available to perform broad analyses. Due to either financial or ethical constraints linked to animal and human *in vivo* studies, *ex vivo* or *in vitro* experiments become more relevant in initial screening research. Additionally, commercially available proteome arrays allow for comprehensive analysis where all experiments are performed in the same conditions. The dozens of analyzed factors allow for complex assessment and predictions regarding unstudied mechanisms. The THP-1 cell line has become a commonly used model to assess the modulation of macrophage activities and represents a competent *in vitro* model for estimation of the immunomodulatory properties of parasite proteins. For example, previous studies have utilized THP-1 cells to examine human monocyte/macrophage stimulation in response to parasite proteins (17, 29–31). A key novel aspect of the present study is that it is the first to comprehensively characterize the impact of the living parasite and its ESP on human THP-1 macrophages. The obtained results highlight the significance of a number of factors concerning the immunomodulatory properties of parasite proteins that have yet to be studied.

ANIMALS AND METHODS

Experimental Animals
Male Lewis rats aged about 3 months at the beginning of the experiment, to be used as experimental hosts, were kept in plastic cages in the animal house facilities of the Institute of Parasitology PAS. They had continuous access to food and water, and natural photoperiod conditions were provided.

Ethics Statement
All experimental procedures used in the present study had been preapproved by the third Local Ethical Committee for Scientific Experiments on Animals in Warsaw, Poland (resolution no. 51/2012, 30th of May 2012).

Cultivation of HD and Collection of ES Products

The HD strain was kept in the Institute of Parasitology PAS (strain WMS). Six-week-old cysticercoids reared in *Tribolium castaneum* beetles were fed in doses of 8–10 to 3-month-old rats (15 male rats). After 6 weeks, coproscopic examination of the rat feces was performed to ascertain the presence of adult parasites. To collect the adult parasites, the rats were euthanized with Thiopental anesthesia (Biochemie GmbH, Austria), administered in 100 mg/kg body weight (b.w.) intraperitoneally (i.p.).

Adult HD were obtained from the small intestine of infected rats and washed few times in PBS at room temperature to remove intestinal debris. The worms were incubated at 37°C in RPMI 1640 culture media containing penicillin and streptomycin (Sigma) for 10 h, with the media changed every 2 h. The harvested media containing ESP were pooled and centrifuged at 5,000 rpm for 15 min and directly placed in an Amicon® Ultra Centrifugal Filters Ultracel-3K (Millipore) to concentrate them. Protein concertation was determined with Bradford protein assay. Prepared ESP samples were stored at −80°C until used.

Cell Culture and Stimulation

The THP-1 human monocyte cell line was purchased from the American Type Culture Collection. Cells were maintained in culture medium (RPMI 1640 supplemented with 10% fetal bovine serum, 2 mM glutamine, 100 U/ml penicillin, 100 µg/ml streptomycin) at 37°C in a humidified atmosphere of 5% CO_2. The cells were seeded into six-well plates at a concentration of 1×10^6/ml in a whole volume of 4.8 ml/well. The cells were

differentiated into macrophages by the addition of 100 ng/ml phorbol 12-myristate 13-acetate (PMA) for 72 h. After differentiation, the cells were washed twice with fresh media w/o PMA and stimulated with ES or whole parasite. For whole parasite stimulation, the cells were maintained in Nunc polystyrene (PS) EasYFlask™ 25 cm² (Thermo Scientific) flasks at the same cell concentration and density per square centimeter. In the case of cells stimulated with parasite antigens and LPS, the cells were first treated with LPS (100 ng/ml), and parasite (one 10-cm worm/10×10^6 cells) or antigens (5 µg/ml) were added after 1 h. After 24 h, the stimulation culture media was collected and cells were washed with sterile PBS. Cells for phosphokinase analysis were lysed with lysis buffer from Proteome Profiler kit (R&D), cells for RNA isolation were directly treated with fenozol supplied with Total RNA kit (A&A Biotechnology) and stored at −80°C until use.

cDNA Synthesis and Real Time PCR Analysis

Total RNAs were isolated from the same number of cells stimulated with ES products or whole parasite according to the kit manufacturer's instructions. First-strand cDNAs were synthesized from 0.7 µg of total isolated RNA using a Maxima™ First Strand cDNA Synthesis Kit for RT-qPCR (Thermo Scientific). qPCR were performed by use of Luminaris Color HiGreen High ROX qPCR master Mix (Thermo Scientific). Reactions were conducted in 10 µl of total volume in StepOne Real-Time PCR System, Applied Biosystems.

Gene-specific primers, presented in **Table 1**, were intron-spanning and purchased from Sigma. Primer sequences were

TABLE 1 | Primers sequences used for Real-Time PCR.

Gene	Name	Forward primer	Reverse primer	Source
TNFα	Tumor necrosis factor alpha	5'CCCATGTTGTAGCAAACCCT	5'CCCTTGAAGAGGACCTGG	sd
IL-1β	Interleukin-1 beta	5'GGACAAGCTGAGGAAGATGC	5'TCGTTATCCCATGTGTCGAA	sd
IL-8	Interleukin-8	5'CAAACCTTTCCACCCCAAAT	5'CTCTGCACCCAGTTTTCCTT	sd
IL-12 p35	Interleukin-12 p35	5'GATGGCCCTGTGCCTTAGTA	5'TCAAGGGAGGATTTTTGTGG	(32)
IL-10	Interleukin-10	5'CCTGGAGGAGGTGATGCCCCA	5'CCTGCTCCACGGCCTTGCTC	sd
TGFβ	Transforming growth factor beta	5'TGCGCTTGAGATCTTCAAA	5'GGGCTAGTCGCACAGAACT	(32)
CCL1	CC chemokine type 1	5'ATACCAGCTCCATCTGCTCC	5'TGCCTCAGCATTTTTCTGTG	sd
CCL3	CC chemokine type 3 (MIP-1α)	5'ACTTTGAGACGAGCAGCCAGTG	5'TTTCTGGACCCACTCCTCACTG	sd
CCL4	CC chemokine type 4 (MIP-1β)	5'GTAGCTGCCTTCTGCTCTCC	5'ACCACAAAGTTGCGAGGAAG	sd
CCL22	CC chemokine type 22	5'ATTACGTCCGTTACCGTCTG	5'TAGGCTCTTCATTGGCTCAG	(32)
CCR7	C-C chemokine receptor type 7	5'GTGGTGGCTCTCCTTGTCAT	5'TGTGGTGTTGTCTCCGATGT	(32)
CXCL11	CXC chemokine type 11 (I-TAC)	5'CCTGGGGTAAAAGCAGTGAA	5'TGGGATTTAGGCATCGTTGT	(32)
CHI3L-1	Chitinase-3-like protein 1	5'GATAGCCTCCAACACCCAGA	5'AATTCGGCCTTCATTTCCTT	(32)
CD36	Cluster of differentiation 36	5'AGATGCAGCCTCATTTCCAC	5'GCCTTGGATGGAAGAACAAA	(32)
CD54	Cluster of differentiation 54 (sICAM)	5'GGCTGGAGCTGTTTGAGAAC	5'AGGAGTCGTTGCCATAGGTG	sd
IDO1	Indoleamine 2,3-dioxygenase 1	5'GCGCTGTTGGAAATAGCTTC	5'CAGGACGTCAAAGCACTGAA	(32)
IRF3	Interferon regulatory factor 3	5'AAGAAGGGTTGCGTTTAGCA	5'TCCCCAACTCCTGAGTTCAC	(32)
Klf4	Krueppel-like factor 4	5'CCCACACAGGTGAGAAACCT	5'ATGTGTAAGGCGAGGTGGTC	(32)
MRC1	Mannose receptor C type 1	5'GGCGGTGACCTCCACAAGTAT	5'ACGAAGCCATTTGGTAAACG	(32)
NFκB p65	Nuclear factor kappa B p65 (RelA)	5'TCTGCTTCCAGGTGACAGTG	5'ATCTTGAGCTCGGCAGTGTT	(32)
PPARc	Peroxisome proliferator-activated receptor c	5'TTCAGAAATGCCTTGCAGTG	5'CCAACAGCTTCTCCTTCTCG	(32)
MHC I	Major histocompatibility complex 1	5'GCAGTTGAGAGCCTACCTGG	5'CTCATGGTCAGAGATGGGGT	sd
MHC II	Major histocompatibility complex 2	5'AGGCAGCATTGAAGTCAGGT	5'CTGTGCAGATTCAGACCGTG	sd
RPL37A	Ribosomal protein L37a	5'ATTGAAATCAGCCAGCACGC	5'AGGAACCACAGTGCCAGATC	(33)
ACTB	β-actin	5'ATTGCCGACAGGATGCAGAA	5'GCTGATCCACATCTGCTGGAA	(33)

sd, self-designed primers.

designed or taken from Jaguin et al. (32). Two reference genes were used (β-actin, RPL37A) (33). All primer pairs were designed to have a melting point of about 64°C. Reaction runs included 2 min at 50°C and 10 min at 95°C followed by 40 cycles of a two-step PCR consisting of a denaturing phase at 95°C for 15 s and a combined annealing and extension phase at 72°C for 30 s. The C_T value of β-actin and RPL37A was subtracted from that of the gene of interest to obtain a ΔC_T value. The ΔC_T value of the least abundant sample at all time points for each gene was subtracted from the ΔC_T value of each sample to obtain a $\Delta\Delta C_T$ value. The gene expression level relative to the calibrator was expressed as $2^{-\Delta\Delta C_T}$ (34).

Phospho-Kinase Arrays

The phospho-antibody array analysis was performed using the Proteome Profiler Human Phospho-Kinase Array Kit from R&D Systems according to the manufacturer's instructions. After a 24-h stimulation period, macrophages were lysed with Lysis Buffer 6 (R&D Systems) and agitated for 30 min at 4°C. Cell lysates were clarified by microcentrifugation at $14,000 \times g$ for 5 min, and the supernatants were subjected to protein assay using a Pierce ™ BCA Protein Assay Kit (Thermo Scientific). Preblocked nitrocellulose membranes of the Human Phospho-Kinase arrays were incubated with ~400 μg (ES/whole parasite stimulation w/o LPS) or ~240 μg (stimulation with LPS) of cellular extract overnight at 4°C on a rocking platform. The membranes were washed three times with 1× Wash Buffer (R&D Systems) to remove the unbound proteins and were then incubated with a mixture of biotinylated detection antibodies and streptavidin-HRP antibodies. Chemiluminescent detection reagents were applied to detect spot densities. Membranes were exposed to X-ray film for 3, 5, and 10 min. Array images were analyzed using image analysis software Quantity One (Biorad).

Cytokine Arrays

The collected culture media from cells stimulated with parasite and ES products were subjected to the Proteome Profiler Human Cytokine Array Panel A (R&D Systems) according to the manufacturer's instructions. Each nitrocellulose membrane contains duplicated spots of 36 different antibodies for anticytokines, chemokines, growth factors, and adhesion proteins. Preblocked nitrocellulose membranes of the Human Cytokine Array were incubated with 1 ml of each culture media and detection antibody cocktail overnight at 4°C on a rocking platform. The membranes were washed three times with 1× Wash Buffer (R&D Systems) to remove unbound proteins. Chemiluminescent detection reagents were applied to detect spot densities. Membranes were exposed to X-ray film for 3, 5, and 10 min. Array images were analyzed using the image analysis software (Quantity One).

ELISAs

Cytokine (TNF-α, IL-1β, IL-6, IL-12p70, IL-10) concentrations were determined using the commercial ELISA kits OptEIA ™ Set Human (BD Biosciences) and DuoSet ELISA (R&D Systems) for IL-23. Supernatants were stored at −80°C until assayed. Experiments yielding supernatants were performed independently in triplicate. Optical densities were read at the appropriate wavelength on a microplate reader, and measurements were calculated as mean ± SE.

Statistical Analysis

ΔC_T values for all genes were normalized to mean C_T of β-actin and RPL37 reference genes. ΔC_T values for treated samples and controls (calibrators) were compared by t-test for independent samples. Differences at $P < 0.05$ were considered as significant. Analyses were performed using Statgraphics Centurion ver. XV (StatPoint Technologies, Warrenton, VA, USA) (***$P < 0.001$, **$P < 0.01$, *$P < 0.05$). The same t-test was used for the analysis of ELISA experiments.

RESULTS

HD and Its ESP Impact on THP-1 Macrophage Gene Expression Levels

Our findings indicate that HD ESP have a significant inhibitory effect on macrophage-originated inflammatory cytokines and chemokines. In order to evaluate the impact of the HD ESP on macrophage activation, their effect on the expression of pro-inflammatory and anti-inflammatory cytokines and chemokines mRNA was investigated. Stimulation of THP-1 macrophages with ESP, with or without LPS, significantly reduced the expression of *TNF-α*, *IL-1β*, *MIP-1α*, *MIP-1β*, *IL-8*, and *TGF-β* (**Figures 1** and **2**). The expression of scavenger receptor *CD36*, major histocompatibility complex (*MHC*) II, *CCR7*, and transcriptions factors *Klf4*, *IRF3*, and *NFκB p65* were also diminished.

The expression of the *CCL1*, *CCL22*, *CXCL11*, and *IL-10* genes differed depending on the addition of LPS to cells stimulated with ESP. *CCL1* and *CCL22* expression was downregulated, but upregulated in the presence of LPS. Additionally, *CHI3L-1*, *MHC I*, *HO-1*, and *IDO 1* expression was also dependent on LPS stimulation (**Figures 1** and **2**). However, treatment with the living parasite triggered a significantly different profile of cytokine expression. The presence of the parasite upregulated the expression of inflammatory cytokines and chemokines such as *TNF-α*, *IL-1β*, *MIP-1α*, *MIP-1β*, and *IL-8* (**Figure 1**). This increase in expression changed after stimulation with the parasite and LPS (**Figure 2**): *IL-8* expression was significantly reduced, and anti-inflammatory cytokines such as *TGF-β* and *IL-10* were upregulated. The expression of the scavenger receptor *CD36* and chemokine receptor *CCR7* were enhanced in cells treated with living parasite and LPS, and downregulated in those treated with parasite only. Analogous effects were noted for transcription factors *Klf4*, *IRF3*, and *NFκB p65*, while *HO-1* levels were comparable in all cells, irrespective of LPS stimulation.

Cytokine and Chemokine Protein Profile after HD and ESP Stimulation

Proteome Profiler cytokine array analysis confirmed inhibited MIP-1α expression in macrophages treated with ESP (ESP/M) and upregulation in macrophages cultured with living HD (HD/M) (**Figure 3**). The cytokine array analysis indicated that

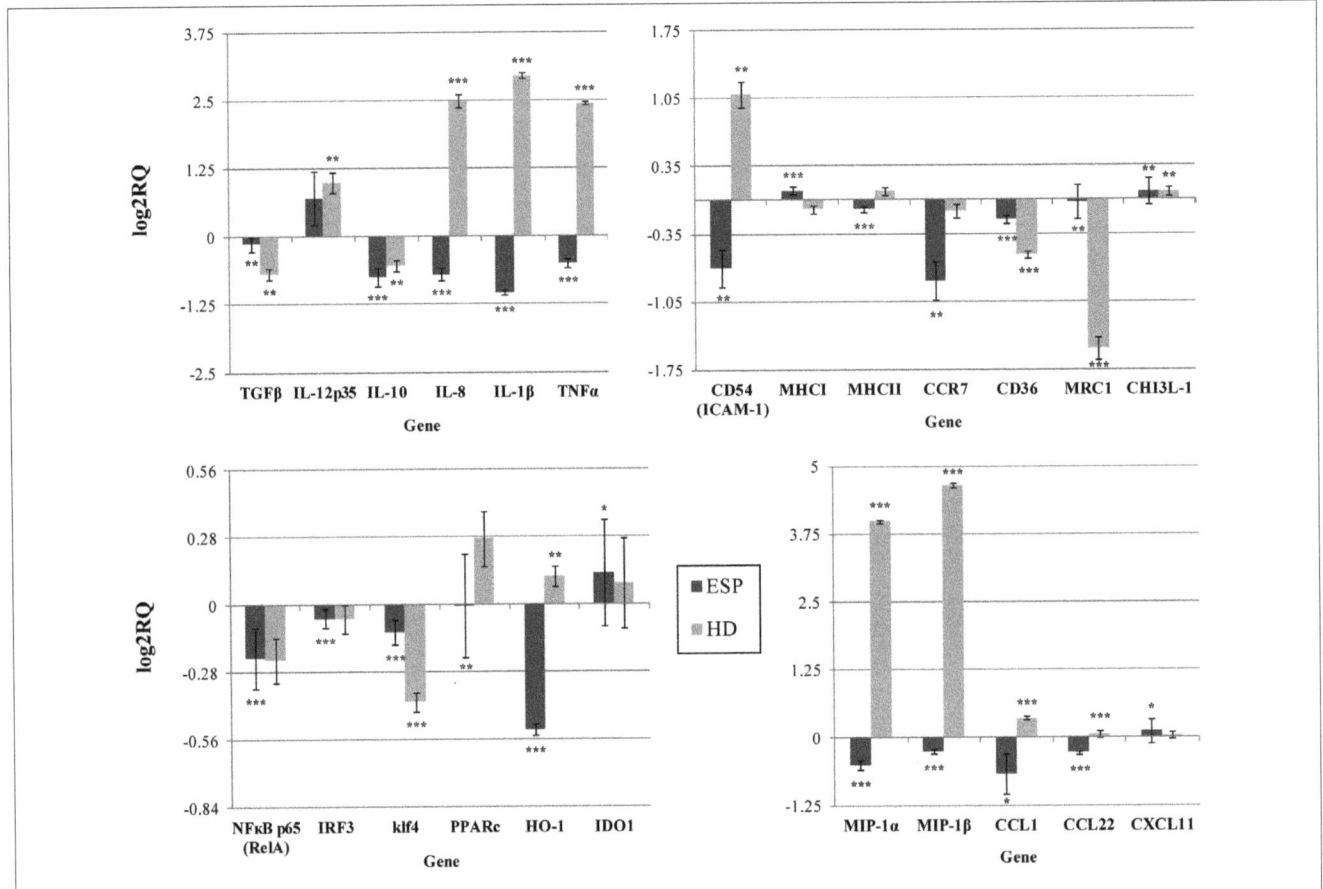

FIGURE 1 | Gene expression profile without LPS. Gene expression analysis was determined by qPCR. The results are calculated relative to control, where excretory/secretory products (ESP) and *Hymenolepis diminuta* (HD) had seperated one. ΔC_T values for all genes were normalized to mean C_T of β-actin and RPL37 housekeeping genes. ΔC_T values for treated samples and controls (calibrators) were compared by t-test for independent samples. Differences at $P < 0.05$ were considered as significant (***$P < 0.001$, **$P < 0.01$, *$P < 0.05$).

ESP and the live parasite have different effects on CXCL1, sICAM-1, and IL-1β levels in the cell culture medium. Additionally, the arrays reveal variations in the secretion of IL-1ra, MIF, and RANTES: MIF level was strongly induced by both types of stimulation, and IL-8, IL-1ra, SERPIN E1, and RANTES were reduced in comparison to unstimulated cells. While ESP stimulation strongly induced the production of CXCL10, minimal induction was seen in cells treated with the living parasite.

The levels of selected cytokines were further investigated with ELISA (**Figure 4**). No significant difference was observed between cells stimulated with ESP or HD, with or without LPS, with regard to TNF-α concentration. While IL-1β secretion rose in macrophages treated with either HD or HD + LPS compared to control cells, no significant changes were observed in those treated with ESP. IL-6 secretion fell in THP-1 macrophages cultured with ESP, ESP + LPS, and HD + LPS, but not in HD. Both the parasite and ESP suppressed the secretion of inflammatory cytokines such as IL-12p70 and IL-23. A similar effect was found for the anti-inflammatory cytokine IL-10, with the exception of the HD cultures, where the IL-10 level found to be higher than controls.

Changes in Kinase Phosphorylation Profiles in THP-1 Macrophages after ESP and HD Treatment

The screening analysis of the phosphorylation profiles of selected kinases in cells suggests that both ESP and HD have a similar effect. Only slight changes were observed when phosphorylation levels of selected kinases were increased more for cells stimulated with ESP than HD: p53 (S392, S46 but not for S15), Akt 1/2/3 (T308), β-catenin, STAT3 (Y705, S727), ERK1/2 (T202/Y204, T185/Y187), Hck (Y411), WNK1, and HSP60 (**Figure 5**; Figure S1 in Supplementary Material).

After LPS stimulation, the phosphorylation levels of p53, Akt1/2/3 T308, ERK1/2, and HSP60 were similar to those noted for cells without LPS (**Figure 6**). However, AMPKα1 (T188), Akt1/2/3 (S473), RSK1/2/3 (S380/S386/S377), Chk-2 (T68), p70S6 (T421/S424), STAT2 (Y689), and STAT6 (Y641) had increased phosphorylation in cells treated with ESP and LPS (**Figure 6**), but not in cells treated with ESP without LPS (**Figure 5**). The β-catenin, STAT3 (Y705, S727), Hck, and WNK1 in ESP + LPS cells demonstrated reduced levels of phosphorylation compared to HD + LPS cells, whereas the opposite was noted

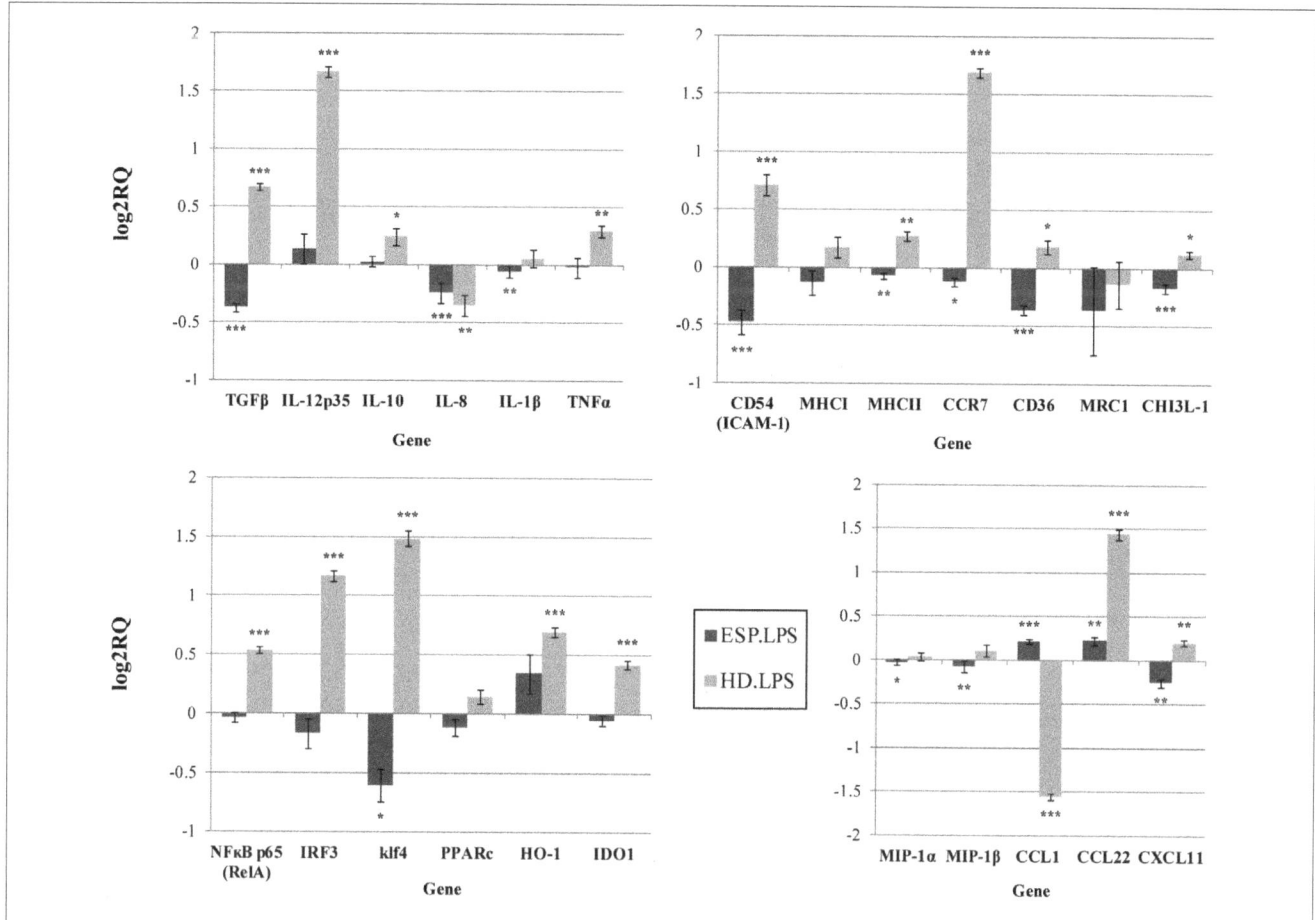

FIGURE 2 | Gene expression profile with LPS. Gene expression analysis was determined by qPCR. The results are calculated relative to the LPS control, where excretory/secretory products (ESP) + LPS and *Hymenolepis diminuta* (HD) + LPS had seperated one. ΔC_T values for all genes were normalized to mean C_T of β-actin and RPL37 housekeeping genes. ΔC_T values for treated samples and controls (calibrators) were compared by *t*-test for independent samples. Differences at $P < 0.05$ were considered as significant (***$P < 0.001$, **$P < 0.01$, *$P < 0.05$).

in cell cultures without LPS. HSP27, CREB c-JUN, JNK, and PYK2 demonstrated higher phosphorylation in HD-treated cells than ESP-treated cells, in which phosphorylation was lower or at the same level as control cells.

DISCUSSION

It is believed that parasitic proteins polarize macrophages principally toward type M2 (35), but the detailed mechanisms underlying this process are as yet unknown. Our results, however, indicate a mixed type of macrophage polarization, which is induced following exposure to HD ESP and the living parasite. Johnston et al. (17) described the general influence of HD antigens on THP-1 macrophages. Similarly to Johnston et al., we also identified anti-inflammatory properties of HD, as we observed inhibition of TNF-α, IL-6, and IL-1β after ESP stimulation. However, in our analyses, stimulation with whole living parasite induced proinflammatory cytokines, whereas Johnston et al. (17) observed no inflammatory stimulation with a high molecular mass extract of HD; there was no IL-1β stimulation, unlike with the HD treatment in our studies. Analysis of ERK1/2

phosphorylation revealed increased levels in ESP and HD, while with *Hd*HMW, there was no effect. Additionally, we also identified inhibition of proinflammatory molecules such as IL-12p70 and IL-23 in ESP and whole parasite treatment, which was not described in the previous reports. These additional results indicate that the anti-inflammatory properties of HD-derived antigens are stronger than was previously known. In relation to previous studies of the influence of the adult tapeworm on macrophages, our results draw attention to several new factors that have not been examined in this respect before. In our studies, we found several markers characteristic for both types of polarization, M1 and M2. This may indicate that parasites evolved mechanisms leading to diminished host reaction to foreign antigens, and this adaptation results in the stimulation of anti-inflammatory pathways in the host organism. Namely, as the host reacts to adult parasite surface antigens, helminths secrete a number of immunomodulatory molecules allowing them to avoid being expelled. This may account for the anti-inflammatory effect of HD ESP. Present data showed distinct cytokine and chemokine expression profiles in the macrophages stimulated with ESP and those with the living HD parasite, some of these results point to

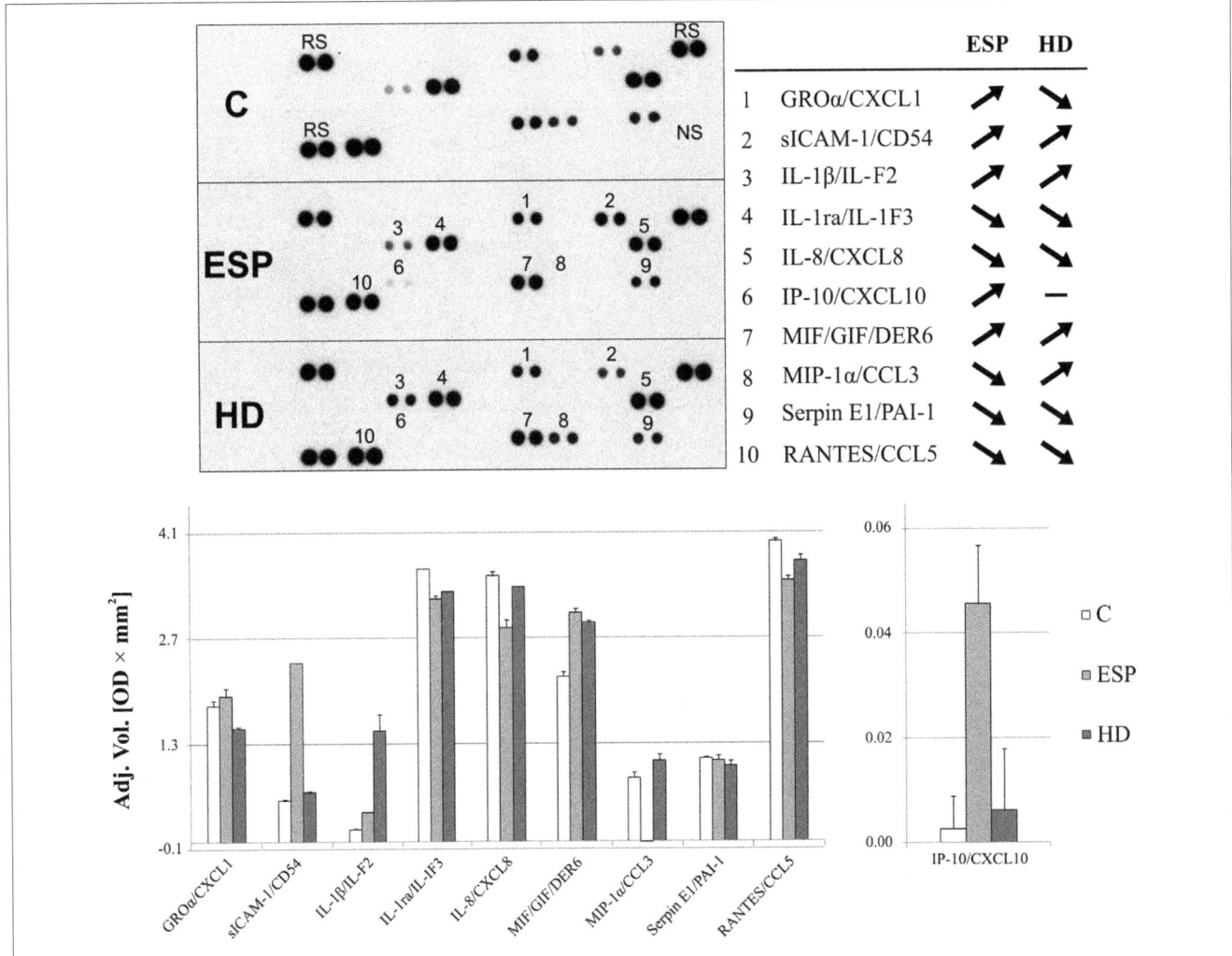

FIGURE 3 | Cytokine array analysis. Secreted protein profile in the supernatants of THP-1 macrophages stimulated 24 h with excretory/secretory products (ESP), *Hymenolepis diminuta* (HD), or unstimulated was determined by Protein Profiler Array Panel A. Supernatants of three independent experiments were mixed then subjected for the assay. As a result, we show the adjusted mean volume (OD × mm²) ± SD of two repeats on membrane. The average intensity of the pixels in background volume was calculated and subtracted from each pixel in all standard and unknown.

new aspects of parasite–host interactions, which are discussed below.

Our results suggest that while ESP, ESP + LPS, and HD + LPS stimulation inhibits the expression of the CXC motif chemokine ligand 8 (*CXCL-8*) gene, the presence of the parasite in the medium enhances the expression level, which is inconsistent with the result of antibody arrays. This might be partially explained by poor protein stability or post-translational mechanisms influencing proteins expression. Some of the factors and mechanisms reflecting the discrepancies in mRNA and protein expression are known and described (36). However, we are unable to establish which mechanism influencing protein abundance was present in our study. The function of IL-8/CXCL8 is to orchestrate the recruitment of neutrophils, basophils, and T-cells, but not monocytes, within inflamed tissues, and is also involved in neutrophil activation (37). The same may be true during the interaction of adult cestodes with host immune cells.

Another chemokine observed by us, which is important for the protective immunity of the host is macrophage inflammatory protein 1 α (MIP-1α/CCL3). The absence of this chemokine greatly impairs the recruitment of monocytes and neutrophils into infected organs. Furthermore, CCL3 induces macrophage activation and the killing of *Escherichia coli*, *Trypanosoma cruzi*, or *Klebsiella pneumonia* (38, 39). Our results indicate that stimulation of macrophages with HD ESP significantly reduces the level of the MIP-1α protein, as well as its gene expression, although different expression is observed when cells are incubated with the whole parasite. While the mRNA level is highly induced, the protein level reflects this trend in lesser extent, which may be associated with the relatively short half-life of CCL3 protein (40).

Our study showed that stimulation with HD and its ESP decreased the level of the CCL5 (RANTES) chemokine, which shares a common receptor with CCL3 and CCL4, namely,

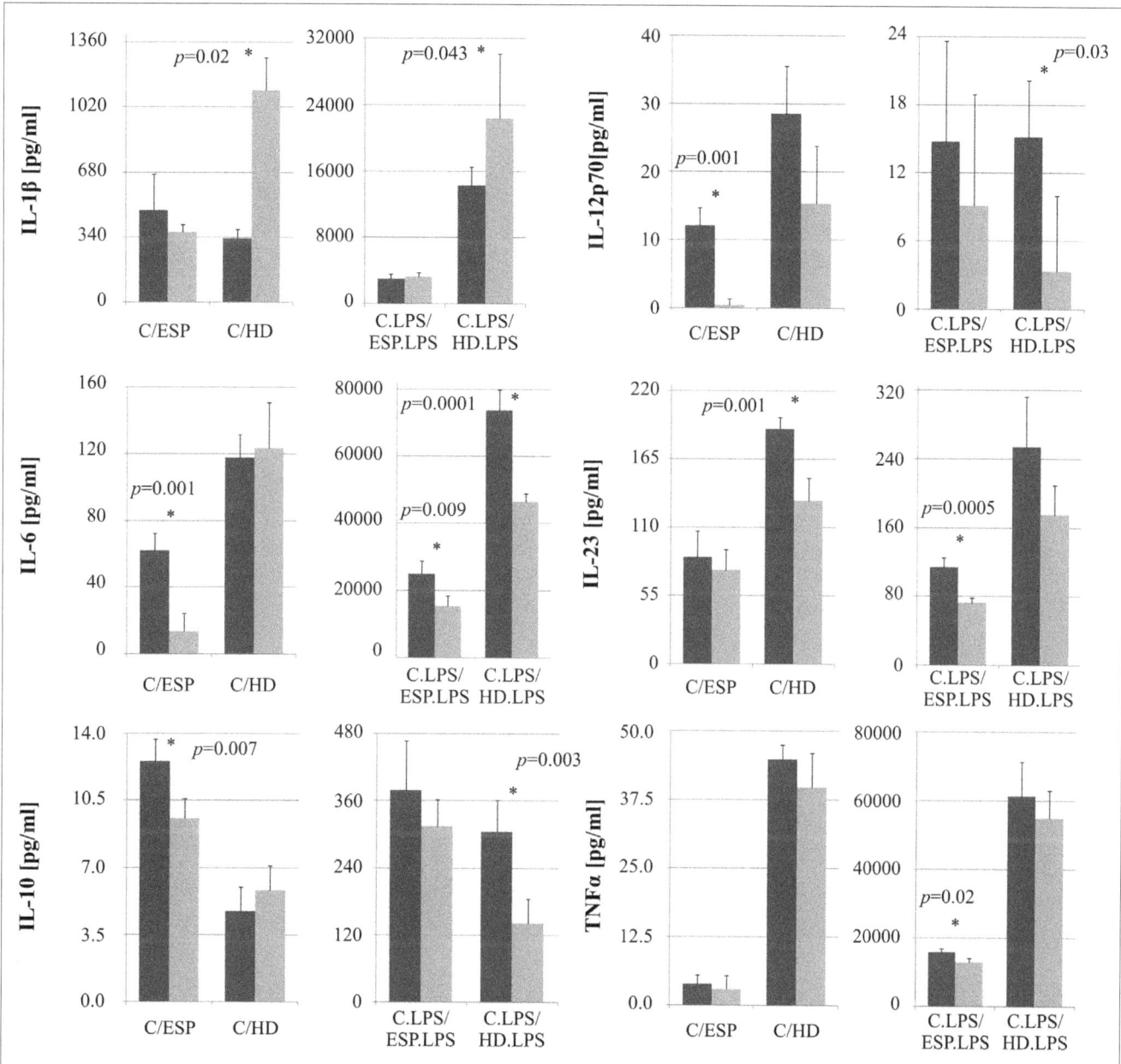

FIGURE 4 | Cytokine ELISA analysis. Cytokine production by monocytes THP-1 differentiated to macrophages and stimulated with *Hymenolepis diminuta* (HD) and ES products of this parasite [excretory/secretory products (ESP)]. Cells (1 × 10⁶/ml) were stimulated *in vitro* for 24 h with whole parasite or 5µg/ml ES products (*light gray bar*) or unstimulated (*dark gray*) or in combination with LPS (100 ng/ml). Results are expressed as a mean ± SD of four independent experiments. Statistical analysis was performed by Student's *t*-test. A value of *P* < 0.05 was considered to be significant.

chemokine receptor 5 (CCR5). RANTES is primarily involved in the migration of monocytes, neutrophils, dendritic cells, and T-cells. For instance, microbial challenge with *Toxoplasma gondii* is able to enhance the production of CCL3, CCL4, and CCL5. These chemokines activate CCR5 and signal the production of IL-12 by CD8α dendritic cells to initiate a Th1 response for clearance of the parasite (41). However, M2 macrophages obtained from mice implanted intraperitoneally with the filarial nematode

Brugia malayi display IL-4-dependent inhibition of the pro-inflammatory chemokines CCL3 and CCL4 (42). During *T. cruzi* infection, CCL5 plays an important protective role in mobilizing B cell populations and is directly able to induce B cell proliferation and IgM secretion (43).

According to the analysis of *CCL1* and *CCL22* gene expression levels in the present study, ESP-stimulated macrophages have the M1 phenotype and those exposed to living HD have

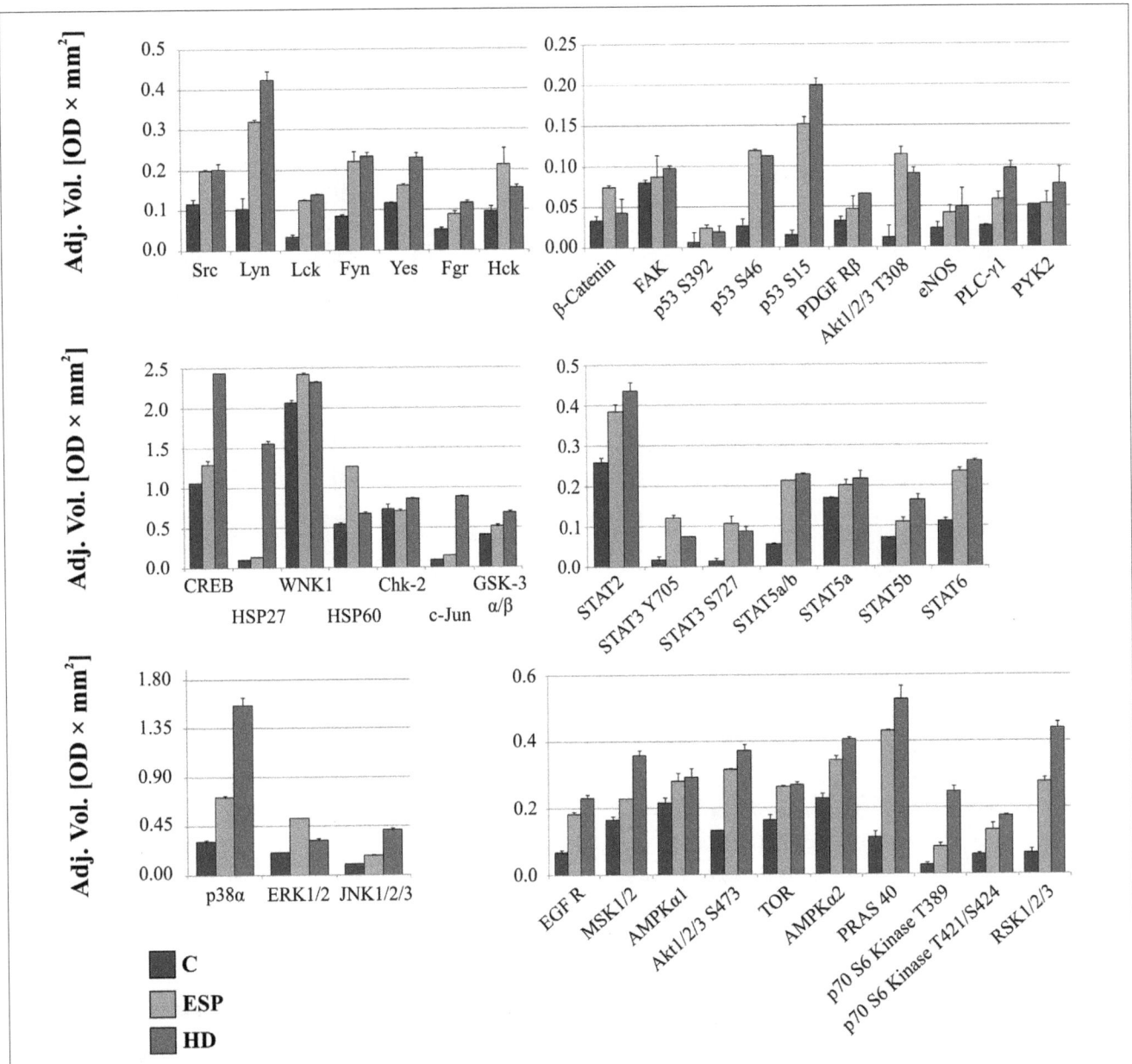

FIGURE 5 | Phospho-kinase analysis without LPS. Changes in signaling proteins phosphorylation profile in macrophages stimulated with excretory/secretory products (ESP) and *Hymenolepis diminuta* (HD) was determined by Proteome Profiler Human Phospho-Kinase Array Kit. Stimulated cells were lysed with usage of special kit buffer and frozen in −80°C until use. Protein concentration was checked and the same amout (400 μg) was used for each analysis. Analysis of control and treated sample has to be performed at one time. As a result we show the adjusted mean volume (OD × mm²) ± SD of two repeats on membrane. The average intensity of the pixels in background volume was calculated and subtracted from each pixel in all standard and unknown.

M2. Stimulation with LPS enhanced the expression of *CCL1* and *CCL22* in ESP/M and reduced *CCL1* in HD/M, whereas *CCL22* was highly induced in the HD/M system. CCL1 is known as an essential chemokine for the maintenance of M2b macrophage properties. CCL22 is a M2 phenotype marker, and together with CCL1, is a Treg-attracting chemokine (44, 45).

The reduced levels of IL-1β, TNF-α, IL-6, IL-23, and IL-12p70 indicate that ESP has superior immunosuppressive properties compared to the whole parasite. Although qPCR analysis confirms that ESP inhibits IL-1β and TNF-α, the ELISA results indicate the opposite for HD stimulation without LPS, where qPCR data indicated higher TNF-α expression. However, it is known that high levels of mRNA expression do not necessarily reflect the amount of protein in a cell. Regulation of gene expression at transcriptional and translational levels (such as alternative splicing, RNA stability influenced by regulatory elements, different half-lifes of the proteins during different conditions or modification) can lead to a weak correlation between

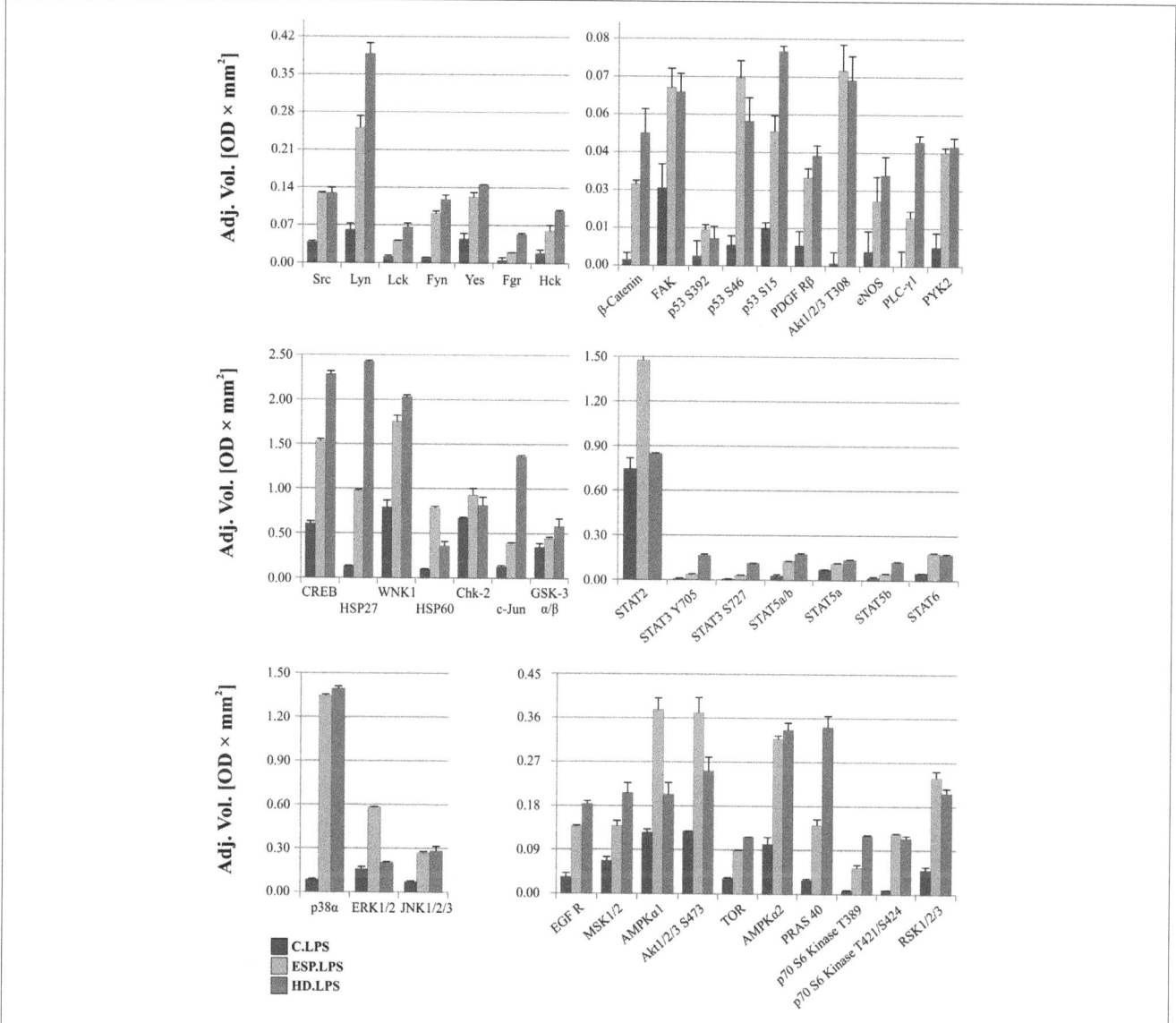

FIGURE 6 | **Phospho-kinase analysis with LPS**. Changes in signaling proteins phosphorylation profile in macrophages stimulated with excretory/secretory products (ESP)/LPS and *Hymenolepis diminuta* (HD)/LPS was determined by Proteome Profiler Human Phospho-Kinase Array Kit. Stimulated cells were lysed with usage of special kit buffer and frozen in −80°C until use. Protein concentration was checked and the same amout (240 μg) was used for each analysis. Analysis of control and treated sample has to be performed at one time. As a result, we show the adjusted mean volume (OD × mm²) ± SD of two repeats on membrane. The average intensity of the pixels in background volume was calculated and subtracted from each pixel in all standard and unknown.

mRNA and protein levels. Despite the progress in methodology, still little is known about specificity of translation regulation, feedback and coupling between regulatory, the roles of miRNAs, RNA-binding proteins, and yet unknown mechanisms of protein abundance regulation (36). For instance, microRNAs can simultaneously downregulate hundreds of genes by inhibiting mRNA translation into protein and thus modulating many cellular processes (46).

Discrepancies are also present regarding the levels of IL-10 and IL-12; the ELISA test examined the p70 subunit, containing p40 and p35, while qPRC examined the p35 subunit. While qPCR indicated greater expression of the IL-12 p35 subunit in all

cases, ELISA analysis revealed inhibition of IL-12 p70. The IL-12 is composed of the p35 (encoded by *Il12a*) and p40 (encoded by *Il12b*) chains and principally activates NK cells and induces CD4+ T lymphocytes to become IFN-γ-producing Th1 cells (47). The p40 chain can also form a dimer with p19 to give rise to IL-23, which is required for Th17 differentiation (48, 49). Similarly, the p35 chain can combine with Epstein–Barr-induced 3 (EBI3) to form IL-35 in induced regulatory T cells (iTr35) and tolerogenic human DCs (50). *T. muris* infection has been shown to induce the expansion of suppressive IL-35-producing CD4+ Foxp3⁻ "Tr35" cells in the murine intestine (51). Our analysis identified high expression of the *IL-12p35* subunit and ELISA validation revealed inhibition of IL-12p70 and IL-23, which may suggest that high expression of p35 is connected with the induction of IL-35, however, this has to be elucidated in future experiments.

Another effect of HD ESP is diminished expression of ICAM-1 (CD54), a transmembrane protein expressed on epithelial cells, endothelial cells, and immune cells such as T cells and macrophages. ICAM-1 enables leukocytes to migrate through the endothelia to the inflammation site (52), participates in immunological synapse formation (53), and it is also implicated in the formation and progression of atherosclerotic lesions (54) and development of experimentally induced intestinal inflammation (55). Interestingly, our study revealed an increased level of sICAM-1 in culture media from ESP-treated cells. This supports the findings of previous studies which suggest that ICAM-1 is not only expressed on the cell membrane (mICAM-1), but is also released as a soluble molecule (sICAM-1), possibly resulting from proteolytic cleavage or alternative RNA splicing (56, 57). This may explain the difference in expression levels of mRNA and protein, especially when there are reports about the presence of separate distinct messenger RNA transcripts coding for mICAM-1 and sICAM-1 (58). The binding of sICAM-1 to LFA-1 is capable of inhibiting lymphocyte attachment to endothelial cells (59); however, the role and functions of soluble ICAM-1 have not yet been completely elucidated. Some analyses have revealed that sICAM-1 plays a role in neutrophil inhibition and macrophage recruitment during inflammation. sICAM-1 can act as a regulator during inflammatory processes. Excessive circulating sICAM-1 in transgenic animals may bind to β2 integrin on the leukocyte and thus decrease its availability for cell–cell interactions (60). All of the above could represent a parasite immune evasion strategy aimed to block leukocyte:endothelial cells interactions.

Recent experiments show that helminth derived molecules may reduce lupus-associated accelerated atherosclerosis in a mouse model (61) and offer strong protection against cholesterol-induced atherosclerosis development (62). Our data show a correlation between the presence of parasite and expression of a macrophage CD36 receptor. This macrophage scavenger receptor is responsible for recognition and internalization of oxidized lipids, and represents a major participant in atherosclerotic foam cell formation (63). In addition, human studies have shown CD36 to be associated with impaired insulin sensitivity (64–66) and pathogenesis of metabolic disorders such as insulin resistance, obesity, and non-alcoholic hepatic steatosis, and an absence of CD36-mediated lipid uptake in muscle or liver is capable of preventing diet-induced lipotoxicity (67–69). Our results show evident reduction in the expression of CD36 in cells stimulated with ESP and HD. For those stimulated with ESP, downregulation was also observed in the presence of LPS. This may indicate one of the mechanisms used by parasites for immunomodulation and could explain the beneficial effects of parasites on atherosclerosis and metabolic disorders.

As HD ESP demonstrated greater downregulation of gene expression for inflammatory cytokines and chemokines, it was decided to analyze the phosphorylation profiles of a range of signaling proteins from various pathways.

Changes in phosphorylation profiles were observed after stimulation with ESP and HD. The phosphorylation level of the following proteins was elevated after HD ESP treatment, irrespective of LPS application: ERK 1/2, Akt 1/2/3 T308, p53 S392, p53 S46, and HSP 60. Studies concerning the lacto-N-fucopentaose III (LNFPIII) carbohydrate moiety present on S. mansoni eggs and ES-62 reveal that sustained ERK activation can suppress Th1-inducing IL-12 production (70). Inhibition of the production of the shared p40 subunit, mediated by ERK, can also cause downregulation of IL-12p70 and IL-23 cytokines, as pretreatment of cells with an ERK pathway inhibitor can reduce IL-12p40 production (71–73).

Another important molecule is β-catenin, a ubiquitously expressed main signal transducer of the canonical Wnt signaling pathway. Activation of the Wnt/β-catenin pathway with Wnt3a in mouse microglial cells leads to the expression and release of the pro-inflammatory cytokines interleukins IL-6, IL-12, and IFNγ (74). In contrast, the Wnt/β-catenin pathway has also been demonstrated to play an anti-inflammatory role in mouse colon epithelial stem cells and macrophages infected with Salmonella (75) or Mycobacterium (76), which indicates that activation of the Wnt/β-catenin pathway downregulates the pro-inflammatory responses to certain bacterial infections (77, 78). Our results show that β-catenin has a higher phosphorylation level in ESP-stimulated cells, which could contribute to the immunosuppressive properties of these proteins.

Significantly greater induction of phosphorylation was observed in c-JUN, HSP27, and CREB after stimulation with HD compared to ESP. Several studies attribute an anti-inflammatory function to HSP27. Stimulation of THP-1 macrophages with recombinant HSP27 resulted in increased NF-κB transcriptional activity and induced the expression of a variety of genes, including the pro-inflammatory factors IL-1β and TNF-α. However, it was also found to increase the expression of anti-inflammatory factors including IL-10 and GM-CSF (79). Our results reveal greater expression of IL-1β at both the gene and protein levels, and increased expression of TNF-α and IL-10 at the gene level following stimulation with HD and LPS. Induction of the inflammatory cytokine might be via the JNK pathway, where c-Jun is a component of the AP-1 transcription factor. This pathway appears to play a significant role in chronic inflammatory diseases involving the expression of specific proteases and cytokines (80, 81); however, the expression of pro-inflammatory cytokines is independent of the JNK/AP-1 signaling cascade in human neutrophils (82). In addition to these findings, HD incubation with cells induces transcription factor CREB, which is known for its role in cell proliferation, differentiation, and survival (83–85). However, recent evidence has revealed its function in immune responses, including inhibiting NF-κB activation, inducing macrophage survival and promoting the proliferation, survival, and regulation of T and B lymphocytes. While some studies identify CREB as a part of the anti-inflammatory immune response (86), others associate it with the pro-inflammatory response (87). As the qPCR analysis revealed upregulation of inflammatory cytokine and chemokine expression in the case of HD treatment, our present results favor a pro-inflammatory response.

STAT2, AMPKα1, and Akt 1/2/3 S473 demonstrated greater phosphorylation following ESP + LPS compared to HD + LPS. While STAT2 may be a novel regulator in the immunosuppressive function of mesenchymal stem cells (88), AMPKα1 is crucial for phagocytosis-induced macrophage skewing from a pro- to anti-inflammatory phenotype at the time of resolution of inflammation (89). Akt is a major metabolic regulator implicated in M2 activation (90, 91). It mediates enhanced glucose consumption in M2 macrophages, which contributes to induction of M2 gene expression (92). These results indicate the induction of an M2 phenotype in the presence of ESP and LPS.

CONCLUSION

We examined several proteins/kinases, which have never been considered in molecular studies devoted to host–parasite interactions, and which are now intended for more in-depth study in subsequent experiments to assess their relevance in the immune response. The information revealed in this report may allow for the discovery of new signaling pathways, or improve our understanding of those already identified. With knowledge of the beneficial effects of parasites on many immune related diseases, such research can contribute to a better understanding of these diseases and the mechanisms underlying them. As shown above, our results suggest the presence of markers for both M1 and M2 macrophage phenotypes; therefore, we can conclude that infections with adult tapeworms induce mixed polarization of macrophages. This may explain the phenomenon where adult cestodes, although attached to the host intestinal epithelium with their adhesive structures and billions of microtriches covering the surface layer of the parasite tegument do not harm the hosts and usually do not induce an inflammatory reaction. We hope our results pave the way for future in-depth studies to find and elucidate novel mechanisms involved in parasite immunomodulation, especially regarding helminth infections that are known to have a considerable influence on a number of serious autoimmune diseases, in which a careful experimental approach is necessary. Bearing in mind that our previous report confirmed that HD ES products have immunogenic properties (18), our next step will be focused on the characteristics and careful analysis of immunomodulatory functions of single identified proteins. This may be essential to establish the molecules involved in the mechanisms of immunomodulation and determine the mechanism of their action.

AUTHOR CONTRIBUTIONS

AZ-D, KB, BS, and DM performed the experiments; designed the experiments; interpreted the data; drafted the manuscript; reviewed and approved the final version of the manuscript; agreed to be accountable for the content of the work.

ACKNOWLEDGMENTS

We wish to thank Dr. Luke J. Norbury (Institute of Parasitology PAS, Poland) for his critical comments and final English proofreading.

FUNDING

Financial support for this study was provided by the National Science Centre, Poland (grant 2012/05/B/NZ6/00769).

REFERENCES

1. Mosser DM, Edwards JP. Exploring the full spectrum of macrophage activation. *Nat Rev Immunol* (2008) 8(12):958–69. doi:10.1038/nri2448
2. Grinberg S, Hasko G, Wu D, Leibovich SJ. Suppression of PLCbeta2 by endotoxin plays a role in the adenosine A(2A) receptor-mediated switch of macrophages from an inflammatory to an angiogenic phenotype. *Am J Pathol* (2009) 175(6):2439–53. doi:10.2353/ajpath.2009.090290
3. Ferrante CJ, Pinhal-Enfield G, Elson G, Cronstein BN, Hasko G, Outram S, et al. The adenosine-dependent angiogenic switch of macrophages to an M2-like phenotype is independent of interleukin-4 receptor alpha (IL-4Ralpha) signaling. *Inflammation* (2013) 36(4):921–31. doi:10.1007/s10753-013-9621-3
4. Murray PJ, Allen JE, Biswas SK, Fisher EA, Gilroy DW, Goerdt S, et al. Macrophage activation and polarization: nomenclature and experimental guidelines. *Immunity* (2014) 41(1):14–20. doi:10.1016/j.immuni.2014.06.008
5. Schultze JL, Freeman T, Hume DA, Latz E. A transcriptional perspective on human macrophage biology. *Semin Immunol* (2015) 27(1):44–50. doi:10.1016/j.smim.2015.02.001
6. Peon AN, Espinoza-Jimenez A, Terrazas LI. Immunoregulation by *Taenia crassiceps* and its antigens. *Biomed Res Int* (2013) 2013:498583. doi:10.1155/2013/498583
7. Baska P, Zawistowska-Deniziak A, Zdziarska AM, Wasyl K, Wisniewski M, Cywinska A, et al. *Fasciola hepatica* – the pilot study of *in vitro* assessing immune response against native and recombinant antigens of the fluke. *Acta Parasitol* (2013) 58(4):453–62. doi:10.2478/s11686-013-0163-5
8. Rehman ZU, Knight JS, Koolaard J, Simpson HV, Pernthaner A. Immunomodulatory effects of adult *Haemonchus contortus* excretory/secretory products on human monocyte-derived dendritic cells. *Parasite Immunol* (2015) 37(12):657–69. doi:10.1111/pim.12288
9. McKay DM. The immune response to and immunomodulation by *Hymenolepis diminuta. Parasitology* (2010) 137(3):385–94. doi:10.1017/S0031182009990886
10. Okada H, Kuhn C, Feillet H, Bach JF. The 'hygiene hypothesis' for autoimmune and allergic diseases: an update. *Clin Exp Immunol* (2010) 160(1):1–9. doi:10.1111/j.1365-2249.2010.04139.x
11. Wiria AE, Sartono E, Supali T, Yazdanbakhsh M. Helminth infections, type-2 immune response, and metabolic syndrome. *PLoS Pathog* (2014) 10(7):e1004140. doi:10.1371/journal.ppat.1004140
12. Weinstock JV, Elliott DE. Helminths and the IBD hygiene hypothesis. *Inflamm Bowel Dis* (2009) 15(1):128–33. doi:10.1002/ibd.20633
13. Maizels RM, Yazdanbakhsh M. Immune regulation by helminth parasites: cellular and molecular mechanisms. *Nat Rev Immunol* (2003) 3(9):733–44. doi:10.1038/nri1183
14. Maizels RM. Parasitic helminth infections and the control of human allergic and autoimmune disorders. *Clin Microbiol Infect* (2016) 22(6):481–6. doi:10.1016/j.cmi.2016.04.024
15. Reyes JL, Fernando MR, Lopes F, Leung G, Mancini NL, Matisz CE, et al. IL-22 restrains tapeworm-mediated protection against experimental colitis

via regulation of IL-25 expression. *PLoS Pathog* (2016) 12(4):e1005481. doi:10.1371/journal.ppat.1005481

16. Graepel R, Leung G, Wang A, Villemaire M, Jirik FR, Sharkey KA, et al. Murine autoimmune arthritis is exaggerated by infection with the rat tapeworm, *Hymenolepis diminuta*. *Int J Parasitol* (2013) 43(7):593–601. doi:10.1016/j.ijpara.2013.02.006

17. Johnston MJ, Wang A, Catarino ME, Ball L, Phan VC, MacDonald JA, et al. Extracts of the rat tapeworm, *Hymenolepis diminuta*, suppress macrophage activation *in vitro* and alleviate chemically induced colitis in mice. *Infect Immun* (2010) 78(3):1364–75. doi:10.1128/IAI.01349-08

18. Bien J, Salamatin R, Sulima A, Savijoki K, Bruce Conn D, Nareaho A, et al. Mass spectrometry analysis of the excretory-secretory (E-S) products of the model cestode *Hymenolepis diminuta* reveals their immunogenic properties and the presence of new E-S proteins in cestodes. *Acta Parasitol* (2016) 61(2):429–42. doi:10.1515/ap-2016-0058

19. Guigas B, Molofsky AB. A worm of one's own: how helminths modulate host adipose tissue function and metabolism. *Trends Parasitol* (2015) 31(9):435–41. doi:10.1016/j.pt.2015.04.008

20. Hussaarts L, Garcia-Tardon N, van Beek L, Heemskerk MM, Haeberlein S, van der Zon GC, et al. Chronic helminth infection and helminth-derived egg antigens promote adipose tissue M2 macrophages and improve insulin sensitivity in obese mice. *FASEB J* (2015) 29(7):3027–39. doi:10.1096/fj.14-266239

21. Reyes JL, Lopes F, Leung G, Mancini NL, Matisz CE, Wang A, et al. Treatment with cestode parasite antigens results in recruitment of CCR2+ myeloid cells, the adoptive transfer of which ameliorates colitis. *Infect Immun* (2016) 84(12):3471–83. doi:10.1128/IAI.00681-16

22. Matisz CE, Leung G, Reyes JL, Wang A, Sharkey KA, McKay DM. Adoptive transfer of helminth antigen-pulsed dendritic cells protects against the development of experimental colitis in mice. *Eur J Immunol* (2015) 45(11):3126–39. doi:10.1002/eji.201545579

23. Lopes F, Reyes JL, Wang A, Leung G, McKay DM. Enteric epithelial cells support growth of *Hymenolepis diminuta in vitro* and trigger TH2-promoting events in a species-specific manner. *Int J Parasitol* (2015) 45(11):691–6. doi:10.1016/j.ijpara.2015.05.004

24. Reyes JL, Wang A, Fernando MR, Graepel R, Leung G, van Rooijen N, et al. Splenic B cells from *Hymenolepis diminuta*-infected mice ameliorate colitis independent of T cells and via cooperation with macrophages. *J Immunol* (2015) 194(1):364–78. doi:10.4049/jimmunol.1400738

25. Gao XX, Wang BX, Fei XF, Zhang J, Gong YJ, Minami M, et al. Effects of polysaccharides (FI0-c) from mycelium of *Ganoderma tsugae* on proinflammatory cytokine production by THP-1 cells and human PBMC (II). *Acta Pharmacol Sin* (2000) 21(12):1186–92.

26. Han XQ, Chung Lap Chan B, Dong CX, Yang YH, Ko CH, Gar-Lee Yue G, et al. Isolation, structure characterization, and immunomodulating activity of a hyperbranched polysaccharide from the fruiting bodies of *Ganoderma sinense*. *J Agric Food Chem* (2012) 60(17):4276–81. doi:10.1021/jf205056u

27. Schroecksnadel S, Jenny M, Fuchs D. Myelomonocytic THP-1 cells for *in vitro* testing of immunomodulatory properties of nanoparticles. *J Biomed Nanotechnol* (2011) 7(1):209–10. doi:10.1166/jbn.2011.1272

28. Chanput W, Mes JJ, Wichers HJ. THP-1 cell line: an *in vitro* cell model for immune modulation approach. *Int Immunopharmacol* (2014) 23(1):37–45. doi:10.1016/j.intimp.2014.08.002

29. Jain SK, Sahu R, Walker LA, Tekwani BL. A parasite rescue and transformation assay for antileishmanial screening against intracellular *Leishmania donovani* amastigotes in THP1 human acute monocytic leukemia cell line. *J Vis Exp* (2012) 30(70):4054. doi:10.3791/4054

30. Gov L, Karimzadeh A, Ueno N, Lodoen MB. Human innate immunity to *Toxoplasma gondii* is mediated by host caspase-1 and ASC and parasite GRA15. *MBio* (2013) 4(4):e255–13. doi:10.1128/mBio.00255-13

31. Baska P, Wisniewski M, Krzyzowska M, Dlugosz E, Zygner W, Gorski P, et al. Molecular cloning and characterisation of *in vitro* immune response against astacin-like metalloprotease Ace-MTP-2 from *Ancylostoma ceylanicum*. *Exp Parasitol* (2013) 133(4):472–82. doi:10.1016/j.exppara.2013.01.006

32. Jaguin M, Houlbert N, Fardel O, Lecureur V. Polarization profiles of human M-CSF-generated macrophages and comparison of M1-markers in classically activated macrophages from GM-CSF and M-CSF origin. *Cell Immunol* (2013) 281(1):51–61. doi:10.1016/j.cellimm.2013.01.010

33. Maess MB, Sendelbach S, Lorkowski S. Selection of reliable reference genes during THP-1 monocyte differentiation into macrophages. *BMC Mol Biol* (2010) 11:90. doi:10.1186/1471-2199-11-90

34. Livak KJ, Schmittgen TD. Analysis of relative gene expression data using real-time quantitative PCR and the 2(-Delta Delta C(T)) method. *Methods* (2001) 25(4):402–8. doi:10.1006/meth.2001.1262

35. Kelly B, O'Neill LA. Metabolic reprogramming in macrophages and dendritic cells in innate immunity. *Cell Res* (2015) 25(7):771–84. doi:10.1038/cr.2015.68

36. Vogel C, Marcotte EM. Insights into the regulation of protein abundance from proteomic and transcriptomic analyses. *Nat Rev Genet* (2012) 13(4):227–32. doi:10.1038/nrg3185

37. Xu L, Kitade H, Ni Y, Ota T. Roles of chemokines and chemokine receptors in obesity-associated insulin resistance and nonalcoholic fatty liver disease. *Biomolecules* (2015) 5(3):1563–79. doi:10.3390/biom5031563

38. Standiford TJ, Kunkel SL, Greenberger MJ, Laichalk LL, Strieter RM. Expression and regulation of chemokines in bacterial pneumonia. *J Leukoc Biol* (1996) 59(1):24–8.

39. Villalta F, Zhang Y, Bibb KE, Kappes JC, Lima MF. The cysteine-cysteine family of chemokines RANTES, MIP-1alpha, and MIP-1beta induce trypanocidal activity in human macrophages via nitric oxide. *Infect Immun* (1998) 66(10):4690–5.

40. Hillyer P, Male D. Expression of chemokines on the surface of different human endothelia. *Immunol Cell Biol* (2005) 83(4):375–82. doi:10.1111/j.1440-1711.2005.01345.x

41. DiPietro LA, Burdick M, Low QE, Kunkel SL, Strieter RM. MIP-1alpha as a critical macrophage chemoattractant in murine wound repair. *J Clin Invest* (1998) 101(8):1693–8. doi:10.1172/JCI1020

42. Loke P, Nair MG, Parkinson J, Guiliano D, Blaxter M, Allen JE. IL-4 dependent alternatively-activated macrophages have a distinctive *in vivo* gene expression phenotype. *BMC Immunol* (2002) 3:7. doi:10.1186/1471-2172-3-7

43. Sullivan NL, Eickhoff CS, Zhang X, Giddings OK, Lane TE, Hoft DF. Importance of the CCR5-CCL5 axis for mucosal *Trypanosoma cruzi* protection and B cell activation. *J Immunol* (2011) 187(3):1358–68. doi:10.4049/jimmunol.1100033

44. Mantovani A, Locati M, Polentarutti N, Vecchi A, Garlanda C. Extracellular and intracellular decoys in the tuning of inflammatory cytokines and Toll-like receptors: the new entry TIR8/SIGIRR. *J Leukoc Biol* (2004) 75(5):738–42. doi:10.1189/jlb.1003473

45. Riezu-Boj JI, Larrea E, Aldabe R, Guembe L, Casares N, Galeano E, et al. Hepatitis C virus induces the expression of CCL17 and CCL22 chemokines that attract regulatory T cells to the site of infection. *J Hepatol* (2011) 54(3):422–31. doi:10.1016/j.jhep.2010.07.014

46. Bartel DP. microRNAs: target recognition and regulatory functions. *Cell* (2009) 136(2):215–33. doi:10.1016/j.cell.2009.01.002

47. Trinchieri G. Interleukin-12: a cytokine produced by antigen-presenting cells with immunoregulatory functions in the generation of T-helper cells type 1 and cytotoxic lymphocytes. *Blood* (1994) 84(12):4008–27.

48. Oppmann B, Lesley R, Blom B, Timans JC, Xu Y, Hunte B, et al. Novel p19 protein engages IL-12p40 to form a cytokine, IL-23, with biological activities similar as well as distinct from IL-12. *Immunity* (2000) 13(5):715–25. doi:10.1016/S1074-7613(00)00070-4

49. Teng MW, Bowman EP, McElwee JJ, Smyth MJ, Casanova JL, Cooper AM, et al. IL-12 and IL-23 cytokines: from discovery to targeted therapies for immune-mediated inflammatory diseases. *Nat Med* (2015) 21(7):719–29. doi:10.1038/nm.3895

50. Dixon KO, van der Kooij SW, Vignali DA, van Kooten C. Human tolerogenic dendritic cells produce IL-35 in the absence of other IL-12 family members. *Eur J Immunol* (2015) 45(6):1736–47. doi:10.1002/eji.201445217

51. Collison LW, Chaturvedi V, Henderson AL, Giacomin PR, Guy C, Bankoti J, et al. IL-35-mediated induction of a potent regulatory T cell population. *Nat Immunol* (2010) 11(12):1093–101. doi:10.1038/ni.1952

52. Muller WA. Mechanisms of leukocyte transendothelial migration. *Annu Rev Pathol* (2011) 6:323–44. doi:10.1146/annurev-pathol-011110-130224

53. Grakoui A, Bromley SK, Sumen C, Davis MM, Shaw AS, Allen PM, et al. The immunological synapse: a molecular machine controlling T cell activation. *Science* (1999) 285(5425):221–7. doi:10.1126/science.285.5425.221

54. Bourdillon MC, Poston RN, Covacho C, Chignier E, Bricca G, McGregor JL. ICAM-1 deficiency reduces atherosclerotic lesions in double-knockout

mice (ApoE(-/-)/ICAM-1(-/-)) fed a fat or a chow diet. *Arterioscler Thromb Vasc Biol* (2000) 20(12):2630–5. doi:10.1161/01.ATV.20.12.2630

55. Bendjelloul F, Maly P, Mandys V, Jirkovska M, Prokesova L, Tuckova L, et al. Intercellular adhesion molecule-1 (ICAM-1) deficiency protects mice against severe forms of experimentally induced colitis. *Clin Exp Immunol* (2000) 119(1):57–63. doi:10.1046/j.1365-2249.2000.01090.x

56. King PD, Sandberg ET, Selvakumar A, Fang P, Beaudet AL, Dupont B. Novel isoforms of murine intercellular adhesion molecule-1 generated by alternative RNA splicing. *J Immunol* (1995) 154(11):6080–93.

57. Budnik A, Grewe M, Gyufko K, Krutmann J. Analysis of the production of soluble ICAM-1 molecules by human cells. *Exp Hematol* (1996) 24(2):352–9.

58. Whiteman SC, Bianco A, Knight RA, Spiteri MA. Human rhinovirus selectively modulates membranous and soluble forms of its intercellular adhesion molecule-1 (ICAM-1) receptor to promote epithelial cell infectivity. *J Biol Chem* (2003) 278(14):11954–61. doi:10.1074/jbc.M205329200

59. Rieckmann P, Michel U, Albrecht M, Bruck W, Wockel L, Felgenhauer K. Soluble forms of intercellular adhesion molecule-1 (ICAM-1) block lymphocyte attachment to cerebral endothelial cells. *J Neuroimmunol* (1995) 60(1–2):9–15. doi:10.1016/0165-5728(95)00047-6

60. Wang HW, Babic AM, Mitchell HA, Liu K, Wagner DD. Elevated soluble ICAM-1 levels induce immune deficiency and increase adiposity in mice. *FASEB J* (2005) 19(8):1018–20. doi:10.1096/fj.04-3094fje

61. Aprahamian TR, Zhong X, Amir S, Binder CJ, Chiang LK, Al-Riyami L, et al. The immunomodulatory parasitic worm product ES-62 reduces lupus-associated accelerated atherosclerosis in a mouse model. *Int J Parasitol* (2015) 45(4):203–7. doi:10.1016/j.ijpara.2014.12.006

62. Wolfs IM, Stoger JL, Goossens P, Pottgens C, Gijbels MJ, Wijnands E, et al. Reprogramming macrophages to an anti-inflammatory phenotype by helminth antigens reduces murine atherosclerosis. *FASEB J* (2014) 28(1):288–99. doi:10.1096/fj.13-235911

63. Park YM. CD36, a scavenger receptor implicated in atherosclerosis. *Exp Mol Med* (2014) 46:e99. doi:10.1038/emm.2014.38

64. An P, Freedman BI, Hanis CL, Chen YD, Weder AB, Schork NJ, et al. Genome-wide linkage scans for fasting glucose, insulin, and insulin resistance in the National Heart, Lung, and Blood Institute Family Blood Pressure Program: evidence of linkages to chromosome 7q36 and 19q13 from meta-analysis. *Diabetes* (2005) 54(3):909–14. doi:10.2337/diabetes.54.3.909

65. Love-Gregory L, Sherva R, Sun L, Wasson J, Schappe T, Doria A, et al. Variants in the CD36 gene associate with the metabolic syndrome and high-density lipoprotein cholesterol. *Hum Mol Genet* (2008) 17(11):1695–704. doi:10.1093/hmg/ddn060

66. Bonen A, Tandon NN, Glatz JF, Luiken JJ, Heigenhauser GJ. The fatty acid transporter FAT/CD36 is upregulated in subcutaneous and visceral adipose tissues in human obesity and type 2 diabetes. *Int J Obes (Lond)* (2006) 30(6):877–83. doi:10.1038/sj.ijo.0803212

67. Goudriaan JR, Dahlmans VE, Teusink B, Ouwens DM, Febbraio M, Maassen JA, et al. CD36 deficiency increases insulin sensitivity in muscle, but induces insulin resistance in the liver in mice. *J Lipid Res* (2003) 44(12):2270–7. doi:10.1194/jlr.M300143-JLR200

68. Hajri T, Han XX, Bonen A, Abumrad NA. Defective fatty acid uptake modulates insulin responsiveness and metabolic responses to diet in CD36-null mice. *J Clin Invest* (2002) 109(10):1381–9. doi:10.1172/JCI0214596

69. Koonen DP, Jacobs RL, Febbraio M, Young ME, Soltys CL, Ong H, et al. Increased hepatic CD36 expression contributes to dyslipidemia associated with diet-induced obesity. *Diabetes* (2007) 56(12):2863–71. doi:10.2337/db07-0907

70. Harnett W, Harnett MM. Helminth-derived immunomodulators: can understanding the worm produce the pill? *Nat Rev Immunol* (2010) 10(4):278–84. doi:10.1038/nri2730

71. Klotz C, Ziegler T, Figueiredo AS, Rausch S, Hepworth MR, Obsivac N, et al. A helminth immunomodulator exploits host signaling events to regulate cytokine production in macrophages. *PLoS Pathog* (2011) 7(1):e1001248. doi:10.1371/journal.ppat.1001248

72. Al-Riyami L, Harnett W. Immunomodulatory properties of ES-62, a phosphorylcholine-containing glycoprotein secreted by *Acanthocheilonema viteae*. *Endocr Metab Immune Disord Drug Targets* (2012) 12(1):45–52. doi:10.2174/187153012799278893

73. Goodridge HS, Harnett W, Liew FY, Harnett MM. Differential regulation of interleukin-12 p40 and p35 induction via Erk mitogen-activated protein kinase-dependent and -independent mechanisms and the implications for bioactive IL-12 and IL-23 responses. *Immunology* (2003) 109(3):415–25. doi:10.1046/j.1365-2567.2003.01689.x

74. Halleskog C, Mulder J, Dahlstrom J, Mackie K, Hortobagyi T, Tanila H, et al. WNT signaling in activated microglia is proinflammatory. *Glia* (2011) 59(1):119–31. doi:10.1002/glia.21081

75. Liu X, Lu R, Wu S, Sun J. *Salmonella* regulation of intestinal stem cells through the Wnt/beta-catenin pathway. *FEBS Lett* (2010) 584(5):911–6. doi:10.1016/j.febslet.2010.01.024

76. Neumann J, Schaale K, Farhat K, Endermann T, Ulmer AJ, Ehlers S, et al. Frizzled1 is a marker of inflammatory macrophages, and its ligand Wnt3a is involved in reprogramming *Mycobacterium tuberculosis*-infected macrophages. *FASEB J* (2010) 24(11):4599–612. doi:10.1096/fj.10-160994

77. Umar S, Sarkar S, Wang Y, Singh P. Functional cross-talk between beta-catenin and NFkappaB signaling pathways in colonic crypts of mice in response to progastrin. *J Biol Chem* (2009) 284(33):22274–84. doi:10.1074/jbc.M109.020941

78. Sun J, Hobert ME, Duan Y, Rao AS, He TC, Chang EB, et al. Crosstalk between NF-kappaB and beta-catenin pathways in bacterial-colonized intestinal epithelial cells. *Am J Physiol Gastrointest Liver Physiol* (2005) 289(1):G129–37. doi:10.1152/ajpgi.00515.2004

79. Salari S, Seibert T, Chen YX, Hu T, Shi C, Zhao X, et al. Extracellular HSP27 acts as a signaling molecule to activate NF-kappaB in macrophages. *Cell Stress Chaperones* (2013) 18(1):53–63. doi:10.1007/s12192-012-0356-0

80. Karin M, Gallagher E. From JNK to pay dirt: jun kinases, their biochemistry, physiology and clinical importance. *IUBMB Life* (2005) 57(4–5):283–95. doi:10.1080/15216540500097111

81. Karin M, Lawrence T, Nizet V. Innate immunity gone awry: linking microbial infections to chronic inflammation and cancer. *Cell* (2006) 124(4):823–35. doi:10.1016/j.cell.2006.02.016

82. Cloutier A, Ear T, Borissevitch O, Larivee P, McDonald PP. Inflammatory cytokine expression is independent of the c-Jun N-terminal kinase/AP-1 signaling cascade in human neutrophils. *J Immunol* (2003) 171(7):3751–61. doi:10.4049/jimmunol.171.7.3751

83. Shaywitz AJ, Greenberg ME. CREB: a stimulus-induced transcription factor activated by a diverse array of extracellular signals. *Annu Rev Biochem* (1999) 68:821–61. doi:10.1146/annurev.biochem.68.1.821

84. Mayr B, Montminy M. Transcriptional regulation by the phosphorylation-dependent factor CREB. *Nat Rev Mol Cell Biol* (2001) 2(8):599–609. doi:10.1038/35085068

85. Sakamoto KM, Frank DA. CREB in the pathophysiology of cancer: implications for targeting transcription factors for cancer therapy. *Clin Cancer Res* (2009) 15(8):2583–7. doi:10.1158/1078-0432.CCR-08-1137

86. Wen AY, Sakamoto KM, Miller LS. The role of the transcription factor CREB in immune function. *J Immunol* (2010) 185(11):6413–9. doi:10.4049/jimmunol.1001829

87. Westbom CM, Shukla A, MacPherson MB, Yasewicz EC, Miller JM, Beuschel SL, et al. CREB-induced inflammation is important for malignant mesothelioma growth. *Am J Pathol* (2014) 184(10):2816–27. doi:10.1016/j.ajpath.2014.06.008

88. Yi T, Lee DS, Jeon MS, Kwon SW, Song SU. Gene expression profile reveals that STAT2 is involved in the immunosuppressive function of human bone marrow-derived mesenchymal stem cells. *Gene* (2012) 497(2):131–9. doi:10.1016/j.gene.2012.01.073

89. Mounier R, Theret M, Arnold L, Cuvellier S, Bultot L, Goransson O, et al. AMPKalpha1 regulates macrophage skewing at the time of resolution of inflammation during skeletal muscle regeneration. *Cell Metab* (2013) 18(2):251–64. doi:10.1016/j.cmet.2013.06.017

90. Byles V, Covarrubias AJ, Ben-Sahra I, Lamming DW, Sabatini DM, Manning BD, et al. The TSC-mTOR pathway regulates macrophage polarization. *Nat Commun* (2013) 4:2834. doi:10.1038/ncomms3834

91. Ruckerl D, Jenkins SJ, Laqtom NN, Gallagher IJ, Sutherland TE, Duncan S, et al. Induction of IL-4Ralpha-dependent microRNAs identifies PI3K/Akt signaling as essential for IL-4-driven murine macrophage proliferation *in vivo*. *Blood* (2012) 120(11):2307–16. doi:10.1182/blood-2012-02-408252

19

Pasteurella multocida Toxin Triggers RANKL-Independent Osteoclastogenesis

Sushmita Chakraborty[1], Bianca Kloos[1], Ulrike Harre[2], Georg Schett[2] and Katharina F. Kubatzky[1]*

[1] Zentrum für Infektiologie, Medizinische Mikrobiologie und Hygiene, Universitätsklinikum Heidelberg, Heidelberg, Germany,
[2] Department of Internal Medicine 3, Institute of Clinical Immunology, University of Erlangen-Nuremberg, Erlangen, Germany

Edited by:
Diana Bahia,
Universidade Federal de Minas
Gerais, Brazil

Reviewed by:
Carl Goodyear,
University of Glasgow, UK
Ricardo Silvestre,
University of Minho, Portugal

*Correspondence:
Katharina F. Kubatzky
kubatzky@uni-heidelberg.de

Bone remodeling is a continuous process to retain the structural integrity and function of the skeleton. A tight coupling is maintained between osteoclast-mediated resorption of old or damaged bones and osteoblast-mediated formation of new bones for bone homeostasis. While osteoblasts differentiate from mesenchymal stem cells, osteoclasts are hematopoietic in origin and derived from myeloid precursor cells. Osteoclast differentiation is driven by two cytokines, cytokine receptor activator of NF-κB ligand (RANKL), and macrophage colony-stimulating factor. Imbalances in the activity of osteoblasts and osteoclasts result in the development of bone disorders. Bacterially caused porcine atrophic rhinitis is characterized by a loss of nasal ventral conche bones and a distortion of the snout. While *Bordetella bronchiseptica* strains cause mild and reversible symptoms, infection of pigs with toxigenic *Pasteurella multocida* strains causes a severe and irreversible decay. The responsible virulence factor *Pasteurella multocida* toxin (PMT) contains a deamidase activity in its catalytical domain that constitutively activates specific heterotrimeric G proteins to induce downstream signaling cascades. While osteoblasts are inhibited by the toxin, osteoclasts are activated, thus skewing bone remodeling toward excessive bone degradation. Still, the mechanism by which PMT interferes with bone homeostasis, and the reason for this unusual target tissue is not yet well understood. Here, we show that PMT has the potential to differentiate bone marrow-derived macrophages into functional osteoclasts. This toxin-mediated differentiation process is independent of RANKL, a cytokine believed to be indispensable for triggering osteoclastogenesis, as addition of osteoprotegerin to PMT-treated macrophages does not show any effect on PMT-induced osteoclast formation. Although RANKL is not a prerequisite, toxin-primed macrophages show enhanced responsiveness to low concentrations of RANKL, suggesting that the PMT-generated microenvironment offers conditions where low concentrations of RANKL lead to an increase in the number of osteoclasts resulting in increased resorption. PMT-mediated release of the osteoclastogenic cytokines such as IL-6 and TNF-α, but not IL-1, supports the differentiation process. Although the production of cytokines and the subsequent activation of signaling cascades are necessary for PMT-mediated differentiation into osteoclasts, they are not sufficient and PMT-induced activation of G protein signaling is essential for efficient osteoclastogenesis.

Keywords: bacterial toxin, osteoclast, immune evasion, pro-inflammatory cytokines, G protein, *Pasteurella multocida* toxin

INTRODUCTION

Bone is a dynamic tissue, which is constantly remodeled to regulate its structural integrity and functions. Bone remodeling is a physiological process to maintain homeostasis and involves the removal of old or damaged bone structures by osteoclasts and the subsequent replacement of new bone by osteoblasts. A tight coupling of bone resorption to bone formation is maintained in normal bone homeostasis. However, bacterial infections can derail this process and result in bone diseases like caries, periodontitis, osteomyelitis, septic arthritis, lyme disease, and atrophic rhinitis (AR) (1). Bacteria can mediate bone damage by (a) releasing substances, which can directly destroy the bone matrix, (b) by activating osteoclasts or inhibiting osteoblasts through their surface or secreted components, (c) by stimulating inflammatory cells, which subsequently activate osteoclasts resulting in enhanced bone resorption, and (d) by invading osteoblasts resulting in inhibition of bone matrix formation and dysregulation of bone remodeling (2).

One of the most potent bacterial toxin involved in bone destruction is *Pasteurella multocida* toxin (PMT), which is produced by capsular type D or some type A strains of *Pasteurella multocida* (3). PMT is the causative agent of porcine AR, an economically important disease, characterized by degeneration of nasal turbinate bones, leading to a shortening or twisting of the snout (4). In humans, few and rare cases of septic arthritis have been reported due to *P. multocida* infection, and these are usually caused by animal bites (5, 6). PMT is a 146 kDa protein and acts as a potent mitogen for cell types such as osteoclasts or fibroblasts (7, 8). In piglets, administration of purified PMT on nasal conche is sufficient to cause AR (9). The toxin mediates AR through interfering with bone biogenesis by promoting osteoclast differentiation and proliferation resulting in increased bone resorption, while inhibiting osteoblast differentiation and bone regeneration on the other hand (10–14). The mode of action of PMT on host cells is through activation of heterotrimeric G protein families $G\alpha_{q/11}$, $G\alpha_{i1-3}$, and $G\alpha_{12/13}$ but not through $G\alpha_s$ activation (15). PMT induces the permanent activation of heterotrimeric G proteins of the host by deamidation of a conserved glutamine residue in the switch II region of the α subunit, critical to maintain the intrinsic GTPase activity of G protein (16). As a consequence of constitutive activation of host G proteins by PMT, signaling cascades such Map kinase, JAK-STAT, or PI3 kinase pathway are activated, which result in mitogenesis, increased survival, and cytoskeletal reorganization (17). We hypothesize that bone destruction is a side-effect of toxin-induced production of cytokines in an attempt to modulate signal transduction pathways of innate and adaptive immune cells to avoid immune recognition (18).

Osteoclasts differentiate from hematopoietic precursors of the monocyte/macrophage lineage through the action of two cytokines, RANKL and macrophage colony-stimulating factor (M-CSF). Upon stimulation of their cognate receptors, a cascade of signaling events is initiated leading to the activation of transcription factors such as nuclear factor of activated T cells, calcineurin dependent 1 (NFATc1), NF-kB, and AP-1, resulting in the fusion of precursor cells and expression of genes for

osteoclast functions (19). Mice deficient in the cytokines M-CSF and RANKL or their cognate receptors display a prominent osteopetrotic phenotype. Osteoprotegerin (OPG) is a soluble receptor that competes with RANK for RANKL and protects the skeleton from excessive bone resorption. Overexpression or administration of OPG in mice results in profound osteopetrosis by reducing osteoclastogenesis (20). Under physiological conditions the RANKL/OPG ratio is balanced to maintain skeletal integrity.

Osteoclastogenic plasticity has been observed across the myeloid lineage ranging from early myeloid precursors to monocytes, macrophages, and dendritic cells, which can differentiate into tartrate-resistant acid phosphatase (TRAP)-positive osteoclasts in the presence of soluble RANKL (sRANKL) and M-CSF (21–23). Studies from our group and others have shown the osteoclastogenic property of PMT in cell lines or heterogeneous precursor populations (24–26); therefore, we decided to investigate the potential of PMT in inducing osteoclastogenesis in a homogeneous population of bone marrow-derived macrophages (BMDMs) to unravel the mechanism of osteoclast differentiation in more detail. We show that PMT drives the differentiation of BMDMs into TRAP-positive cells independent of RANKL-RANK signaling, as treatment of PMT-stimulated BMDMs with OPG did not abrogate osteoclast formation. In addition, our investigation shows that PMT-induced osteoclastogenesis is modulated by cytokines and their downstream signaling pathways that are necessary but not sufficient for efficient differentiation of macrophages into osteoclasts.

MATERIALS AND METHODS

Ethics Statement

All animal studies were approved by the Regierungspräsidium Karlsruhe, Germany.

Mice

C57BL/6 wild-type mice were purchased from Janvier Labs (Le Genest St. Isle, France), and IL-1R-deficient mice were obtained from Jackson Laboratories (Bar Harbor, ME, USA). Mice were maintained under SPF conditions in accordance with the German policies on animal welfare.

Reagents

Tissue culture reagents were purchased from Biochrom GmbH (Berlin, Germany), PAA laboratories, PAN biotech (Aidenbach, Germany), Merck and Sigma, respectively. Antibodies against phosphorylated c-Jun (p-c-Jun) (Ser63), p-NF-kB (Ser536), phosphorylated STAT-3 (Tyr705), HistonH3, and β-actin were purchased from Cell Signaling Technology (Frankfurt, Germany). Antibodies against NFATc1 and Gq were procured from Santa Cruz Biotechnology (Heidelberg, Germany). An antibody that recognizes the Q209E modification of $G\alpha_q$ was a gift of Prof. S. Kamitami (Osaka, Japan). Secondary HRP-linked antibodies were obtained from Cell Signaling Technology (anti-rabbit IgG, anti-Mouse IgG) or Santa Cruz (anti-rat IgG). FITC-conjugated anti-CD11b, FITC-conjugated anti-CD80, FITC-conjugated anti-CD40 were from BD Biosciences (Heidelberg, Germany). PE-conjugated anti-CD86 and PECy7-conjugated anti-MHCII were purchased from eBioscience (Frankfurt, Germany). PECy7-conjugated anti-F4/80 and PE-conjugated anti-RANK and the corresponding isotype control were purchased from BioLegend (San Diego, CA, USA).

Rat anti-mouse Interleukin 6 receptor (IL-6R) antibody was a gift from Chugai Pharmaceutical, Tokyo, Japan. Etanercept (Enbrel; Pfizer) used in blocking experiment of TNF-α was generously provided by Prof. G. Schett (Erlangen, Germany) and Prof. H. Lorenz (Heidelberg, Germany). The Anti-IL-6 antibody for blocking IL-6 signaling was obtained from BioXCell (West Lebanon, USA).

PCR primers were purchased from Apara (Denzlingen, Germany) or Biomol (Hamburg, Germany). Recombinant PMT and the catalytically inactive mutant PMT^{C1165S} were kindly provided by Prof. Klaus Aktories (Freiburg).

Differentiation of BMDMs

Bone marrow (BM) cells were isolated from the femur and tibia of 6–12 weeks old C57BL/6 mice. These cells were then used to generate BMDMs, using L929-cell conditioned medium (LCCM) as a source of granulocyte/M-CSF. On day 1, BM cells were resuspended in 20 ml of complete medium (DMEM supplemented with 10% FBS, 100 U/ml penicillin, 100 μg/ml streptomycin, and 50 μM 2-mercaptoethanol). On day 2, non-adherent cells were collected by flushing the petri dish several times. These cells were then resuspended in complete medium containing 30% LCCM. On day 4, 30% LCCM was again added and incubated for an additional 3 days. To obtain the BMDM, the supernatants were discarded and the attached cells were collected in 10 ml of complete medium and centrifuged at 200 g for 5 min.

Stimulation

Cells were stimulated with 1 nM PMT or 20–100 ng/ml rec. mouse sRANKL, 25 ng/ml rec. mouse M-CSF 100 ng/ml OPG (all from R&D Systems, Abington, UK). For cytokine experiments, cells were treated with 700 pg/ml TNF-α (eBioscience), 2,500 pg/ml IL-6, and 900 pg/ml IL-1β, respectively (Miltenyi, Bergisch-Gladbach, Germany).

Quantitative Real-time PCR

A total of 1×10^6 BMDM cells were seeded per well in a 6-well plate and then stimulated as indicated. RNA was extracted using High-Pure RNA isolation kit (Roche), according to the manufacturers' protocol. cDNA was prepared by using Revert Aid First strand cDNA synthesis kit (Thermo Scientific). Quantitative RT-PCR was performed using SYBR Green Rox mix (Thermo Scientific) with the primers listed in **Table 1**. RT-PCR was performed using the 7900 HT Fast Real-Time PCR System (AB Applied Biosystems). An initial denaturation step of 10 min at 95°C was common for all genes, but the following cycle for annealing and amplification was different. For *Nfatc1, Il1a, Acp5, Ocstamp* (40 cycles at 95°C for 15 s and at 58°C for 1 min); *calcr* (50 cycles at 95°C for 10 s and at 58°C for 45 s); *Tnf, Il6,* and *Il1b* (40 cycles at 95°C for 15 s and at 60°C for 1 min); *Ctsk, Atp6vod2, Oscar* (40 cycles at 95°C for 15 s,

TABLE 1 | Primer sequences.

Gene	Forward	Reverse
Acp5	5'-TTC CAG GAG ACC TTT GAG GA-3'	5'-GGT AGT AAG GGC TGG GGA AG-3'
Oscar	5'-AGG GAA ACC TCA TCC GTT TG-3'	5'-GAG CCG GAA ATA AGG CAC AG-3'
Ctsk	5'-AGG GAA GCA AGC ACT GGA TA-3'	5'-GCT GGC TGG AAT CAC ATC TT-3'
Nfatc1	5'-GGG TCA GTG TGA CCG AGG AT-3'	5'-GGA AGT CAG AAG TGG GTG GA-3'
Ocstamp	5'-TGG GCC TCC ATA TGA CCT CGA GTA G-3'	5'-TCA AAG GCT TGT AAA TTG GAG GAG T-3'
Atp6v0d2	5'-TCA GAT CTC TTC AAG GCT GTG CTG-3'	5'-GTG CCA AAT GAG TTC AGA GTG ATG-3'
Rps29	5'-AGC CGA CTC GTT CCT TTC TC-3'	5'-CGT ATT TGC GGA TCA GAC C-3'
Tnf	5'-AGC CCC CAG TCT GTA TCC TT-3'	5'-CTC CCT TTG CAG AAC TCA GG-3'
Il6	5'-CCG GAG AG GAGA CTT CAC AG-3'	5'-TTC TGC AAG TGC ATC ATC GT-3'
Il1b	5'-ACT CAT TGT GGC TGT GGA GAA G-3'	5'-GCC GTC TTT CAT TAC ACA GGA-3'
Il1a	5'-CGG GTG ACA GTA TCA GCA AC-3'	5'-GAC AAA CTT CTG CCT GAC GA-3'
calcr	5'-AGA GTG AAA AGG CGG AAT CT-3'	5'-TTT GTA CTG AGC ATC CAG CA-3'
Tnfrsf11a	5'-CGA GGA AGA TTC CCA CAG AG-3'	5'-CAG TGA AGT CAC AGC CCT CA-3'

60°C for 30 s, and 72°C for 30 s). As normalization control *Rsp29* was used, relative expression (rE) was calculated as rE = $1/(2^{ΔCt})$.

ELISA

Supernatants of stimulated cells were harvested and analyzed for IL-6 and TNF-α using mouse IL-6 ELISA MAX™ Standard Set and mouse TNF-α ELISA MAX™ Standard Set, respectively (BioLegend, San Diego, CA, USA). A TecanGENios Pro plate reader (Tecan, Crailsheim, Deutschland) was used for quantification. Results were analyzed using the Magellan5 software.

Determination of Bone Resorption Pit Area

A total of 2.5×10^5 cells were plated in 1 ml of complete medium in 24-well plates and treated as mentioned. On day 3, cells were transferred to a 96-well plate containing bovine cortical bone slices (http://Boneslices.com, Jelling, Denmark) and cultivated for 15–21 days. For measurement of resorbed bone, bone slices were washed with phosphate buffer saline (PBS), incubated in 5% sodium hypochlorite for 1–2 h, washed thoroughly with water, and stained with 0.1% toluidine blue. The pits developed a blue to purple color. The resorbed area was calculated from micro images with Adobe® Photoshop® CS5.

TRAP Staining

A total of 2.5×10^5 cells were plated in 1 ml of complete medium in 24-well plates and treated as described in figure legends. Cells were then fixed and stained using Acid Phosphatase, Leukocyte (TRAP) Kit (Sigma, St. Louis, MO, USA). TRAP-positive cells with three or more nuclei were scored as osteoclasts.

Western Blot Analysis

A total of 1×10^6 cells were stimulated in 2 ml of complete medium in a 6-well format as indicated with PMT. For the lysis, cells were washed twice with ice-cold PBS and collected in ice-cold PBS by scrapping. Cell pellets were lysed in 200 µl of 1× NP40 buffer, freshly supplemented with a Phosphatase and Protease-Inhibitor Cocktail (Roche). Lysates were separated by SDS-PAGE (4–20% gradient polyacrylamid gel, Anamed). Proteins were transferred to nitrocellulose membrane *via* semi-dry western blot, blocked in TBST (5% BSA) for 1 h at RT before the membranes were incubated with the primary antibody, diluted as suggested by Cell Signaling Technology over night at 4°C. After 1 h incubation with the secondary antibody (HRP-coupled), protein bands were detected by enhanced chemiluminescence.

Nuclear Extract Preparation

A total of 1.2×10^7 BMDMs were stimulated as indicated in the figure legend in 10 ml of complete medium. Nuclear extracts were prepared using a nuclear extraction kit (Active Motif). Briefly cells were washed and then collected by scrapping in ice-cold PBS containing phosphatase inhibitor. Cells were then centrifuged at 500 g for 10 min at 4°C. Pellet was resuspended in 500 µl of 1× hypotonic buffer and incubated for 15 min on ice. Then 25 µl of detergent was added and vortexed for 10 s. Cells were then centrifuged at 14,000 g for 1 min at 4°C. Supernatant containing the cytoplasmic fraction was transferred into new microcentrifuge tubes. The pellet was washed with PBS twice, resuspended in 50 µl of complete lysis buffer, vortexed, and incubated for 30 min rocking on ice. After that the samples were centrifuged at 14,000 g for 10 min, and the supernatant containing the nuclear fraction was collected. Western Blot analysis was performed with the nuclear fraction.

Cathepsin K Activity Assay

A total of 5×10^5 BMDMs were seeded per well in a six-well plate. Cells were then stimulated with either PMT or M-CSF or M-CSF+sRANKL for 6 days, and cathepsin K Activity Assay was performed (Abcam). Cells were lysed in 200 µl of cathepsin K cell lysis buffer and incubated on ice for 10 min. Cell debris were centrifuged at 14,000 g for 5 min, and the supernatants were removed. The amount of protein in the lysates was determined with a BCA Assay (Pierce), and the amount was adjusted to 3 µg of protein in 50 µl of lysis buffer per well of a 96-well plate. Fifty microliters of cathepsin K reaction buffer was added to each well. The Ac-LR-AFC substrate was added to a final concentration of 140 µM, and the plate was incubated for 2 h at 37°C. Fluorescence was measured with a microplate reader (FLUOstar OPTIMA; BMG LABTECH) with an excitation of 355 nm and an emission of 520 nm.

FACS Analysis

For FACS analysis 1×10^6 cells were used per sample. Cells were blocked for 15 min in PBS, 2% BSA on ice in a total volume of 100 µl before staining the cells for 1 h on ice with the appropriate antibody or incubated in the corresponding isotype control diluted as suggested by manufacturers. Surface expression of RANK (R12-31), CD11b (M1/70), CD80 (16-10A1), CD86 (GL-1), F4/80 (BM8), CD40 (3/23), and MHCII (M5/114.15.2) was quantified by flow cytometry on a FACSCanto cytometer (BD Biosciences, Heidelberg, Germany). For RANK expression, the mean fluorescence intensity was recorded and the values were corrected for differences in basal fluorescence of unstained cells. Overlays were generated using Flowing Software 2.5.

Phagocytosis Assay

5×10^5 cells were incubated in 100 µl of complete medium at 37°C or at 4°C with green fluorescent latex beads (diameter: 1 µm) diluted 1:100 for 1 h. Afterward, cells were washed five times with ice-cold PBS, and phagocytosis was measured by performing flow cytometry (FACSCanto cytometer, BD Biosciences, Heidelberg, Germany).

Cytokine and Inhibitor Experiments

A total of 2.5×10^5 cells were plated in 1 ml of complete medium in 24-well plates and stimulated according as detailed in the figure legends. For the analysis of cytokine-mediated effects, cytokines were added to the appropriate stimuli or were added alone as cytokine mix directly afterward. Half of the medium was changed after 1½ days cytokines were replenished. After 10–12 days, TRAP staining was performed. To address the role of IL-1, TNF-α and IL-6 signaling in PMT-induced osteoclastogenesis, BMDMs were generated from wt-mice and IL1-R-deficient mice and were treated with the inhibitor etanercept (Enbrel; Pfizer) with a concentration of 120 µg/ml and a neutralization antibody for murine IL-6 (anti m IL-6, BioXCell) at 105 µg/ml for 1 h prior to stimulation. Cells were stimulated with PMT and M-CSF, and the inhibitors were again added 12 h after stimulation. Forty-eight hours after addition of the inhibitors, half of the medium was changed and stimulated or inhibited, respectively. After 10–12 days, a TRAP stain was performed.

Statistical Analyses

Data are presented as means ± SD. Comparison between the groups was performed by employing a Student's *t*-test. *p* values ≤0.05 were considered statistically significant. Multiple-group comparisons were analyzed by analysis of variance.

RESULTS

PMT Triggers Osteoclast Formation in Mouse BMDMs

Pasteurella multocida toxin was shown to have the potential to differentiate osteoclast precursors into osteoclasts in pig, mouse, and rat models using heterogeneous and mostly ill-defined precursor populations (11, 25, 27). As none of these studies checked the osteoclastogenic potential of PMT on a homogenous population of macrophages devoid of stromal or osteoblastic contamination, we decided to investigate the effect of PMT in BMDMs as a model system. After an initial characterization of the population for the expression of typical macrophage markers that allow to distinguish macrophages from monocytes (Figure S1 in Supplementary Material) (28), macrophages were washed and stimulated with PMT, in the absence of M-CSF or other cytokines. PMT treatment resulted in TRAP-positive osteoclast formation with PMT comparable to stimulation with M-CSF/sRANKL (**Figure 1A**; Figure S2 in Supplementary Material). We then tested whether the observed osteoclastogenesis in BMDMs was due to a direct effect of PMT. As PMT is known to deamidate the alpha-subunit of heterotrimeric Gq, we checked the kinetics of deamidation of Gq in BMDMs (29). We observed a very rapid uptake of the toxin resulting in the deamidation of Gq already after 1 h of PMT stimulation that persisted for at least 72 h (**Figure 1B**). We next compared the expression of osteoclast markers in PMT-treated macrophages along with M-CSF/sRANKL-treated macrophages. Real-time qPCR analysis revealed that PMT stimulation in macrophages induced expression of NFATc1 (*Nfatc1*), acid phosphatase 5, tartrate-resistant (*Acp5*, TRAP), cathepsin K (*Ctsk*), d2 isoform of vacuolar (H+) ATPase (v-ATPase) V0 domain (*Atp6v0d2*, ATP6v0d2), osteoclast associated receptor (*Oscar*, OSCAR), calcitonin receptor, and osteoclast stimulatory transmembrane protein (*Ocstamp*, OC-STAMP) (**Figure 1C**). We further checked the activation of transcription factors essential for driving osteoclastogenesis and observed an activation of NFATc1, c-Jun, and NF-kB after PMT stimulation (**Figure 1D**). Next, we sought to examine whether PMT-induced osteoclasts were able to resorb cortical bovine slices. Indeed, PMT-stimulated osteoclasts derived from pure macrophages were able to efficiently resorb bone matrix (**Figures 1E,F**). Further characterization revealed significant cathepsin K activity in PMT-treated cells, suggesting that PMT induces differentiation of macrophages into functional osteoclasts (**Figure 1G**). Together, these observations suggest that PMT induces direct differentiation of BMDMs into functional osteoclasts.

PMT Induces Differentiation of Macrophages into Osteoclasts in a RANKL-Independent Manner

Osteoprotegerin is a soluble secreted protein of the TNF receptor superfamily that is also known as osteoclast inhibitor factor (30). Both OPG and RANK are receptors for RANKL. OPG is an antagonistic endogenous receptor that inhibits osteoclastogenesis. The OPG–RANKL complex counterbalances the effect of the RANK–RANKL complex, thus playing an important role in maintaining bone homeostasis.

As PMT induces differentiation of macrophages into osteoclasts, we wanted to check the effect of OPG on PMT-induced osteoclastogenesis. We treated macrophages with OPG along with M-CSF/sRANKL or PMT, respectively, and then assessed osteoclast formation. Concomitant treatment with OPG completely abrogated osteoclast formation after M-CSF/sRANKL stimulation but failed to show any significant effect on PMT-mediated osteoclast formation (**Figure 2A**). To exclude that PMT might outcompete OPG at the concentration used (1 nM), we gradually decreased PMT concentrations to as low as 0.01 nM, but no variation in the number of TRAP-positive osteoclasts was found (Figure S3 in Supplementary Material). Furthermore, we checked the effect of OPG on PMT-induced signaling pathways and transcription factors. In the presence of OPG, PMT was still

able to activate NFATc1, NF-kB, and c-Jun (**Figure 2B**). We validated this observation by checking the expression of osteoclast markers induced by PMT in the presence of OPG (**Figure 2C**). OPG treatment of macrophages along with PMT did not alter the expression of *Nfatc1*, *Acp5*, *Oscar*, and *Ctsk* but upregulated the expression of *Ocstamp* and *Atp6v0d2* compared to PMT-treated macrophages. Next, we checked the bone resorptive potential of osteoclasts derived from PMT treated with OPG. Again, we did not observe any loss of function of PMT-generated osteoclasts in the presence of OPG, suggesting that OPG treatment does not interfere with PMT-mediated osteoclastogenesis (**Figures 2D,E**). Together, these data prove that PMT-induced differentiation of macrophages into osteoclasts is RANKL-RANK signaling independent.

PMT Primes Macrophages for Enhanced Response to RANKL Stimulation

As PMT-mediated osteoclast formation was not blocked by the RANKL inhibitor OPG, we wanted to see if PMT could influence RANKL-mediated signaling pathways. To check if PMT might augment the response of macrophages at low concentrations of RANKL, we stimulated macrophages with PMT for 1 day, then carefully washed the macrophages with culture medium and re-stimulated them with a low amount of sRANKL (20 ng) in the presence of M-CSF. We observed a marked increase in TRAP-positive cells when the cells had been pretreated with PMT compared to the treatment with sRANKL/M-CSF or PMT, alone (**Figure 3A**). This resulted in an enhanced number of resorption pits (**Figures 3B,C**). These observations demonstrate that PMT increases the responsiveness of the stimulated progenitor cells for RANKL-mediated osteoclastogenesis.

We next determined whether PMT modulates the expression of RANK in order to increase the responsiveness of RANKL. We observed a slight increase in surface expression but not gene expression of RANK at 24 h of treatment with PMT compared to M-CSF-treated samples (**Figure 3D**; Figures S4A,B in Supplementary Material). However, we wondered, if there were other effects involved in the increased activity of cells pretreated with PMT, given the strong effect of the PMT pretreatment on pit formation. As we had seen before that PMT is a strong inducer of pro-inflammatory genes, we investigated whether these cytokines influence the process of differentiation induced by PMT in macrophages.

PMT-Induced TNF-Alpha and IL-6 Are Important Modulators of Its Osteoclastogenic Potential

We have previously shown that PMT is a strong inducer of NF-kB, subsequently causing the induction of pro-inflammatory genes (18). This inflammatory reaction occurs independently of TLR4 or the inflammasome through G-protein-mediated RhoA signaling (31). Therefore, we investigated whether pro-inflammatory cytokines influence PMT-mediated osteoclast differentiation. TNF-α is secreted by various cell types, including macrophages, and is one of the most potent osteoclastogenic cytokines produced during inflammation. It plays an important role in the pathogenesis of rheumatoid arthritis and other forms of chronic inflammatory osteolysis (32, 33). Therefore, we investigated the kinetics of expression and secretion of TNF-α after PMT stimulation in macrophages. We observed that PMT induces the expression of TNF-α starting after 1 h of stimulation, with a maximum

induction after 12 h of stimulation (**Figure 4A**) and continuous secretion of TNF-α for at least 72 h (**Figure 4B**).

We thus questioned, whether blocking of TNF-α signaling by addition of etanercept, a TNF-α antibody, affects PMT-induced downstream signaling pathways. Addition of etanercept prior

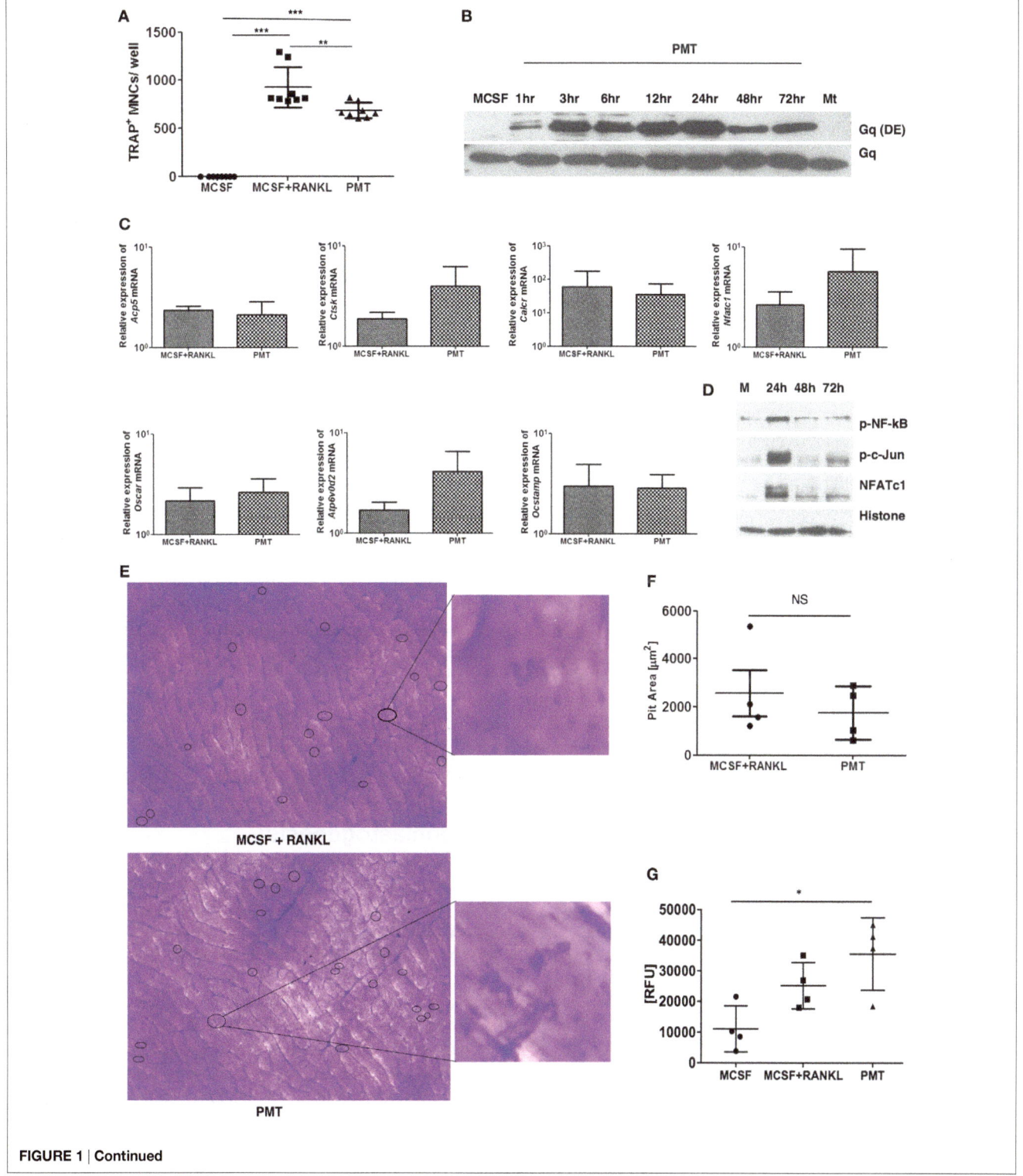

FIGURE 1 | Continued

FIGURE 1 | Continued

Pasteurella multocida* toxin (PMT) induces differentiation of bone marrow-derived macrophages (BMDMs) into osteoclasts. (A)** BMDMs were stimulated with standard concentrations of macrophage colony-stimulating factor (M-CSF), M-CSF/sRANKL, and PMT for 6–10 days as described in Section "Materials and Methods." Cells were then fixed and stained for tartrate-resistant acid phosphatase (TRAP) activity and TRAP+ multinucleated cells (MNCs) were counted. The indicated SD was obtained from four experiments (mean ± SD; $n = 4$). Statistical analysis was performed using analysis of variance (ANOVA) followed by Bonferroni's multiple comparison test ($p \leq 0.0005$; **$p \leq 0.005$). **(B)** To investigate the kinetics of PMT-mediated $G\alpha_q$ deamidation, cells were stimulated for the indicated time points with 1 nM PMT prior to lysis. As a control, cells were either stimulated with M-CSF or a catalytically inactive mutant of PMT (PMT^{C1165S}, Mt) for the longest time point. The immunoblot was probed with an antibody detecting the deamidated form of $G\alpha_q$ (Q209E) or total $G\alpha_q$ ($n = 3$). **(C)** Quantitative RT-PCR analysis of gene expression of *Acp5* (TRAP), *Ctsk* (cathepsin K), calcitonin receptor (*Calcr*), *Nfatc1*, *Oscar*, *Atp6v0d2*, and *Ocstamp* in BMDMs treated either with M-CSF, M-CSF/sRANKL, or PMT; the results were normalized to *Rps29* expression. Cells were stimulated for 12 h to check the expression of *Nfatc1* and *Ocstamp*; for 24 h to check the expression of *Oscar*; for 48 h to check the expression of *Acp5*, *Ctsk*, and *Atp6v0d2*; and for 72 h to check the expression of *Calcr*. The data are presented as fold change relative to the expression of M-CSF-treated cells at the same time point. The indicated SD was obtained from three or more experiments (mean ± SD; $n \geq 3$). No significant difference was observed comparing gene expression of PMT-treated samples with M-CSF/sRANKL-treated sample using a paired Student's *t*-test. **(D)** Immunoblot of phosphorylated NF-κB (p-NF-κB), phosphorylated c-Jun (p-c-Jun), NFATc1, and histone in nuclear extracts from BMDM treated either with PMT or M-CSF (M) for the indicated time points; histone was used as a loading control ($n = 3$). **(E)** Representative photographs of resorption pits (left panel) and magnified image of a resorption pit (right panel) induced by M-CSF + RANKL or PMT. Resorption pit is marked with circle. **(F)** The diagram represents the pit area that was calculated by deducting the pit area value of M-CSF + RANKL or PMT, respectively, from MCSF-treated wells ($n = 4$). Resorption pit pictures were evaluated in a blinded fashion, and false-positive pits were excluded by marking similar structures in M-CSF-treated samples. No significant difference was observed between PMT-treated samples with M-CSF/sRANKL-treated samples using a paired Student's *t*-test (ns, not significant). **(G)** Cells were stimulated with M-CSF, M-CSF/sRANKL, and PMT and lysed as described in Section "Materials and Methods." Cathepsin K activity was analyzed by fluorescence detection measured in duplicates ($n = 4$). Statistical analysis was performed using one-way ANOVA comparing cells stimulated with M-CSF/sRANKL and PMT to the M-CSF sample (*$p \leq 0.05$).

to PMT stimulation in macrophages abrogated p-c-Jun and decreased NFATc1 and NF-kB levels in nuclear extracts (**Figure 4C**). The number of TRAP-positive cells was significantly reduced (approximately 63%) (**Figure 4D**), and osteoclast-specific gene expression was impaired (**Figure 4E**). Collectively, these results suggest that PMT-induced TNF-α production increases its osteoclastogenic potential.

In a previous study, we showed that PMT strongly induces interleukin-6 expression of immune cells (34). In mice, overexpression of IL-6 leads to arthritis while absence of IL-6 prevents formation of arthritis in an experimental murine system (35, 36). In addition, enhanced IL-6 levels are observed in the joints and serum of rheumatoid arthritis patients (37). This prompted us to check the expression and secretion of IL-6 in BMDM. We observed expression of IL-6 after 1 h until 72 h of stimulation with PMT (**Figure 5A**) and the 100-fold increased expression of IL-6 resulted in an enhanced and sustained IL-6 secretion (**Figure 5B**). These observations suggest that in addition to TNF-α, IL-6 may help PMT in driving osteoclastogenesis.

We next evaluated the effect of a murine anti-IL-6R antibody (MR-16) on PMT-induced NFATc1, c-Jun, and NF-kB activation. Anti-IL-6R antibody treatment reduced PMT-induced activation of these transcription factors (**Figure 5C**) and, as a consequence, osteoclast formation (**Figure 5D**), and expression of osteoclast-specific genes was reduced (**Figure 5E**). Together these observations suggest that PMT-mediated osteoclast formation relies, at least in part, on the osteoclastogenic effects of IL-6 and TNF-α.

PMT Induces Osteoclastogenesis in Absence of IL-1 Receptor

Like IL-6 and TNF-α, IL-1α and IL-1β are important osteoclastogenic cytokines (38). Mice deficient in IL-1α, IL-1β, and IL-1α/β suppress arthritis in a mouse model, suggesting that both forms of IL-1 are required for inflammatory bone loss (39). We observed

strongly enhanced expression of both, *Il1a* and *Il1b* after 12 h of PMT treatment that lasted until 72 h (**Figures 6A,B**). As IL-1α and IL-1β transduce signals by binding to IL-1R, we investigated the effect of PMT on BMDM cells derived from IL-1R knockout mice (40). Macrophages deficient in IL-1R showed no differences in the activation of NF-kB, p-c-Jun and NFATc1, respectively (**Figure 6C**) and IL-1R-deficient BMDM differentiated into TRAP-positive cells similar to wild-type macrophages (**Figure 6D**). However, when we tested the cells for their ability to resorb bone, IL-1R-deficient osteoclasts treated with PMT seemed to act slightly less efficient, although the difference did not reach statistical significance (**Figures 6E,F**). Also, we observed comparable expression of osteoclast marker genes after PMT stimulation in IL-1R knockout macrophages (**Figure 6G**). In summary, these observations suggest that PMT can differentiate macrophages into functional osteoclasts in the absence of IL-1R.

Cytokines Are Necessary but Not Sufficient for PMT-Induced Osteoclastogenesis

While it is generally accepted that RANKL is necessary to induce the differentiation of osteoclasts from precursor cells, some publications raise the question whether RANKL-independent osteoclastogenesis can occur under physiological or pathological conditions. It was described that sustained stimulation with cytokines allows osteoclast formation (41, 42). Therefore, we investigated whether the induction of cytokines by PMT is sufficient to induce OC formation. BMDM were treated with the amount of IL-6, TNF-α, and IL-1β that had been measured in the ELISA experiments. Quantification of TRAP-positive cells showed that the addition of cytokines alone did not induce osteoclast formation, neither in the presence nor absence of a supporting M-CSF stimulus (**Figure 7A**). In addition, we did not observe any synergistic effect of those cytokines on M-CSF/

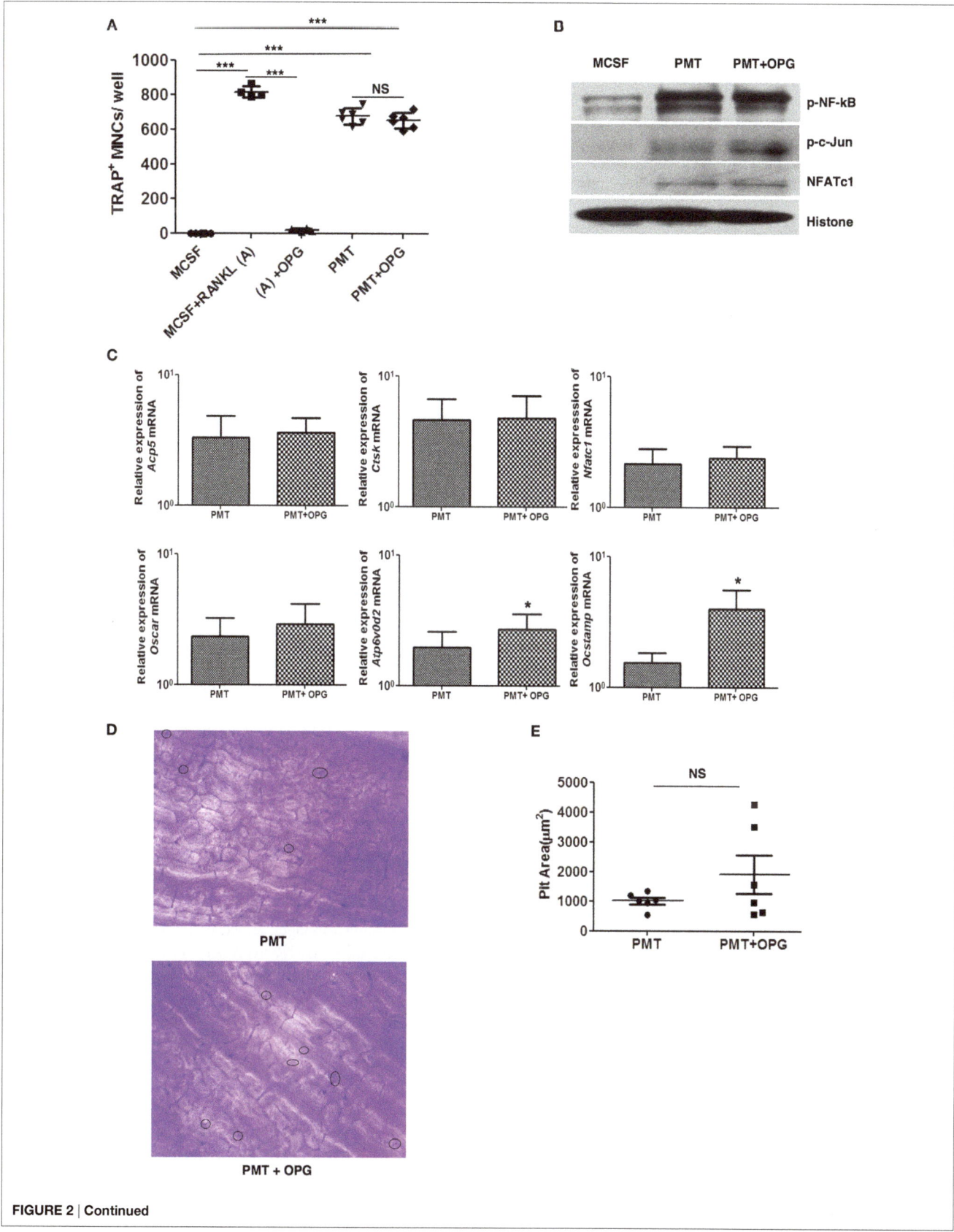

FIGURE 2 | Continued

FIGURE 2 | Continued

Pasteurella multocida toxin (PMT)-mediated osteoclastogenesis is unaltered with osteoprotegerin (OPG) treatment. **(A)** Bone marrow-derived macrophages (BMDMs) were stimulated either with PMT, PMT/OPG, M-CSF/sRANKL, M-CSF/sRANKL/OPG, or macrophage colony-stimulating factor (M-CSF) alone for 6–10 days (PMT 1 nM; OPG 100 ng/ml; M-CSF 25 ng/ml; sRANKL 100 ng/ml). Multinucleated TRAP+ cells were counted per well. The indicated SD was obtained from two or more experiments measured in duplicates (mean ± SD; $n \geq 2$). Statistical analysis was performed using ANOVA followed by Bonferroni's multiple comparison test (***$p \leq 0.0005$; NS, not significant). **(B)** BMDMs were treated with M-CSF, PMT, or with PMT/OPG for 24 h, and nuclear extracts were prepared. The immunoblot shows phosphorylated NF-κB (p-NF-κB), phosphorylated c-Jun (p-c-Jun), NFATc1, and histone, which was used as a loading control ($n = 2$). **(C)** Quantitative RT-PCR analysis of gene expression of Acp5, Ctsk, Ocstamp, Nfatc1, Oscar, and Atp6v0d2 in BMDMs treated either with M-CSF, PMT/OPG, or PMT; values were normalized to Rps29 expression. Data are presented fold change relative to the gene expression of M-CSF-treated cells at the same time point. Cells were stimulated for 12 h to check the expression of Nfatc1 and Ocstamp; for 24 h to check the expression of Oscar; and for 48 h to check the expression of Acp5, Ctsk, and Atp6v0d2. The indicated SD was obtained from three or more experiments (mean ± SD; $n \geq 3$). Statistical analysis was performed using a paired Student's t-test comparing gene expression of PMT/OPG-treated cells to the PMT-treated cells (*$p \leq 0.05$). **(D)** Representative picture of bone resorption assays performed with PMT-stimulated cells with or without OPG. Resorption pits are marked with circles. **(E)** The histogram represents the calculated pit area that was obtained by subtracting the pit area value of PMT or PMT/OPG wells from M-CSF-treated conditions ($n \geq 3$). The resorption pit pictures were evaluated in a blinded fashion and the false-positive pits were excluded by marking similar structures in M-CSF-treated samples. Statistical analysis was performed using a paired Student's t-test comparing PMT + OPG treatment with PMT-treated sample (NS, not significant).

sRANKL-treated cells. To verify that the cytokine amounts used were appropriate, we checked the activity of specific transcription factors (**Figure 7B**). Although there was a strong cytokine-specific activation of STAT-3 and NF-κB comparable to that induced by PMT, this activity was insufficient to induce osteoclast formation.

We next studied the effect of an inhibition of all three osteoclastogenic cytokines investigated in osteoclast formation (**Figure 7C**). Simultaneous inhibition of IL-6 and TNF-α substantially decreased osteoclast formation by 72% for both WT and IL-1R-deficient mice compared to 63% for etanercept alone and 49% for IL-6R inhibition. These data imply that while cytokines are needed to successfully induce PMT-mediated osteoclast formation, the presence of cytokines alone is not sufficient and that other G-protein-related signaling pathways are required.

DISCUSSION

Pasteurella multocida toxin is a bacterial protein toxin that is known to manipulate host cell signaling cascades which for many cell types result in the activation of mitogenic pathways and the production of cytokines (17). In AR PMT signaling influences bone degeneration by accelerating osteoclast formation and by inhibiting osteoblast differentiation and function. In this study, we show that PMT stimulation of macrophages leads to the induction of osteoclast markers similar to M-CSF/sRANKL stimulation resulting in the differentiation of macrophages into functional osteoclasts. Additionally, we show that PMT-derived osteoclasts are able to resorb bone similar to classical osteoclasts derived from M-CSF/sRANKL stimulation. Strack et al. have recently discussed that PMT-induced differentiation of monocytes into osteoclasts might be RANKL-independent, however, they did not show any experimental proof of their speculation (27). Our data show for the first time that PMT-induced osteoclast formation does not depend on RANKL-RANK signaling. OPG treatment with PMT did not alter the activation of essential transcription factors or the expression of osteoclast marker genes, and the resulting osteoclasts were able to resorb bone without any loss of function.

Additionally, we show for the first time that the PMT-induced inflammatory cytokines such as TNF-α, IL-6, and IL-1 play a central role in mediating differentiation of macrophages into osteoclasts. TNF-α alone is a weak inducer of osteoclastogenesis

in mouse BMDM (43), but has been reported to induce functional osteoclast formation along with IL-6 (44) or IL-1 (45). Recently, Yarilina et al. have shown that prolonged TNF-α exposure of human macrophages can trigger differentiation of macrophages into osteoclasts (46). We also observe a sustained production of TNF-α after PMT stimulation in macrophages, and this continuous exposure of macrophages to TNF-α may additionally help PMT in inducing osteoclastogenesis. In accordance, our *in vitro* study using etanercept strongly reduced PMT-induced osteoclast formation (**Figures 4C–E**), suggesting that TNF-α plays a crucial role in PMT-mediated osteoclastogenesis. PMT stimulation also caused the continuous increased production of IL-6. In rheumatoid arthritis, overproduction of IL-6 is observed and there is a correlation between elevated IL-6 level and clinical indices (47, 48). When we blocked IL-6 signaling, a decrease in PMT-induced osteoclast formation in macrophages was observed (**Figures 5C–E**). A similar observation was made by Axmann et al. where IL-6R blockade directly suppressed M-CSF/RANKL-induced osteoclast formation (49).

Similar to TNF-α and IL-6, we observed a high expression of IL-1α and β after PMT stimulation of macrophages. However, our observations in IL-1R knockout mice suggest that PMT can induce differentiation of macrophages into functional osteoclast even in the absence of IL-1R signaling, suggesting that IL-1α and IL-1β do not directly contribute to PMT-induced osteoclast formation in macrophages. However, under physiological conditions, where other IL-1R expressing cells are available, PMT-induced IL-1 production from macrophages may contribute in the progression of AR by either inhibiting osteoblast-mediated bone formation (50) or by inducing the expression of factors supporting osteoclastogenesis such as prostaglandin E2 or RANKL by osteoblasts/stromal cells (51, 52).

The current literature suggests that under physiological conditions the RANKL/RANK axis drives functional osteoclast formation. Pathological conditions, such as cancer or auto-inflammatory bone diseases, are characterized by the presence of a specific microenvironment, where the increased expression of inflammatory cytokines can provide an additional stimulus that further potentiates the generation of bone-resorbing osteoclasts (42, 53). Our data also suggest that the PMT-mediated effect on osteoclasts is probably due to two signaling cascades. While

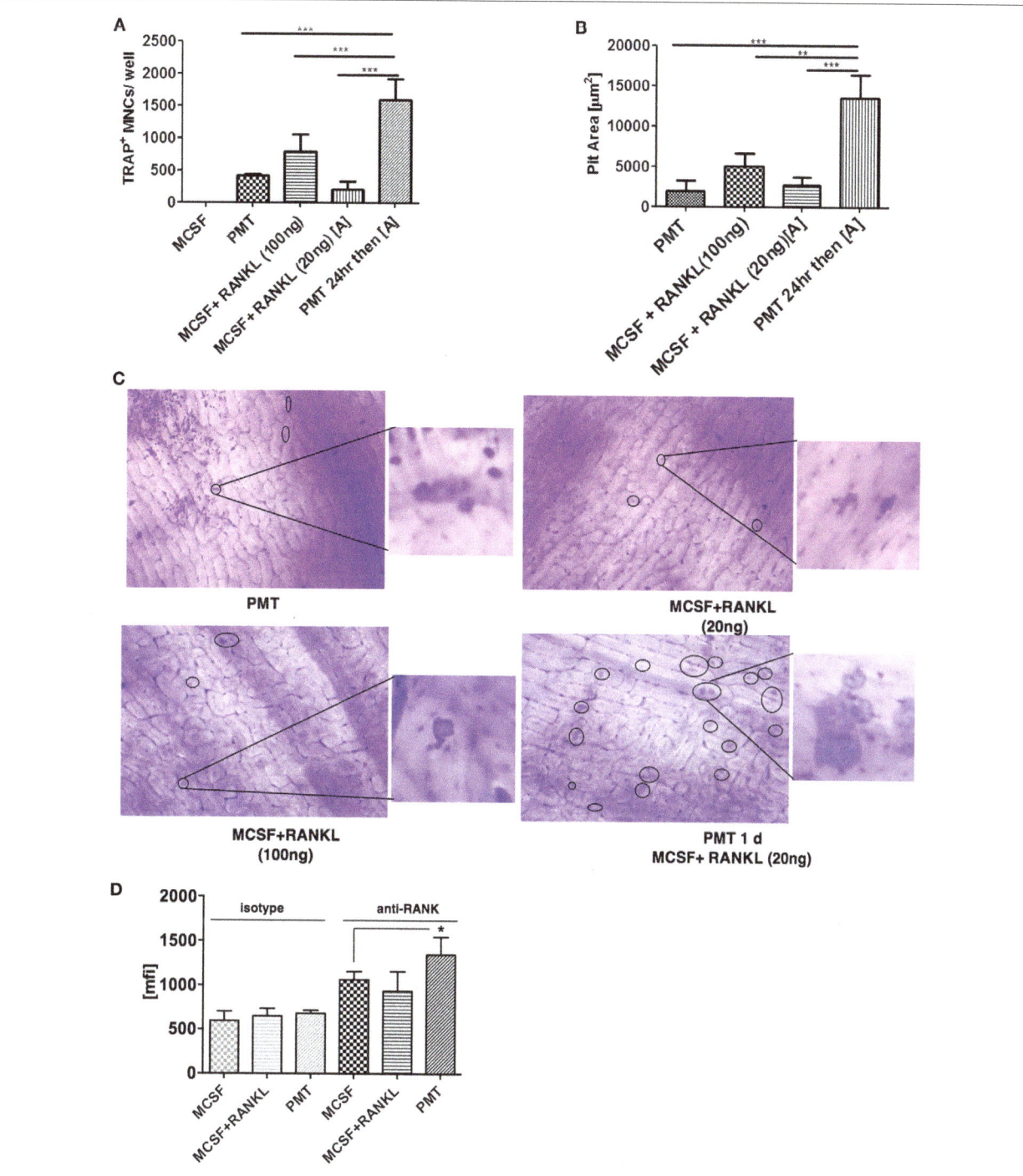

FIGURE 3 | *Pasteurella multocida* **toxin (PMT) pre-treatment augments M-CSF/RANKL-induced osteoclastogenesis in macrophages. (A)** Bone marrow-derived macrophages were treated with stimuli as indicated for 10 days using high (100 ng/ml) and low (20 ng/ml) concentrations of RANKL. TRAP+ multinucleated cells (MNCs) were counted (mean ± SD; $n = 2$). Statistical analysis was performed using analysis of variance (ANOVA) followed by Bonferroni's multiple comparison test comparing TRAP+ cells treated with PMT for 1 day before switching to low M-CSF/sRANKL concentrations with samples treated only with PMT, M-CSF/sRANKL (high), or M-CSF/sRANKL (low), respectively (***$p \leq 0.0005$). **(B)** Graph of the calculated pit area that was obtained by deducting the pit area value of macrophage colony-stimulating factor (M-CSF)-treated conditions from the respective samples (mean ± SD; $n = 3$). Resorption pit pictures were evaluated in a blinded fashion, and false-positive pits were excluded by marking similar structures in M-CSF-treated samples. Statistical analysis was performed using ANOVA followed by Bonferroni's multiple comparison test comparing pit areas in samples treated with PMT for 1 day before switching to M-CSF/sRANKL (low) with PMT, M-CSF/sRANKL (high), or M-CSF/sRANKL (low) samples (**$p \leq 0.005$; ***$p \leq 0.0005$). **(C)** Representative photographs of resorption pits (left panel) and magnified image of a resorption pit (right panel). Resorption pit is marked with circle. **(D)** FACS analysis of RANK surface expression after 1 day of treatment with M-CSF/ sRANKL or PMT (mean ± SD; $n = 3$). Statistical analysis of RANK expression was performed using ANOVA followed by Bonferroni's multiple comparison test (*$p \leq 0.05$).

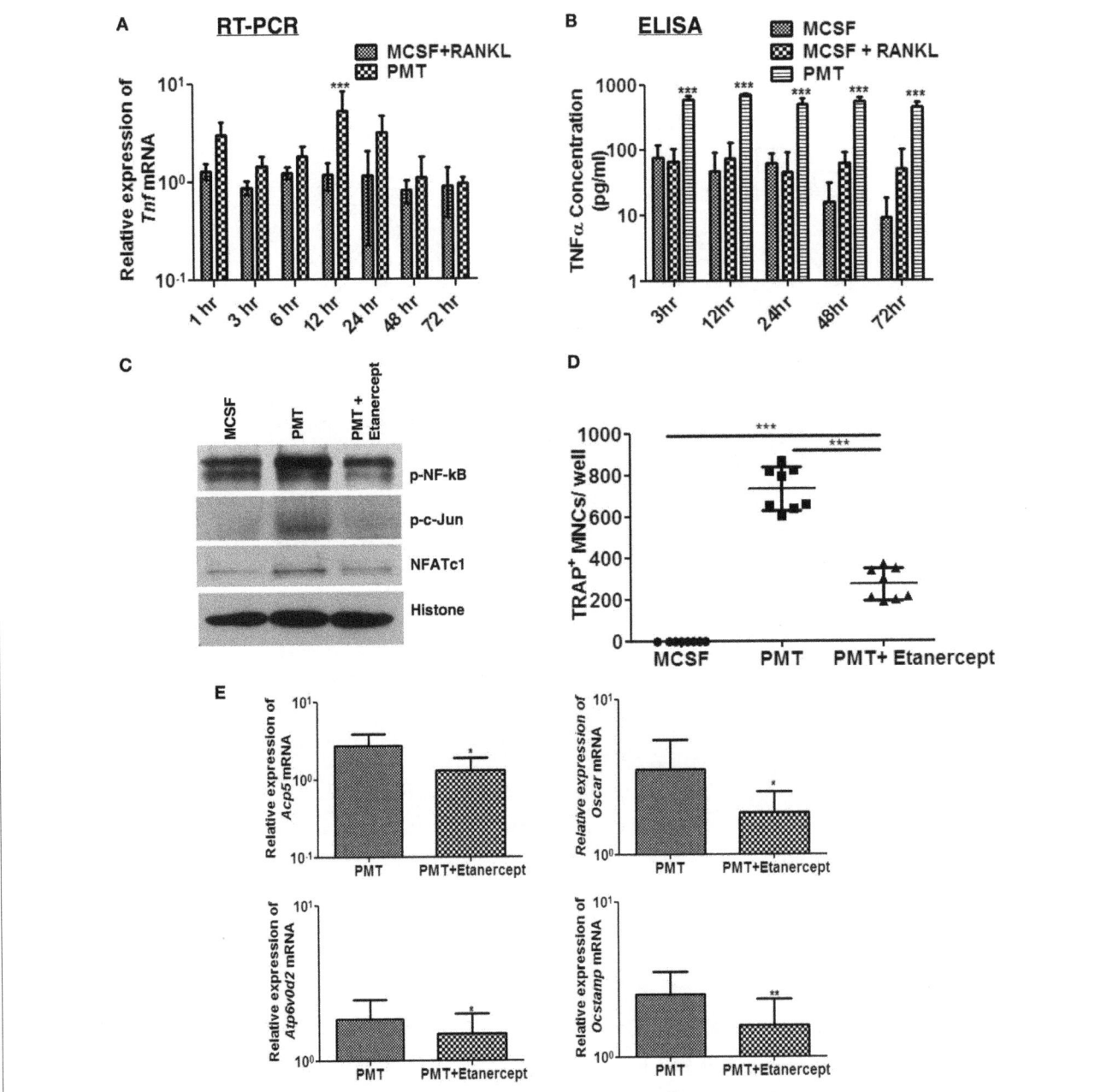

FIGURE 4 | TNF-α modulates *Pasteurella multocida* toxin (PMT)-induced osteoclastogenesis. (A,B) Bone marrow-derived macrophages (BMDMs) were stimulated with standard concentration of macrophage colony-stimulating factor (M-CSF), M-CSF/sRANKL, or PMT for the indicated the time points.
(A) Quantitative RT-PCR analysis of *Tnf* normalized to *Rps29* expression. Data are presented fold change relative to the expression of M-CSF-treated cells at the same time point (mean ± SD; n = 3). **(B)** ELISA of TNF-α production (mean ± SD; n = 3). Statistical analysis was done by two-way analysis of variance (ANOVA) (***$p \leq 0.001$). **(C)** BMDMs were stimulated with PMT, PMT/etanercept (120 μg/ml) or M-CSF for 24 h, and nuclear extracts were prepared. Immunoblots of nuclear phosphorylated NF-κB (p-NF-κB), c-Jun (p-c-Jun), NFATc1, and histone were performed (n = 2). **(D)** BMDMs were stimulated with PMT, PMT/etanercept (120 μg/ml), or M-CSF for 10 days. Multinucleated TRAP⁺ cells were counted per well. The indicated SD was obtained from four experiments (mean ± SD; n = 4). Statistical analysis was performed using ANOVA followed by Bonferroni's multiple comparison test comparing TRAP⁺ cells in PMT/etanercept-treated wells to the PMT- and M-CSF-treated wells; TRAP⁺ cells in PMT-treated wells to the M-CSF-treated wells (***$p \leq 0.0005$). **(E)** Quantitative RT-PCR analysis of gene expression of *Acp5*, *Oscar*, *Atp6v0d2*, and *Ocstamp* in BMDMs treated either with PMT, PMT/etanercept, or M-CSF; data were normalized to *Rps29* expression. Cells were stimulated for 12 h to check the expression of *Ocstamp*; for 24 h to check the expression of *Oscar*; and for 48 h to check the expression of *Acp5* and *Atp6v0d2*. The data are presented as fold change relative to the expression of M-CSF-treated cells at the same time point. The indicated SD was obtained from three or more experiments (mean ± SD; n ≥ 3). Statistical analysis was performed using a paired Student's *t*-test comparing gene expression of PMT/etanercept-treated cells to the PMT-treated cells (*$p \leq 0.05$; **$p \leq 0.005$).

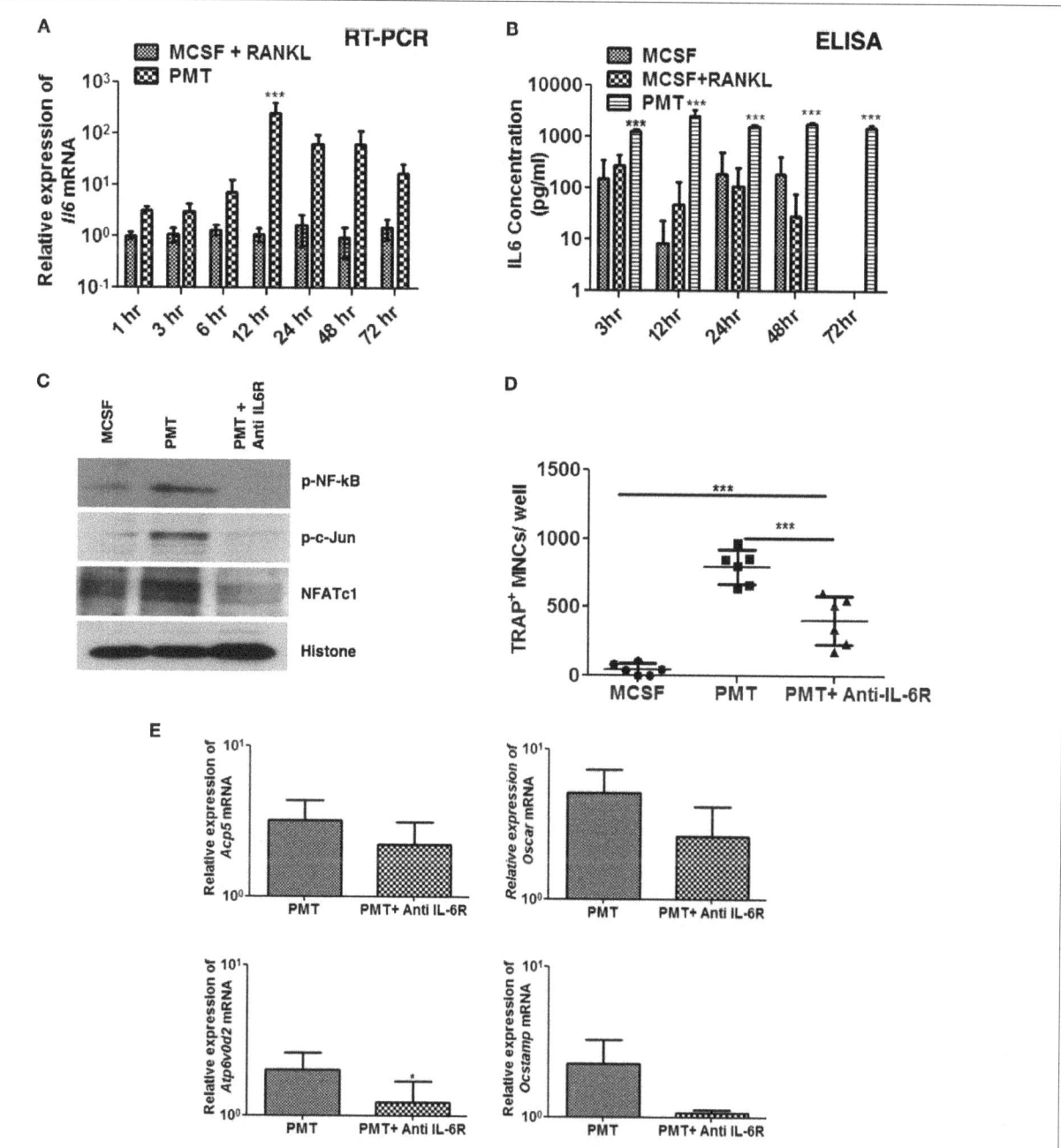

FIGURE 5 | *Pasteurella multocida* **toxin (PMT)-mediated osteoclastogenesis is inhibited with blockade of Interleukin 6 receptor (IL-6R). (A,B)** Bone marrow-derived macrophages (BMDMs) were stimulated with standard concentrations of macrophage colony-stimulating factor (M-CSF), M-CSF/sRANKL, or PMT for the indicated time points. **(A)** Quantitative RT-PCR analysis of *Il6* mRNA; normalized to *Rps29* expression. Data are presented fold change relative to the expression of M-CSF-treated cells at the same time point (mean ± SD; n = 3). **(B)** ELISA of IL-6 production (mean ± SD; n = 3). Statistical analysis was done by analysis of variance (ANOVA) (***$p \leq 0.001$). **(C)** BMDMs were stimulated with PMT, PMT/anti IL-6 R (105 µg/ml), or M-CSF for 24 h, and nuclear extracts were prepared. A representative immunoblot of the phosphorylated forms of NF-κB (p-NF-κB) and c-Jun (p-c-Jun), as well as NFATc1 and histone are shown; histone was used as a loading control (n = 2). **(D)** BMDMs were stimulated with PMT, PMT/Anti IL-6R (105 µg/ml), or M-CSF for 10 days. Multinucleated TRAP⁺ cells were counted per well. The indicated SD was obtained from three experiments (mean ± SD; $n \geq 3$). Statistical analysis was performed using ANOVA followed by Bonferroni's multiple comparison test comparing TRAP⁺ cells in PMT/Anti IL-6R-treated wells to the PMT- and M-CSF-treated wells; TRAP⁺ cells in PMT-treated wells to the M-CSF-treated wells (***$p \leq 0.0005$). Statistical analysis was performed using a paired Student's t-test comparing (***$p \leq 0.0005$). **(E)** Quantitative RT-PCR analysis of gene expression of *Acp5*, *Oscar*, *Atp6v0d2*, and *Ocstamp* in BMDMs treated either with PMT, PMT/Anti-IL6 R (105 µg/ml), or M-CSF alone; values were normalized to *Rps29* expression (mean ± SD; n = 3). Cells were stimulated for 12 h to check the expression of *Ocstamp*; for 24 h to check the expression of *Oscar*; and for 48 h to check the expression of *Acp5* and *Atp6v0d2*. The data are presented as fold change relative to the expression of M-CSF-treated cells at the same time point. Statistical analysis was performed using a paired Student's t-test comparing gene expression of PMT/Anti IL-6R-treated cells with PMT-treated cells (*$p \leq 0.05$).

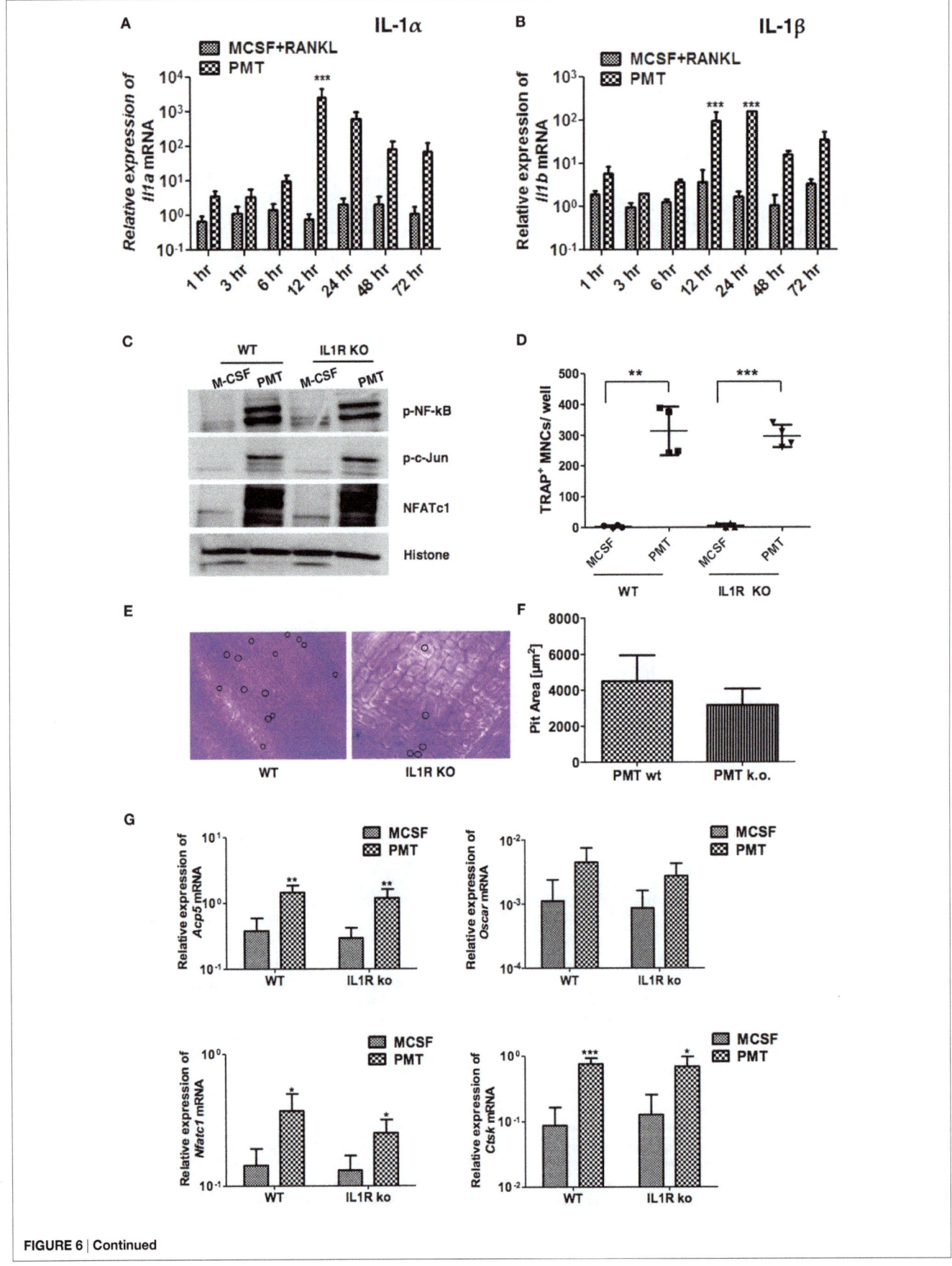

FIGURE 6 | Continued

FIGURE 6 | Continued

Absence of interleukin 1 receptor (IL-1R) fails to block *Pasteurella multocida* toxin (PMT)-induced osteoclast formation. (A,B) Bone marrow-derived macrophages (BMDMs) were stimulated as described using standard concentrations of macrophage colony-stimulating factor (M-CSF), M-CSF/sRANKL, or PMT for the indicated time points. Graphs show quantitative RT-PCR analysis of *Il1a* and *Il1b* mRNA levels normalized to *Rps29* expression. The data are presented as fold change relative to the expression of M-CSF-treated cells at the same time point (mean ± SD; $n \geq 2$). Statistical analysis was done by two-way ANOVA (***$p \leq 0.001$). **(C)** BMDMs from wt and IL-1R-deficient mice were stimulated with PMT or M-CSF for 24 h before nuclear extracts were prepared. Immunoblots of phosphorylated NF-κB (p-NF-κB), c-Jun (p-c-Jun), NFATc1, and histone were performed. Histone served as a loading control ($n = 4$). **(D)** BMDMs from wt and IL-1R-deficient mice were stimulated with PMT and M-CSF for 10 days. Multinucleated TRAP$^+$ cells were counted; the indicated SD was obtained from four experiments (mean ± SD; $n \geq 3$). Statistical analysis was performed using a paired Student's t-test comparing TRAP$^+$ cells in PMT-treated wells to the M-CSF-treated wells (**$p \leq 0.005$; ***$p \leq 0.0005$). **(E)** Representative pictures of bone resorption assays using osteoclasts from wt and IL-1R-deficient mice are shown. Resorption pits are marked with a circle. **(F)** Bone resorption assay of BMDMs derived from wt and IL-1R-deficient mice and differentiated into osteoclasts with PMT. The data presented show the pit area that was calculated by subtracting the pit area value of the M-CSF from the PMT-treated condition ($n = 6$). The resorption pit pictures were evaluated in a blinded fashion and false-positive pits were excluded by marking similar structures in M-CSF-treated samples. **(G)** Quantitative RT-PCR analysis of gene expression of *Acp5*, *Oscar*, *Nfatc1*, and *Ctsk* in BMDMs derived from wild-type mice or IL-1R-deficient mice, treated with M-CSF or PMT, normalized to *Rps29* expression. Cells were stimulated for 12 h to check the expression of *Nfatc1*; for 24 h to check the expression of *Oscar*; and for 48 h to check the expression of *Acp5* and *Ctsk*. The data are presented as fold change relative to the expression of M-CSF-treated cells at the same time point. The indicated SD was obtained from four experiments (mean ± SD; $n = 4$). Statistical analysis was performed using unpaired Student's t-test comparing gene expression to the M-CSF-treated samples (*$p \leq 0.05$; **$p \leq 0.005$; ***$p \leq 0.0005$).

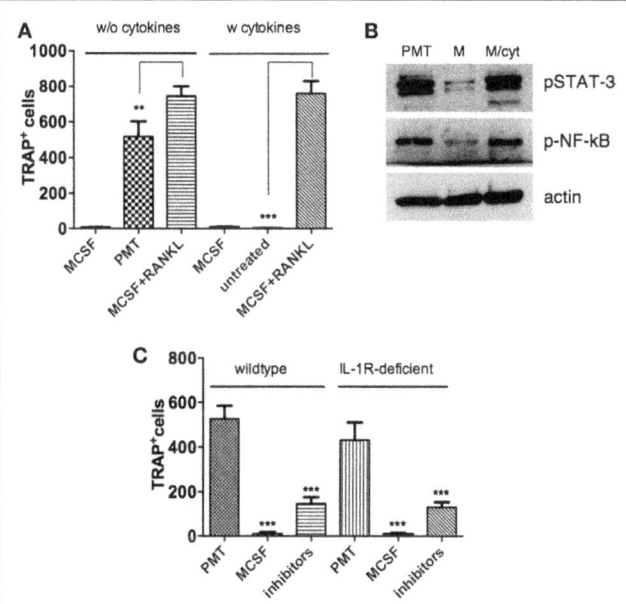

FIGURE 7 | Cytokines alone are not sufficient to mimic *Pasteurella multocida* toxin (PMT)-stimulated osteoclast differentiation. (A) Bone marrow-derived macrophages (BMDMs) were stimulated with macrophage colony-stimulating factor (M-CSF) or M-CSF/sRANKL with or without addition of cytokines: TNF-α (700 pg/ml), IL-6 (2,500 pg/ml), and IL-1β (900 pg/ml) and PMT alone for 10–12 days and multinucleated TRAP$^+$ cells were counted. The indicated SD was obtained from three experiments (mean ± SD). Statistical analysis was performed using one-way analysis of variance (ANOVA) comparing the difference of tartrate-resistant acid phosphatase (TRAP)-positive cells derived from PMT or cytokine treatment compared to M-CSF/sRANKL-treated control cells. **(B)** BMDMs were stimulated with PMT, M-CSF, or M-CSF with cytokines **(A)** for 1½ days. Total lysates were prepared and immunoblots for phosphorylated STAT-3 (p-STAT-3), phosphorylated NF-κB (p-NF-κB), and β-actin were performed ($n = 3$). **(C)** BMDMs from wt and IL-1R-deficient mice were stimulated with PMT, M-CSF, or PMT in combination with etanercept (120 µg/ml) and the IL-6 neutralizing antibody (105 µg/ml) for 10–12 days. Multinucleated TRAP$^+$ cells were quantified and the indicated SD was obtained from three experiments (mean ± SD; $n = 3$). Statistical analysis was performed using one-way ANOVA comparing the decrease of TRAP-positive cells in inhibitor-treated versus PMT-treated control cells.

PMT-induced cytokine production seems to be necessary to allow efficient osteoclast formation, additional PMT-mediated signaling is required. Strack et al. recently showed that NFATc1 is activated by PMT downstream of the activated Gαq subunit (27). In addition, NFATc1 overexpression is known to be sufficient for the induction of osteoclastic genes, even in the absence of RANKL (54). Interestingly, NFATc1 expression was not effectively reduced in our inhibitor experiments, and thus, we hypothesize that Gq-dependent expression of NFATc1 could be the additional signal needed for osteoclast formation. This would also explain why inhibition of cytokines did not abrogate osteoclast formation completely, as there is still a robust Gq-mediated induction of NFATc1 transcriptional activity even in the absence of IL-6 and TNF-α signaling. Collectively, our data suggest that PMT-induced osteoclastogenesis is dependent on cytokines and Gq-mediated signaling but independent of RANKL/RANK axis. A recent article speculates about the possibility of osteoclast subtypes based on large number of studies showing the existence of various subsets of myeloid cells under physiological or pathological state (55). Therefore, we suggest that osteoclasts generated with PMT are distinct from RANKL-derived osteoclasts and are likely to represent a separate subset of osteoclasts; however, detailed future investigations are required for validating our assumption.

Although GPCR signaling is recognized as an important player for many human pathologies, including cardiovascular diseases, inflammation, and cancer, the importance of G proteins in auto-inflammatory bone diseases has not been addressed yet. We recently showed that PMT triggers the differentiation of naïve T cells into a osteoclastogenesis-promoting Th17 phenotype as a consequence of constitutive Gq activation and the resulting downstream activation of STAT-3 (56, 57). This seems to be in contrast with findings of Liu et al. who report that a lack of Gq triggers Th17 differentiation in human lymphocytes (58, 59). However, their data compared mRNA levels but not cellular Gq activity, which for GTPases is a more relevant readout. In support of our findings, other scientists suggest that activation of G proteins by AlF4 treatment activates osteoclast differentiation (60) and that elevated expression of Gq in osteoblasts increases osteoclastogenesis in a transgenic mouse model (61). New therapeutic tools to inhibit G protein signaling are currently being characterized, and it remains to be seen whether they will provide helpful tools in the treatment of auto-inflammatory diseases such as RA as well (62).

AUTHOR CONTRIBUTIONS

SC carried out experiments, participated in the design of the study, and drafted the manuscript; BK performed experiments; UH and GS carried out bone resorption studies; and KK generated FACS data, participated in study design, and drafted the manuscript.

ACKNOWLEDGMENTS

The authors thank Rosalie Deutsch for her technical assistance.

We thank Konrad Bode and Maren Bechberger for IL-1R-deficient animals.

FUNDING

This work was supported by the Deutsche Forschungsgemeinschaft (DFG) through grants to KK (HI 1747/4-2) and GS (SCHE 1583/12-1) within the priority program SPP1468 Immunobone. We acknowledge the financial support of the Deutsche Forschungsgemeinschaft and the Ruprecht-Karls-Universität Heidelberg within the funding programme Open Access Publishing

REFERENCES

1. Henderson B, Nair SP. Hard labour: bacterial infection of the skeleton. *Trends Microbiol* (2003) 11:570–7.
2. Nair SP, Meghji S, Wilson M, Reddi K, White P, Henderson B. Bacterially induced bone destruction: mechanisms and misconceptions. *Infect Immun* (1996) 64:2371–80.
3. Frandsen PL, Foged NT, Petersen SK, Bording A. Characterization of toxin from different strains of *Pasteurella multocida* serotype A and D. *Zentralbl Veterinarmed B* (1991) 38:345–52.
4. Horiguchi Y. Swine atrophic rhinitis caused by *Pasteurella multocida* toxin and bordetella dermonecrotic toxin. *Curr Top Microbiol Immunol* (2012) 361:113–29. doi:10.1007/82_2012_206
5. Chevalier X, Martigny J, Avouac B, Larget-Piet B. Report of 4 cases of *Pasteurella multocida* septic arthritis. *J Rheumatol* (1991) 18:1890–2.
6. Kumar A, Kannampuzha P. Septic arthritis due to *Pasteurella multocida*. *South Med J* (1992) 85:329–30.
7. Rozengurt E, Higgins T, Chanter N, Lax AJ, Staddon JM. *Pasteurella multocida* toxin: potent mitogen for cultured fibroblasts. *Proc Natl Acad Sci U S A* (1990) 87:123–7.
8. Mullan PB, Lax AJ. *Pasteurella multocida* toxin is a mitogen for bone cells in primary culture. *Infect Immun* (1996) 64:959–65.
9. Martineau-Doize B, Frantz JC, Martineau GP. Effects of purified *Pasteurella multocida* dermonecrotoxin on cartilage and bone of the nasal ventral conchae of the piglet. *Anat Rec* (1990) 228:237–46.
10. Felix R, Fleisch H, Frandsen PL. Effect of *Pasteurella multocida* toxin on bone resorption in vitro. *Infect Immun* (1992) 60:4984–8.
11. Martineau-Doize B, Caya I, Gagne S, Jutras I, Dumas G. Effects of *Pasteurella multocida* toxin on the osteoclast population of the rat. *J Comp Pathol* (1993) 108:81–91.
12. Sterner-Kock A, Lanske B, Uberschar S, Atkinson MJ. Effects of the *Pasteurella multocida* toxin on osteoblastic cells in vitro. *Vet Pathol* (1995) 32:274–9.
13. Mullan PB, Lax AJ. *Pasteurella multocida* toxin stimulates bone resorption by osteoclasts via interaction with osteoblasts. *Calcif Tissue Int* (1998) 63:340–5.
14. Siegert P, Schmidt G, Papatheodorou P, Wieland T, Aktories K, Orth JH. *Pasteurella multocida* toxin prevents osteoblast differentiation by transactivation of the MAP-kinase cascade via the Galpha(q/11) – p63RhoGEF – RhoA axis. *PLoS Pathog* (2013) 9:e1003385. doi:10.1371/journal.ppat.1003385
15. Orth JH, Fester I, Siegert P, Weise M, Lanner U, Kamitani S, et al. Substrate specificity of *Pasteurella multocida* toxin for alpha subunits of heterotrimeric G proteins. *FASEB J* (2013) 27:832–42. doi:10.1096/fj.12-213900
16. Orth JH, Preuss I, Fester I, Schlosser A, Wilson BA, Aktories K. *Pasteurella multocida* toxin activation of heterotrimeric G proteins by deamidation. *Proc Natl Acad Sci U S A* (2009) 106:7179–84. doi:10.1073/pnas.0900160106
17. Wilson BA, Ho M. Cellular and molecular action of the mitogenic protein-deamidating toxin from *Pasteurella multocida*. *FEBS J* (2011) 278:4616–32. doi:10.1111/j.1742-4658.2011.08158.x
18. Kubatzky KF, Kloos B, Hildebrand D. Signaling cascades of *Pasteurella multocida* toxin in immune evasion. *Toxins (Basel)* (2013) 5:1664–81. doi:10.3390/toxins5091664
19. Teitelbaum SL. Bone resorption by osteoclasts. *Science* (2000) 289:1504–8.
20. Simonet WS, Lacey DL, Dunstan CR, Kelley M, Chang MS, Luthy R, et al. Osteoprotegerin: a novel secreted protein involved in the regulation of bone density. *Cell* (1997) 89:309–19.
21. Takeshita S, Kaji K, Kudo A. Identification and characterization of the new osteoclast progenitor with macrophage phenotypes being able to differentiate into mature osteoclasts. *J Bone Miner Res* (2000) 15:1477–88. doi:10.1359/jbmr.2000.15.8.1477
22. Rivollier A, Mazzorana M, Tebib J, Piperno M, Aitsiselmi T, Rabourdin-Combe C, et al. Immature dendritic cell transdifferentiation into osteoclasts: a novel pathway sustained by the rheumatoid arthritis microenvironment. *Blood* (2004) 104:4029–37. doi:10.1182/blood-2004-01-0041
23. Gallois A, Lachuer J, Yvert G, Wierinckx A, Brunet F, Rabourdin-Combe C, et al. Genome-wide expression analyses establish dendritic cells as a new osteoclast precursor able to generate bone-resorbing cells more efficiently than monocytes. *J Bone Miner Res* (2010) 25:661–72. doi:10.1359/jbmr.090829
24. Jutras I, Martineau-Doize B. Stimulation of osteoclast-like cell formation by *Pasteurella multocida* toxin from hemopoietic progenitor cells in mouse bone marrow cultures. *Can J Vet Res* (1996) 60:34–9.
25. Gwaltney SM, Galvin RJ, Register KB, Rimler RB, Ackermann MR. Effects of *Pasteurella multocida* toxin on porcine bone marrow cell differentiation into osteoclasts and osteoblasts. *Vet Pathol* (1997) 34:421–30.
26. Kloos B, Chakraborty S, Lindner SG, Noack K, Harre U, Schett G, et al. *Pasteurella multocida* toxin-induced osteoclastogenesis requires mTOR activation. *Cell Commun Signal* (2015) 13:40. doi:10.1186/s12964-015-0117-7
27. Strack J, Heni H, Gilsbach R, Hein L, Aktories K, Orth JH. Noncanonical G-protein-dependent modulation of osteoclast differentiation and bone resorption mediated by *Pasteurella multocida* toxin. *MBio* (2014) 5:e02190. doi:10.1128/mBio.02190-14
28. Francke A, Herold J, Weinert S, Strasser RH, Braun-Dullaeus RC. Generation of mature murine monocytes from heterogeneous bone marrow and description of their properties. *J Histochem Cytochem* (2011) 59:813–25. doi:10.1369/0022155411416007
29. Kamitani S, Ao S, Toshima H, Tachibana T, Hashimoto M, Kitadokoro K, et al. Enzymatic actions of *Pasteurella multocida* toxin detected by monoclonal antibodies recognizing the deamidated alpha subunit of the heterotrimeric GTPase Gq. *FEBS J* (2011) 278:2702–12. doi:10.1111/j.1742-4658.2011.08197.x
30. Yasuda H, Shima N, Nakagawa N, Mochizuki SI, Yano K, Fujise N, et al. Identity of osteoclastogenesis inhibitory factor (OCIF) and osteoprotegerin (OPG): a mechanism by which OPG/OCIF inhibits osteoclastogenesis in vitro. *Endocrinology* (1998) 139:1329–37.
31. Hildebrand D, Bode KA, Riess D, Cerny D, Waldhuber A, Rommler F, et al. Granzyme A produces bioactive IL-1beta through a nonapoptotic inflammasome-independent pathway. *Cell Rep* (2014) 9:910–7. doi:10.1016/j.celrep.2014.10.003

32. Feldmann M, Brennan FM, Maini RN. Role of cytokines in rheumatoid arthritis. *Annu Rev Immunol* (1996) 14:397 440.

33. Kitaura H, Zhou P, Kim HJ, Novack DV, Ross FP, Teitelbaum SL. M-CSF mediates TNF-induced inflammatory osteolysis. *J Clin Invest* (2005) 115:3418–27. doi:10.1172/JCI26132

34. Hildebrand D, Heeg K, Kubatzky KF. *Pasteurella multocida* toxin-stimulated osteoclast differentiation is B cell dependent. *Infect Immun* (2011) 79:220–8. doi:10.1128/IAI.00565-10

35. Alonzi T, Fattori E, Lazzaro D, Costa P, Probert L, Kollias G, et al. Interleukin 6 is required for the development of collagen-induced arthritis. *J Exp Med* (1998) 187:461–8.

36. Sasai M, Saeki Y, Ohshima S, Nishioka K, Mima T, Tanaka T, et al. Delayed onset and reduced severity of collagen-induced arthritis in interleukin-6-deficient mice. *Arthritis Rheum* (1999) 42:1635–43.

37. Nishimoto N, Kishimoto T. Interleukin 6: from bench to bedside. *Nat Clin Pract Rheumatol* (2006) 2:619–26. doi:10.1038/ncprheum0338

38. Dinarello CA. Interleukin-1. *Cytokine Growth Factor Rev* (1997) 8:253–65.

39. Saijo S, Asano M, Horai R, Yamamoto H, Iwakura Y. Suppression of autoimmune arthritis in interleukin-1-deficient mice in which T cell activation is impaired due to low levels of CD40 ligand and OX40 expression on T cells. *Arthritis Rheum* (2002) 46:533–44.

40. Glaccum MB, Stocking KL, Charrier K, Smith JL, Willis CR, Maliszewski C, et al. Phenotypic and functional characterization of mice that lack the type I receptor for IL-1. *J Immunol* (1997) 159:3364–71.

41. Kobayashi K, Takahashi N, Jimi E, Udagawa N, Takami M, Kotake S, et al. Tumor necrosis factor alpha stimulates osteoclast differentiation by a mechanism independent of the ODF/RANKL-RANK interaction. *J Exp Med* (2000) 191:275–86.

42. Sabokbar A, Mahoney DJ, Hemingway F, Athanasou NA. Non-canonical (RANKL-independent) pathways of osteoclast differentiation and their role in musculoskeletal diseases. *Clin Rev Allergy Immunol* (2015) 51(1):16–26. doi:10.1007/s12016-015-8523-6

43. Lam J, Takeshita S, Barker JE, Kanagawa O, Ross FP, Teitelbaum SL. TNF-alpha induces osteoclastogenesis by direct stimulation of macrophages exposed to permissive levels of RANK ligand. *J Clin Invest* (2000) 106:1481–8. doi:10.1172/JCI11176

44. Yokota K, Sato K, Miyazaki T, Kitaura H, Kayama H, Miyoshi F, et al. Combination of tumor necrosis factor alpha and interleukin-6 induces mouse osteoclast-like cells with bone resorption activity both in vitro and in vivo. *Arthritis Rheumatol* (2014) 66:121–9. doi:10.1002/art.38218

45. Kim N, Kadono Y, Takami M, Lee J, Lee SH, Okada F, et al. Osteoclast differentiation independent of the TRANCE-RANK-TRAF6 axis. *J Exp Med* (2005) 202:589–95. doi:10.1084/jem.20050978

46. Yarilina A, Xu K, Chen J, Ivashkiv LB. TNF activates calcium-nuclear factor of activated T cells (NFAT)c1 signaling pathways in human macrophages. *Proc Natl Acad Sci U S A* (2011) 108:1573–8. doi:10.1073/pnas.1010030108

47. Madhok R, Crilly A, Watson J, Capell HA. Serum interleukin 6 levels in rheumatoid arthritis: correlations with clinical and laboratory indices of disease activity. *Ann Rheum Dis* (1993) 52:232–4.

48. Sack U, Kinne RW, Marx T, Heppt P, Bender S, Emmrich F. Interleukin-6 in synovial fluid is closely associated with chronic synovitis in rheumatoid arthritis. *Rheumatol Int* (1993) 13:45–51.

49. Axmann R, Bohm C, Kronke G, Zwerina J, Smolen J, Schett G. Inhibition of interleukin-6 receptor directly blocks osteoclast formation in vitro and in vivo. *Arthritis Rheum* (2009) 60:2747–56. doi:10.1002/art.24781

50. Stashenko P, Dewhirst FE, Rooney ML, Desjardins LA, Heeley JD. Interleukin-1 beta is a potent inhibitor of bone formation in vitro. *J Bone Miner Res* (1987) 2:559–65.

51. Akatsu T, Takahashi N, Udagawa N, Imamura K, Yamaguchi A, Sato K, et al. Role of prostaglandins in interleukin-1-induced bone resorption in mice in vitro. *J Bone Miner Res* (1991) 6:183–9.

52. Hofbauer LC, Lacey DL, Dunstan CR, Spelsberg TC, Riggs BL, Khosla S. Interleukin-1beta and tumor necrosis factor-alpha, but not interleukin-6, stimulate osteoprotegerin ligand gene expression in human osteoblastic cells. *Bone* (1999) 25:255–9.

53. Feng X, Teitelbaum SL. Osteoclasts: new insights. *Bone Res* (2013) 1:11–26. doi:10.4248/BR201301003

54. Takayanagi H. The role of NFAT in osteoclast formation. *Ann N Y Acad Sci* (2007) 1116:227–37. doi:10.1196/annals.1402.071

55. Novack DV. Inflammatory osteoclasts, a different breed of bone eaters? *Arthritis Rheumatol* (2016) 68(12):2834–6. doi:10.1002/art.39835

56. Orth JH, Aktories K, Kubatzky KF. Modulation of host cell gene expression through activation of STAT transcription factors by *Pasteurella multocida* toxin. *J Biol Chem* (2007) 282:3050–7. doi:10.1074/jbc.M609018200

57. Hildebrand D, Heeg K, Kubatzky KF. *Pasteurella multocida* toxin manipulates T cell differentiation. *Front Microbiol* (2015) 6:1273. doi:10.3389/fmicb.2015.01273

58. Wang Y, Li Y, He Y, Sun Y, Sun W, Xie Q, et al. Expression of G protein alphaq subunit is decreased in lymphocytes from patients with rheumatoid arthritis and is correlated with disease activity. *Scand J Immunol* (2012) 75:203–9. doi:10.1111/j.1365-3083.2011.02635.x

59. Liu Y, Wang D, Li F, Shi G. Galphaq controls rheumatoid arthritis via regulation of Th17 differentiation. *Immunol Cell Biol* (2015) 93:616–24. doi:10.1038/icb.2015.13

60. Park B, Yang YM, Choi BJ, Kim MS, Shin DM. Activation of G proteins by aluminum fluoride enhances RANKL-mediated osteoclastogenesis. *Korean J Physiol Pharmacol* (2013) 17:427–33. doi:10.4196/kjpp.2013.17.5.427

61. Dela Cruz A, Grynpas MD, Mitchell J. Elevated galpha11 expression in osteoblast lineage cells promotes osteoclastogenesis and leads to enhanced trabecular bone accrual in response to pamidronate. *Am J Physiol Endocrinol Metab* (2016) 310:E811–20. doi:10.1152/ajpendo.00049.2016

62. Jo M, Jung ST. Engineering therapeutic antibodies targeting G-protein-coupled receptors. *Exp Mol Med* (2016) 48:e207. doi:10.1038/emm.2015.105

20

Regulatory Dynamics of Cell Differentiation Revealed by True Time Series from Multinucleate Single Cells

Anna Pretschner[1], Sophie Pabel[1], Markus Haas[1], Monika Heiner[2] and Wolfgang Marwan[1]*

[1] *Magdeburg Centre for Systems Biology and Institute of Biology, Otto von Guericke University, Magdeburg, Germany,*
[2] *Computer Science Institute, Brandenburg University of Technology Cottbus-Senftenberg, Cottbus, Germany*

Edited by:
Jianhua Xing,
University of Pittsburgh, United States

Reviewed by:
Shou-Wen Wang,
Harvard Medical School,
United States
Junyue Cao,
The Rockefeller University,
United States

*Correspondence:
Wolfgang Marwan
wolfgang.marwan@ovgu.de

Dynamics of cell fate decisions are commonly investigated by inferring temporal sequences of gene expression states by assembling snapshots of individual cells where each cell is measured once. Ordering cells according to minimal differences in expression patterns and assuming that differentiation occurs by a sequence of irreversible steps, yields unidirectional, eventually branching Markov chains with a single source node. In an alternative approach, we used multi-nucleate cells to follow gene expression taking true time series. Assembling state machines, each made from single-cell trajectories, gives a network of highly structured Markov chains of states with different source and sink nodes including cycles, revealing essential information on the dynamics of regulatory events. We argue that the obtained networks depict aspects of the Waddington landscape of cell differentiation and characterize them as reachability graphs that provide the basis for the reconstruction of the underlying gene regulatory network.

Keywords: single cell time series, gene regulatory network, Petri net, Markov chain, systems biology, Waddington landscape

INTRODUCTION

Single-cell analyses revealed complex dynamics of gene regulation in differentiating cells (Spiller et al., 2010; Junker and van Oudenaarden, 2014; Paul et al., 2015; Marr et al., 2016; Plass et al., 2018). It is believed that dynamic effects possibly superimposed by stochastic fluctuations in gene expression levels may play crucial roles in cell fate choice, commitment, and reprogramming (Graf and Enver, 2009; Huang et al., 2009; Zhou and Huang, 2011; FerrellJr., 2012; Il Joo et al., 2018; Bornholdt and Kauffman, 2019). Changes in gene expression over time have not been directly measured in single mammalian cells as cells are - for technical reasons - sacrificed during the analysis procedure and hence can be measured only once. Instead, algorithms have been developed to infer the gene expression trajectory of a typical cell in pseudo-time from static snapshots of gene expression states in a cell population, resulting in Markov chains of states (Bendall et al., 2014; Cannoodt et al., 2016; Chen et al., 2019; Saelens et al., 2019; Setty et al., 2019). Most trajectory inference algorithms are based on the assumption that differentiation is unidirectional (Bendall et al., 2014; Haghverdi et al., 2016; Street et al., 2018; Saelens et al., 2019) and that the probability of transiting from one state to the next similar state is independent of the individual history of a cell (Setty et al., 2019). The inference of trajectories has been used to create pseudo-time series for differentiation (Marco et al., 2014; Moignard et al., 2015; Shin et al., 2015; Macaulay et al., 2016), cell cycle (Kafri et al., 2013), and the response to perturbation (Gaublomme Jellert et al., 2015). As any given distribution of expression patterns could result from multiple dynamics, the reconstruction of trajectories from snapshots faces fundamental limits (Weinreb et al., 2018). Even though regulatory mechanisms cannot be directly and rigorously inferred from snapshots (Weinreb et al., 2018), dynamic analyses may be of immediate importance to resolve competing views on basic mechanisms and the role of stochasticity in cell fate decisions (Moris et al., 2016).

True single cell time series can be obtained in *Physarum polycephalum* by taking multiple samples of one and the same giant cell. *Physarum* belongs to the amoebozoa group of organisms. It has a complex, prototypical eukaryote genome (Schaap et al., 2016) and forms different cell types during its life cycle (Alexopoulos and Mims, 1979).

Giant, multi-nucleate cells, so-called plasmodia provide a source of macroscopic amounts of homogeneous protoplasm with a naturally synchronous population of nuclei, which is continually mixed by vigorous shuttle-streaming (Guttes and Guttes, 1961, 1964; Rusch et al., 1966; Dove et al., 1986). The differentiation of a plasmodium into fruiting bodies involves extensive remodeling of signal transduction and transcription factor networks with alterations at the transcriptional, translational, and post-translational level (Glöckner and Marwan, 2017).

In starving plasmodial cells, the formation of fruiting bodies can be experimentally triggered by a brief pulse of far-red light received by phytochrome as photoreceptor (Starostzik and Marwan, 1995b; Lamparter and Marwan, 2001; Schaap et al., 2016). Retrieving small samples of the same plasmodial cell before and at different time points after an inductive light pulse allows to follow how gene expression changes over real time. Because cell cycle, cell fate choice, and development are synchronous throughout the plasmodium (Rusch et al., 1966; Starostzik and Marwan, 1995a; Hoffmann et al., 2012; Walter et al., 2013; Rätzel and Marwan, 2015), single-cell gene expression trajectories can indeed be constructed from time series. By assembling finite state machines made from trajectories we have constructed Petri net models for the state transitions that predict Markov chains as variable developmental routes to differentiation (Werthmann and Marwan, 2017; Rätzel et al., 2020) which may be considered as trajectories through the Waddington landscape (Waddington, 1957; Huang et al., 2009). These Petri nets also predict reversible and irreversible steps, commitment points, and meta-stable states in cells responding to a differentiation stimulus. However, the computational approach for the construction of Petri nets from time series has been originally developed with data sets of a coarse resolution in time and the structural resolution of the nets was accordingly limited. Nevertheless, the approach turned out to be useful for capturing the dynamics of the process. For this paper, we developed a method for retrieving smaller samples from even larger plasmodial cells and showed that these cells provide a homogeneous source for samples to

be taken. This allowed us to considerably improve the time resolution as compared to previous studies. Sampling cells at higher time resolution, allowed the construction of Petri nets with enhanced structural and dynamic resolution. Structural complexity, highly connected nodes, parallel pathways, reversible reactions, and Petri net places representing meta-stable states in the developmental network, as revealed by the new data sets, characterize the differentiation response as complex and dynamic in contrast to a smooth, continuous process. We describe the graph properties of the Waddington Petri nets and conclude that the gene expression dynamics revealed by our analysis most likely emerge from the non-linear dynamic behavior of the underlying regulatory network rather than from stochastic fluctuations in the concentration of regulatory molecules.

MATERIALS AND METHODS

Plasmodial Strain, Growth of Cells, Sample Preparation, and Gene Expression Analysis

Sporulation-competent plasmodial cells of wild type strain LU897 × LU898 (Starostzik and Marwan, 1998) were obtained as previously described (Starostzik and Marwan, 1998; Rätzel et al., 2020). A total of 2.8 gram of plasmodial mass was applied to a 14 cm Ø Petri dish that contained 90 ml of semi-rich Golderer agar (Golderer et al., 2001), based on a salt solution of 0.01% (w/v) niacin, 0.01% (w/v) niacinamide, 0.1% (w/v) $CaCO_3$, and 0.14 mM $CuCl_2$, supplemented with 5 g peptone from meat (Sigma Aldrich), 0.75 g yeast extract (Becton, Dickinson & Co.), and 3.9 mM glucose per liter, adjusted to pH 4.6 with concentrated HCl. After starvation for 7 days at 22°C in complete darkness, sporulation was induced with a 15 min pulse of far-red light ($\lambda \geq 700$ nm, 13 W/m^2) (Starostzik and Marwan, 1998). Before and at 1-h time intervals after the start of the far-red pulse, samples were taken in duplicate at arbitrarily chosen but distant positions on the plate. Each sample was obtained by picking an agar plug of 1.13 cm^2 with the cut bulb of a disposable Pasteur pipette (EA62.1; Carl Roth, Karlsruhe, Germany). The plasmodial mass on the agar plug was scraped off with a pipet tip and, by cutting the tip, transferred into a vial of glass beads immersed in liquid nitrogen (**Figure 1**). After extraction of RNA and removal of contaminating DNA (Marquardt et al., 2017), the relative abundance of the mRNAs of 35 genes, differentiation marker and reference genes (Hoffmann et al., 2012; **Supplementary Table 2**) was analyzed by gene expression profiling (GeXP), a multiplex RT-PCR method (Hayashi et al., 2007) as previously described (Rätzel and Marwan, 2015; Marquardt et al., 2017).

Data Analysis Pipeline and Automated Generation of Petri Nets

To correct for differences in the concentration of total RNA and in the efficiency of the RT-PCR reaction, the gene expression values were normalized to the median of the estimated relative concentrations of mRNAs of the 35 genes in each RNA sample. Each normalized expression value was subsequently normalized

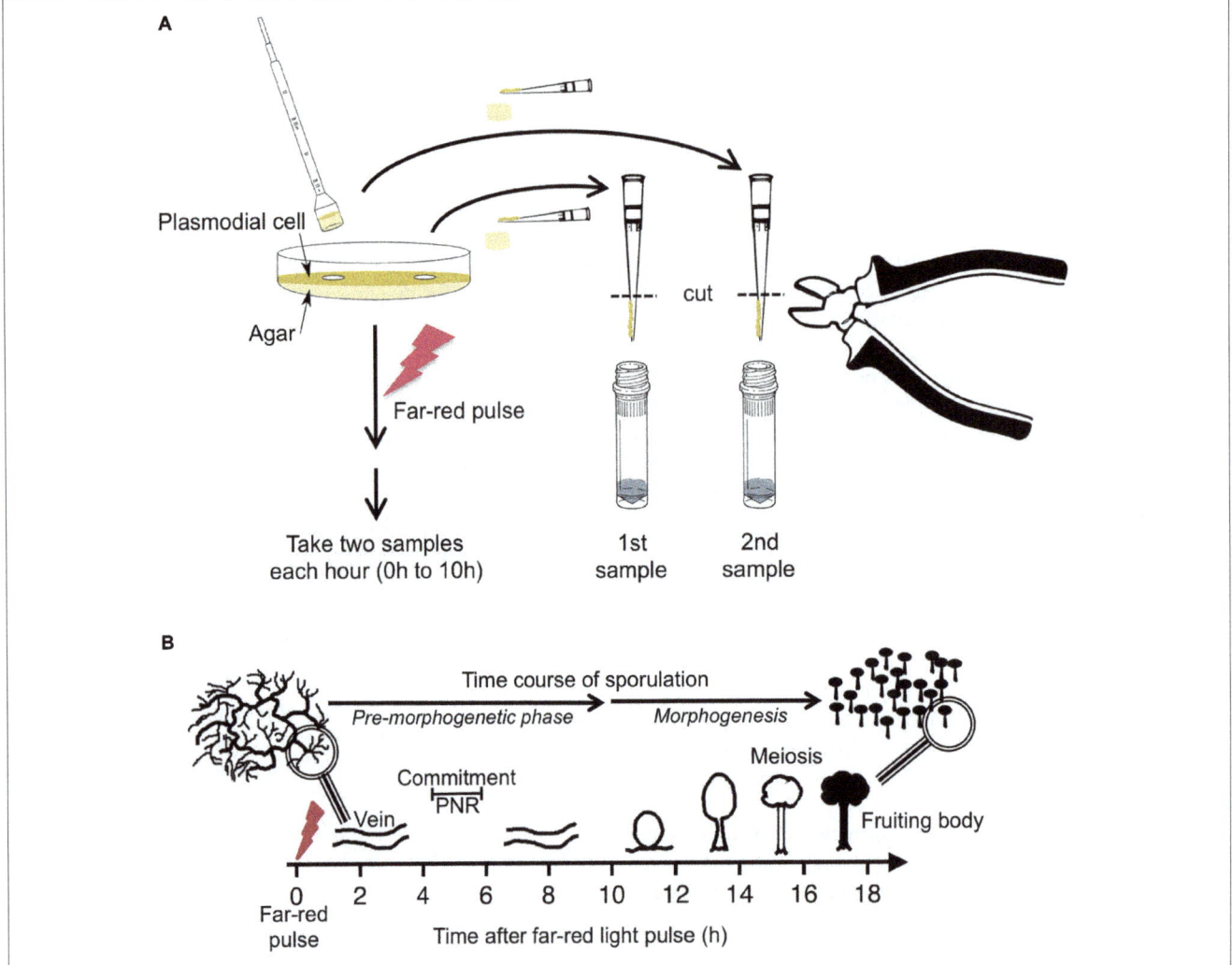

FIGURE 1 | Experimental protocol for taking time series by repeated sampling of individual plasmodial cells and time course of light-induced sporulation. **(A)** Each Petri dish contained one individual plasmodial cell supported by an agar substratum. Before and at 1-h time intervals after stimulation of the cell with a pulse of far-red light, samples were taken in duplicate by picking an agar plug at arbitrarily chosen but distant positions on the plate. The cell mass was scraped off from the agar plug with a pipet tip and transferred into a vial containing glass beads and liquid nitrogen. After purification of RNA, the gene expression pattern was estimated twice in each sample, each with two independent multiplex RT-PCR reactions (see section "Materials and Methods" for details). For dark controls, the far-red light stimulus was omitted. **(B)** Time frame of light-induced sporulation of a plasmodial cell. At about four to 6 h after stimulation with a pulse of far-red light, the cell is irreversibly committed to sporulation by crossing the point of no return (PNR) while there is no obvious change in the plasmodial morphology. Morphogenesis then starts at about 11 h after the stimulus by the formation of nodules that subsequently culminate to form the fruiting bodies. Panel B was taken from Glöckner and Marwan (2017).

to the geometric mean of all values obtained for a given gene, and this was performed separately for each gene.

Data were analyzed and processed with a revised and extended pipeline written in R (R Core Team., 2016), based on the previously described script (Rätzel et al., 2020). The normalized gene expression data were clustered and significant clusters were determined with the help of the Simprof algorithm (Clarke et al., 2008) as provided by the clustsig package (Whitaker and Christman, 2014). Expression patterns were visualized in the form of a heatmap generated by the heatmap.2 function, provided as part of the gplots package (Warnes et al., 2016). Changes in gene expression over time were visualized by multidimensional

scaling based on Euclidean distance (Gower, 1966) with the help of the *cmdscale* function provided as part of the stats package v3.5.1 (R Core Team., 2016). Petri nets were constructed from single cell trajectories of gene expression as previously described (Rätzel et al., 2020). Each trajectory is a temporal sequence of gene expression states, where each state corresponds to a Simprof significant cluster. Petri nets specified in ANDL format (Abstract Net Description Language) (Heiner et al., 2013) were imported into Snoopy (Rohr et al., 2010) and graphically displayed by running the Sugiyama layout algorithm (Sugiyama et al., 1981). Petri net places, each representing a gene expression state, were colored according to the relative temporal stability

Regulatory Dynamics of Cell Differentiation Revealed by True Time Series from Multinucleate...

197

of the expression state or according to the relative frequency with which each gene expression state occurred. Petri net transitions, corresponding to transits between states were colored according to the frequency with which each transit occurred or according to the data subset in which the transit occurred. These parameters were computed and coloring was performed by automatic editing of Snoopy files encoded in xml format, again with the help of a R script. The raw data and the complete computational pipeline used in this study is provided as part of the **Supplementary Material**.

RESULTS

Even Large Plasmodia Provide a Source of Homogeneous Cell Material for True Time Series Analysis of Gene Expression

In previous studies we have shown that the gene expression pattern in samples taken at the same time from different sites of a plasmodium covering a standard Petri dish (9 cm Ø) did not change within the limits of accuracy of the measurements. Accordingly, repeated sampling of the same plasmodial cell yields true time series (Rätzel, 2015; Werthmann and Marwan, 2017). To allow more samples to be taken without consuming too much of the plasmodial mass, we now prepared plasmodia on 14 cm Ø Petri dishes, increasing the surface area covered by the plasmodial mass by 2.4-fold, and took smaller samples by punching agar plugs of 1.13 cm^2 per sample, to harvest a small portion of the initial total plasmodial mass. To test whether the homogeneity in gene expression is impaired or even lost in the larger plasmodia, we took 9 or 16 samples at the same time from approximately evenly spread sites of a plasmodium, and estimated the gene expression pattern twice in the RNA of each sample, to obtain one technical replicate of each measurement (Aselmeyer, 2019; Driesch, 2019). This allowed to estimate the biological variation in gene expression within a plasmodium as compared to the technical accuracy of the measurements. In order to correct for potential differences in the efficiency of the RT-PCR between reactions, the expression value for each gene was normalized to the median of the expression values of all genes measured in the sample (for details see Materials and Methods). To estimate the technical accuracy of the measurements, the relative deviation of first and second measurement from the mean of the two measurements was estimated for each assayed gene in each of the retrieved plasmodial samples. The frequency distribution of all values was almost symmetrical with a tail consisting of a small number of low values, obviously as a result of inefficient RT-PCR reactions. To estimate the degree of homogeneity in gene expression within a plasmodial cell, we asked to which extent the expression values for the 9 or 16 samples taken from the same plasmodium deviated from their median. To restrict the influence of technical artifacts on the result, we considered the subset of the data where first and the second measurement of the same plasmodial sample deviated not more than two-fold from the mean of the two values. In three of the total of 46 analyzed plasmodia (30 far-red stimulated; 16 dark controls), individual

samples deviated from the rest of the samples of the same plasmodium by more than a factor of two. As errors in sample preparation could not be ruled out, these three plasmodia were excluded and the remaining data set of 43 plasmodia (28 far-red stimulated; 15 dark controls) was analyzed taking the mean of 1st and 2nd measurement for each gene in each sample. Among the total of 7160 values, 98% of the symmetric frequency distribution (**Supplementary Figure 1**) were between 0.48- and 2.10-fold deviation of the median of all values of the respective plasmodium (**Supplementary Table 1**). There was no obvious difference between dark controls and far-red stimulated plasmodia which were measured at 6 h after the light pulse when genes were already differentially regulated (**Supplementary Figure 1** and **Supplementary Table 1**), indicating that even during the period where the mRNA abundance changed in time, the homogeneity in gene expression levels is maintained. Visual inspection of outliers within the distribution did not reveal any candidates for specific genes that might be inhomogeneously expressed.

In summary, the gene expression values throughout a plasmodium deviated not more than approximately two-fold from the median of all samples from the same plasmodium and were thus within the limits of the technical accuracy of the measurements, even under conditions were genes were in the process of being up- or down-regulated. These differences measured between samples were minor as compared to the differential regulation where the expression level of genes changed in the order of ten to more than hundred-fold (**Supplementary Figure 2**). These results are consistent with the results of the time-series experiment, where for each time point, two samples were retrieved and analyzed from the same plasmodial cell (see below).

Sampling of Plasmodia at 1 h Time Interval

As the assayed genes were evenly expressed and changed evenly in time throughout the large plasmodia, at least within the limits of accuracy of the measurements, we took time series at 1 h time intervals. In order to assay, in each experiment, for the homogeneity and synchrony in gene expression throughout the plasmodium, we took two samples at each time point from different, arbitrarily chosen but distant sites of the plasmodium (**Figure 1**). In far-red stimulated plasmodia, the first samples (referred to as the 0 h samples) were taken at the start of the experiment, i.e., immediately before application of the 15 min pulse of far-red light. All subsequent samples were taken at 1 h time intervals until 10 h after the start of the experiment [At 5 to 6 h after the far-red pulse cells have passed the commitment point, while visible morphogenesis starts several hours later by entering the transient nodulation stage at about 11 h after the pulse (Hoffmann et al., 2012)]. In the dark controls, the far-red stimulus was omitted. Gene expression in each plasmodial sample taken at a given time point was analyzed twice by GeXP-RT-PCR, where the measurement and the corresponding technical replicate are referred to as 1st and 2nd measurement for sample #1, and 3rd and 4th measurement for sample #2, respectively. Data were normalized as described above.

The technical quality of the measurements was estimated separately for the two data sets, each comprising the data of the samples collected at the 11 time points of the time series. For each plasmodial sample, the relative deviation of the two measurements (1st and 2nd, or 3rd and 4th) from the mean of the two measurements was estimated. The frequency distributions of the deviations and corresponding quantil values indicated that the technical qualities of 1st and 2nd, as well as 3rd and 4th measurement were virtually identical with 95% of the values differing less than a factor of two from each other (**Figure 2** and **Table 1**).

The degree of spatial variability of gene expression within a plasmodium was estimated by combining the data sets for the first and the second sample of a plasmodium taken at each time point of the time series. The frequency distribution of the deviation of each measurement from the mean of 1st, 2nd, 3rd, and 4th measurement of the two samples taken from each plasmodium at any time point was virtually identical to the frequency distributions obtained for the technical replicates, indicating that gene expression within the analyzed plasmodia varied at maximum within the limits of accuracy of the measurements (within a factor of 2 in 95% of the samples). This conclusion is based on the comparison of the quantile distributions of the data sets (**Figure 2** and **Table 1**) considering a total of 36,540 data points.

Multi-Dimensional Scaling Analysis

With this data set, we investigated how expression changes as a function of time in the individual plasmodial cells. The gene expression pattern of a plasmodial cell at a given time point was obtained as the mean of the four expression values of each gene measured in the two plasmodial samples picked at that time point.

For visual representation of the data set and of single-cell trajectories of gene expression, we performed multidimensional scaling (MDS) to obtain a data point for the expression pattern of each cell at each time point. Single-cell trajectories of gene expression are shown in **Figure 3**. Notably, the gene expression patterns of un-stimulated cells (dark controls) changed as a function of time with the highest variability along coordinate 2 of the MDS plot **Figures 3A,B**. Trajectories of far-red stimulated cells (**Figures 3C,D**) moved from the left side to the right side of the plot, while the shape of individual trajectories varied to a certain extent, indicating that the response of the cells was similar though not identical. Obviously, the trajectories of six of the eight far-red-stimulated plasmodia of experiment #1 traversed a considerably larger area of the MDS plot (**Figure 3C**) as compared to the other stimulated cells, indicating a larger variation in gene expression during the response to the stimulus. The extent of variation is accordingly obvious when the bulk of data points is placed in the same plot (**Figure 4A**). To search for genes that may account for the scattering along coordinate 2, we visually inspected the individual time series displayed in the form of a heat map (**Supplementary Figure 2**). In addition to the genes that were clearly up- or down-regulated in response to the stimulus, the messages of four genes, *hstA*, *nhpA*, *pcnA*, and *uchA*,

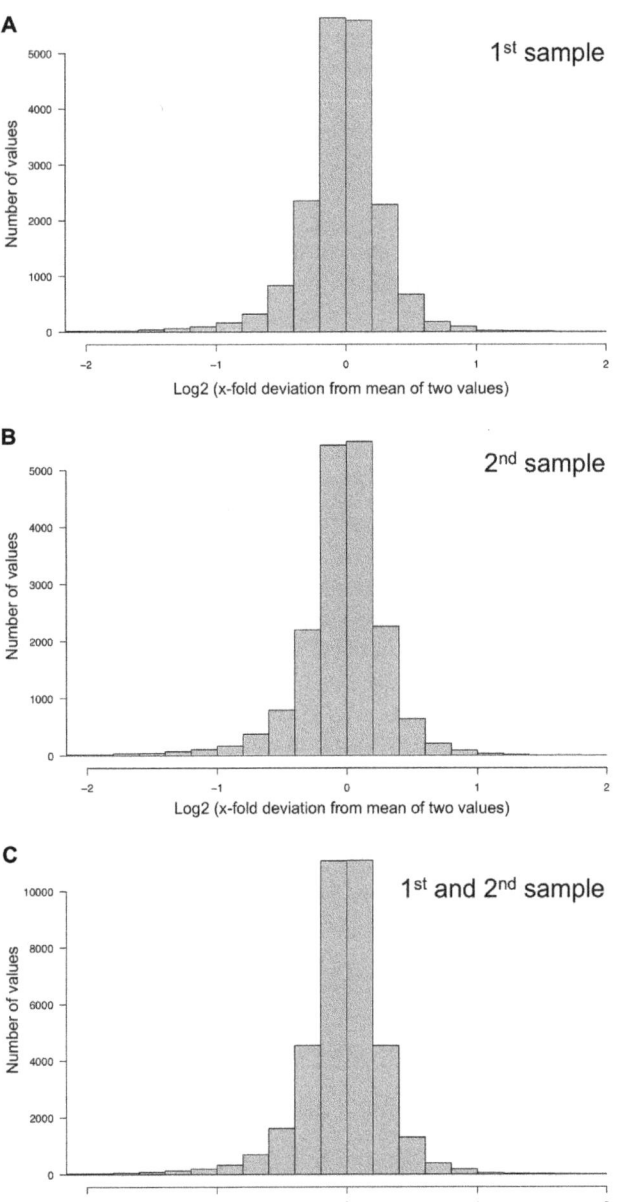

FIGURE 2 | Technical accuracy of measurements and homogeneity of gene expression within a plasmodial cell as determined by technical and biological replicates, taken in experiments #1 and #2 (**Table 2**). **(A,B)** Technical accuracy of measurements of gene expression. The concentration of the mRNAs of the set of 35 genes (**Supplementary Table 2**) was determined twice by RT-PCR for each RNA sample. The frequency distributions display the Log2 of the x-fold deviation of each expression value of each gene from the mean of the two values obtained by technical replication. Panels **(A,B)** show the results obtained for each of the two biological samples [**(A)**, sample #1; **(B)**, sample #2], that both were simultaneously taken from the same plasmodial cell at any time point during the experiments. **(C)** Combination of the data sets shown in panels **(A,B)**. This frequency distribution shows the deviation of each measurement from the mean of four values, obtained by twice measuring each of the two biological samples simultaneously taken from the same plasmodium at any time point of the experiments. The figure represents the complete data set of 36,540 data points that was analyzed in the present study.

TABLE 1 | Quantil distributions of the x-fold deviation (x) of a value from the mean of two or four values, characterizing the reproducibility of measurements as estimated through technical and biological replicates, respectively.

Measurements	1st and 2nd		3rd and 4th		1st to 4th	
Percent of values	Quantile (Log2 (x))	Quantile (x)	Quantile (Log2 (x))	Quantile (x)	Quantile (Log2 (x))	Quantile (x)
1%	−1.308	0.404	−1.383	0.383	−1.349	0.392
5%	−0.555	0.681	−0.592	0.663	−0.574	0.672
25%	−0.169	0.890	−0.169	0.890	−0.169	0.890
50%	−0.010	0.993	−0.008	0.995	−0.009	0.994
75%	0.137	1.100	0.142	1.103	0.139	1.101
95%	0.414	1.333	0.415	1.334	0.415	1.333
99%	0.750	1.682	0.746	1.677	0.747	1.679

The table quantitatively characterizes the frequency distributions shown in **Figure 2**.

in the following called *pcnA*-group genes, changed over time in some of the plasmodia, but there was no obvious consistent relationship to the time point of stimulus application. When only genes were included in the analysis that were clearly up- or down-regulated in response to the stimulus (**Supplementary Figure 2**, see also **Figure 8**), the data points of the MDS plot were indeed less scattered (**Figure 4B**). A qualitatively similar result was obtained plotting the up-regulated and the down-regulated genes separately (**Figures 4C,D**), with some more variation in the expression of the down-regulated genes. However, expression of the *pcnA*-group over time (**Figure 4E**) was clearly different from the up- or down-regulated genes. Expression of the pcnA-group genes was different between cells from experiments #1 and #2 as seen from the trajectories of the cells (**Supplementary Figure 3**), suggesting that ongoing internal processes in cells of experiment #1 might even influence their response to the light stimulus. Indeed, according to the corresponding MDS plots of the bulk data points (**Supplementary Figure 4**) the response of the up- and down-regulated genes was less uniformly in cells of experiment #1 (**Supplementary Figure 4C**) as compared to those of experiment #2 (**Supplementary Figure 4D**).

Construction and Graph Properties of Waddington Landscape Petri Nets

For a further analysis, we performed hierarchical clustering of the expression data for all assayed 35 genes, differentiation marker and reference genes (**Supplementary Table 2**), and identified significantly different clusters of expression patterns with the help of the Simprof algorithm (Clarke et al., 2008; **Supplementary Figure 5**). The temporal sequences of gene expression patterns classified as Simprof significant clusters defined a trajectory for each individual cell and revealed significant differences between cell trajectories (**Table 2**). To relate gene expression states and trajectories we constructed a Petri net (bipartite graph) as previously described (Werthmann and Marwan, 2017; Rätzel et al., 2020), by representing each gene expression state by a *place* and the temporal transit between two states by a *transition* (**Figure 5**). A single token marking one place of the Petri net indicates the current gene expression state of a cell. The token moves from its place to a downstream place when the transition, connecting the two places through directed arcs, fires. As each

transition is connected to exactly two places (one pre-place and one post-place), tokens are neither formed nor destroyed when moving through the net, so the gene expression state of the cell remains unequivocally defined at any time. The coherent Petri net obtained this way represents a state machine predicting possible developmental trajectories in terms of Markov chains of gene expression states (Rätzel et al., 2020).

The basic modeling principles are summarized in **Table 3**. We observe the following structural properties of the model which we call '*Waddington landscape Petri net*':

- Each transition has exactly one pre-place and one post-place.
- There are places having more than one post-transition. These post-transitions are in conflict. But, because every transition has exactly one preplace, each conflict is a free choice conflict, meaning the token is free to choose which route to take, predicting a corresponding free choice for the cell (see Discussion).
- There are places having more than one pre-transition, i.e., alternative paths may re-join. Thus, the Petri net structure does not form a tree.
- There are cycles: a cell may switch back to previous states or oscillate between states as defined by the expression patterns of the set of observed genes.

For technical reasons we add immediate transitions starting alternative trajectories, in order to get a statistical distribution of states in which the experiments have started or will start with a given probability.

In contrast to most state-of-the-art pseudo-time series approaches found in the literature (Saelens et al., 2019), the structure of the Waddington landscape Petri net is not restricted to a partial order, meaning it is neither restricted to a directed acyclic graph nor to a tree. Instead we obtain what is known in Petri net theory as '*state machine*,' also called in other communities 'finite state machine' or 'finite automata,' which may involve cycles.

A state machine with one token and its reachability graph, or Markov chain for stochastic Petri nets, are isomorph (i.e., have the same structure, there is a 1-to-1 correspondence); to put it differently: our (stochastic) Petri net represents the

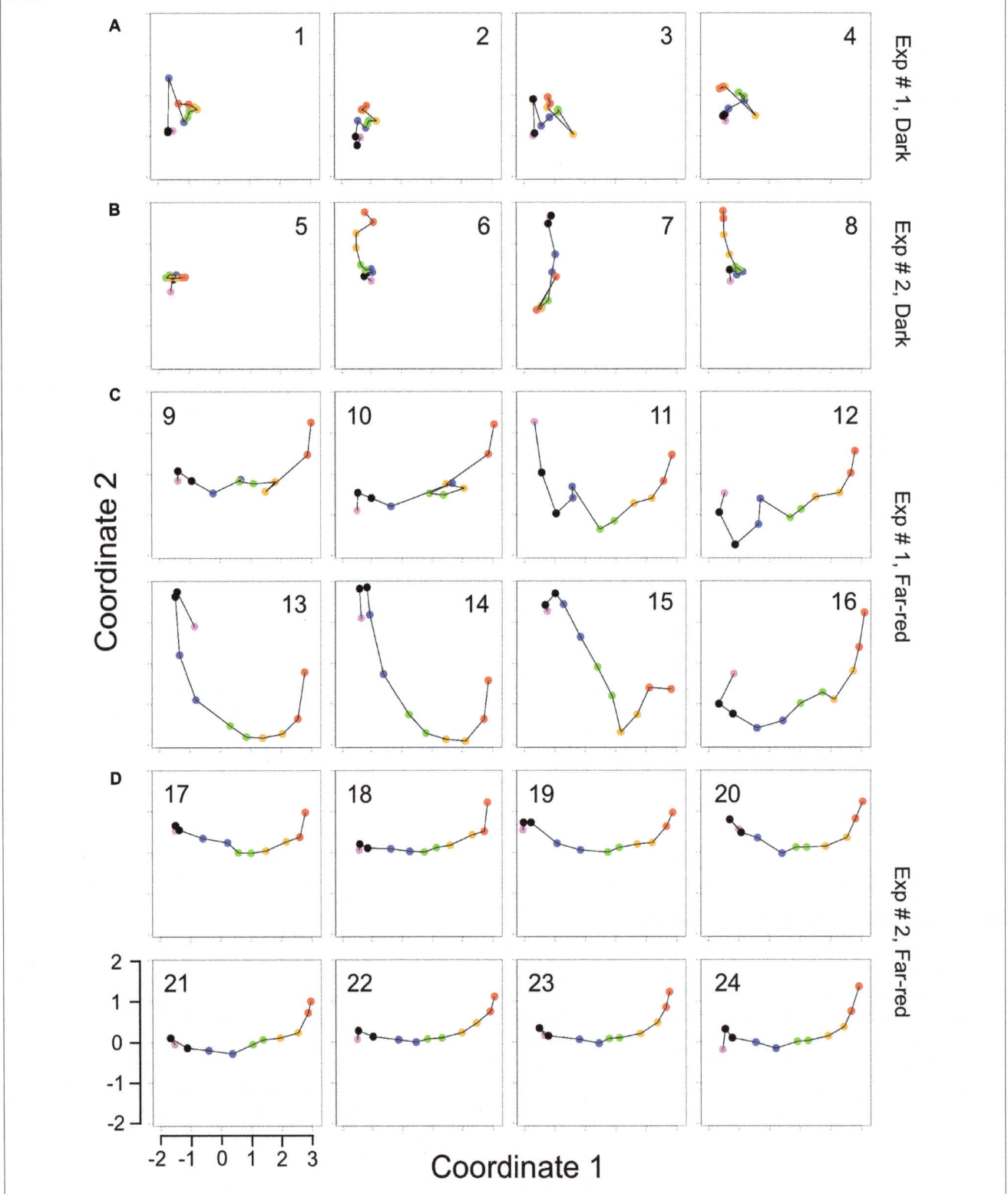

FIGURE 3 | Single cell trajectories of gene expression displayed after multidimensional scaling (MDS) of the expression patterns. Panels **(A,B)** show the trajectories of unstimulated cells of experiment #1 and experiment #2, respectively. Panels **(C,D)** show the trajectories of far-red stimulated cells of the two experiments. Each data point represents the gene expression pattern of a cell at a given time point at 1 h time intervals. The start position (0 h) of each trajectory is encoded in pink and the endpoint (10 h) in red. The number displayed in each subpanel refers to the ID number of the plasmodial cell (P1 to P24) as listed in **Table 2**. All cells of the dark controls did not sporulate while all far-red irradiated cells sporulated. All plots are displayed at the same scale.

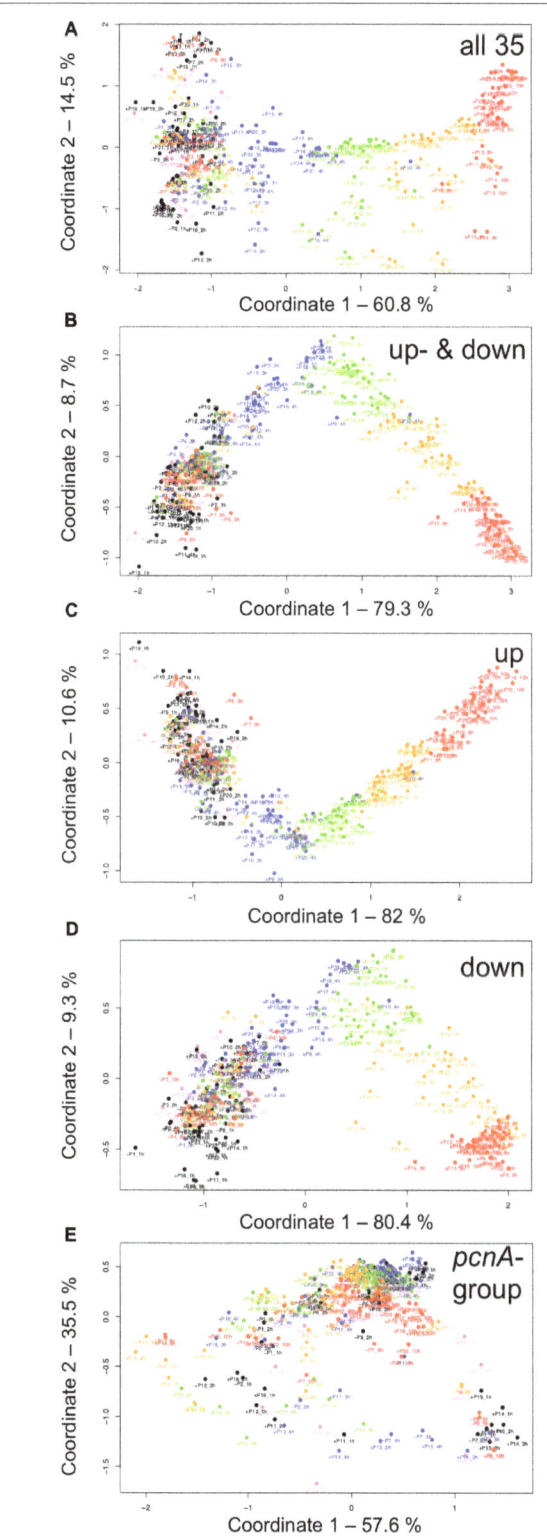

FIGURE 4 | Continued

of an individual cell at a given time point. Time is encoded by color (0 h, pink; 1 h, 2 h, black; 3 h, 4 h, blue; 5 h, 6 h, green; 7 h, 8 h, ocher; 9 h, 10 h, red). Developmental destiny, cell number (as assigned in **Table 2**), and time is given for each data point. The label + P12_1 h, for example, indicates that the data point refers to the expression pattern of plasmodium number 12 as measured at 1 h after the start of the experiment (corresponding to the onset of the far-red light stimulus in light-stimulated cells) and that the plasmodium had sporulated (+) in response to the stimulus (+, sporulated; -, not sporulated). The percent of variance is given for each coordinate.

Markov chain of states the cells assume in the course of their developmental trajectory and accordingly on their walk through the Waddington landscape. We assume that the Petri net represents the corresponding region of the Waddington landscape predicting possible developmental paths a single cell can follow, which of course yields a state machine.

Representing Markov chains as Petri nets comes with a couple of advantages.

First, Petri nets are equipped with the concept of T-invariants, which belong to the standard body of Petri net theory from very early on (Lautenbach, 1973). We consider T-invariants as crucial in terms of biological interpretation of the generated net structures (Sackmann et al., 2006; Heiner, 2009). The computation of T-invariants is rather straightforward for state machines; due to their simple structure it holds:

- each cycle in a state machine defines a T-invariant, and
- each elementary cycle (no repetition of transitions) is a minimal T-invariant.

Second, modeling the differentiation-inducing stimuli, what we have not done so far, would turn some of the free choice conflicts into non-free choice conflicts, which involves, technically speaking, leaving the state machine net class. To unequivocally identify transits that are stimulus-dependent, we need a higher data density which we will hopefully achieve in one of our next experiments. With stimulus-dependent transitions, the constructed Petri nets and their Markov chains do not coincide anymore, instead the Markov chains as well as the reachability graph are directly derived from the Petri nets and may be analyzed by standard algorithms. Finally, our Petri net approach paves the way for the actual ultimate goal of our future work - reconstructing the underlying gene regulatory networks based on the reachability graphs encoded by the Waddington landscape Petri nets.

Characterization of Petri Nets Constructed From Gene Expression Data

Figure 6 displays a Petri net assembled from cell trajectories considering the set of 35 genes. The graphical representation laid out using the Sugiyama algorithm emphasizes the directionality of concurrent processes (Sugiyama et al., 1981). The net indicates that cells started in different states (connected to C0; see Legend to **Figure 5** for details) and, after stimulation by far-red light, proceeded to a small set of terminal states (places C71, C76,

FIGURE 4 | Gene expression patterns of all analyzed cells displayed for different sub sets of genes. Multidimensional scaling was performed for the complete set of 35 genes **(A)**, the subset of the up- and the down-regulated genes **(B)**, or exclusively for up-regulated **(C)**, down-regulated **(D)** or the *pcnA*-group of genes **(E)**. Each data point represents the expression pattern

(Continued)

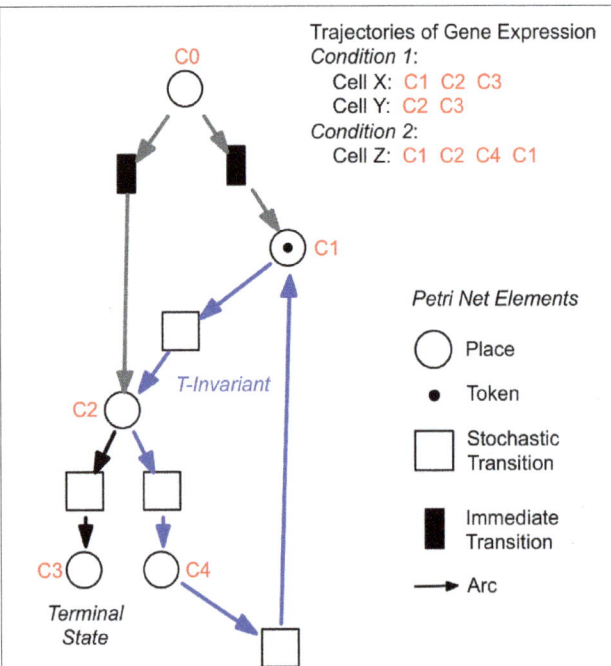

Trajectories of Gene Expression
Condition 1:
Cell X: C1 C2 C3
Cell Y: C2 C3
Condition 2:
Cell Z: C1 C2 C4 C1

Petri Net Elements

○ Place

• Token

▢ Stochastic Transition

■ Immediate Transition

→ Arc

FIGURE 5 | Construction of Petri nets from single cell trajectories of gene expression. Petri nets are directed, bipartite graphs with two types of nodes, places and transitions, that are connected by arcs (see symbols, lower right). Petri nets are used in this work to model state machines, as exemplified in the following. Any gene expression state of a cell, as defined by its assignment to a Simprof significant cluster of gene expression patterns, is represented by a corresponding place (drawn as a circle). Any transit between two states is mediated by a transition (drawn as a rectangle). The current gene expression state of a cell is indicated by one token which marks the respective place. When a place contains a token, one of its post-transitions can fire to move the token into its post-place. (A post-transition of a place is a downstream transition which is immediately connected to that place, as indicated by a directed arc). Because each transition of the Petri nets as they are used here, has exactly one pre-place (one incoming arc) and one post-place (one outgoing arc), and because all arc weights are one, tokens can neither be produced nor destroyed, and the state of gene expression remains unequivocally defined. A transit cycle, i.e., the ensemble of reactions that bring a subsystem back to the state from which it started, is called transition-invariant (T-invariant) (Sackmann et al., 2006). The arcs that contribute to the T-invariant of the Petri net displayed in the figure are highlighted in blue. Petri nets, as they are used in this paper, contain one additional place C0, which does not represent a gene expression state. For simulation, C0 defines the initial gene expression state of the cell by randomly delivering its token to one of the places that are connected to C0 through so-called *immediate transitions* (filled in black) that fire immediately when the simulation starts [for details see (Rätzel et al., 2020)]. Connection to C0 also graphically highlights the places representing those gene expression states in which cell trajectories started. In the example shown, the cell trajectory started in a gene expression state assigned to Simprof cluster C1. The token can move to C2 where it randomly moves to either C3 (a terminal state in this example) or to C4, from which it may return to C1 and possibly continue.

differently. Each place is colored according to the relative frequency of its corresponding gene expression state, indicating that some states occurred more frequently than others. From this representation it is obvious that cells of experiments #1 and #2 form different branches, in part projecting onto different terminal states, reflecting accordingly different developmental trajectories to commitment and sporulation. **Figure 6B** displays the same Petri net, but with a different color coding. Here, transitions are colored according to how frequent the corresponding transits occurred, indicating that some paths were more frequently taken than others. Places are colored according to their relative stability, defined as the average residence time of a cell in the respective state (see Methods for details). Coloring indicates that cells reached a terminal state through states of different stability, e.g., meta-stable intermediates.

Un-stimulated cells (**Table 2**, dark controls) spontaneously switched between significantly different states of gene expression. Their trajectories gave three disconnected Petri nets (**Figure 7A**; the three nets were connected to C0 for technical reasons, see legend to **Figure 5**).

Petri nets of **Figures 6A,B**, **7A** indicate that the expression pattern in both, stimulated and un-stimulated cells developed predominantly in forward directions while there were some transits back to previous states creating so-called transition-invariants (T-Invariants; **Figure 5**). We asked whether stimulus-independent temporal expression differences like those observed in the subset of the *pcnA*-group of genes might have added to this directedness. Therefore, we constructed a Petri net from trajectories based on significant clusters, this time exclusively clustering the subset of up- and down-regulated genes. Basic features found in the Petri nets of up- and down-regulated genes were similar to the ones found in the Petri nets for the full set of 35 genes: Trajectories formed parallel main branches, there were intermediate nodes of different stability, of different connectedness, and hence states that occurred with different frequency (**Supplementary Figure 6**). In contrast, there was a high number of minimal T-invariants that heavily involved places representing gene expression states that occurred in un-stimulated cells. A Petri net built by considering only transits that occurred in un-stimulated cells (**Figure 7B**) was nearly covered with T-Invariants, indicating spontaneous, reversible alterations in the expression of up- and/or down-regulated genes. Again, states of gene expression displayed different stability. Considerable variation in the expression level of the up-and down-regulated genes is even most obvious from the heat map of initial states from which trajectories emerged and of terminal states that were observed during the experiment (**Figure 8**). The high density of T-invariants in the un-stimulated cells suggests that similarly, the T-invariants involving places corresponding to light-stimulated cells are due to gene expression changes that do spontaneously occur before cells are caught by a new attractor formed in response to the far-red stimulus (see Discussion).

Figure 7 also shows that selecting sets or subsets of genes for hierarchical clustering and subsequent Petri net construction may yield Petri nets of different structure delivering accordingly non-redundant information on corresponding subsets. This is also shown in **Table 4** for the set of 35 genes and for subsets, the

C77, C78, C79, C95) via multiple, more or less highly connected intermediate states. To facilitate the interpretation of the Petri net, we have colored the places and transitions according to different criteria. In **Figure 6A**, the transitions being specific to cells of experiments #1 or #2 and for dark controls or light-stimulated cells in the respective experiments, are colored down-regulated, up-regulated, down- and up-regulated, and the *pcnA*-group of genes. Except for the *pcnA*-group, the number of places per gene was approximately the same. The number of transitions per gene however became less with more genes considered. This suggests that up-regulation, down-regulation and even expression of the *pcnA*-group of genes are at least partly coordinated or co-regulated processes. The average number of minimal T-invariants per gene compared for the different groups of genes (**Table 4**) suggests that reversibility observed for the subsets vanishes when more genes are considered, obviously due to combinatorial effects.

Single Cell Trajectories Reveal Qualitatively Different Patterns in Differential Gene Regulation

We have argued that Petri net modeling disentangles the complex response of cell reprogramming (Rätzel et al., 2020) and predicts feasible developmental pathways through the Waddington landscape, resulting in significantly distinct single cell trajectories. To reveal similarities and differences of the expression kinetics of individual genes, we plot the geometric mean of the concentration values of the mRNA in a cluster

TABLE 2 | Single cell trajectories of gene expression.

Experiment	Cell	0.0 h	1.0 h	2.0 h	3.0 h	4.0 h	5.0 h	6.0 h	7.0 h	8.0 h	9.0 h	10.0 h
Exp #1, Dark	P1	60	56	59	26	35	34	34	28	28	32	31
Exp #1, Dark	P2	57	57	60	63	35	35	35	14	32	32	32
Exp #1, Dark	P3	59	56	26	36	34	34	28	13	31	32	32
Exp #1, Dark	P4	37	37	37	33	28	28	28	15	29	30	30
Exp #2, Dark	P5	24	52	52	52	49	52	52	52	48	47	48
Exp #2, Dark	P6	25	45	45	44	44	43	43	21	7	5	1
Exp #2, Dark	P7	3	3	3	11	12	63	63	60	60	26	61
Exp #2, Dark	P8	46	45	45	42	42	42	43	21	7	6	1
Exp #1, Far-red	P9	33	23	27	16	101	111	105	87	88	70	79
Exp #1, Far-red	P10	36	23	27	16	88	111	105	89	80	75	78
Exp #1, Far-red	P11	10	53	54	54	15	98	100	89	91	93	71
Exp #1, Far-red	P12	53	55	54	54	15	100	100	89	91	92	71
Exp #1, Far-red	P13	8	4	4	12	53	69	69	68	66	65	95
Exp #1, Far-red	P14	22	4	2	9	53	69	69	68	66	65	95
Exp #1, Far-red	P15	22	4	2	9	96	96	69	67	64	94	95
Exp #1, Far-red	P16	62	55	58	54	97	99	89	90	93	75	78
Exp #2, Far-red	P17	50	40	45	19	104	103	105	87	82	72	79
Exp #2, Far-red	P18	51	41	25	19	104	102	110	85	81	72	77
Exp #2, Far-red	P19	20	20	20	17	18	102	109	83	82	72	77
Exp #2, Far-red	P20	44	40	45	19	103	108	110	84	82	73	77
Exp #2, Far-red	P21	51	39	45	19	103	106	110	84	82	74	76
Exp #2, Far-red	P22	52	39	45	19	103	107	110	86	81	73	76
Exp #2, Far-red	P23	46	39	45	19	103	107	110	86	81	73	77
Exp #2, Far-red	P24	24	38	45	19	104	108	110	84	81	74	77

Trajectories are displayed as temporal sequences of gene expression states. Each state is given by the cluster ID number to which it was assigned by the Simprof algorithm. The two experiments, Exp #1 and Exp #2, were performed on two different days, respectively, with the same strain (LU897 × LU898) and under virtually identical experimental conditions. All cells of the dark controls did not sporulate while all far-red irradiated cells sporulated.

TABLE 3 | Terminology, basic modeling principles, and Petri net elements.

Place	Cellular state as defined by a significant cluster of gene expression patterns.
Transition	Transit, i.e., reprogramming step as defined by a discrete change in the state of gene expression.
Source place (no predecessor node)	The cellular state in which an experiment (recording of a time series) starts. There can be various source nodes, depending on the particular state in which a cell is in the moment when the experiment starts.
Sink node (no successor node)	Place with no outgoing arc, indicating a terminal state of gene expression reached at the end of the experiment. There can be multiple sink nodes.
Conflict	Forward branching places modeling bifurcation.
Token	The token indicates the cell and its current gene expression state; there is always just a single token.
Path	A single path from a source node to a sink node represents a possible developmental trajectory of a single cell. But a trajectory may not necessarily involve a source node and/or a sink node.
Petri net	The entire Petri net gives, for the genes analyzed, that part of the Waddington landscape through which cells passed and accordingly all corresponding developmental trajectories a single cell may undergo.

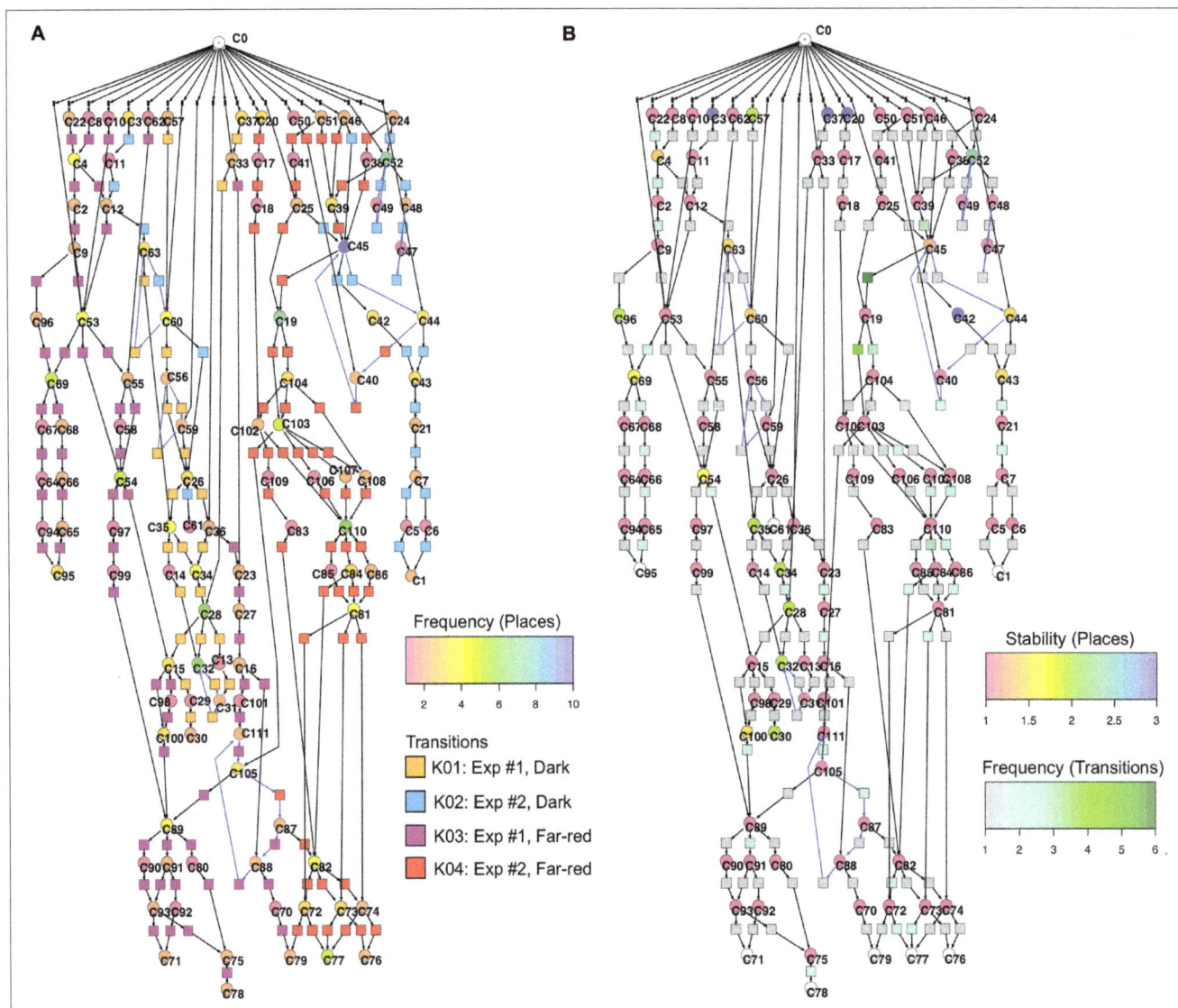

FIGURE 6 | Two copies of the same Petri net, automatically constructed from the single cell trajectories of gene expression as displayed in **Table 2**. Places and transitions in the two nets were colored according to different criteria. Arcs as part of T-invariants are highlighted in blue. **(A)** As indicated by the rainbow color key, places are colored according to the relative frequency with which respective gene expression states occurred in the data set. Transitions are colored according to whether corresponding transits occurred in experiments #1 or #2, and whether cells were far-red stimulated or un-stimulated (dark controls), respectively, as indicated by the panel lower right. **(B)** Places are color coded according to the relative stability of the states of gene expression they represent. This relative stability indicates how long a cell on average resided in a certain state. The color of transitions indicates, in absolute numbers, how frequent a corresponding transit occurred in the data set.

logarithmically, normalized to its concentration at the start of the experiment ($t = 0$ h) as a function of time for any single cell trajectory. In this kind of plot, the time course of the mRNA of each gene starts at the same point, while the slope of the curve indicates the x-fold change in mRNA abundance over time. Plotting subsets of genes suggests that trajectories through different regions of the Petri net of **Figure 6** indeed emerge from qualitatively different expression kinetics, and that genes are also differently regulated relative to each other when different trajectories are compared. In the example shown in **Figure 9**, *pldA* is early up-regulated in quite a number of trajectories, followed by *pwiA* and finally by *ligA* and *rgsA*

that appear strongly correlated at least in some of the plots. Qualitatively different patterns of regulation relative to each other are also evident for the three phospholipase D-encoding genes (**Supplementary Figure 7**). The *pldA* gene is up-regulated while *pldB* and *pldC* are down-regulated. In some of the trajectories, the initial change in the concentration of the *pldA* and *pldC* mRNAs is inverse as compared to the overall time course. A more comprehensive representation with more genes displayed makes similarities and differences between trajectories even more obvious (**Supplementary Figure 8**). Here, we observe a phenomenon, which is also seen in **Table 2**, namely that cells remain in a certain state for some time. This occurs

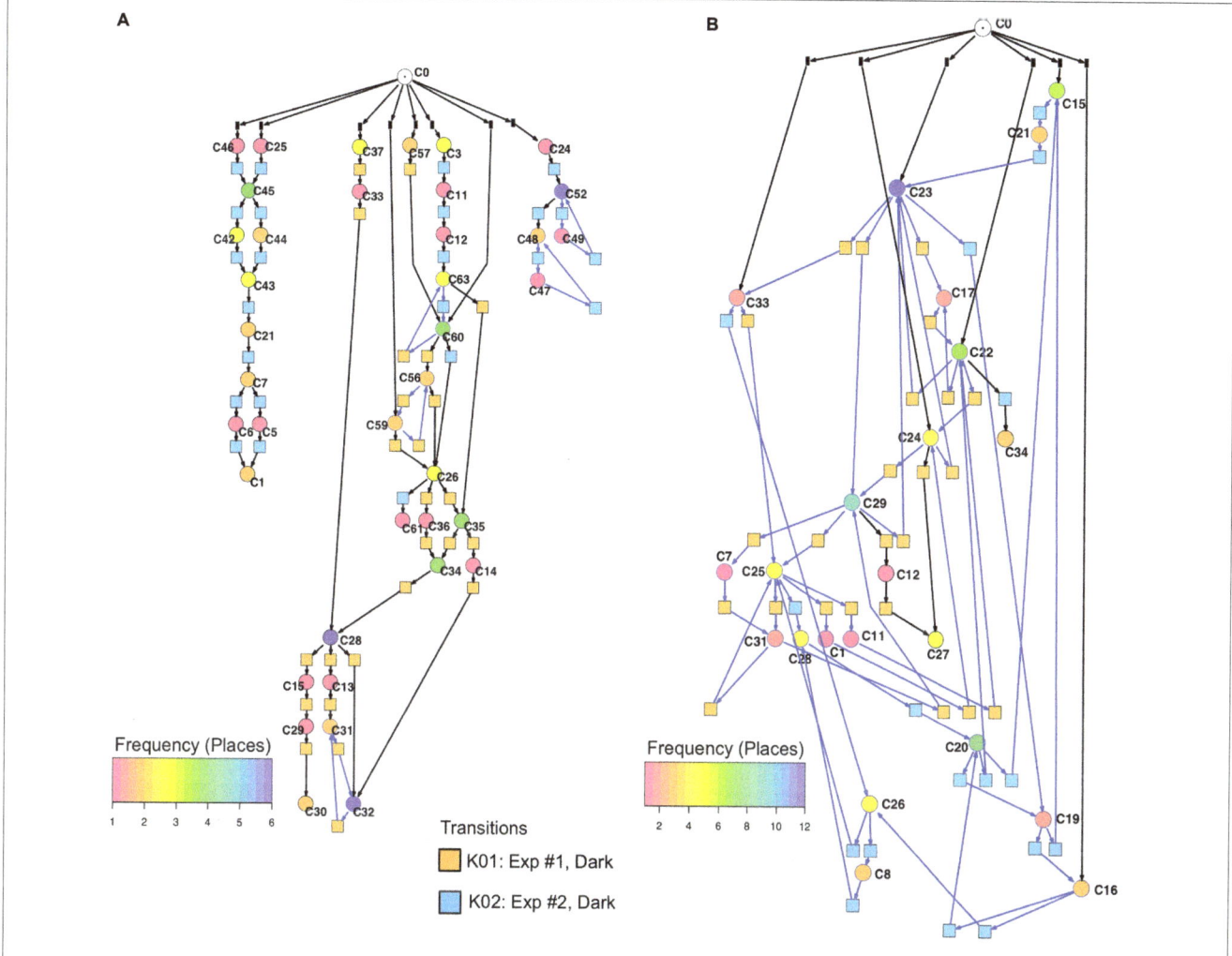

FIGURE 7 | Petri nets constructed for different sets of genes from the gene expression trajectories of un-stimulated cells. **(A)** Trajectories considering the complete set of 35 genes yield three disconnected Petri nets containing a low number of T-invariants (as indicated by arcs highlighted in blue). **(B)** When only those genes are considered that are up- or down-regulated in response to far-red stimulation (omitting the *pcnA*-group of genes), the trajectories of un-stimulated cells give one coherent Petri net which is almost covered with T-invariants. Color coding of places indicates the relative frequencies of states of gene expression in un-stimulated cells. Differences in gene expression patterns of un-stimulated cells between the cells of experiment #1 and #2 are indicated by the appearance of experiment-specific transits and accordingly differently colored transitions.

predominantly in the unstimulated cells but is also seen in some of the light stimulated cells, e.g., in those that proceed to state C95 (**Figures 6A,B**). Cells seemingly are trapped in a meta-stable state (e.g., C96, C69, C100, etc.; **Figure 6B**) for some time until the developmental program proceeds. We presumably will need more data to see whether this is an artifact which occurs by the discretization of gene expression through clustering. Conversely, discretization might help to identify tipping points for the differential regulation of gene expression as the plots in **Supplementary Figure 8** suggest.

DISCUSSION

We have analyzed the gene expression dynamics in response to a differentiation-inducing stimulus pulse in true time by repeatedly

taking samples of large, multinucleate plasmodial cells. Control experiments have demonstrated that the gene expression patterns in samples simultaneously retrieved from different sites of a large plasmodial cell did not deviate within the range of the technical accuracy of the measurements. This again confirms that the plasmodial cytoplasm, at least at the level of macroscopic sampling, can be considered as a homogeneous reaction volume.

The injuries caused by multiple sampling of a plasmodial cell heal spontaneously and cutting the plasmodial mass neither induces nor prevents sporulation (Starostzik and Marwan, 1994, 1995b, 1998; Rätzel and Marwan, 2015). This is also confirmed for the large plasmodia used in this study through the dark controls that did not sporulate (**Table 2** and **Supplementary Figure 1**), while the far-red light induced plasmodia sporulated. The changes in gene expression patterns observed in the dark controls seem to be spontaneous and not caused by repeated

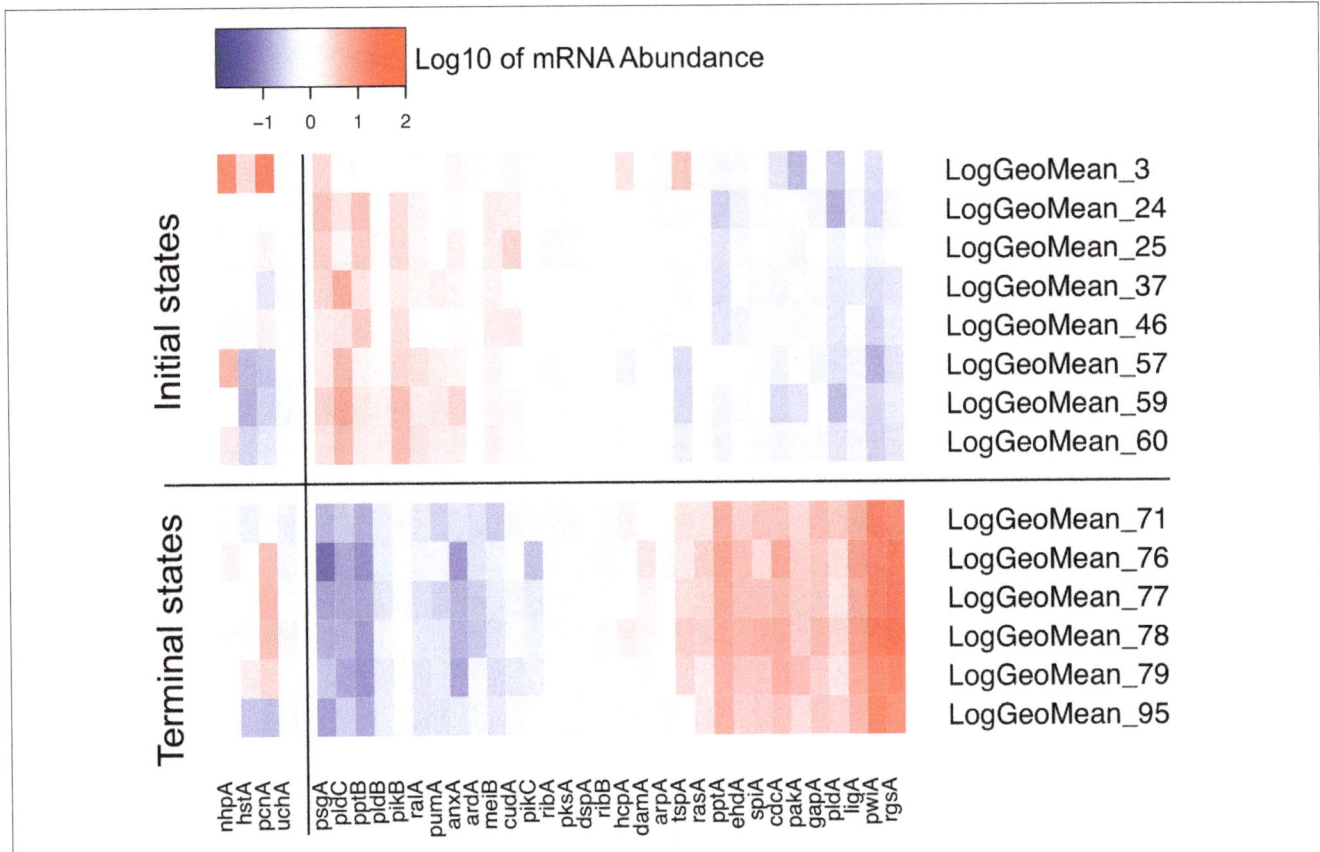

FIGURE 8 | Heat map visualizing the variability of initial and terminal states of gene expression. The initial states displayed in this panel are the start points of trajectories of un-stimulated cells as listed in **Table 2**. The terminal states are endpoints of trajectories of far-red stimulated cells corresponding to terminal places of the Petri net of **Figure 6**. Color-coded expression values correspond to the logarithm to the base 10 of the geometric mean of all expression values of a respective cluster with its ID number indicated on the right side of the panel. For clarity, expression of the *pcnA*-group of genes is displayed as a separate block.

sampling for the following reasons. First, plasmodia are in different states of gene expression already at the start of the experiments (**Figure 8** and **Table 2**), i.e., at the moment before the first sample is taken, so sampling cannot be the reason for this heterogeneity. Second, the trajectories of unstimulated cells developed in different directions although the sampling procedure was the same in all of these cells (**Figures 3A,B**), indicating that the shift in gene expression was not directed with respect to the start of the experiment. Third, the Petri net considering the differential regulation of the subset of light-regulated genes, the expression of which changed to some extent even in unstimulated cells, is almost completely covered with T-invariants (**Figure 7B**), indicating no directed change in the expression of this set of light-regulated genes, neither with respect to the start of the experiment nor with respect to the initial states in which the cells resided before the first sample was taken. We cannot exclude that certain genes might be differentially regulated in response to injury, e.g., genes involved in membrane biosynthesis, as leaks in the membrane readily heal. However, there seems to be no systematic effect on the differential expression of the genes analyzed in the present study.

In contrast to a typical mammalian cell, which has a relatively small cytoplasmic volume, while many genes are present in two copies only, the plasmodial cell contains many millions of nuclei. These nuclei are suspended in a large cytoplasmic volume which continually mixes by the vigorous shuttle streaming. The T-invariants in the Petri nets of **Figure 7B** and **Supplementary Figure 6** indicating the up- and down-regulation of genes were due to changes in the mRNA concentration that occurred at distantly located sites of the plasmodium at the same time. Hence, it seems unlikely that these changes are stochastic gene expression noise. Instead, the T-invariants, presumably do reflect the (non-linear) dynamics of the system. Non-linear, switch-like behavior, bifurcations, multistability, or oscillations all certainly do have fundamental biological and functional implications, and the T-invariant analysis of true single cell time series can help to identify them.

Gene expression states of the cells were defined by hierarchical clustering and discretized by assigning each gene expression pattern to a Simprof significant cluster (Clarke et al., 2008; Rätzel et al., 2020). Trajectories of subsequent discrete states were then assembled into a state machine implemented as a Petri net. In the Petri net, each gene expression state is represented by a place and each transit between two states is represented by a transition.

The Petri net, as it has been defined in this and previous studies depends on the data pre-processing by clustering of the data

(Clarke et al., 2008; Werthmann and Marwan, 2017; Rätzel et al., 2020). It models gene expression trajectories as Markov chains (Gagniuc, 2017), which assumes that each subsequent state only depends on the current state of a cell and not on its previous states, i.e., it does not depend on the individual history of a cell. This assumption is commonly made by computing pseudo-time series from snapshots of individual mammalian cells (Bendall et al., 2014; Shin et al., 2015; Haghverdi et al., 2016; Marr et al., 2016; Street et al., 2018; Weinreb et al., 2018; Chen et al., 2019; Setty et al., 2019). It follows the principle of parsimony in making not more assumptions than necessary and giving the simplest possible explanation for an observed phenomenon. Practically this means that any path which a token can take through the Petri net, by stochastic firing of the transitions, translates into a feasible trajectory of an individual cell. Hence, firing of a transition does only depend on the marking of the pre-place of this transition and not on the identity of any upstream places from which the token originally came. Defining the cell's state of gene expression by measuring more genes might well diversify places and hence change the structure of the Petri net. This has been demonstrated by constructing nets from subsets of genes. In the examples provided, the structure of the Petri net changed and the number of T-invariants increased drastically upon reduction of the number of considered genes (**Table 4**).

If the structure of the Petri net does depend on the set of genes analyzed, what is its actual value? The actual value is that it reveals the behavior of states defined by sets or subsets of genes. Limiting the analysis to the chosen subset of up- and down-regulated genes, as we have done here, revealed extensive on- and off-switching of the genes in unstimulated cells that are differentially regulated in response to a differentiation-inducing stimulus. This became immediately obvious through structural analysis of the Petri net by determining the number of minimal T-invariants. Displaying the net in Sugiyama representation revealed another phenomenon with respect to this subset of genes. The light stimulus caused directed development toward a small number of terminal states reducing the overall number of alternative states in which the cells resided. This suggests that a cellular attractor is formed in response to the stimulus causing the commitment to differentiation.

Coloring the transitions of the Petri net according to the frequency by which transits occurred, allows identification and visualization of main paths, i.e., paths which the system preferably took. Coloring places according to the relative stability of the states they represent indicated metastable states that were not necessarily identical to highly connected places. Places having many pre-transitions (many incoming arcs) represent states, the system is likely to assume, like a corrie in the metaphor of the Waddington landscape, through which the system will pass. Places having many post-transitions (many out-going arcs) represent branching points from which the system has multiple options to proceed.

Our analysis has confirmed former observations (Rätzel et al., 2020), now at considerably larger resolution in time, that unstimulated cells spontaneously and reversibly change their expression pattern. These changes involved the expression of genes that are differentially regulated in response to a

differentiation-inducing stimulus. Spontaneous switching of gene expression patterns is at least one reason why stimulated cells started their way to commitment and differentiation from quite different states, indicating substantial heterogeneity in the population of cells. In other words, cells can start differentiation while being in various different states. The differentiation-inducing stimulus then collects or focusses these cells onto a narrow set of states like an attractor of a dynamic system would do. This phenomenon is graphically revealed by the funnel- or cone-like appearance of the Petri net in the Sugiyama layout (**Supplementary Figure 6**).

The response of a cell to a differentiation-inducing stimulus seems to depend on the cell's current internal state. **Figure 6A** revealed distinct main branches (visible through transitions of different color) for cells from experiment #1 as compared to experiment #2, suggesting that the response of the cell in terms of its developmental pathway did indeed depend on the initial physiological or gene expression state in which the cell resided while receiving the stimulus. The cells proceeded to slightly different terminal states that however might belong to the same cellular attractor.

One might be tempted to suspect a certain structure in the list of subsequently recorded trajectories (**Table 2**). Changes in the initial state of the plasmodia until the time of stimulus application might have occurred as the experiment proceeded. Similarities in subsequently recorded trajectories may be by chance and we cannot draw any final conclusion because the number of analyzed plasmodia is by far too low. With more plasmodia analyzed and more genes measured, we might discover that the individual history of a cell indeed matters, meaning that the Markov assumption is wrong. Even if this should be the case, the Petri net representation would still be valid, however with the firing probability of certain transitions depending on which path the token came from. Technically, this dependency could be implemented in the form of a colored Petri net. In this context it is trivial and at the same time

TABLE 4 | Number and relative frequency of places, transitions, and T-invariants of Petri nets constructed for different subsets of genes (see legend to **Supplementary Figure 4**) from the data of cells listed in **Table 2**.

Gene set	Down	Up	Up and down	All 35	*pcnA* etc.
Genes	10	10	20	35	4
Places	35	44	69	111	24
Transitions	105	114	144	159	59
P/Gene	3.5	4.4	3.5	3.2	6.0
T/Gene	10.5	11.4	7.2	4.5	14.8
Time points	11	11	11	11	11
Cells	24	24	24	24	24
P/(Genes × tps × cells)	0.013	0.017	0.013	0.012	0.023
T/(Genes × tps × cells)	0.040	0.043	0.027	0.017	0.056
T/P	3.00	2.59	2.09	1.43	2.46
T-Inv	4,413	1,371	1,063	7	732
T-Inv/Gene	441	137	53	0.20	183

P, places; T, transitions; tps, time points; T-Inv, T-invariants.

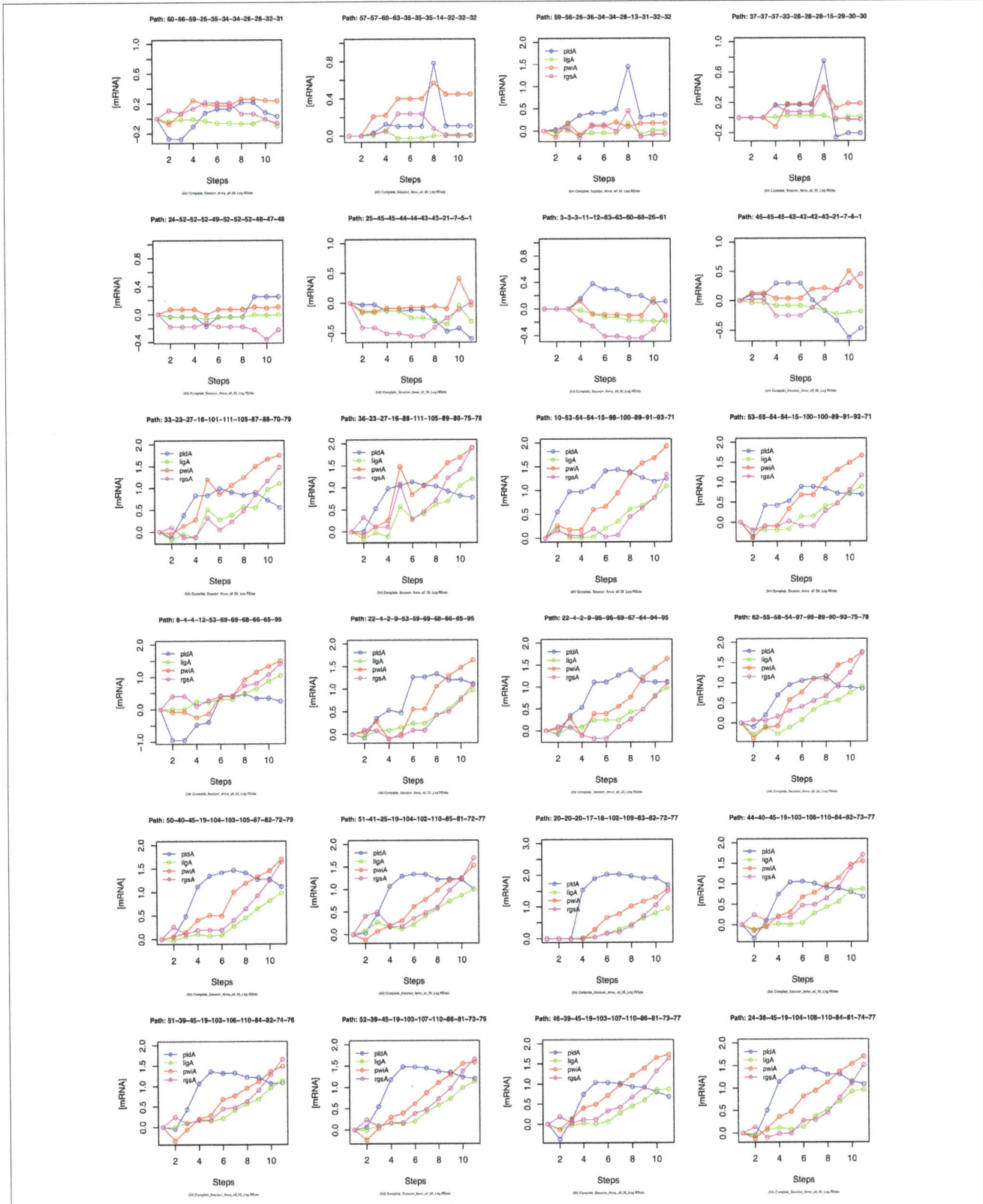

FIGURE 9 | Gene expression kinetics as derived from single cell trajectories. Each panel represents the trajectory of one single cell as characterized by subsequent states of expression of the same, arbitrarily chosen set of genes. For each gene, the logarithm to the base 10 of the geometric mean of the expression values of each cluster was plotted against time. Plotting the geometric mean instead of individual expression values results in discretization of the data while comparing different trajectories.

important to note that the identity of a place (i.e., state) is always defined by the set of genes that have been measured. Measuring more genes might split any place into more places or even into a separate Petri net with an increased overall number of places. This holds not only for states defined by gene expression but also for cellular states defined by the covalent modification of proteins, *etc.*

We have previously argued that the Petri net depicts aspects of the topology of the Waddington landscape (Waddington, 1957; Huang et al., 2009) with respect to and limited to the set of observed genes (or measured molecular entities) (Werthmann and Marwan, 2017; Rätzel et al., 2020). Then, the token in the Petri net corresponds to the marble rolling down the Waddington landscape as developmental processes unfold. Each Petri net place represents, albeit implicitly, a significantly distinct gene expression state. Despite this implicit representation, the temporal information on each gene for each cell trajectory is available and we have used this information to reveal the temporal hierarchy of differentially regulated genes. The

Petri net representation disentangles accordingly the complex gene expression response and identifies alternative regulatory programs or routes. Using this information, the underlying regulatory network can be inferred by applying appropriate algorithms (Marwan et al., 2008; Durzinsky et al. 2011, 2013).
This is a possible next step to go.

AUTHOR CONTRIBUTIONS

AP and SP performed single cell time-series experiments and gene expression analyses and evaluated their results. MHa supervised the experimental work and developed the sample preparation method together with SP. MHe essentially contributed the analysis of the graph-theoretical properties of Petri nets and wrote a corresponding section of the manuscript. WM conceived and supervised the study, performed computational analyses including the automated generation of the Petri nets and wrote the manuscript. All authors read and approved the final version of the manuscript.

REFERENCES

Alexopoulos, C. J., and Mims, C. W. (1979). *Introductory Mycology*, 3rd Edn. New York,NY: John Wiley Sons.

Aselmeyer, F. (2019). *Analyse des Genexpressionsmusters an unterschiedlichen Stellen von Plasmodien von Physarum polycephalum (II)*.Bachelor Thesis Magdeburg: Otto von Guericke Universität.

Bendall, S. C., Davis, K. L., Amir, E.-A. D., Tadmor, M. D., Simonds, E. F., Chen, T. J., et al. (2014). Single-cell trajectory detection uncovers progression and regulatory coordination in human B cell development. *Cell* 157, 714–725. doi: 10.1016/j.cell.2014.04.005

Bornholdt, S., and Kauffman, S. (2019). Ensembles, dynamics, and cell types: Revisiting the statistical mechanics perspective on cellular regulation. *J. Theor. Biol.* 467, 15–22. doi: 10.1016/j.jtbi.2019.01.036

Cannoodt, R., Saelens, W., and Saeys, Y. (2016). Computational methods for trajectory inference from single-cell transcriptomics. *Eur. J. Immunol.* 46, 2496–2506. doi: 10.1002/eji.201646347

Chen, H., Albergante, L., Hsu, J. Y., Lareau, C. A., Lo Bosco, G., Guan, J., et al. (2019). Single-cell trajectories reconstruction, exploration and mapping of omics data with STREAM. *Nat. Comm.* 10:1903.

Clarke, K., Somerfield, P., and Gorley, R. (2008). Testing of null hypotheses in exploratory community analyses: similarity profiles and biota-environment linkage. *J. Exp. Mar. Biol. Ecol.* 366, 56–69. doi: 10.1016/j.jembe.2008.07.009

Dove, W. F., Dee, J., Hatano, S., Haugli, F. B., and Wohlfarth-Bottermann, K.-E. (eds) (1986). *The Molecular Biology of Physarum polycephalum*. New York,NY: Plenum Press.

Driesch, J. (2019). *Analyse des Genexpressionsmusters an unterschiedlichen Stellen von Plasmodien von Physarum polycephalum (I)*.Bachelor Thesis, Magdeburg: Otto von Guericke Universität.

Durzinsky, M., Marwan, W., and Wagler, A. (2013). Reconstruction of extended Petri nets from time-series data by using logical control functions. *J. Math. Biol.* 66, 203–223. doi: 10.1007/s00285-012-0511-3

Durzinsky, M., Wagler, A., and Marwan, W. (2011). Reconstruction of extended Petri nets from time series data and its application to signal transduction and to gene regulatory networks. *BMC Syst. Biol.* 5:113. doi: 10.1186/1752-0509-5-113

Ferrell, J. E Jr. (2012). Bistability, bifurcations, and Waddington's epigenetic landscape. *Curr. Biol.* 22, R458–R466.

Gagniuc, P. A. (2017). *Markov Chains: From Theory to Implementation and Experimentation*. New York, NY: John Wiley & Sons.

Gaublomme Jellert, T., Yosef, N., Lee, Y., Gertner Rona, S., Yang, Li, V., et al. (2015). Single-cell genomics unveils critical regulators of Th17 cell pathogenicity. *Cell* 163, 1400–1412.

Glöckner, G., and Marwan, W. (2017). Transcriptome reprogramming during developmental switching in *Physarum polycephalum* involves extensive remodelling of intracellular signaling networks. *Sci. Rep.* 7:12304.

Golderer, G., Werner, E. R., Leitner, S., Gröbner, P., and Werner-Felmayer, G. (2001). Nitric oxide synthase is induced in sporulation of *Physarum polycephalum*. *Genes Devel.* 15, 1299–1309. doi: 10.1101/gad.890501

Gower, J. C. (1966). Some distance properties of latent root and vector methods used in multivariate analysis. *Biometrika* 53, 325–328. doi: 10.2307/2333639

Graf, T., and Enver, T. (2009). Forcing cells to change lineages. *Nature* 462, 587–594. doi: 10.1038/nature08533

Guttes, E., and Guttes, S. (1961). Synchronous mitosis in starved plasmodia of the myxomycete *Physarum polycephalum*. *Feder. Proc.* 20:419.

Guttes, E., and Guttes, S. (1964). Mitotic synchrony in the plasmodia of *Physarum polycephalum* and mitotic synchronisation by coalescence of microplasmodia. *Meth. Cell Physiol.* 1, 43–54. doi: 10.1016/s0091-679x(08)62085-3

Haghverdi, L., Büttner, M., Wolf, F. A., Buettner, F., and Theis, F. J. (2016). Diffusion pseudotime robustly reconstructs lineage branching. *Nat. Methods* 13:845. doi: 10.1038/nmeth.3971

Hayashi, E., Aoyama, N., Wu, Y., Chi, H. C., Boyer, S. K., and Still, D. W. (2007). *Multiplexed, quantitative gene expression analysis for lettuce seed germination on GenomeLabTM GeXP genetic analysis system*. *Beckman Coulter Application Information A-10295A*. Available online at https://ls.beckmancoulter.co.jp/files/appli_note/A_10295A.pdf (accessed December 16, 2020).

Heiner, M. (2009). "Understanding Network Behavior by Structured Representations of Transition Invariants," in *Algorithmic Bioprocesses. Natural Computing Series*, eds A. Condon, D. Harel, J. Kok, A. Salomaa, and E. Winfree, (Berlin: Springer), 367–389. doi: 10.1007/978-3-540-88869-7_19

Heiner, M., Rohr, C., and Schwarick, M. (2013). "MARCIE – Model Checking and Reachability Analysis Done Efficiently," in *Application and Theory of Petri Nets and Concurrency. PETRI NETS 2013. Lecture Notes in Computer Science*, Vol. 7927, eds J. M. Colom, and J. Desel, (Berlin: Springer).

Hoffmann, X.-K., Tesmer, J., Souquet, M., and Marwan, W. (2012). Futile attempts to differentiate provide molecular evidence for individual differences within a population of cells during cellular reprogramming. *FEMS Microbiol. Lett.* 329, 78–86. doi: 10.1111/j.1574-6968.2012.02506.x

Huang, S., Ernberg, I., and Kauffman, S. (2009). Cancer attractors: a systems view of tumors from a gene network dynamics and developmental perspective. *Semin. Cell Dev. Biol.* 20, 869–876. doi: 10.1016/j.semcdb.2009.07.003

Il Joo, J., Zhou, J. X., Huang, S., and Cho, K.-H. (2018). Determining relative dynamic stability of cell states using boolean network model. *Sci. Rep.* 8:12077.

Junker, J. P., and van Oudenaarden, A. (2014). Every cell is special: Genome-wide studies add a new dimension to single-cell biology. *Cell* 157, 8–11. doi: 10.1016/j.cell.2014.02.010

Kafri, R., Levy, J., Ginzberg, M. B., Oh, S., Lahav, G., and Kirschner, M. W. (2013). Dynamics extracted from fixed cells reveal feedback linking cell growth to cell cycle. *Nature* 494, 1–4. doi: 10.1201/b14602-2

Lamparter, T., and Marwan, W. (2001). Spectroscopic detection of a phytochrome-like photoreceptor in the myxomycete *Physarum polycephalum* and the kinetic mechanism for the photocontrol of sporulation by P$_{fr}$. *Photochem. Photobiol.* 73, 697–702. doi: 10.1562/0031-8655(2001)073<0697:sdoapl>2.0.co;2

Lautenbach, K. (1973). Exact liveness conditions of a Petri net class. GMD Report 82, Bonn (in German).

Macaulay, I. C., Svensson, V., Labalette, C., Ferreira, L., Hamey, F., Voet, T., et al. (2016). Single-cell RNA-sequencing reveals a continuous spectrum of differentiation in hematopoietic cells. *Cell Rep.* 14, 966–977. doi: 10.1016/j.celrep.2015.12.082

Marco, E., Karp, R. L., Guo, G., Robson, P., Hart, A. H., Trippa, L., et al. (2014). Bifurcation analysis of single-cell gene expression data reveals epigenetic landscape. *Proc. Natl. Acad. Sci. U S A* 111, E5643–E5650.

Marquardt, P., Werthmann, B., Raetzel, V., Haas, M., and Marwan, W. (2017). Quantifying 35 transcripts in a single tube: Model-based calibration of the GeXP RT-PCR assay. *bioRxiv.* doi: 10.1101/159723

Marr, C., Zhou, J. X., and Huang, S. (2016). Single-cell gene expression profiling and cell state dynamics: collecting data, correlating data points and connecting the dots. *Curr. Opin. Biotechnol.* 39, 207–214. doi: 10.1016/j.copbio.2016.04.015

Marwan, W., Wagler, A., and Weismantel, R. (2008). A mathematical approach to solve the network reconstruction problem. *Math. Meth. Oper. Res.* 67, 117–132. doi: 10.1007/s00186-007-0178-5

Moignard, V., Woodhouse, S., Haghverdi, L., Lilly, A. J., Tanaka, Y., Wilkinson, A. C., et al. (2015). Decoding the regulatory network of early blood development from single-cell gene expression measurements. *Nat. Biotechnol.* 33, 269–276. doi: 10.1038/nbt.3154

Moris, N., Pina, C., and Arias, A. M. (2016). Transition states and cell fate decisions in epigenetic landscapes. *Nat. Rev. Genet.* 7, 693–703. doi: 10.1038/nrg.2016.98

Paul, F., Arkin, Ya, Giladi, A., Jaitin Diego, A., Kenigsberg, E., et al. (2015). Transcriptional heterogeneity and lineage commitment in myeloid progenitors. *Cell* 163, 1663–1677. doi: 10.1016/j.cell.2015.11.013

Plass, M., Solana, J., Wolf, F. A., Ayoub, S., Misios, A., Glažar, P., et al. (2018). Cell type atlas and lineage tree of a whole complex animal by single-cell transcriptomics. *Science* 360:eaaq1723. doi: 10.1126/science.aaq1723

R Core Team. (2016). *R: A language and environment for statistical computing* Vienna: R Foundation for Statistical Computing.

Rätzel, V. (2015). *Dynamische Fließgleichgewichte und ihre Übergänge in Reaktionsnetzwerken: Experimenteller Nachweis der Quasi-potential-Landschaft der zellulären Reprogrammierung.* Magdeburg: Otto von Guericke University.

Rätzel, V., and Marwan, W. (2015). Gene expression kinetics in individual plasmodial cells reveal alternative programs of differential regulation during commitment and differentiation. *Dev. Growth Differ.* 57, 408–420. doi: 10.1111/dgd.12220

Rätzel, V., Werthmann, B., Haas, M., Strube, J., and Marwan, W. (2020). Disentangling a complex response in cell reprogramming and probing the Waddington landscape by automatic construction of Petri nets. *BioSystems* 189:104092. doi: 10.1016/j.biosystems.2019.104092

Rohr, C., Marwan, W., and Heiner, M. (2010). Snoopy–a unifying Petri net framework to investigate biomolecular networks. *Bioinformatics* 26, 974–975. doi: 10.1093/bioinformatics/btq050

Rusch, H. P., Sachsenmaier, W., Behrens, K., and Gruter, V. (1966). Synchronization of mitosis by the fusion of the plasmodia of *Physarum polycephalum J. Cell Biol.* 31, 204–209.

Sackmann, A., Heiner, M., and Koch, I. (2006). Application of Petri net based analysis techniques to signal transduction pathways. *BMC Bioinform.* 7:482.

Saelens, W., Cannoodt, R., Todorov, H., and Saeys, Y. (2019). A comparison of single-cell trajectory inference methods. *Nat. Biotechnol.* 37, 547–554.

Schaap, P., Barrantes, I., Minx, P., Sasaki, N., Anderson, R. W., Bénard, M., et al. (2016). The *Physarum polycephalum* genome reveals extensive use of prokaryotic two-component and metazoan-type tyrosine kinase signaling. *Genome Biol. Evol.* 8, 109–125.

Setty, M., Kiseliovas, V., Levine, J., Gayoso, A., Mazutis, L., and Pe'er, D. (2019). Characterization of cell fate probabilities in single-cell data with Palantir. *Nat. Biotechnol.* 37, 451–460.

Shin, J., Berg, D. A., Zhu, Y., Shin, J. Y., Song, J., Bonaguidi, M. A., et al. (2015). Single-cell RNA-seq with Waterfall reveals molecular cascades underlying adult neurogenesis. *Stem Cell* 17, 360–372.

Spiller, D. G., Wood, C. D., Rand, D. A., and White, M. R. H. (2010). Measurement of single-cell dynamics. *Nature* 465, 736–745.

Starostzik, C., and Marwan, W. (1994). Time-resolved detection of three intracellular signals controlling photomorphogenesis in *Physarum polycephalum. J. Bacteriol.* 176, 5541–5543.

Starostzik, C., and Marwan, W. (1995a). Functional mapping of the branched signal transduction pathway that controls sporulation in *Physarum polycephalum Photochem. Photobiol.* 62, 930–933.

Starostzik, C., and Marwan, W. (1995b). A photoreceptor with characteristics of phytochrome triggers sporulation in the true slime mould *Physarum polycephalum. FEBS Lett.* 370, 146–148. doi: 10.1016/0014-5793(95)00820-y

Starostzik, C., and Marwan, W. (1998). Kinetic analysis of a signal transduction pathway by time-resolved somatic complementation of mutants. *J. Exp. Biol.* 201, 1991–1999.

Street, K., Risso, D., Fletcher, R. B., Das, D., Ngai, J., Yosef, N., et al. (2018). Slingshot: cell lineage and pseudotime inference for single-cell transcriptomics. *BMC Genomics* 19:477.

Sugiyama, K., Tagawa, S., and Toda, M. (1981). Methods for visual understanding of hierarchical system structures. *IEEE Transac. Syst. Man Cyber.* 11, 109–125. doi: 10.1109/tsmc.1981.4308636

Waddington, C. H. (1957). *The Strategy of the Genes; a Discussion of Some Aspects of Theoretical Biology.* London: Allen & Unwin.

Walter, P., Hoffmann, X.-K., Ebeling, B., Haas, M., and Marwan, W. (2013). Switch-like reprogramming of gene expression after fusion of multinucleate plasmodial cells of two *Physarum polycephalum* sporulation mutants. *Biochem. Biophys. Res. Comm.* 435, 88–93.

Warnes, G. R., Bolker, B., Bonebakker, L., Gentleman, R., Huber, W., Liaw, A., et al. (2016). *gplots: Various R Programming Tools for Plotting Data. R package version 3.0.*

Weinreb, C., Wolock, S., Tusi, B. K., Socolovsky, M., and Klein, A. M. (2018). Fundamental limits on dynamic inference from single-cell snapshots. *Proc. Natl. Acad. Sci.* 115, E2467–E2476.

Werthmann, B., and Marwan, W. (2017). Developmental switching in *Physarum polycephalum*: Petri net analysis of single cell trajectories of gene expression indicates responsiveness and genetic plasticity of the Waddington quasipotential landscape. *J. Phys. D Appl. Phys.* 50:464003. doi: 10.1088/1361-6463/aa8e2b

Whitaker, D., and Christman, M. (2014). *clustsig: Significant Cluster Analysis. R package version 1.1.*

Zhou, J. X., and Huang, S. (2011). Understanding gene circuits at cell-fate branch points for rational cell reprogramming. *Trends Genet.* 27, 55–62.

21

P2X and P2Y Receptor Signaling in Red Blood Cells

Ronald Sluyter [1, 2, 3]*

School of Biological Sciences, University of Wollongong, Wollongong, NSW, Australia, [2] Centre for Medical and Molecular Bioscience, University of Wollongong, Wollongong, NSW, Australia, [3] Illawarra Health and Medical Research Institute, Wollongong, NSW, Australia

[1]

Edited by:
Lello Zolla,
Tuscia University, Italy

Reviewed by:
Peizhong Mao,
Oregon Health and Science University,
USA
Helle Praetorius,
Aarhus University, Denmark

*Correspondence:
Ronald Sluyter
rsluyter@uow.edu.au

Purinergic signaling involves the activation of cell surface P1 and P2 receptors by extracellular nucleosides and nucleotides such as adenosine and adenosine triphosphate (ATP), respectively. P2 receptors comprise P2X and P2Y receptors, and have well-established roles in leukocyte and platelet biology. Emerging evidence indicates important roles for these receptors in red blood cells. P2 receptor activation stimulates a number of signaling pathways in progenitor red blood cells resulting in microparticle release, reactive oxygen species formation, and apoptosis. Likewise, activation of P2 receptors in mature red blood cells stimulates signaling pathways mediating volume regulation, eicosanoid release, phosphatidylserine exposure, hemolysis, impaired ATP release, and susceptibility or resistance to infection. This review summarizes the distribution of P2 receptors in red blood cells, and outlines the functions of P2 receptor signaling in these cells and its implications in red blood cell biology.

Keywords: erythrocyte, red blood cell, adenosine triphosphate, purinergic receptor, P2X1 receptor, P2X7 receptor, P2Y1 receptor, P2Y13 receptor

INTRODUCTION

It is well-established that extracellular adenosine triphosphate (ATP) and other nucleotides function through cell surface purinergic receptors to mediate numerous signaling events in all cell types (Burnstock and Knight, 2004). Purinergic receptors that respond to extracellular nucleotides are termed P2 receptors, and comprise P2X and P2Y receptor subtypes (Burnstock and Kennedy, 1985). P2X receptors are trimeric ATP-gated cation channels that mediate the rapid flux of Na^+, K^+, and Ca^{2+}, with some members also mediating the rapid flux of organic ions (Kaczmarek-Hajek et al., 2012). In mammals, seven P2X receptor subunits exist (P2X1–P2X7), which combine to form either homomeric or heteromeric receptors (Kaczmarek-Hajek et al., 2012). P2Y receptors are G protein-coupled receptors and modulate various signaling events including adenylyl cyclase, phospholipase C, and ion channel activation (Abbracchio et al., 2006). To date eight P2Y receptors have been identified in mammals (P2Y1, P2Y2, P2Y4, P2Y6, and P2Y11–P2Y14). Unlike P2X receptors, some P2Y receptor subtypes are preferentially activated by nucleotides other than ATP, such as P2Y2 and P2Y13, which are preferentially activated by uridine triphosphate (UTP) and adenosine diphosphate (ADP), respectively. Furthermore, ADP is an agonist of many P2Y receptor subtypes (Abbracchio et al., 2006).

P2 receptors are present on all blood cells (Burnstock, 2015). In particular, P2X7 has well-established roles on leukocytes (Bartlett et al., 2014), while P2Y1 and P2Y12 have well-defined functions on platelets (Gachet, 2008). P2 receptors also play important roles in hematopoietic stem cells (Rossi et al., 2012). Collectively, P2 receptor activation contributes to inflammation

(Idzko et al., 2014a), and vascular and blood disease (Idzko et al., 2014b), as evidenced by studies of P2 receptor-deficient mice (Labasi et al., 2002; Stachon et al., 2014). Moreover, it is becoming apparent that P2 receptors have important roles in red blood cells (RBCs), a salient point given the importance of ATP release from RBCs within the vasculature (Sprague and Ellsworth, 2012). This review aims to provide an overview of the distribution of P2 receptors on RBC progenitors and mature RBCs (erythrocytes), and to outline the functions of P2 receptor signaling in these cell types. Other aspects relevant to purinergic signaling in RBCs including ATP release, ectonucleotidases and P1 receptors have been subject to earlier reviews (Huber, 2012; Burnstock, 2015).

DISTRIBUTION OF P2 RECEPTORS IN PROGENITOR RED BLOOD CELLS

P2 receptors have been identified in progenitor RBCs from humans and mice. RT-PCR of human erythroid progenitors, derived by culture of CD34[+] cells, reveals mRNA for P2X1, P2X4, P2X7, and P2Y1, but not P2Y2, P2Y4, and P2Y6 (Hoffman et al., 2004). Further, quantitative PCR reveals high amounts of P2Y13 mRNA, and lower amounts of P2Y1 and P2Y12 in human reticulocytes (Wang et al., 2005). RT-PCR, and immunoblotting and immunolabeling reveal P2X7 mRNA and protein, respectively in murine erythroleukemic (MEL) cells (Constantinescu et al., 2010), a model of progenitor RBCs (Friend et al., 1971). Finally, RT-PCR demonstrates P2Y1, P2Y2, and P2Y12, but not P2Y4, mRNA in murine bone marrow erythroblasts (Paredes-Gamero et al., 2006).

P2X RECEPTOR FUNCTION IN PROGENITOR RED BLOOD CELLS

Evidence for functional P2X receptors in progenitor RBCs is limited to P2X7, and then only in MEL cells. Over 30 years ago, ATP was shown to induce Na^+, K^+, and Ca^{2+} fluxes, and death in MEL cells (Chahwala and Cantley, 1984), although the role of purinergic receptors in these processes was not considered at the time. Subsequently, it was demonstrated that P2X7 activation mediates ATP-induced rapid dye uptake and apoptosis in MEL cells (Constantinescu et al., 2010). A role for P2X7 in ATP-induced Na^+, K^+, and Ca^{2+} fluxes was not examined, but this study indicates that the initial ATP-induced cation fluxes observed in the earlier study (Chahwala and Cantley, 1984) were most probably mediated by P2X7. P2X7 activation in MEL cells induces rapid phosphatidylserine (PS) exposure, microparticle release, apoptosis (Constantinescu et al., 2010) and reactive oxygen species formation (Wang and Sluyter, 2013). In contrast to other cell types (Bartlett et al., 2013), reactive oxygen species formation is not essential for P2X7-induced apoptosis in MEL cells, but requires p38 mitogen-activated protein kinase and caspase activation (Wang and Sluyter, 2013). The role of P2X7 activation in progenitor RBCs remains to be determined, but may cause the removal of damaged RBC progenitors to prevent development of anemia, leukemia, or autoimmunity.

P2Y RECEPTOR FUNCTION IN PROGENITOR RED BLOOD CELLS

The presence of functional P2Y receptors in progenitor RBCs is mainly limited to P2Y1. ATP, ADP, and UTP induce release of Ca^{2+} from intracellular stores within murine bone marrow erythroblasts suggesting the presence of functional P2Y receptors in these cells (Paredes-Gamero et al., 2006). Although the identity of the receptors responsible for the ATP- and UTP-induced responses were not resolved, the ADP-induced release of intracellular Ca^{2+} was caused by P2Y1 activation (Paredes-Gamero et al., 2006). Functional P2Y1 may also be present in in human progenitor RBCs, but direct evidence is sparse. ADP, an agonist of P2Y1 but also other P2Y receptors (Abbracchio et al., 2006), can cause the release of intracellular Ca^{2+} within human erythroid progenitors generated from peripheral blood (Porzig et al., 1995). The physiological roles of P2Y1 activation in progenitor RBCs remain to be explored.

DISTRIBUTION OF P2 RECEPTORS IN RED BLOOD CELLS

P2 receptors have been identified in RBCs from various species. Quantitative PCR reveals high amounts of P2Y13 mRNA, low amounts of P2X1, P2X4, P2X7, and P2Y2, and even lower amounts of P2Y1, P2Y4, P2Y6, P2Y11, and P2Y12 in human RBCs (Wang et al., 2005). Immunoblotting demonstrates the presence of P2X1 and P2X7 protein in human, canine and murine RBCs (Sluyter et al., 2007a; Skals et al., 2009), and P2Y1 protein in human RBCs (Tanneur et al., 2006). Immunolabeling confirms the presence of P2X7 protein in RBCs from humans (Sluyter et al., 2004) and dogs (Sluyter et al., 2007a), as well as P2Y1 (Tanneur et al., 2006), and to a lesser extent P2X2 (Sluyter et al., 2004) and P2Y2 (Tanneur et al., 2004) in human RBCs.

P2X RECEPTOR FUNCTION IN RED BLOOD CELLS

P2X1 and P2X7 mediate bacterial toxin-induced lysis of RBCs from various species. Both P2X1 and P2X7, but not P2Y1 or P2Y2, mediate *Escherichia coli* α-hemolysin-induced lysis of human, murine and equine RBCs (Skals et al., 2009). This effect is primarily mediated by P2X1 in murine RBCs, but P2X7 in human RBCs (Skals et al., 2009) suggesting that these receptors are differentially expressed in RBCs from these two species. In contrast, python RBCs are resistant to α-hemolysin (Larsen et al., 2011). α-Hemolysin also induces cell shrinkage of and PS exposure on human RBCs, and the subsequent phagocytosis of these cells by human THP-1 monocytes (Fagerberg et al., 2013) implying that RBCs exposed to this toxin can be cleared from the circulation. Through the use of the P2X1 antagonist MRS2159 and P2X7 antagonists, this study also indicated that both P2X1 and P2X7 mediate α-hemolysin-induced PS exposure (Fagerberg et al., 2013). However, MRS2159 is also a potent antagonist of human P2X7 (Sophocleous et al., 2015) leaving open the possibility that P2X7, but not P2X1, mediates this event in human RBCs.

P2X1 and P2X7 also mediate *Staphylococcus aureus* α-toxin-induced lysis of murine and equine RBCs (Skals et al., 2011) and *Aggregatibacter actinomycetemcomitans* leukotoxin A-induced lysis of human RBCs (Munksgaard et al., 2012). The latter study also demonstrated that leukotoxin A induced shrinkage of and PS exposure on RBCs (Munksgaard et al., 2012), although the role of P2X1 or P2X7 in these processes was not elucidated. P2X7 activation also mediates *Actinobacillus pleuropneumoniae* ApxIA toxin-induced lysis of ovine RBCs (Masin et al., 2013). In contrast, P2X receptor activation was not required for *Bordetlla pertussis* adenylate cyclase toxin-induced lysis of ovine RBCs (Masin et al., 2013). Collectively, the authors concluded that involvement of P2X receptor activation in hemolysis could be regulated by toxin pore size, with ApxIA hemolysin forming larger pores (∼2.4 nm) than adenylate cyclase toxin (∼0.7 nm) (Masin et al., 2013). Finally, *E. coli* shiga toxin can induce microvesicle release from human RBCs; a process blocked by broad-spectrum P2 receptor antagonists (Arvidsson et al., 2015), however the specific P2 receptors involved remain unknown.

Complement can also induce lysis of human, murine, and ovine RBCs via P2X1 and P2X7 activation (Hejl et al., 2013). Notably, the P2 receptor antagonist suramin was originally shown to impair complement-mediated lysis of human and guinea pig RBCs, although this effect appeared to be due to suramin directly binding complement components (Fong and Good, 1972). Nevertheless, a role for P2X receptor activation in this early study cannot be excluded.

Bacterial toxin-induced and complement-induced hemolysis involves the release of ATP acting on P2X receptors in an autocrine or paracrine fashion (**Figure 1A**). ATP scavenging enzymes impair α-hemolysin, α-toxin and leukotoxin A-induced lysis of RBCs (Skals et al., 2009; Munksgaard et al., 2012), and complement-induced hemolysis (Hejl et al., 2013) supporting the concept that released ATP activates P2X receptors. Originally it was thought that hemichannel pannexin-1, which can mediate ATP release from RBCs (Locovei et al., 2006), was responsible for the above ATP release (Skals et al., 2011), as pannexin-1 antagonists prevented toxin-induced hemolysis (Skals et al., 2009, 2011; Munksgaard et al., 2012). However, recent findings indicate that α-hemolysin and leukotoxin A induce ATP release from RBCs by forming toxin pores rather than via pannexin-1 (Skals et al., 2014). Thus, current evidence suggests that bacterial toxins directly from pores in RBCs to allow ATP release, which then acts on P2X1 and P2X7 to mediate hemolysis. The mechanism by which complement causes ATP release remains to be resolved. However, recent data indicates that ligation of complement receptor 1 on human RBCs mediates ATP release (Melhorn et al., 2013). Both bacterial toxin-induced and complement-induced hemolysis pose potential health problems during certain bacterial infections and in diseases associated with prolonged complement activation.

Direct evidence for functional P2X7 in RBCs was first demonstrated for human RBCs. P2X7 activation mediates Na$^+$ and Rb$^+$ (K$^+$) fluxes, as well as choline$^+$ uptake in human RBCs (Sluyter et al., 2004; Stevenson et al., 2009). Moreover,

P2X7 activation can induce Rb$^+$ efflux and choline$^+$ uptake in canine RBCs (Sluyter et al., 2007a; Shemon et al., 2008; Stevenson et al., 2009). The ability of ATP to induce cation fluxes in canine RBCs was first observed in 1972 (Parker and Snow, 1972) and subsequently by others that same decade (Elford, 1975; Romualdez et al., 1976). These early investigations did not attribute this effect to purinergic signaling, despite the establishment of this concept by Burnstock also in 1972 (Burnstock, 1972). Nevertheless, it is evident from the initial observations (Parker and Snow, 1972) that ATP induced Na$^+$ or K$^+$ fluxes in canine RBCs in a manner characteristic of P2X7 activation, and with a time course and order of magnitude near identical to that of ATP-induced Rb$^+$ fluxes in canine RBCs observed some 30 years later (Sluyter et al., 2007a). Thus, this original observation that ATP mediates cation fluxes in canine RBCs (Parker and Snow, 1972) remains one of the earliest known reports of functional P2X7 in any cell type. Notably, relative P2X7 activity in canine RBCs is up to 100-fold greater than that observed in human RBCs (Sluyter et al., 2007a; Stevenson et al., 2009). This increased P2X7 activity in canine RBCs corresponds to increased amounts of P2X7 in canine RBCs compared to human RBCs (Sluyter et al., 2007a). The physiological significance of this observation remains unknown, as does the relative amount or activity of P2X7 on RBCs between other species.

P2X7 activation induces PS exposure in human RBCs (**Figure 1B**) either freshly isolated from peripheral blood (Sluyter et al., 2007b) or following cold storage for up to 6 weeks (Sophocleous et al., 2015). Notably, the amount of P2X7-induced PS exposure varies between donors (Sophocleous et al., 2015). This is mostly likely due to single nucleotide polymorphisms in the *P2RX7* gene that code for loss or gain of P2X7 function (Sluyter and Stokes, 2011). Consistent with this concept, ATP-induced cation fluxes and PS exposure are reduced in RBCs from subjects coding loss-of-function *P2RX7* gene mutations (Sluyter et al., 2004, 2007b), while gain-of-function mutations correspond with augmented ATP-induced cation fluxes in RBCs (Stokes et al., 2010). ATP can also induce PS exposure and hemolysis in canine RBCs (Sluyter et al., 2007a), but direct evidence for P2X7 in these processes is lacking. Furthermore, mutations that alter receptor function are found in the *P2RX7* gene of dogs (Spildrejorde et al., 2014), but it remains to be determined if these mutations alter P2X7-mediated events in canine RBCs. The physiological significance of P2X7-mediated PS exposure in RBCs remains unknown, but the propensity of human RBCs to undergo PS externalization does not change with *in vitro* or *in vivo* aging (Sophocleous et al., 2015). This suggests that P2X7-mediated PS exposure in RBCs does not play a role in the normal removal of senescent RBCs, but perhaps in the removal of RBCs following cell stress or damage, or in diseased states.

P2X7 activation induces epoxyeicosatrienoic acid (EET) release from rat RBCs (Jiang et al., 2007). This release of EETs is partly dependent on phospholipase A$_2$ stimulation, but not hemolysis (Jiang et al., 2007). In combination with earlier data (Jiang et al., 2005), EETs released downstream of P2X7 activation represent both EETs generated from stored

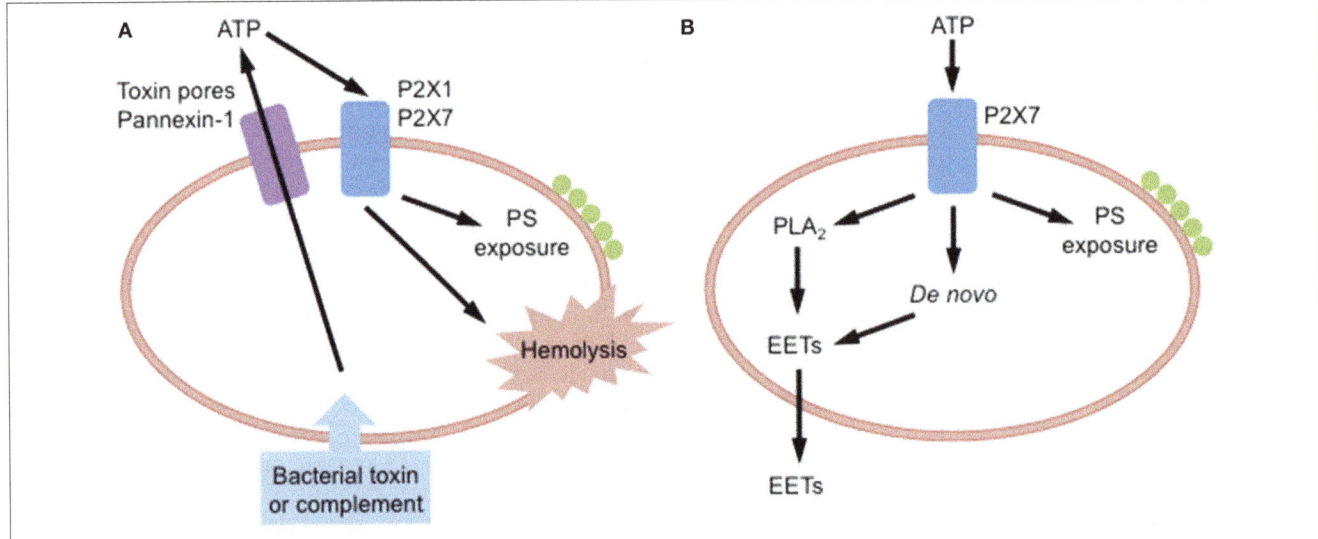

FIGURE 1 | P2X receptor activation in red blood cells. (A) Binding of bacterial toxins or complement to red blood cells causes adenosine triphosphate (ATP) release via toxin pores and pannexin-1. Released ATP can activate P2X1 and P2X7 receptors on these cells to induce phosphatidylserine (PS) exposure and hemolysis. **(B)** Extracellular ATP can activate P2X7 receptors on red blood cells to induce PS exposure, or formation of epoxyeicosatrienoic acids (EETs), via phospholipase A_2 (PLA_2) acting on stored phospholipids or via *de novo* synthesis, and subsequent EETs release.

phospholipids and from *de novo* synthesis (Jiang et al., 2007) (**Figure 1B**). EETs are eicosanoids that mediate a variety of functions within the circulation including vasodilation (Jiang et al., 2010), thus P2X7-mediated EET release may amplify the circulatory responses mediated by extracellular ATP (Jiang et al., 2007, 2010). It remains to be determined if P2X7 activation can induce EET release from RBCs of other species.

Functional P2X receptors have been reported in non-mammalian RBCs. During hypotonic swelling, ATP is released from *Necturus* salamander RBCs to stimulate regulatory volume decrease in these cells (Light et al., 1999). Pharmacological approaches indicated that this receptor is most likely a P2X2 homolog (Light et al., 2001), while other studies showed that activation of this P2X2-like receptor mediates Ca^{2+} influx during hypotonic swelling of *Necturus* RBCs (Light et al., 2003). ATP release also regulates volume decreases during hypotonic swelling of skate RBCs (Goldstein et al., 2003), but direct evidence for P2 receptors in this process is lacking. In contrast, a P2X-like receptor stimulates regulatory volume decrease in alligator cells (Wormser et al., 2011). Activation of this receptor stimulates Ca^{2+} influx to activate phospholipase A_2 and arachidonic acid release to increase K^+ permeability and volume recovery (Wormser et al., 2011). At present there is no evidence that P2X receptor activation stimulates regulatory volume decrease in mammalian RBCs. Finally, functional P2X receptors have been identified in RBCs from other reptiles. Activation of P2X-like receptors in RBCs from Iguania lizards causes an influx of Ca^{2+} (Bagnaresi et al., 2007; Beraldo and Garcia, 2007). In contrast, RBCs from Scleroglossa lizards do not appear to express functional P2X receptors, but rather a P2Y4-like receptor that causes intracellular Ca^{2+} release following activation (Sartorello and Garcia, 2005).

P2Y RECEPTOR FUNCTION IN RED BLOOD CELLS

The first direct evidence for functional P2 receptors in RBCs was established through a series of studies demonstrating the presence of P2Y1 in turkey RBCs (see Boyer et al., 1996). A P2Y receptor was initially identified in membranes of turkey RBCs (Harden et al., 1988) and then in whole turkey RBCs (Berrie et al., 1989; Boyer et al., 1989). Combined, these studies showed that activation of this receptor stimulates phosphatidylinositol 4,5-biphophate hydrolysis and phospholipase C activation (Harden et al., 1988; Berrie et al., 1989; Boyer et al., 1989). Subsequent cloning identified this receptor as the turkey homolog of human and chick P2Y1 (Filtz et al., 1994).

Functional P2Y1 is also present in human and murine RBCs, where it plays a role in promoting malaria parasite development. *Plasmodium* infection of human or murine RBCs results in ATP release (Tanneur et al., 2006; Akkaya et al., 2009) and the subsequent activation of P2Y1 to open an osmolyte permeability pathway (Tanneur et al., 2006), which potentially promotes parasite development through the supply of nutrients and removal of metabolic waste products (Kirk, 2001). Similar findings where also observed with oxidized RBCs suggesting that parasite-derived oxidative stress is involved in the induction of this P2Y1-induced osmolyte permeability pathway (Tanneur et al., 2006) (**Figure 2A**). Studies of P2 receptor activation in malaria-infected RBCs however are complicated by evidence that *Plasmodium* malaria parasites also express functional P2 receptors (Levano-Garcia et al., 2010; da Cruz et al., 2012).

RBCs may express functional P2Y12, but direct evidence is limited. In human RBCs, the P2Y12 antagonist ticagrelor inhibits adenosine uptake (van Giezen et al., 2012) and induces the release of ATP, which can be subsequently degraded to adenosine

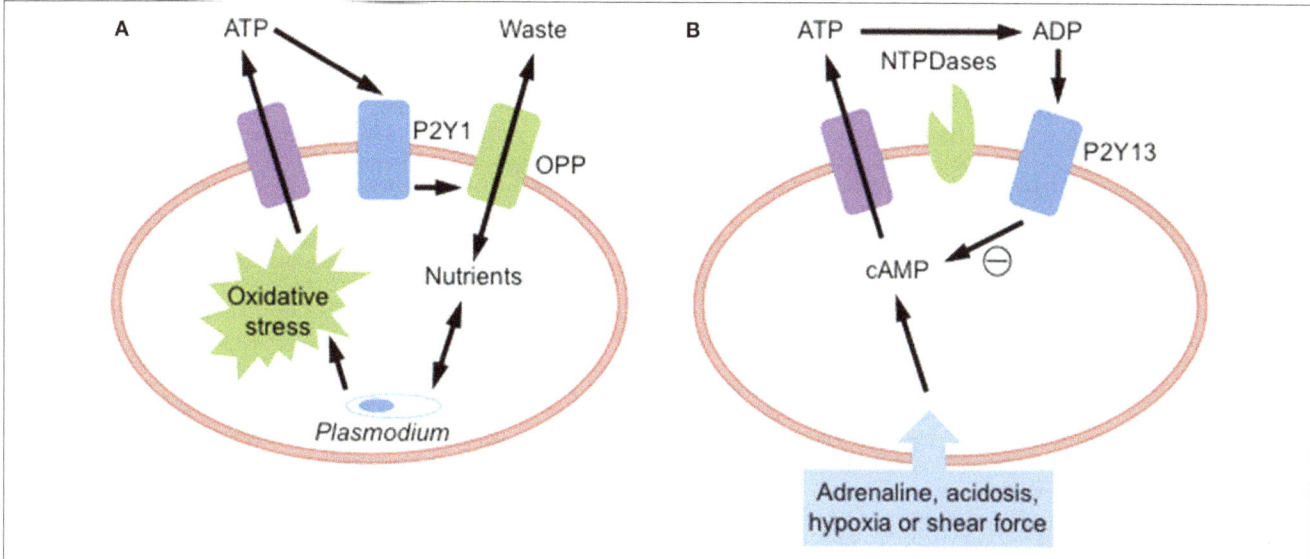

FIGURE 2 | P2Y receptor activation in red blood cells. (A) *Plasmodium* malarial parasite infection of red blood cells causes oxidative stress to induce adenosine triphosphate (ATP) release, which activates P2Y1 receptors to open an osmolyte permeability pathway (OPP) to facilitate parasite growth by supplying incoming nutrients and removing outgoing metabolic waste products. **(B)** Cellular stress (acidosis, adrenaline, hypoxia, or shear force) of red blood cells increases intracellular cyclic adenosine monophosphate (cAMP) to cause ATP release. Released ATP can be degraded by ectonucleotidases (NTPDases) to adenosine diphosphate (ADP), which then activates P2Y13 receptors to reduce cAMP and prevent further ATP release.

(Ohman et al., 2012). Further, ticagrelor augments cardiac blood flow in dogs (van Giezen et al., 2012) indirectly suggesting that P2Y12 may be present on canine RBCs. Combined these studies suggested that ticagrelor may provide cardiovascular benefits in addition to ADP-induced platelet aggregation.

Functional P2Y13 is present on RBCs, where it negatively regulates ATP release from these cells (Wang et al., 2005). Activation of this receptor by ADP impairs the release of ATP from human RBCs (**Figure 2B**). Moreover, intracoronary injection of the P2Y13 agonist 2-methylthio-ADP into pigs reduces the amount of circulating ATP (Wang et al., 2005). Further evidence defining a role for this receptor in this feedback mechanism is wanting.

CONCLUSIONS

Various P2 receptors are present in progenitor and mature RBCs. Evidence for functional P2 receptors in primary progenitor RBCs remains to be fully explored, but studies of MEL cells indicate that P2X7 can mediate microparticle release, reactive oxygen species formation, and apoptosis. A larger body of evidence is available for the presence of functional P2 receptors in mature RBCs, with P2X1, P2X7, P2Y1, and P2Y13 being the major P2 receptor subtypes present. In RBCs, P2X1 and P2X7 mediate ATP-induced PS exposure, hemolysis, and eicosanoid release. P2Y1 facilitates malaria parasite development within RBCs, while P2Y13 functions to negatively regulate ATP release from RBCs. Despite these findings, further investigations are required to fully define the role of P2 receptors in RBCs.

AUTHOR CONTRIBUTIONS

RS conceived and wrote the manuscript, and prepared the figures.

ACKNOWLEDGMENTS

The American Kennel Club Canine Health Foundation, the University of Wollongong, the Centre for Medical and Molecular Bioscience, and the Illawarra Health and Medical Research Institute currently support the laboratory of RS.

REFERENCES

Abbracchio, M. P., Burnstock, G., Boeynaems, J. M., Barnard, E. A., Boyer, J. L., Kennedy, C., et al. (2006). International Union of Pharmacology LVIII: update on the P2Y G protein-coupled nucleotide receptors: from molecular mechanisms and pathophysiology to therapy. *Pharmacol. Rev.* 58, 281–341. doi: 10.1124/pr.58.3.3

Akkaya, C., Shumilina, E., Bobballa, D., Brand, V. B., Mahmud, H., Lang, F., et al. (2009). The *Plasmodium falciparum*-induced anion channel of human erythrocytes is an ATP-release pathway. *Pflugers Arch.* 457, 1035–1047. doi: 10.1007/s00424-008-] 0572-8

Arvidsson, I., Ståhl, A. L., Hedström, M. M., Kristoffersson, A. C., Rylander, C., Westman, J. S., et al. (2015). Shiga toxin-induced complement-mediated hemolysis and release of complement-coated red blood cell-derived microvesicles in hemolytic uremic syndrome. *J. Immunol.* 194, 2309–2318. doi: 10.4049/jimmunol.1402470

Bagnaresi, P., Rodrigues, M. T., and Garcia, C. R. (2007). Calcium signaling in lizard red blood cells. *Comp. Biochem. Physiol. A Mol. Integr. Physiol.* 147, 779–787. doi: 10.1016/j.cbpa.2006.09.015

Bartlett, R., Stokes, L., and Sluyter, R. (2014). The P2X7 receptor channel: recent developments and the use of P2X7 antagonists in models of disease. *Pharmacol. Rev.* 66, 638–675. doi: 10.1124/pr.113.008003

Bartlett, R., Yerbury, J. J., and Sluyter, R. (2013). P2X7 receptor activation induces reactive oxygen species formation and cell death in murine EOC13 microglia. *Mediators Inflamm.* 2013:271813. doi: 10.1155/2013/271813

Beraldo, F. H., and Garcia, C. R. (2007). Divergent calcium signaling in RBCs from *Tropidurus torquatus* (*Squamata-Tropiduridae*) strengthen classification in lizard evolution. *BMC Physiol.* 7:7. doi: 10.1186/1472-6793-7-7

Berrie, C. P., Hawkins, P. T., Stephens, L. R., Harden, T. K., and Downes, C. P. (1989). Phosphatidylinositol 4,5-bisphosphate hydrolysis in turkey erythrocytes is regulated by P_{2Y} purinoceptors. *Mol. Pharmacol.* 35, 526–532.

Boyer, J. L., Downes, C. P., and Harden, T. K. (1989). Kinetics of activation of phospholipase C by P_{2Y} purinergic receptor agonists and guanine nucleotides. *J. Biol. Chem.* 264, 884–890.

Boyer, J. L., Schachter, J. B., Sromek, S. M., Palmer, R. K., Jacobson, K. A., Nicholas, R. A., et al. (1996). Avian and human homologues of the P2Y, receptor: pharmacological, signaling, and molecular properties. *Drug Develop. Res.* 39, 253–261.

Burnstock, G. (1972). Purinergic nerves. *Pharmacol. Rev.* 24, 509–581.

Burnstock, G. (2015). Blood cells: an historical account of the roles of purinergic signalling. *Purinergic Signal.* doi: 10.1007/s11302-015-9462-7. [Epub ahead of print].

Burnstock, G., and Kennedy, C. (1985). Is there a basis for distinguishing two types of P2-purinoceptor? *Gen. Pharmacol.* 16, 433–440. doi: 10.1016/0306-3623(85)90001-1

Burnstock, G., and Knight, G. E. (2004). Cellular distribution and functions of P2 receptor subtypes in different systems. *Int. Rev. Cytol.* 240, 31–304. doi: 10.1016/S0074-7696(04)40002-3

Chahwala, S. B., and Cantley, L. C. (1984). Extracellular ATP induces ion fluxes and inhibits growth of Friend erythroleukemia cells. *J. Biol. Chem.* 259, 13717–13722.

Constantinescu, P., Wang, B., Kovacevic, K., Jalilian, I., Bosman, G. J., Wiley, J. S., et al. (2010). P2X7 receptor activation induces cell death and microparticle release in murine erythroleukemia cells. *Biochim. Biophys. Acta* 1798, 1797–1804. doi: 10.1016/j.bbamem.2010.06.002

da Cruz, L. N., Juliano, M. A., Budu, A., Juliano, L., Holder, A. A., Blackman, M. J., et al. (2012). Extracellular ATP triggers proteolysis and cytosolic Ca^{2+} rise in *Plasmodium berghei* and *Plasmodium yoelii* malaria parasites. *Malar. J.* 11:69. doi: 10.1186/1475-2875-11-69

Elford, B. C. (1975). Independent routes for Na transport across dog red cell membranes. *Nature* 256, 580–582. doi: 10.1038/256580a0

Fagerberg, S. K., Skals, M., Leipziger, J., and Praetorius, H. A. (2013). P2X receptor-dependent erythrocyte damage by alpha-hemolysin from *Escherichia coli* triggers phagocytosis by THP-1 cells. *Toxins (Basel)* 5, 472–487. doi: 10.3390/toxins5030472

Filtz, T. M., Li, Q., Boyer, J. L., Nicholas, R. A., and Harden, T. K. (1994). Expression of a cloned P2Y purinergic receptor that couples to phospholipase C. *Mol. Pharmacol.* 46, 8–14.

Fong, J. S., and Good, R. A. (1972). Suramin—a potent reversible and competitive inhibitor of complement systems. *Clin. Exp. Immunol.* 10, 127–138.

Friend, C., Scher, W., Holland, J. G., and Sato, T. (1971). Hemoglobin synthesis in murine virus-induced leukemic cells *in vitro*: stimulation of erythroid differentiation by dimethyl sulfoxide. *Proc. Natl. Acad. Sci. U.S.A.* 68, 378–382. doi: 10.1073/pnas.68.2.378

Gachet, C. (2008). P2 receptors, platelet function and pharmacological implications. *Thromb. Haemost.* 99, 466–472. doi: 10.1160/th07-11-0673

Goldstein, L., Koomoa, D. L., and Musch, M. W. (2003). ATP release from hypotonically stressed skate RBC: potential role in osmolyte channel regulation. *J. Exp. Zool. A Comp. Exp. Biol.* 296, 160–163. doi: 10.1002/jez.a.10228

Harden, T. K., Hawkins, P. T., Stephens, L., Boyer, J. L., and Downes, C. P. (1988). Phosphoinositide hydrolysis by guanosine 5′-[gamma-thio]triphosphate-activated phospholipase C of turkey erythrocyte membranes. *Biochem. J.* 252, 583–593. doi: 10.1042/bj2520583

Hejl, J. L., Skals, M., Leipziger, J., and Praetorius, H. A. (2013). P2X receptor stimulation amplifies complement-induced haemolysis. *Pflugers Arch.* 465, 529–541. doi: 10.1007/s00424-012-1174-z

Hoffman, J. F., Dodson, A., Wickrema, A., and Dib-Hajj, S. D. (2004). Tetrodotoxin-sensitive Na^+ channels and muscarinic and purinergic receptors identified in human erythroid progenitor cells and red blood cell ghosts. *Proc. Natl. Acad. Sci. U.S.A.* 101, 12370–12374. doi: 10.1073/pnas.0404228101

Huber, S. M. (2012). Purinoceptor signaling in malaria-infected erythrocytes. *Microbes Infect.* 14, 779–786. doi: 10.1016/j.micinf.2012.04.009

Idzko, M., Ferrari, D., and Eltzschig, H. K. (2014a). Nucleotide signalling during inflammation. *Nature* 509, 310–317. doi: 10.1038/nature13085

Idzko, M., Ferrari, D., Riegel, A. K., and Eltzschig, H. K. (2014b). Extracellular nucleotide and nucleoside signaling in vascular and blood disease. *Blood* 124, 1029–1037. doi: 10.1182/blood-2013-09-402560

Jiang, H., Anderson, G. D., and McGiff, J. C. (2010). Red blood cells (RBCs), epoxyeicosatrienoic acids (EETs) and adenosine triphosphate (ATP). *Pharmacol. Rep.* 62, 468–474. doi: 10.1016/S1734-1140(10)70302-9

Jiang, H., Quilley, J., Reddy, L. M., Falck, J. R., Wong, P. Y., and McGiff, J. C. (2005). Red blood cells: reservoirs of cis- and trans-epoxyeicosatrienoic acids. *Prostaglandins Other Lipid Mediat.* 75, 65–78. doi: 10.1016/j.prostaglandins.2004.10.003

Jiang, H., Zhu, A. G., Mamczur, M., Falck, J. R., Lerea, K. M., and McGiff, J. C. (2007). Stimulation of rat erythrocyte $P2X_7$ receptor induces the release of epoxyeicosatrienoic acids. *Br. J. Pharmacol.* 151, 1033–1040. doi: 10.1038/sj.bjp.0707311

Kaczmarek-Hájek, K., Lörinczi, É., Hausmann, R., and Nicke, A. (2012). Molecular and functional properties of P2X receptors—recent progress and persisting challenges. *Purinergic Signal.* 8, 375–417. doi: 10.1007/s11302-012-9314-7

Kirk, K. (2001). Membrane transport in the malaria-infected erythrocyte. *Physiol. Rev.* 81, 495–537.

Labasi, J. M., Petrushova, N., Donovan, C., McCurdy, S., Lira, P., Payette, M. M., et al. (2002). Absence of the P2X7 receptor alters leukocyte function and attenuates an inflammatory response. *J. Immunol.* 168, 6436–6445. doi: 10.4049/jimmunol.168.12.6436

Larsen, C. K., Skals, M., Wang, T., Cheema, M. U., Leipziger, J., and Praetorius, H. A. (2011). Python erythrocytes are resistant to alpha-hemolysin from *Escherichia coli*. *J. Membr. Biol.* 244, 131–140. doi: 10.1007/s00232-011-9406-2

Levano-Garcia, J., Dluzewski, A. R., Markus, R. P., and Garcia, C. R. (2010). Purinergic signalling is involved in the malaria parasite *Plasmodium falciparum* invasion to red blood cells. *Purinergic Signal.* 6, 365–372. doi: 10.1007/s11302-010-9202-y

Light, D. B., Attwood, A. J., Siegel, C., and Baumann, N. L. (2003). Cell swelling increases intracellular calcium in *Necturus* erythrocytes. *J. Cell. Sci.* 116, 101–109. doi: 10.1242/jcs.00202

Light, D. B., Capes, T. L., Gronau, R. T., and Adler, M. R. (1999). Extracellular ATP stimulates volume decrease in *Necturus* red blood cells. *Am. J. Physiol.* 277, C480–C491.

Light, D. B., Dahlstrom, P. K., Gronau, R. T., and Baumann, N. L. (2001). Extracellular ATP activates a P2 receptor in *Necturus* erythrocytes during hypotonic swelling. *J. Membr. Biol.* 182, 193–202. doi: 10.1007/s0023201-0043-z

Locovei, S., Bao, L., and Dahl, G. (2006). Pannexin 1 in erythrocytes: function without a gap. *Proc. Natl. Acad. Sci. U.S.A.* 103, 7655–7659. doi: 10.1073/pnas.0601037103

Masin, J., Fiser, R., Linhartova, I., Osicka, R., Bumba, L., Hewlett, E. L., et al. (2013). Differences in purinergic amplification of osmotic cell lysis by the pore-forming RTX toxins Bordetella pertussis CyaA and Actinobacillus pleuropneumoniae ApxIA: the role of pore size. *Infect. Immun.* 81, 4571–4582. doi: 10.1128/IAI.00711-13

Melhorn, M. I., Brodsky, A. S., Estanislau, J., Khoory, J. A., Illigens, B., Hamachi, I., et al. (2013). CR1-mediated ATP release by human red blood cells promotes CR1 clustering and modulates the immune transfer process. *J. Biol. Chem.* 288, 31139–31153. doi: 10.1074/jbc.M113.486035

Munksgaard, P. S., Vorup-Jensen, T., Reinholdt, J., Soderstrom, C. M., Poulsen, K., Leipziger, J., et al. (2012). Leukotoxin from *Aggregatibacter actinomycetemcomitans* causes shrinkage and P2X receptor-dependent lysis of human erythrocytes. *Cell. Microbiol.* 14, 1904–1920. doi: 10.1111/cmi.12021

Ohman, J., Kudira, R., Albinsson, S., Olde, B., and Erlinge, D. (2012). Ticagrelor induces adenosine triphosphate release from human red blood cells. *Biochem. Biophys. Res. Commun.* 418, 754–758. doi: 10.1016/j.bbrc.2012.01.093

Paredes-Gamero, E. J., Craveiro, R. B., Pesquero, J. B., França, J. P., Oshiro, M. E., and Ferreira, A. T. (2006). Activation of P2Y1 receptor triggers two calcium signaling pathways in bone marrow erythroblasts. *Eur. J. Pharmacol.* 534, 30–38. doi: 10.1016/j.ejphar.2006.01.010

Parker, J. C., and Snow, R. L. (1972). Influence of external ATP on permeability and metabolism of dog red blood cells. *Am. J. Physiol.* 223, 888–893.

Porzig, H., Gutknecht, R., Kostova, G., and Thalmeier, K. (1995). G-protein-coupled receptors in normal human erythroid progenitor cells. *Naunyn Schmiedebergs Arch. Pharmacol.* 353, 11–20. doi: 10.1007/BF00168910

Romualdez, A., Volpi, M., and Sha'afi, R. I. (1976). Effect of exogenous ATP on sodium transport in mammalian red cells. *J. Cell. Physiol.* 87, 297–306. doi: 10.1002/jcp.1040870305

Rossi, L., Salvestrini, V., Ferrari, D., Di Virgilio, F., and Lemoli, R. M. (2012). The sixth sense: hematopoietic stem cells detect danger through purinergic signaling. *Blood* 120, 2365–2375. doi: 10.1182/blood-2012-04-422378

Sartorello, R., and Garcia, C. R. (2005). Activation of a P2Y4-like purinoceptor triggers an increase in cytosolic [Ca^{2+}] in the red blood cells of the lizard *Ameiva ameiva* (*Squamata, Teiidae*). *Braz. J. Med. Biol. Res.* 38, 5–10. doi: 10.1590/S0100-879X2005000100002

Shemon, A. N., Sluyter, R., Stokes, L., Manley, P. W., and Wiley, J. S. (2008). Inhibition of the human P2X7 receptor by a novel protein tyrosine kinase antagonist. *Biochem. Biophys. Res. Commun.* 365, 515–520. doi: 10.1016/j.bbrc.2007.11.008

Skals, M., Bjaelde, R. G., Reinholdt, J., Poulsen, K., Vad, B. S., Otzen, D. E., et al. (2014). Bacterial RTX toxins allow acute ATP release from human erythrocytes directly through the toxin pore. *J. Biol. Chem.* 289, 19098–19109. doi: 10.1074/jbc.M114.571414

Skals, M., Jorgensen, N. R., Leipziger, J., and Praetorius, H. A. (2009). α-Hemolysin from *Escherichia coli* uses endogenous amplification through P2X receptor activation to induce hemolysis. *Proc. Natl. Acad. Sci. U.S.A.* 106, 4030–4035. doi: 10.1073/pnas.0807044106

Skals, M., Leipziger, J., and Praetorius, H. A. (2011). Haemolysis induced by α-toxin from *Staphylococcus aureus* requires P2X receptor activation. *Pflugers Arch.* 462, 669–679. doi: 10.1007/s00424-011-1010-x

Sluyter, R., Shemon, A. N., Barden, J. A., and Wiley, J. S. (2004). Extracellular ATP increases cation fluxes in human erythrocytes by activation of the P2X7 receptor. *J. Biol. Chem.* 279, 44749–44755. doi: 10.1074/jbc.M405631200

Sluyter, R., Shemon, A. N., Hughes, W. E., Stevenson, R. O., Georgiou, J. G., Eslick, G. D., et al. (2007a). Canine erythrocytes express the P2X7 receptor: greatly increased function compared with human erythrocytes. *Am. J. Physiol. Regul. Integr. Comp. Physiol.* 293, R2090–R2098. doi: 10.1152/ajpregu.00166.2007

Sluyter, R., Shemon, A. N., and Wiley, J. S. (2007b). P2X7 receptor activation causes phosphatidylserine exposure in human erythrocytes. *Biochem. Biophys. Res. Commun.* 355, 169–173. doi: 10.1016/j.bbrc.2007.01.124

Sluyter, R., and Stokes, L. (2011). Significance of P2X7 receptor variants to human health and disease. *Recent Pat. DNA Gene Seq.* 5, 41–54. doi: 10.2174/187221511794839219

Sophocleous, R. A., Mullany, P. R., Winter, K. M., Marks, D. C., and Sluyter, R. (2015). Propensity of red blood cells to undergo P2X7 receptor-mediated phosphatidylserine exposure does not alter during *in vivo* or *ex vivo* aging. *Transfusion* 55, 1946–1954. doi: 10.1111/trf.13101

Spildrejorde, M., Bartlett, R., Stokes, L., Jalilian, I., Peranec, M., Sluyter, V., et al. (2014). R270C polymorphism leads to loss of function of the canine P2X7 receptor. *Physiol. Genomics* 46, 512–522. doi: 10.1152/physiolgenomics.00195.2013

Sprague, R. S., and Ellsworth, M. L. (2012). Erythrocyte-derived ATP and perfusion distribution: role of intracellular and intercellular communication. *Microcirculation* 19, 430–439. doi: 10.1111/j.1549-8719.2011.00158.x

Stachon, P., Peikert, A., Michel, N. A., Hergeth, S., Marchini, T., Wolf, D., et al. (2014). P2Y6 deficiency limits vascular inflammation and atherosclerosis in mice. *Arterioscler. Thromb. Vasc. Biol.* 34, 2237–2245. doi: 10.1161/ATVBAHA.114.303585

Stevenson, R. O., Taylor, R. M., Wiley, J. S., and Sluyter, R. (2009). The P2X7 receptor mediates the uptake of organic cations in canine erythrocytes and mononuclear leukocytes: comparison to equivalent human cell types. *Purinergic Signal.* 5, 385–394. doi: 10.1007/s11302-009-9163-1

Stokes, L., Fuller, S. J., Sluyter, R., Skarratt, K. K., Gu, B. J., and Wiley, J. S. (2010). Two haplotypes of the P2X7 receptor containing the Ala-348 to Thr polymorphism exhibit a gain-of-function effect and enhanced interleukin-1β secretion. *FASEB J.* 24, 2916–2927. doi: 10.1096/fj.09-150862

Tanneur, V., Duranton, C., Brand, V. B., Sandu, C. D., Akkaya, C., Gachet, C., et al. (2006). Purinoceptors are involved in the induction of an osmolyte permeability in malaria-infected and oxidized human erythrocytes. *FASEB J.* 20, 133–135. doi: 10.1096/fj.04-3371fje

Tanneur, V., Duranton, C., Brand, V. B., Sandu, C. D., Gachet, C., Sluyter, R., et al. (2004). Oxidation-induced activation of an organic osmolyte permeability in human erythrocytes involves purinoceptor signalling. *Pflugers Arch.* 447, S135.

van Giezen, J. J., Sidaway, J., Glaves, P., Kirk, I., and Björkman, J. A. (2012). Ticagrelor inhibits adenosine uptake *in vitro* and enhances adenosine-mediated hyperemia responses in a canine model. *J. Cardiovasc. Pharmacol. Ther.* 17, 164–172. doi: 10.1177/1074248411410883

Wang, B., and Sluyter, R. (2013). P2X7 receptor activation induces reactive oxygen species formation in erythroid cells. *Purinergic Signal.* 9, 101–112. doi: 10.1007/s11302-012-9335-2

Wang, L., Olivecrona, G., Götberg, M., Olsson, M. L., Winzell, M. S., and Erlinge, D. (2005). ADP acting on P2Y13 receptors is a negative feedback pathway for ATP release from human red blood cells. *Circ. Res.* 96, 189–196. doi: 10.1161/01.RES.0000153670.07559.E4

Wormser, C., Pore, S. A., Elperin, A. B., Silverman, L. N., and Light, D. B. (2011). Potentiation of regulatory volume decrease by a P2-like receptor and arachidonic acid in American alligator erythrocytes. *J. Membr. Biol.* 242, 75–87. doi: 10.1007/s00232-011-9377-3

22

Circulating microRNAs as Potential Biomarkers of Infectious Disease

Carolina N. Correia[1], Nicolas C. Nalpas[1†], Kirsten E. McLoughlin[1], John A. Browne[1], Stephen V. Gordon[2,3], David E. MacHugh[1,3]* and Ronan G. Shaughnessy[2]

[1] Animal Genomics Laboratory, UCD School of Agriculture and Food Science, University College Dublin, Dublin, Ireland, [2] UCD School of Veterinary Medicine, University College Dublin, Dublin, Ireland, [3] University College Dublin, UCD Conway Institute of Biomolecular and Biomedical Research, Dublin, Ireland

Edited by:
Diana Bahia,
Universidade Federal de Minas
Gerais, Brazil

Reviewed by:
Giulia Carla Marchetti,
University of Milan, Italy
Katherine J. Siddle,
Harvard University, USA

*Correspondence:
David E. MacHugh
david.machugh@ucd.ie

†Present address:
Nicolas C. Nalpas,
Quantitative Proteomics and
Proteome Center Tübingen,
Interfaculty Institute for Cell Biology,
University of Tübingen, Tübingen,
Germany

microRNAs (miRNAs) are a class of small non-coding endogenous RNA molecules that regulate a wide range of biological processes by post-transcriptionally regulating gene expression. Thousands of these molecules have been discovered to date, and multiple miRNAs have been shown to coordinately fine-tune cellular processes key to organismal development, homeostasis, neurobiology, immunobiology, and control of infection. The fundamental regulatory role of miRNAs in a variety of biological processes suggests that differential expression of these transcripts may be exploited as a novel source of molecular biomarkers for many different disease pathologies or abnormalities. This has been emphasized by the recent discovery of remarkably stable miRNAs in mammalian biofluids, which may originate from intracellular processes elsewhere in the body. The potential of circulating miRNAs as biomarkers of disease has mainly been demonstrated for various types of cancer. More recently, however, attention has focused on the use of circulating miRNAs as diagnostic/prognostic biomarkers of infectious disease; for example, human tuberculosis caused by infection with *Mycobacterium tuberculosis*, sepsis caused by multiple infectious agents, and viral hepatitis. Here, we review these developments and discuss prospects and challenges for translating circulating miRNA into novel diagnostics for infectious disease.

Keywords: biomarker, diagnostic, infection, transcriptomics, microRNA, serum, plasma

WHAT MAKES A GOOD BIOMARKER?

According to the working group of the National Institutes of Health Director's Initiative on Biomarkers and Surrogate Endpoints, a biomarker is "a characteristic that is objectively measured and evaluated as an indicator of normal biological processes, pathogenic processes, or pharmacologic responses to a therapeutic intervention" (1). A simpler but broader definition of biomarkers as objective, quantifiable characteristics of biological processes has also been emphasized by Strimbu and Tavel (2). The ideal biomarker has high specificity and sensitivity, is detectable by minimally invasive sampling procedures, and its concentration should be indicative of a disease state (1–5). Diagnostic biomarkers can be used to evaluate disease status, prognostic biomarkers are informative of disease outcome, and predictive biomarkers help determine treatment efficacy when experimental groups are compared to controls (4, 6).

In recent years, high-throughput sequencing (HTS) technologies have enabled simultaneous screening of thousands of potential transcriptional biomarkers, which facilitates both discovery of specific host disease expression biosignatures (7–9) and new insight on host–pathogen interaction and immunobiology (10–12). Host biomarkers may also help evaluate vaccine efficacy in both humans and domestic animals (13, 14), as well as provide information on the molecular mechanisms underlying latent infections (15) and drug resistance in pathogens (16).

CANONICAL BIOGENESIS AND IMMUNOLOGICAL FUNCTIONS OF microRNAs (miRNAs)

The role of miRNAs in post-transcriptional regulation of gene expression was discovered in 1993 through analyses of the *lin-4* locus in the roundworm *Caenorhabditis elegans*. Two contemporaneous studies showed that an RNA transcript from *lin-4* repressed translation of the lin-14 messenger RNA (mRNA), thereby exerting temporal developmental control on a diverse range of cell lineages (17, 18). Since then, it has been demonstrated that eukaryotic organisms contain hundreds to thousands of these small non-coding regulatory RNA molecules (19). Many miRNAs are evolutionarily conserved across divergent metazoan taxa (20, 21), highlighting the extensive roles that these small RNAs play in the regulatory networks and pathways governing complex biological processes such as cell fate specification, and innate and adaptive immunity (22–24).

Canonical biogenesis of miRNA in mammalian cells starts with transcription of a long RNA molecule called the primary-miRNA (pri-miRNA) by RNA polymerase II (25). Within the nucleus, pri-miRNA undergoes cleavage by the microprocessor complex, which consists of a Drosha ribonuclease III and the RNA-binding DGCR8 microprocessor complex subunit protein (26, 27). The intermediate product is a precursor-miRNA (pre-miRNA) hairpin of ~70 nucleotides in length that is transported to the cytoplasm by the exportin-5 protein (28). An additional cleavage occurs near the pre-miRNA terminal loop through the action of endoribonuclease Dicer (29). The final product is an 18–25 nucleotide double-stranded RNA with short 3′ overhangs that binds to argonaute (AGO) proteins and is loaded into the RNA-induced silencing complex (RISC) by the RISC-loading complex (RLC), which is formed by endoribonuclease Dicer, RLC subunit TARBP2, and AGO1–4 proteins (30). One strand of the RNA duplex, the mature miRNA, remains within the RLC and is used as a guide by the RISC for complementary nucleotide base pairing with a target mRNA (31). The second strand is known as miRNA* (or passenger strand) and is normally degraded after its release from the RLC. Further details on canonical biogenesis (32) and the processes driving mature miRNA strand selection (33, 34) have been extensively reviewed elsewhere. The development of HTS technologies has facilitated high-resolution miRNA-sequencing (miRNA-seq), revealing the existence of multiple functional mature variants that are termed isomiRs (35–37). In addition, non-canonical pathways have been identified as alternative mechanisms of miRNA biogenesis (38, 39).

Dysregulation of intracellular miRNAs during disease was first reported in 2002, with evidence that miR-15 and miR-16 were tumor suppressors for chronic lymphocyte leukemia (40). Shortly afterward, it was shown that higher let-7 expression levels were associated with a better prognosis for lung cancer survival (41). Notably, the cancer research literature has highlighted miRNAs as

powerful classifiers for disease onset and patient survival (42–45), as well as tumor driver mutations (46–48). These, and several other studies that followed, laid the groundwork for research that focuses on exploring the potential of miRNAs as biomarkers and therapeutic gene targets.

In silico analyses suggest that at least two-thirds of mammalian mRNAs are regulated by miRNAs (22, 23, 49); therefore, it is perhaps unsurprising that these non-coding transcripts have emerged as important molecular fine-tuners of the host immune response during infection (50–54). For example, multiple miRNAs are known to regulate the toll-like receptor 4 (TLR-4) pathway in the host innate immune response (23, 55, 56) and are also essential for optimal T cell activation and differentiation (57–62). More specifically, mice lacking miR-155 show diminished immune responses against infections with *Citrobacter rodentium* (63), *Salmonella typhimurium* (64), and *Listeria monocytogenes* (65). miR-155 has also been found to be increased in peripheral monocytes of chronic hepatitis C (CHC)-infected patients following *in vitro* stimulation with lipopolysaccharide (LPS) (66), and in murine bone marrow-derived macrophages stimulated with LPS plus interferon-γ (IFN-γ) (67). miR-146 is another important miRNA that exhibits increased expression in immune cells following TLR activation by bacterial pathogens (68). Moreover, members of the miR-146 family were found to form distinct expression profiles in human monocyte-derived macrophage cells infected with *Mycobacterium tuberculosis* (69) and *M. bovis*-infected bovine alveolar macrophages (70). The immunoregulatory roles of miRNAs in different cells involved with the host response to bacterial infections has been comprehensively reviewed (71, 72).

Collectively, these studies highlight the importance of post-transcriptional regulation of gene expression mediated by intracellular miRNAs in mammalian infection and immunity processes. A growing number of public databases provide information on miRNA–disease relationships (73), and informative reviews on this topic have been published (22, 23, 49, 52, 74).

Circulating miRNAs

So far, we have shown examples of intracellular miRNAs with immunological roles; however, there is a growing consensus that immune and non-immune cells routinely and actively release miRNAs into extracellular environments (75–77). Commonly associated with RNA-binding proteins, high-density lipoprotein particles or enclosed within lipid vesicles (**Figure 1**), miRNAs have been found to be extremely stable in extracellular fluids of mammals, such as blood plasma, serum, urine, saliva, and semen (78–80). miRNAs released by a human THP-1 monocyte cell line may be taken up by recipient cells in an alternative means of cell-to-cell communication (81). Wang and colleagues have shown that nucleophosmin, an RNA-binding protein involved with nuclear export of ribosomes, mediates export and protection of circulating miRNAs against degradation in several human cell lines (HepG2, A549, T98, and BSEA2B) immediately after serum deprivation, which is suggestive of an active response to stress (82). Active release of extracellular circulating miRNAs supports the hypothesis that they may act as "hormones" in cell-to-cell communication (82, 83).

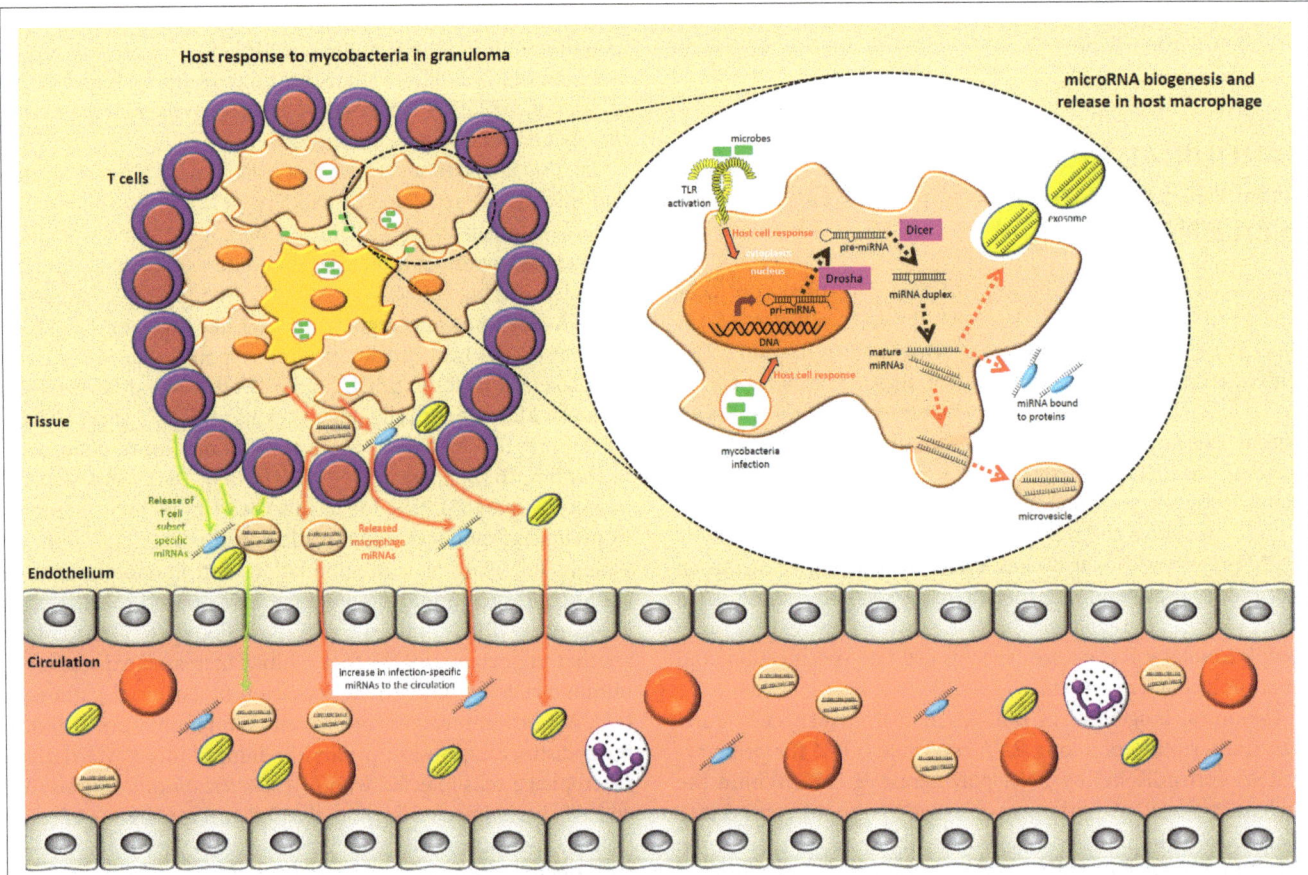

FIGURE 1 | A tuberculosis lung granuloma demonstrates how specific circulating microRNAs (miRNAs) may arise during an infection process. Mycobacterial pathogen-associated molecular patterns are recognized by toll-like receptors (TLRs) and other pattern recognition receptors, which result in the upregulation of primary-miRNAs in macrophages. These transcripts are subsequently cleaved in the nucleus and cytoplasm by Drosha and Dicer, respectively, resulting in 21–25 nucleotide mature miRNAs that act to fine-tune intracellular immune processes. Specific pathways and components of the immune response may be regulated by different miRNA subsets. Concurrently, the surrounding T lymphocytes involved in granuloma formation/maintenance upregulate T cell subset-specific miRNAs as a means of modulating the type of adaptive immune response. Mature miRNAs generated in macrophages and T cells may also be released into the extracellular environment within exosomes, heterogeneous microvesicles, or in association with high-density lipoprotein, LDL, or other protein complexes. Subsequently, by means not yet fully understood, these extracellular miRNAs move from local sites of infection to the circulatory system. This process can therefore give rise to infection-specific circulating miRNA expression signatures that can readily be accessed from multiple biological fluids (e.g., serum, plasma, or sputum).

Further work is required to fully understand how the release of extracellular miRNAs and uptake by target cell populations influences biomolecular signaling networks. Regardless of their precise functions, the main utility of miRNAs in the field of diagnostics and prognostics is based on the premise that different miRNA expression signatures are linked to different pathological states (**Figure 1**). With this in mind, it is noteworthy that a number of infectious diseases have been the focus of recent studies to assess circulating miRNAs as biomarkers.

CHALLENGES FOR ACCURATE DETECTION OF CIRCULATING miRNAs IN BIOFLUIDS

The observation that extracellular nucleic acids (both DNA and RNA) are present in vertebrate bodily fluids was first recorded almost 70 years ago (84), but their potential as biomarkers for disease states was not fully realized until the 1990s (85). In turn, detection of extracellular miRNAs was first reported in 2008 when placental miRNAs were observed in maternal plasma (86). In the same year, circulating miRNAs were also described in blood serum (87, 88) and plasma samples collected from cancer patients (88). In this regard, the potential of circulating miRNAs as non-invasive diagnostic and prognostic biomarkers of disease status in biological fluids was first realized in the field of cancer biology, particularly because techniques for cancer diagnosis and prognosis still primarily rely on invasive tissue biopsies (89–91), and the establishment of new circulating protein biomarkers has not been able to meet the demand (92). The marked stability of circulating miRNAs in body fluids, which are still viable after repeated cycles of freeze–thawing and long-term storage of frozen samples (88, 93–95), makes them attractive biomarker candidates for diagnosis or prognosis of complex diseases. However, the main challenges for profiling circulating miRNAs are biases introduced during pre-analytical and analytical steps that are described below.

Biological Fluids and RNA Extraction

Analysis of circulating miRNAs is normally performed on peripheral blood plasma or serum, and to a lesser extent on sputum, urine, breast milk, saliva, semen, and cerebrospinal fluid (CSF). The choice of starting material can significantly impact the expression profiles that are generated; in particular, because each biofluid can be enriched for a distinct set of miRNAs (78, 96, 97).

The miRNA fraction in biological fluids typically represents a very low proportion of total RNA. Investigations using serum and plasma samples have demonstrated that different protocols for pre- and post-storage sample processing can impact the quality of subsequent RNA extraction (96, 98, 99). For example, collection of peripheral blood in heparin-coated tubes can inhibit downstream laboratory steps that are based on polymerase chain reaction (PCR) protocols (100, 101). In addition, special attention is required to avoid contamination with intracellular miRNAs originating from blood components such as platelets and erythrocytes, which can introduce significant bias in circulating miRNA expression profiles (96, 98, 99, 102). It is also important that circulating miRNAs with low GC content are not lost when performing phenol-based RNA isolation, a problem that can be overcome by using small RNA extraction kits customized for specific biofluids (103).

Expression Profiling Methods for Circulating miRNAs

The biochemistry and molecular structure of miRNAs can cause difficulties for accurate transcriptional profiling and quantification (104, 105). Consequently, various established techniques for mRNA detection have been modified to improve miRNA detection, irrespective of tissue type (106–108).

Reverse transcription quantitative real-time PCR (RT-qPCR) is currently the most widely used method for miRNA profiling: it provides excellent sensitivity, high sample throughput, and the capacity for moderate multiplexing of targets (109). A number of strategies can be used with miRNA RT-qPCR, including (1) reverse transcription using stem loop primers as implemented in TaqMan™ MicroRNA Assays (110); (2) incorporation of locked nucleic acids (LNAs) (111) in primer sequences to reduce melting temperature (T_m) differences in primer–target duplexes, as used for miRCURY LNA™ Universal RT microRNA PCR; and (3) approaches that enzymatically incorporate a poly(A) tail to miRNAs prior to the reverse transcription step, which facilitates hybridization with a poly(T) sequence linked to a universal reverse primer (109). Regardless of these technical developments, RT-qPCR-based methods cannot identify novel miRNAs and, importantly, special attention is required for the design of standardized internal controls (104, 112–114).

Hybridization-based methods normally rely on DNA capture probes that are immobilized on a microarray platform such that fluorescent signal intensities can be quantified to estimate expression of individual miRNAs. Commercially available and cost-effective microarray assays include miRCURY LNA™

microRNA Arrays (Exiqon), GeneChip® miRNA Arrays (Affymetrix), and SurePrint miRNA Microarrays (Agilent) (115). However, due to lower specificity and reduced dynamic range compared to other methods, microarrays often require additional validation *via* RT-qPCR (116, 117). A relatively new hybridization-based method that does not require a PCR amplification step or direct labeling of target miRNAs is the nCounter® miRNA Expression Assay developed by NanoString Technologies (118). This approach has comparable sensitivity to RT-qPCR, is high-throughput, and also facilitates multiplexing of up to 800 distinct miRNA variant targets in the same assay. miRNA profiling in serum (119), peripheral blood (120), and aortic tissue (121) provide examples of studies that have used this technology. However, like RT-qPCR, it is again important to note that hybridization-based profiling methods cannot be used to identify novel miRNA variants.

Unlike the methods discussed above, HTS technologies in the form of miRNA-seq can be used for discovery-focused global expression profiling of the whole miRNA transcriptome (miRNome) from a particular biological sample (122). In addition, miRNA-seq approaches can identify with high accuracy, novel mature miRNAs, sequence variants (specific isomiRs or particular miRNA family members), and also pre-miRNAs (37, 123, 124). The rapid adoption of HTS for miRNA profiling has been driven by significant increases in sample throughput, a wide range of laboratory methods for different applications, and a thriving ecosystem of open-source software for data analysis and interpretation (9, 125, 126). However, it is important to note that technical biases inherent to different sequencing technologies (e.g., Illumina®, ABI SOLiD®, and Ion Torrent™) may generate reads that are not bona fide miRNAs (5, 127, 128).

Finally, in addition to established methods described here, emerging biosensor approaches for miRNA profiling have been reviewed in detail elsewhere and are beyond the scope this review (129).

Data Normalization

Transcriptomics experiments are characteristically "noisy," therefore, appropriate normalization is critical to minimize technical variation that may compromise interpretation of results. A range of methods have been successfully used for this purpose in mRNA and intracellular miRNA transcriptomics. Manufacturers' instructions for data normalization vary greatly depending on the platform used, and a consensus on how circulating miRNA data should be normalized is yet to emerge (114, 130).

There is significant debate concerning the optimal strategy for normalization of circulating miRNAs with RT-qPCR assays. Methods currently used include (1) normalization to small nucleolar RNAs as reference genes, (2) normalization to an external spike-in synthetic oligonucleotide, (3) normalization to specific miRNAs, and (4) the global mean normalization method (131). Most studies report the use of small nucleolar RNA genes as reference genes, such as the small nucleolar RNA, C/D box 44 gene (*SNORD44*), the small nucleolar RNA, C/D box 48 gene (*SNORD48*), and the RNA, U6 small nuclear 6, pseudogene (*RNU6-6P*). However, there is growing evidence that these genes may be unsuitable due to significant variability in expression among individual serum samples (130, 132, 133). Use of a synthetic RNA spike-in as a reference gene has also been criticized because this method only accounts for specific components of technical variation introduced, for example, during RNA extraction or reverse transcription (134). When identifying circulating miR-NAs to serve as reference genes, normally those that do not vary significantly among biological replicates are selected. Marabita et al. (114) advise caution when selecting endogenous reference controls and recommend the use of data-driven approaches for this purpose. The global mean normalization method does not require a reference gene and appears to be the most robust option but should only be applied when simultaneously profiling hundreds of miRNAs (131).

Conflicting reports on the efficiency of methods for statistical normalization are also problematic for miRNA-seq data. Tam and collaborators evaluated a range of available methods and recommended the use of trimmed mean of M-values (TMM) (135) and upper quartile scaling (136) for count normalization in comprehensive miRNA profiling studies (137). Conversely, Garmire and Subramaniam (138) did not support the use of TMM but strongly recommended the application of locally weighted regression (139, 140) or quantile normalization (QN) (141) instead. However, a parallel benchmarking study published soon afterward came to the opposite conclusion, recommending TMM over QN (142). Finally, a rebuttal to Garmire and Subramaniam highlighted several drawbacks with their data analysis and evaluation of the TMM method (143).

In summary, it is imperative that rigorous independent benchmarking studies are performed to systematically evaluate normalization methods proposed for miRNA profiling. With these challenges in mind, in the next section we review studies that have assessed the usefulness of circulating miRNAs as biomarkers for selected bacterial and viral infections.

DIFFERENTIAL EXPRESSION OF CIRCULATING miRNAs IN SPECIFIC INFECTIOUS DISEASES

Human Tuberculosis (TB)

Human TB, caused by *M. tuberculosis*, continues to be a significant global health problem with 9.6 million new cases and 1.5 million deaths in 2014 (144). Classical methods for TB diagnostics in clinical settings include smear microscopy and mycobacterial culture. The former is the most used test in middle- and low-income countries, but its sensitivity is highly variable (20–60%), and the latter can take up to 8 weeks to yield results (145).

Diagnostic tests based on molecular methods represent a significant improvement in turnaround time and accuracy. Nonetheless, most molecular-based platform assays in use today are costly and have not been designed to be used in lower tiers of the health-care system (145, 146). According to Pai and Schito, one of the highest priorities for TB diagnostics is the development of a point-of-care non-sputum-based test capable of detecting all forms of TB, including extra pulmonary TB. Improved methods for distinction between active and latent TB are also urgently required (146).

Human TB was one of the first infectious diseases to be targeted for development of new diagnostics based on circulating serum or plasma miRNAs. Using a human miRNA microarray platform (Exiqon miRCURY™ LNA), Fu and colleagues were able to detect 92 differentially expressed miRNAs in serum from patients with active pulmonary TB compared to healthy individuals (147). However, it is important to note that an appropriate correction procedure for multiple statistical tests was not used in this study. Notwithstanding this, RT-qPCR validation demonstrated that circulating miR-93* and miR-29a were significantly upregulated in serum from the TB cases. In addition, miR-29a was also shown to be differentially expressed in sputum samples from TB patients compared to healthy controls (HC). A follow-up study using the same miRNA expression microarray, but with sputum samples from active pulmonary TB cases and HC, also found that miR-29a was upregulated in sputum from TB patients (148). However, inconsistencies were observed between the results obtained for circulating serum miRNAs by Fu and coworkers and those obtained by Yi and colleagues for sputum miRNAs. In particular, the 2 sets of 10 miRNAs that showed the most increased or decreased expression in TB patients were different for each body fluid (147, 148).

Parallel work using a different miRNA expression platform (Applied Biosystems TaqMan® Low Density Array Human MicroRNA Panel) and a comparable statistical approach identified a total of 97 differentially expressed miRNAs in serum samples from active pulmonary TB patients compared to HC (149). Following RT-qPCR validation and receiver operating characteristic (ROC) curve analysis, a panel of three miRNAs (miR-361-5p, miR-889, and miR-576-3p) was shown to differentiate TB patients from HC with moderate sensitivity and specificity. Further evaluation of the specificity of this panel of miRNAs for diagnosis of pulmonary TB was performed using RT-qPCR analysis of serum from pediatric patients infected with enterovirus, varicella-zoster virus, or *Bordetella pertussis*. All three miRNAs exhibited significant differences between the TB patient group and the other microbial infection groups, leading Qi and colleagues to propose this set of miRNAs as the starting point for a biosignature of human TB (149).

A comparative study of the diagnostic potential of a small panel of circulating serum miRNAs for pulmonary TB, lung cancer, and pneumonia was undertaken by Abd-El-Fattah et al. (150). Using RT-qPCR, these workers examined expression of four miRNAs (miR-21, miR-155, miR-182, and miR-197) in serum from pulmonary TB, lung cancer, and pneumonia patient groups compared to a HC group. They observed that all four miRNAs were significantly differentially expressed between lung cancer patients and HC, three miRNAs (miR-21, miR-155, and miR-197) distinguished pneumonia patients from controls, but only one miRNA (miR-197) was significantly differentially expressed between the pulmonary TB group and the control group.

Two independent miRNA-seq studies of circulating serum miRNAs for diagnosis of active pulmonary TB revealed distinct panels of miRNAs as potential expression biomarkers of disease. The first study (151) showed that six circulating serum miR-NAs (miR-378, miR-483-5p, miR-22, miR-29c, miR-101, and miR-320b) could serve as a distinct biosignature of pulmonary TB compared to HC, and importantly, groups of pneumonia, lung cancer, and chronic obstructive pulmonary disease patients. Furthermore, ROC curve analysis demonstrated that a six-miRNA biosignature could discriminate pulmonary TB patients from HC with a sensitivity of 95.0% and a specificity of 91.8% (151). In the second study (152), miRNA-seq was used to identify a total of 30 circulating serum miRNAs that were differentially expressed (24 increased and 6 decreased) in active pulmonary TB patients compared to 3 different control groups (latent TB infection, BCG-vaccinated, and HC). However, only 1 of these 30 circulating serum miRNAs (miR-22) was also detected in the earlier study by Zhang et al. (151).

Natural killer (NK) cells are effector lymphocytes that represent an important component of the innate immune system; they are able to rapidly target and kill virus-infected and tumorigenic cells in the absence of antibodies (153). Zhang et al. (154) observed decreased expression of circulating serum miR-155 in TB patients when compared to HC. From a functional perspective, levels of miRNA-155 were also inversely associated with cytotoxicity of NK cells isolated from the TB patients, which suggested that miR-155 may be used as an indicator of NK cell activity in TB patients (154).

The human TB studies described in this review have used serum and sputum as a source of circulating miRNAs and multiple transcriptomics technologies (miRNA-seq, microarray, and RT-qPCR) and data normalization methods, which together may contribute to the discordance among the results obtained by different researchers. **Table 1** provides summary information on circulating miRNA biomarker studies for diagnosis and prognosis of human TB.

Sepsis

Sepsis is a subtype of systemic inflammatory response syndrome (SIRS), which is caused by an immune response triggered by various microbial infections. The causative agent is most commonly a bacterial pathogen, but it can also be triggered by infections involving fungi, viruses, or parasites (155). Sepsis is a major burden on health-care systems and of greatest concern in intensive care units (ICUs), where delayed diagnosis is a major cause of mortality. Consequently, in recent years there has been a concerted effort to develop circulating miRNA biomarkers for sepsis diagnosis and prognosis (156, 157).

Plasma levels of miR-150 have been shown to correlate with those of TNF-α, IL-10, and IL-18, which are important immune response markers. More specifically, the ratio of miR-150/IL-18 has been suggested as a useful indicator of sepsis (158). miR-150 was also shown to exhibit increased expression in plasma from septic shock patients and was an independent predictor of mortality (159). Contrary to these results, circulating serum miR-150 levels could not be used to differentiate between critical illness patients and healthy individuals. However, although circulating miR-150 had no association with common markers of inflammation, it was independently correlated with unfavorable prognosis for patients (160).

It has previously been proposed that decreased expression of circulating miR-146a serves as an indicator of sepsis in both serum (161) and plasma (162). In addition, miR-223, which also exhibits decreased expression in plasma during sepsis, has been shown to display a greater capacity to distinguish sepsis from non-infectious SIRS than miR-146a (161). However, using more stringent statistical methods, a recent study demonstrated that miR-146a and miR-223 neither exhibited differential expression in plasma samples of sepsis and septic shock patients nor were they correlated with markers of inflammation, disease progression, or mortality (159). In addition, a comprehensive animal and clinical study has demonstrated that miR-223 serum levels do not correspond to the presence of sepsis in murine models or in a large cohort of ICU patients and do not reflect clinical outcome for critically ill patients (163). Taken together, these results constitute good evidence that circulating serum miR-223 cannot be used as a biomarker for sepsis.

TABLE 1 | Circulating microRNAs (miRNAs) profiled in selected human tuberculosis studies.

Platform/assay	Biological fluid	Notable miRNAs detected (arrows indicate direction of expression)	Data normalization	Reference
miRCURY LNA array (Exiqon) Gene Amp PCR system 9700 (Applied Biosystems)	Serum and sputum	miR-93*↑, miR-29a↑	Median normalization U6 snRNA	Fu et al. (147)
miRCURY LNA array (Exiqon) GeneAmp PCR System 9700 (Applied Biosystems)	Sputum	miR-3179↑, miR-147↑, miR-19b-2*↓, miR-29a↑	Median normalization U6 snRNA	Yi et al. (148)
TaqMan Low Density array (Applied Biosystems) TaqMan RT-qPCR (Applied Biosystems)	Serum	miR-361-5p↑, miR-889↑, miR-576-3p↑	cel-miR-238 miR-16	Qi et al. (149)
7500 Real-Time PCR system (Applied Biosystems)	Serum	miR-197↑	SNORD68	Abd-El-Fattah et al. (150)
Solexa Small RNA-seq (Illumina) SYBR green RT-qPCR assay	Serum	miR-378↑, miR-483-5p↑, miR-22↑, miR-29c↑, miR-101↓, miR-320b↓	No information provided concerning miRNA-sequencing normalization method miR-16	Zhang et al. (151)
Solexa Small RNA-seq (Illumina) TaqMan RT-qPCR (Applied Biosystems)	Serum	miR-516b↑, miR-486-5p↓, miR-196b↑, miR-376c↑	Total copy number of each sample was normalized to 100,000 cel-miR-238	Zhang et al. (152)
SYBR green RT-qPCR assay	Serum	miR-155↓	U6 snRNA	Zhang et al. (154)

In a murine model of polymicrobial sepsis, circulating serum miR-133a, miR-155, miR-150, and miR-193b* displayed increased expression compared to baseline measurements (164). When extended to humans using large cohorts of ICU patients and HC, serum miR-133a also exhibited increased expression and displayed an increasing trend with disease severity. In addition, miR-133a was correlated not only with markers of inflammation and bacterial infection but also with renal and hepatic damage, cholestasis, and liver biosynthetic capacity. This work therefore supports further evaluation of miR-133a as a useful marker for the clinical state of critically ill patients (164). Another recent study using a murine model that was extended to human patients also revealed that miR-122 displayed increased expression independent of the presence of infection or sepsis in human ICU patients (165).

It has been shown that circulating serum miR-297 is upregulated in non-surviving sepsis patients when compared to survivors, whereas miR-574-5p is downregulated. In addition, the combination of sepsis stage, sequential organ failure assessment (SOFA) score and miR-574-5p expression was identified as an excellent predictor for patient survival from sepsis (166). Finally, another comprehensive study investigated aberrantly expressed serum miRNAs, demonstrating that a combination of four miRNAs (miR-15a, miR-16, miR-193b*, and miR483-5p) and three clinical indicators (SOFA score, acute physiology and chronic health evaluation score, and sepsis stage) can be used as a good predictor for mortality by sepsis (167).

Table 2 provides a summary of methodologies and notable miRNAs profiled for the sepsis studies discussed in this section.

Viral Hepatitis
Hepatitis B
According to the WHO, an estimated 240 million people are chronically infected with hepatitis B virus (HBV), and more than 686,000 people die every year due to complications of HBV infection (168). This viral infection attacks the liver and presents as acute or chronic disease. In comparison to other regions of the world, sub-Saharan Africa and East Asia show high endemicity of HBV infection (169). Additionally, chronic hepatitis B (CHB)

infection is a major cause of liver cirrhosis and hepatocellular carcinoma (HCC), and reliable indicators of disease progression are urgently needed.

It has been shown that the occurrence of specific circulating miRNAs in blood serum of HBV-infected individuals increases with disease severity: 37 miRNAs in HC, 77 in chronic asymptomatic carriers, 101 in CHB, and 135 in HBV-associated acute-on-chronic liver failure (170). Circulating serum miR-210 (171) and miR-124 (172) are among the miRNAs implicated as being increased in conjunction with disease severity. Markers for liver fibrosis in HBV-infected patients have also been examined, with miR-345-3p, miR-371a-5p, and miR-2861 reported as positive indicators of fibrosis, whereas miR-486-3p and miR-497-5p exhibited lower expression at all stages of fibrosis when compared to non-fibrosis CHB patients (173).

Many published studies have suggested groups of circulating miRNAs that could distinguish CHB patients at early stages of HCC from those without the presence of cancer, such as plasma miR-122, miR-223, miR-26a, miR-27a, miR-192, miR-21, and miR-801 (174); plasma miR-28-5p, miR-30a-5p, miR-30e-3p, miR-378a-3p, miR-574-3p, and let-7c (175); serum miR-222, miR-223, and mir-21 (176); serum miR-206, miR-141-3p, miR-433-3p, miR-1228-5p, miR-199a-5p, miR-122-5p, miR-192-5p, and miR-26a-5p (177); and exosomal serum miR-221, miR-222, miR-224, and miR-18a (178). Furthermore, miR-150 (179) and miR-18a (180) have been independently profiled in serum of HBV-HCC patients and found to exhibit significantly higher expression in these groups when compared to CHB samples.

An miRNA consistently reported in hepatitis infection studies is miR-122. Significant higher levels of this miRNA in plasma (181, 182) and serum (170, 183–185) samples of HBV-infected patients have been observed, and hence miR-122 abundance has been suggested as a potential disease signature. miR-122 abundance was also positively correlated with current markers of viral activity in HBV-infected patients (170, 185, 186), but conflicting reports have been published regarding its correlation to degree of liver injury (170, 181, 186). It is also important to note that miR-122 was significantly upregulated in a murine model for alcohol- and chemical-induced liver diseases (181) and non-alcoholic

TABLE 2 | Circulating microRNAs (miRNAs) profiled in selected sepsis studies.

Platform/assay	Biological fluid	Notable miRNAs detected (arrows indicate direction of expression)	Data normalization	Reference
miRNA microarray (Agilent Technologies)	Plasma	miR-486↑, miR-182↑, miR-150↓, miR-342-5p↓	Median normalization	Vasilescu et al. (158)
RT-qPCR TaqMan MicroRNA assays (Applied Biosystems)			U6B snRNA	
RT-qPCR TaqMan MicroRNA assays (Applied Biosystems)	Plasma	miR-150↑, miR-146a NS, miR-223 NS	cel-miR-39	Puskarich et al. (159)
RT-qPCR miScript system (Qiagen)	Serum	miR-146a↓, miR-223↓	mmu-miR-295	Wang et al. (161)
RT-qPCR miScript system (Qiagen)	Serum	miR-133a↑	SV40	Tacke et al. (164)
RT-qPCR miScript system (Qiagen)	Serum	miR-122↑	SV40	Roderburg et al. (165)
GeneChip miRNA 1.0 arrays (Affymetrix)	Serum	miR-297↑, miR-574-5p↓	5S rRNA	Wang et al. (166)
RT-qPCR miRcute (Tiangen Biotech Company)				
Solexa Small RNA-seq (Illumina)	Serum	miR-193b*↑, miR-15↑, miR-122↑, miR-483-5p↑, miR-16↓, miR-223↓	U6 snRNA	Wang et al. (167)
RT-qPCR TaqMan MicroRNA assays (Applied Biosystems)				

steatohepatitis (184), which might implicate it as an unreliable biomarker.

miR-122 was also investigated as a component of miRNA panels that aim to offer a more robust biosignature of hepatitis B. For example, an assessment of CHB and healthy children during a period of 6 years revealed that miR-122-5p, miR-122-3p, miR-99a-5p, and miR-125b-5p could be used to monitor pathological status (187, 188). Furthermore, the combination of miR-122, miR-let7c, miR-23b, and miR-150 was able to distinguish either HBV or occult HBV-infected patients from HC (189).

Other miRNAs have also been reported as potential biomarkers for HBV-infected patients: miR-375, miR-10a, miR-223, and miR-423 distinguished HBV-infected patients from HC, and the combination of miR-920 and miR-423 was able to differentiate between HBV- and hepatitis C virus (HCV)-infected individuals (190).

Table 3 provides a summary of methodologies and notable miRNAs profiled for the HBV infection studies discussed in this section.

Hepatitis C

Similar to HBV, the HCV can also give rise to both acute and chronic infection, possibly leading to progressive liver disease, cirrhosis, and hepatocellular cancer. Considering that miR-122 is a liver-specific miRNA (191), it is therefore not surprising that the majority of circulating miRNA studies have focused on miR-122 as a potential biomarker for hepatic pathologies. As previously observed in HBV studies, expression of circulating miR-122 has been found to be significantly higher in chronic HCV patients when compared to healthy cohorts (192–195). Panels of miRNAs that contained miR-122 also resulted in positive indicators for the presence of either CHC (66) or HCV (196). Positive correlation with current markers of HCV infection has been seen for miR-122 (66, 192) and miR-122-5p (196). However, miR-122 levels were also elevated in non-alcoholic fatty liver disease patients (193), and decreased levels were observed in one study with advanced stage fibrosis CHC patients (197).

Interestingly, high levels of serum miR-122 may predict favorable virological responses to therapy in pretreatment pegylated IFN alpha/ribavirin (pegIFN/RBV) patients of Asian ethnicity (198), but not for those of African and Caucasian ethnicities (199).

In a longitudinal study using plasma samples from non-HCV-infected injection drug users who eventually acquired the infection, miR-122 and miR-885-5p were increased in abundance during acute infection, whereas miR-494 and miR-411 were decreased in expression. Also, in an independent cohort of individuals, all but miR-411 were validated (200). Furthermore, miR-122 and miR-885-5p levels remained elevated during viremia and returned to preinfection levels after infection resolution (200).

Considering other miRNA species, miR-571 has been associated with HCV-related cirrhosis progression (201); miR-20a and miR-92a serum levels were elevated in HCV-infected fibrosis patients, and miR–92a expression was significantly reduced after infection resolution (202); circulating serum miR-320c, miR-134, and miR-483-5p were shown to be significantly increased in expression for HCV-infected patients when compared to HC (203).

Early detection of HCC is also a major concern for HCV patients. Several published studies have investigated the usefulness of circulating miRNAs as a less-invasive diagnostic method. Serum levels of miR–16 were lower in HCC patients

TABLE 3 | Circulating microRNAs (miRNAs) profiled in selected hepatitis B studies.

Platform/assay	Biological fluid	Notable miRNAs detected (arrows indicate direction of expression)	Data normalization	Reference
RT-qPCR TaqMan MicroRNA assays (Applied Biosystems)	Serum	miR-122↑, miR-16↑, miR-223↑, miR-19b↑, miR-20a↑, miR-92a↑, miR-106a↑, let-7b↑, miR-194↑	U6 snRNA	Ji et al. (170)
RT-qPCR TaqMan MicroRNA assays (Applied Biosystems)	Serum	miR-210↑	cel-miR-39	Song et al. (171)
RT-qPCR miRcute (Tiangen Biotech Company)	Serum	miR-124↑	5S rRNA	Wang et al. (172)
miRNA microarray (Agilent Technologies)	Plasma	miR-4695-5p↑, miR-486-3p↓, miR-497-5p↓	Quantile normalization	Zhang et al. (173)
SYBR Green I-based RT-qPCR with individual miRNA-specific primers (Applied Biosystems)	Plasma	miR-122↑	U6 snRNA	Zhang et al. (181)
Thunderbird SYBR qPCR mix (Toyobo, Japan)	Plasma	miR-122↑	No information provided	Zhang et al. (182)
SYBR Green PCR Master Mixture (Takara)	Serum	miR-122↑	miR-181a	Xu et al. (183)
RT-qPCR TaqMan MicroRNA assays (Applied Biosystems)	Serum	miR-122↑	No information provided	Waidmann et al. (186)
RT-qPCR miRNA arrays and individual assays (Exiqon)	Plasma	miR-122-5p↑, miR-122-3p↑, miR-99a-5p↑, miR-125b-5p↑	Global mean normalization, U6, and geometric mean normalization	Winther et al. (187)
TaqMan probe-based RT-qPCR (Applied Biosystems)	Serum	miR-122↑, miR-let7c↑, miR-23b↑, miR-150↑	Plant MIR-168	Chen et al. (189)
Solexa Small RNA-seq (Illumina) TaqMan probe-based RT-qPCR (Applied Biosystems)	Serum	miR-375↑, miR-10a↑, miR-223↑, miR-423↑	Plant MIR-168	Li et al. (190)

when compared to either HCV (204) or chronic liver diseases (CLD) groups (205). miR-21 serum levels were elevated in CHC and CHC-associated HCC patients, in comparison with HC (206). In plasma samples, miR-21 levels were also significantly higher for HCC patients when compared to chronic hepatitis (B or C types) and healthy groups (207). Lastly, miR-199a exhibited moderate power to distinguish HCC patients from CLD groups (205).

Although there are many published studies describing the use of circulating miRNAs as biomarkers of hepatitis B (**Table 3**) or hepatitis C (**Table 4**), specificity and lack of independent validation remain significant problems hindering adoption of circulating miRNAs as useful biomarkers for hepatitis.

Other Infectious Diseases

Pertussis

Pertussis, also known as whooping cough, is a respiratory infection caused by *B. pertussis*. A panel of five circulating miRNAs was observed to be upregulated in serum samples from infected patients (miR-202, miR-342-5p, miR-206, miR-487b, and miR-576-5p), showing high sensitivity and specificity for differentiation of pertussis patients and HC. In addition, analysis of this miRNA panel in samples from patients with a range of other microbial infections (*M. tuberculosis*, enterovirus, varicella-zoster virus, mumps virus, and measles virus) demonstrated that the expression signature for pertussis disease was distinct and unambiguous (208).

Human Immunodeficiency Virus (HIV)-Associated Neurological Disorders (HAND)

Cognitive, motor, and behavioral impairments that affect individuals infected with the HIV are collectively referred to as HAND (209, 210). An miRNA pairwise approach has demonstrated the potential use of two pairs of plasma miRNAs as biomarkers for cognitive-impaired HIV-positive individuals: miR-495-3p in combination with let-7b-5p, miR-151a-5p, or miR-744-5p; and miR-376a-3p/miR-16-5p (211).

It is recognized that early detection of HAND would facilitate better treatment choices and fewer sequelae caused by neuronal damage. However, this phenomenon has been difficult to investigate in patient cohorts; therefore, it has required the use of animal models such as the macaque (*Macaca nemestrina*) simian immunodeficiency virus (SIV) model of HIV (212). A combination of six circulating plasma miRNAs (miR-125b, miR-34a, miR-21, miR-1233, miR-130b, and miR-146a) could be used to predict the development of central nervous system disease in a macaque/SIV model, when animal samples from pre- and post-infection were compared to HC (212). Expression of circulating miRNAs in CSF of HIV-encephalitis (HIVE) patients has been compared to HIV-positive patients without signs of HAND, and also to HIV-negative individuals. Overall, decreased expression of miRNAs was observed between HIV-positive and HIV-negative groups, whereas between HIVE and HIV-negative no changes in expression were observed. General increased expression was only observed when HIVE and HIV-positive groups were compared, with miR-19b-2*, miR-937, and miR-362-5p displaying the largest fold changes (213).

Hand, Foot and Mouth Disease (HFMD)

Human enterovirus 71 (EV71) and coxsackievirus A16 (CVA16) are the most common pathogens responsible for HFMD. More than 500,000 cases, including 176 fatal ones, have been reported in China since an outbreak in 2008 (214). Levels of eight circulating serum miRNAs (miR-148a, miR-143, miR-324-3p, miR-628-3p, miR-206, miR-140-5p, miR-455-5p, and miR-362-3p) were significantly higher in sera of patients with enteroviral infections (215). The combination of six miRNAs (miR-148a, miR-143, miR-324-3p, miR-628-3p, miR-140-5p, and miR-362-3p) generated

TABLE 4 | Circulating microRNAs (miRNAs) profiled in selected hepatitis C studies.

Platform/assay	Biological fluid	Notable miRNAs detected (arrows indicate direction of expression)	Data normalization	Reference
RT-qPCR TaqMan MicroRNA assays (Applied Biosystems)	Serum	miR-122↑	No information provided	Bihrer et al. (192)
RT-qPCR TaqMan MicroRNA assays (Applied Biosystems)	Serum	miR-122↑, miR-34a↑, miR-16↑	cel-miR-238	Cermelli et al. (193)
RT-qPCR TaqMan MicroRNA assays (Applied Biosystems)	Serum	miR-122↑, miR-192↑	Normalized for initial serum input	van der Meer et al. (194)
RT-qPCR TaqMan MicroRNA assays (Applied Biosystems)	Serum	miR-122↑	cel-miR-39	Wang et al. (195)
RT-qPCR TaqMan MicroRNA assays (Applied Biosystems)	Serum	miR-122↑, miR-155↑	cel-miR-39	Bala et al. (66)
miRNA PCR arrays / Individual RT-qPCR assays	Serum	miR-122↑, miR-134↑, miR-424-3p↑, miR-629-5p↑	U6 snRNA	Zhang et al. (196)
RT-qPCR TaqMan MicroRNA assays	Serum	miR-122↑	cel-miR-39	Su et al. (198)
TaqMan RT-qPCR OpenArray chips / Individual TaqMan RT-qPCR assays	Plasma	miR-122↑, miR-885-5p↑, miR-494↓	Quantile normalization / ath-miR-159a	El-Diwany et al. (200)
miScript miRNA PCR array / Individual TaqMan RT-qPCR assays	Serum and plasma	miR-20a↑, miR-92a↑	cel-miR-39	Shrivastava et al. (202)
miRNA microarray (Agilent)	Serum	miR-134↑, miR-320c↑, miR-483-5p↑	Percentile shift normalization	Shwetha et al. (203)

a biosignature that could distinguish enteroviral patients and HC. In addition, a panel comprising miR-143, miR-324-3p, and miR-545 had moderate ability in discriminating patients infected with CVA16 from those with EV71 (215). Circulating exosomal miRNAs (miR-671-5p, miR-16-5p, and miR-150-3p) have also been observed to be differentially expressed in serum samples from both mild and extremely severe cases of HFMD when compared to that from healthy individuals (216). Lastly, a signature of eight miRNAs (miR-494, miR-29b-3p, miR-551a, miR-606, miR-876-5p, miR-30c-5p, miR-221-3p, and miR-150-5p) was identified in serum of children infected with EV71 (119). Furthermore, the results presented by Wang and collaborators suggested that upregulation of miR-876-5p is a specific response to severe EV71 infection.

Varicella

Varicella, commonly known as chickenpox, is caused by the varicella-zoster virus. Aberrant serum miRNA expression in non-vaccinated children that contracted varicella revealed a panel of five miRNAs (miR-197, miR-629, miR-363, miR-132, and miR-122) that could differentiate, with moderate sensitivity

and specificity, varicella patients from HC, and also varicella patients from patients with three other microbial infections (*B. pertussis*, measles virus, and enterovirus) (217).

Influenza

Influenza A viruses are the causative agents of influenza in birds and mammals. A panel of 14 circulating miRNAs was observed to be aberrantly expressed in whole blood samples from patients infected with the H1N1 strain of influenza virus A (218). Further analyses showed that six of these miRNAs (miR-1260, miR-335*, miR-664, miR-26a, miR-576-3p, and miR-628-3p) had similar expression signatures in human A549 and Madin–Darby canine kidney (MDCK) cells infected with H1N1 *in vitro*. In addition, examination of MDCK supernatant exosomes indicated that only miR-576-3p was not detectable (218). Also, evaluation of serum samples from influenza A/H1N1 patients demonstrated that critically ill patients exhibited elevated expression of miR-150, when compared to those presenting a milder form of the disease (219).

Avian influenza A (H7N9) has been recently detected in China and was associated with fatal cases. Circulating serum miR-17,

TABLE 5 | Circulating microRNAs (miRNAs) profiled in selected infectious disease studies.

Pathology	Platform/assay	Biological fluid	Notable miRNAs detected (arrows indicate direction of expression)	Data normalization	Reference
Pertussis (human)	RT-qPCR TaqMan Array Human miRNA panel (Applied Biosystems)	Serum	miR-202↑, miR-342-5p↑ miR-206↑, miR-487b↑, miR-576-5p↑	cel-miR-238	Ge et al. (208)
Varicella (human)	RT-qPCR TaqMan Array Human miRNA panel (Applied Biosystems)	Serum	miR-197↑, miR-629↑, miR-363↑, miR-132↑, miR-122↑	cel-miR-238	Qi et al. (217)
Influenza H1N1 (human)	miRCURY LNA microRNA Arrays (Exiqon) RT-qPCR TaqMan	Whole blood	miR-1260↑, miR-335↑*, miR-664↑, miR-26a↓, miR-576-3p↑, miR-628-3p↓	Normalized to endogenous controls and the spike-in control 18S rRNA	Tambyah et al. (218)
Influenza A/H1N1 virus (human)	RT-qPCR TaqMan Array Human miRNA panel (Applied Biosystems)	Serum	miR-150↑	U6 snRNA	Moran et al. (219)
Avian Influenza A H7N9 (human)	RT-qPCR TaqMan Array Human miRNA panel (Applied Biosystems)	Serum	miR-17↑, miR-20a↑, miR-106a↑, miR-376c↑	cel-miR-238	Zhu et al. (220)
Hand, foot and mouth disease (human)	RT-qPCR TaqMan Array Human miRNA panel (Applied Biosystems)	Serum	miR-148a↑, miR-143↑, miR-324-3p↑, miR-628-3p↑, miR-140-5p↑, miR-362-3p↑	cel-miR-238	Cui et al. (215)
Hand, foot and mouth disease (human)	miRNA microarrays (Agilent Technologies) RT-qPCR	Serum exosomes	miR-671-5p↓, miR-16-5p↑, miR-150-3p↓	Standard Agilent normalization miR-642a-3p	Jia et al. (216)
Hand, foot and mouth disease (human)	nCounter® miRNA Expression assays (NanoString) RT-qPCR TaqMan assays (Applied Biosystems)	Serum	miR-494↑, miR-29b-3p↑, miR-551a↓, miR-606↓, miR-876-5p↑, miR-30c-5p↑, miR-221-3p↓, miR-150-5p↓	Geometric mean of top 100 miRNAs U6 snRNA	Wang et al. (119)
Human immunodeficiency virus (HIV)-associated neurological disorders (human)	RT-qPCR microRNA panels (Exiqon)	Plasma	miR-151a-5p↑, miR-194-5p↑, miR-19b-1-5p↑	miR-23a-3p and miR-23b-3p	Kadri et al. (211)
HIV-encephalitis (human)	RT-qPCR microRNA panels (Exiqon)	Cerebrospinal fluid	miR-19b-2*↑, miR-937↑, miR-362-5p↑	miR-622 and miR-1266	Pacifici et al. (213)
Staphylococcus aureus-induced mastitis (bovine)	TruSeq small RNA sequencing (Illumina)	Bovine milk exosomes	miR-142-5p↑, miR-223↑	Upper quantile normalization	Sun et al. (222)

miR-20a, miR-106a, and miR-376c were significantly increased in expression for patients compared to healthy individuals (220).

Infectious Diseases of Veterinary Importance

Staphylococcus aureus is most common causative agent of contagious bovine mastitis, which represents a significant economic burden to the dairy industry (221). Evaluation of milk exosomes from *S. aureus*-infected Holstein cows indicated miR-142-5p and miR-223 as potential biomarkers for this infection in mammary glands (222).

It is important to note that the use of age-matched subjects may be critical for experiments designed to detect circulating miRNAs as biomarkers of infection in young animals. Consequently, results from such experiments should be interpreted with caution since they may be confounded by expression of miRNAs associated with developmental processes. This issue has been addressed recently by Farrell and colleagues using serum samples collected from calves at preinfection and 6 months after infection with *Mycobacterium avium* subspecies *paratuberculosis* (MAP). Expression profiling of circulating miRNAs *via* miRNA-seq showed differential expression of miR-205 and miR-432, but a signature of infection could not be identified (223). On the other hand, analysis of biobanked bovine serum samples from experimental infections with MAP (stored at −20°C for 10–15 years) revealed that the circulating miRNA profile was remarkably similar in composition to the profile from fresh sera (<1 year at −80°C) (94). miRNAs have also been shown to correlate with season (summer, after the calves were born; fall, at weaning; and the following spring) when serum from *Mycoplasma bovis*-infected calves was profiled, but similar to the MAP studies described above a strong signature of infection was not observed (224).

Studies focusing on miRNAs in animals of veterinary relevance are still relatively few and most are restricted to cellular miRNAs. However, unlike human studies where cohort sizes may be limited, where age, gender, and ethnicity profiles may differ, and where patient histories cannot be readily assessed, veterinary studies can leverage sophisticated experimental designs that may help uncover nuances of circulating miRNA expression profiles that better represent disease status in infected individuals.

A summary of the results obtained for studies discussed in Sections "Other Infectious Diseases" and "Infectious Diseases of Veterinary Importance" are presented in **Table 5**.

CONCLUSION AND FUTURE PERSPECTIVES FOR CIRCULATING miRNAs AS BIOMARKERS OF INFECTION

Robustness to adverse sampling and storage conditions, especially for body fluids, is the most compelling reason that circulating miRNAs have significant potential as ancillary minimally invasive biomarkers for a wide range of pathologies. Thus far, research work has focused on identifying biomarkers and/or biosignatures for several diseases; however, studies that use systematic validation and independent confirmation are still relatively rare. In this regard, inconsistent results have been described across multiple studies that may be attributable to underpowered experimental design, lack of validation in subjects with pathologies that elicit similar immune responses, absence of standardized reference genes for normalization, and inappropriate statistical methodologies (e.g., no correction for multiple testing). Therefore, throughout the workflow, from sample collection and handling to the downstream bioinformatics, standardization of analytical methods will be key to establishing circulating miRNAs as robust and reliable biomarkers in clinical settings.

AUTHOR CONTRIBUTIONS

CC and RS reviewed the relevant literature and drafted the initial manuscript. CC, RS, and DM prepared the figure. NN, KM, and JB contributed additional material to the manuscript. DM and SG supervised the study and oversaw preparation, editing, and revision of the manuscript. All the authors have read and approved the final manuscript.

ACKNOWLEDGMENTS

The authors wish to thank Damien Farrell, David Magee, Kate Killick, and Kevin Rue-Albrecht for fruitful discussions.

FUNDING

The study was founded by the Brazilian Science Without Borders Programme (CAPES grant no. BEX-13070-13-4), Science Foundation Ireland (grant nos. SFI/08/IN.1/B2038 and SFI/15/IA/3154), and the Department of Agriculture, Food and the Marine (grant no. RSF11/S/141 – ICONMAP).

REFERENCES

1. Biomarkers Definitions Working Group. Biomarkers and surrogate endpoints: preferred definitions and conceptual framework. *Clin Pharmacol Ther* (2001) 69:89–95. doi:10.1067/mcp.2001.113989
2. Strimbu K, Tavel JA. What are biomarkers? *Curr Opin HIV AIDS* (2010) 5:463–6. doi:10.1097/COH.0b013e32833ed177
3. Etheridge A, Lee I, Hood L, Galas D, Wang K. Extracellular microRNA: a new source of biomarkers. *Mutat Res* (2011) 717:85–90. doi:10.1016/j.mrfmmm.2011.03.004
4. Nalejska E, Maczynska E, Lewandowska MA. Prognostic and predictive biomarkers: tools in personalized oncology. *Mol Diagn Ther* (2014) 18:273–84. doi:10.1007/s40291-013-0077-9

5. Buschmann D, Haberberger A, Kirchner B, Spornraft M, Riedmaier I, Schelling G, et al. Toward reliable biomarker signatures in the age of liquid biopsies – how to standardize the small RNA-Seq workflow. *Nucleic Acids Res* (2016) 44:5995–6018. doi:10.1093/nar/gkw545
6. Ballman KV. Biomarker: predictive or prognostic? *J Clin Oncol* (2015) 33:3968–71. doi:10.1200/JCO.2015.63.3651
7. Ramilo O, Allman W, Chung W, Mejias A, Ardura M, Glaser C, et al. Gene expression patterns in blood leukocytes discriminate patients with acute infections. *Blood* (2007) 109:2066–77. doi:10.1182/blood-2006-02-002477
8. Mejias A, Ramilo O. Transcriptional profiling in infectious diseases: ready for prime time? *J Infect* (2014) 68(Suppl 1):S94–9. doi:10.1016/j.jinf.2013.09.018

9. Weiner J, Kaufmann SH. High-throughput and computational approaches for diagnostic and prognostic host tuberculosis biomarkers. *Int J Infect Dis* (2017) (in press). doi:10.1016/j.ijid.2016.10.017

10. Mohr S, Liew CC. The peripheral-blood transcriptome: new insights into disease and risk assessment. *Trends Mol Med* (2007) 13:422–32. doi:10.1016/j.molmed.2007.08.003

11. Blankley S, Berry MP, Graham CM, Bloom CI, Lipman M, O'Garra A. The application of transcriptional blood signatures to enhance our understanding of the host response to infection: the example of tuberculosis. *Philos Trans R Soc Lond B Biol Sci* (2014) 369:20130427. doi:10.1098/rstb.2013.0427

12. Chaussabel D. Assessment of immune status using blood transcriptomics and potential implications for global health. *Semin Immunol* (2015) 27:58–66. doi:10.1016/j.smim.2015.03.002

13. Kaufmann SH, Fortune S, Pepponi I, Ruhwald M, Schrager LK, Ottenhoff TH. TB biomarkers, TB correlates and human challenge models: new tools for improving assessment of new TB vaccines. *Tuberculosis (Edinb)* (2016) 99(Suppl 1):S8–11. doi:10.1016/j.tube.2016.05.010

14. Vordermeier HM, Jones GJ, Buddle BM, Hewinson RG, Villarreal-Ramos B. Bovine tuberculosis in cattle: vaccines, diva tests, and host biomarker discovery. *Annu Rev Anim Biosci* (2016) 4:87–109. doi:10.1146/annurev-animal-021815-111311

15. Esterhuyse MM, Weiner J III, Caron E, Loxton AG, Iannaccone M, Wagman C, et al. Epigenetics and proteomics join transcriptomics in the quest for tuberculosis biomarkers. *MBio* (2015) 6:e1187–1115. doi:10.1128/mBio.01187-15

16. Wallis RS, Peppard T. Early biomarkers and regulatory innovation in multidrug-resistant tuberculosis. *Clin Infect Dis* (2015) 61(Suppl 3):S160–3. doi:10.1093/cid/civ612

17. Lee RC, Feinbaum RL, Ambros V. The *C. elegans* heterochronic gene lin-4 encodes small RNAs with antisense complementarity to lin-14. *Cell* (1993) 75:843–54. doi:10.1016/0092-8674(93)90529-Y

18. Wightman B, Ha I, Ruvkun G. Posttranscriptional regulation of the heterochronic gene *lin-14* by lin-4 mediates temporal pattern formation in *C. elegans*. *Cell* (1993) 75:855–62. doi:10.1016/0092-8674(93)90530-4

19. Kozomara A, Griffiths-Jones S. miRBase: annotating high confidence microRNAs using deep sequencing data. *Nucleic Acids Res* (2014) 42:D68–73. doi:10.1093/nar/gkt1181

20. Wheeler BM, Heimberg AM, Moy VN, Sperling EA, Holstein TW, Heber S, et al. The deep evolution of metazoan microRNAs. *Evol Dev* (2009) 11:50–68. doi:10.1111/j.1525-142X.2008.00302.x

21. Berezikov E. Evolution of microRNA diversity and regulation in animals. *Nat Rev Genet* (2011) 12:846–60. doi:10.1038/nrg3079

22. O'Connell RM, Rao DS, Chaudhuri AA, Baltimore D. Physiological and pathological roles for microRNAs in the immune system. *Nat Rev Immunol* (2010) 10:111–22. doi:10.1038/nri2708

23. O'Neill LA, Sheedy FJ, McCoy CE. microRNAs: the fine-tuners of toll-like receptor signalling. *Nat Rev Immunol* (2011) 11:163–75. doi:10.1038/nri2957

24. Shenoy A, Blelloch RH. Regulation of microRNA function in somatic stem cell proliferation and differentiation. *Nat Rev Mol Cell Biol* (2014) 15:565–76. doi:10.1038/nrm3854

25. Lee Y, Kim M, Han J, Yeom KH, Lee S, Baek SH, et al. microRNA genes are transcribed by RNA polymerase II. *EMBO J* (2004) 23:4051–60. doi:10.1038/sj.emboj.7600385

26. Lee Y, Ahn C, Han J, Choi H, Kim J, Yim J, et al. The nuclear RNase III Drosha initiates microRNA processing. *Nature* (2003) 425:415–9. doi:10.1038/nature01957

27. Han J, Lee Y, Yeom KH, Kim YK, Jin H, Kim VN. The Drosha-DGCR8 complex in primary microRNA processing. *Genes Dev* (2004) 18:3016–27. doi:10.1101/gad.1262504

28. Lund E, Guttinger S, Calado A, Dahlberg JE, Kutay U. Nuclear export of microRNA precursors. *Science* (2004) 303:95–8. doi:10.1126/science.1090599

29. Zhang HD, Kolb FA, Brondani V, Billy E, Filipowicz W. Human Dicer preferentially cleaves dsRNAs at their termini without a requirement for ATP. *EMBO J* (2002) 21:5875–85. doi:10.1093/emboj/cdf582

30. Chendrimada TP, Gregory RI, Kumaraswamy E, Norman J, Cooch N, Nishikura K, et al. TRBP recruits the Dicer complex to Ago2 for microRNA processing and gene silencing *Nature* (2005) 436:740–4. doi:10.1038/nature03868

31. Lewis BP, Shih IH, Jones-Rhoades MW, Bartel DP, Burge CB. Prediction of mammalian microRNA targets. *Cell* (2003) 115:787–98. doi:10.1016/s0092-8674(03)01018-3

32. Schwarz DS, Hutvagner G, Du T, Xu Z, Aronin N, Zamore PD. Asymmetry in the assembly of the RNAi enzyme complex. *Cell* (2003) 115:199–208. doi:10.1016/s0092-8674(03)00759-1

33. Kim Y, Yeo J, Lee JH, Cho J, Seo D, Kim JS, et al. Deletion of human tarbp2 reveals cellular microRNA targets and cell-cycle function of TRBP. *Cell Rep* (2014) 9:1061–74. doi:10.1016/j.celrep.2014.09.039

34. Wilson RC, Tambe A, Kidwell MA, Noland CL, Schneider CP, Doudna JA. Dicer-TRBP complex formation ensures accurate mammalian microRNA biogenesis. *Mol Cell* (2015) 57:397–407. doi:10.1016/j.molcel.2014.11.030

35. Morin RD, O'Connor MD, Griffith M, Kuchenbauer F, Delaney A, Prabhu AL, et al. Application of massively parallel sequencing to microRNA profiling and discovery in human embryonic stem cells. *Genome Res* (2008) 18:610–21. doi:10.1101/gr.7179508

36. Neilsen CT, Goodall GJ, Bracken CP. IsomiRs – the overlooked repertoire in the dynamic microRNAome. *Trends Genet* (2012) 28:544–9. doi:10.1016/j.tig.2012.07.005

37. Guo L, Chen F. A challenge for miRNA: multiple isomiRs in miRNAomics. *Gene* (2014) 544:1–7. doi:10.1016/j.gene.2014.04.039

38. Okamura K, Hagen JW, Duan H, Tyler DM, Lai EC. The mirtron pathway generates microRNA-class regulatory RNAs in *Drosophila*. *Cell* (2007) 130:89–100. doi:10.1016/j.cell.2007.06.028

39. Ruby JG, Jan CH, Bartel DP. Intronic microRNA precursors that bypass Drosha processing. *Nature* (2007) 448:83–6. doi:10.1038/nature05983

40. Calin GA, Dumitru CD, Shimizu M, Bichi R, Zupo S, Noch E, et al. Frequent deletions and down-regulation of micro-RNA genes miR15 and miR16 at 13q14 in chronic lymphocytic leukemia. *Proc Natl Acad Sci U S A* (2002) 99:15524–9. doi:10.1073/pnas.242606799

41. Takamizawa J, Konishi H, Yanagisawa K, Tomida S, Osada H, Endoh H, et al. Reduced expression of the let-7 microRNAs in human lung cancers in association with shortened postoperative survival. *Cancer Res* (2004) 64:3753–6. doi:10.1158/0008-5472.CAN-04-0637

42. Bianchi F, Nicassio F, Marzi M, Belloni E, Dall'Olio V, Bernard L, et al. A serum circulating miRNA diagnostic test to identify asymptomatic high-risk individuals with early stage lung cancer. *EMBO Mol Med* (2011) 3:495–503. doi:10.1002/emmm.201100154

43. Boeri M, Verri C, Conte D, Roz L, Modena P, Facchinetti F, et al. microRNA signatures in tissues and plasma predict development and prognosis of computed tomography detected lung cancer. *Proc Natl Acad Sci U S A* (2011) 108:3713–8. doi:10.1073/pnas.1100048108

44. Ciesla M, Skrzypek K, Kozakowska M, Loboda A, Jozkowicz A, Dulak J. microRNAs as biomarkers of disease onset. *Anal Bioanal Chem* (2011) 401:2051–61. doi:10.1007/s00216-011-5001-8

45. Tan X, Qin W, Zhang L, Hang J, Li B, Zhang C, et al. A 5-microRNA signature for lung squamous cell carcinoma diagnosis and hsa-miR-31 for prognosis. *Clin Cancer Res* (2011) 17:6802–11. doi:10.1158/1078-0432.CCR-11-0419

46. Gasparini P, Cascione L, Landi L, Carasi S, Lovat F, Tibaldi C, et al. microRNA classifiers are powerful diagnostic/prognostic tools in ALK-, EGFR-, and KRAS-driven lung cancers. *Proc Natl Acad Sci U S A* (2015) 112:14924–9. doi:10.1073/pnas.1520329112

47. Du F, Yuan P, Zhao ZT, Yang Z, Wang T, Zhao JD, et al. A miRNA-based signature predicts development of disease recurrence in HER2 positive breast cancer after adjuvant trastuzumab-based treatment. *Sci Rep* (2016) 6:33825. doi:10.1038/srep33825

48. Kurozumi S, Yamaguchi Y, Kurosumi M, Ohira M, Matsumoto H, Horiguchi J. Recent trends in microRNA research into breast cancer with particular focus on the associations between microRNAs and intrinsic subtypes. *J Hum Genet* (2017) 62:15–24. doi:10.1038/jhg.2016.89

49. Baltimore D, Boldin MP, O'Connell RM, Rao DS, Taganov KD. microRNAs: new regulators of immune cell development and function. *Nat Immunol* (2008) 9:839–45. doi:10.1038/ni.f.209

50. Pauley KM, Chan EK. microRNAs and their emerging roles in immunology. *Ann N Y Acad Sci* (2008) 1143:226–39. doi:10.1196/annals.1443.009

51. Lu LF, Liston A. microRNA in the immune system, microRNA as an immune system. *Immunology* (2009) 127:291–8. doi:10.1111/j.1365-2567.2009.03092.x

52. O'Connell RM, Rao DS, Baltimore D. microRNA regulation of inflammatory responses. *Annu Rev Immunol* (2012) 30:295–312. doi:10.1146/annurev-immunol-020711-075013

53. Chen CZ, Schaffert S, Fragoso R, Loh C. Regulation of immune responses and tolerance: the microRNA perspective. *Immunol Rev* (2013) 253:112–28. doi:10.1111/imr.12060

54. Zhu S, Pan W, Qian Y. microRNA in immunity and autoimmunity. *J Mol Med (Berl)* (2013) 91:1039–50. doi:10.1007/s00109-013-1043-z

55. Yang L, Seki E. Toll-like receptors in liver fibrosis: cellular crosstalk and mechanisms. *Front Physiol* (2012) 3:138. doi:10.3389/fphys.2012.00138

56. He X, Jing Z, Cheng G. microRNAs: new regulators of toll-like receptor signalling pathways. *Biomed Res Int* (2014) 2014:945169. doi:10.1155/2014/945169

57. Liston A, Linterman M, Lu LF. microRNA in the adaptive immune system, in sickness and in health. *J Clin Immunol* (2010) 30:339–46. doi:10.1007/s10875-010-9378-5

58. Dooley J, Linterman MA, Liston A. microRNA regulation of T-cell development. *Immunol Rev* (2013) 253:53–64. doi:10.1111/imr.12049

59. Jeker LT, Bluestone JA. microRNA regulation of T-cell differentiation and function. *Immunol Rev* (2013) 253:65–81. doi:10.1111/imr.12061

60. Podshivalova K, Salomon DR. microRNA regulation of T-lymphocyte immunity: modulation of molecular networks responsible for T-cell activation, differentiation, and development. *Crit Rev Immunol* (2013) 33:435–76. doi:10.1615/CritRevImmunol.2013006858

61. Tang X, Tang R, Xu Y, Wang Q, Hou Y, Shen S, et al. microRNA networks in regulatory T cells. *J Physiol Biochem* (2014) 70:869–75. doi:10.1007/s13105-014-0348-x

62. Amado T, Schmolka N, Metwally H, Silva-Santos B, Gomes AQ. Cross-regulation between cytokine and microRNA pathways in T cells. *Eur J Immunol* (2015) 45:1584–95. doi:10.1002/eji.201545487

63. Clare S, John V, Walker AW, Hill JL, Abreu-Goodger C, Hale C, et al. Enhanced susceptibility to *Citrobacter rodentium* infection in microRNA-155-deficient mice. *Infect Immun* (2013) 81:723–32. doi:10.1128/IAI.00969-12

64. Rodriguez A, Vigorito E, Clare S, Warren MV, Couttet P, Soond DR, et al. Requirement of bic/microRNA-155 for normal immune function. *Science* (2007) 316:608–11. doi:10.1126/science.1139253

65. Lind EF, Elford AR, Ohashi PS. micro-RNA 155 is required for optimal CD8+ T cell responses to acute viral and intracellular bacterial challenges. *J Immunol* (2013) 190:1210–6. doi:10.4049/jimmunol.1202700

66. Bala S, Tilahun Y, Taha O, Alao H, Kodys K, Catalano D, et al. Increased microRNA-155 expression in the serum and peripheral monocytes in chronic HCV infection. *J Transl Med* (2012) 10:151. doi:10.1186/1479-5876-10-151

67. Jablonski KA, Gaudet AD, Amici SA, Popovich PG, Guerau-de-Arellano M. Control of the inflammatory macrophage transcriptional signature by miR-155. *PLoS One* (2016) 11:e0159724. doi:10.1371/journal.pone.0159724

68. Staedel C, Darfeuille F. microRNAs and bacterial infection. *Cell Microbiol* (2013) 15:1496–507. doi:10.1111/cmi.12159

69. Furci L, Schena E, Miotto P, Cirillo DM. Alteration of human macrophages microRNA expression profile upon infection with *Mycobacterium tuberculosis*. *Int J Mycobacteriol* (2013) 2:128–34. doi:10.1016/j.ijmyco.2013.04.006

70. Vegh P, Magee DA, Nalpas NC, Bryan K, McCabe MS, Browne JA, et al. microRNA profiling of the bovine alveolar macrophage response to *Mycobacterium bovis* infection suggests pathogen survival is enhanced by microRNA regulation of endocytosis and lysosome trafficking. *Tuberculosis* (2015) 95:60–7. doi:10.1016/j.tube.2014.10.011

71. Eulalio A, Schulte L, Vogel J. The mammalian microRNA response to bacterial infections. *RNA Biol* (2012) 9:742–50. doi:10.4161/rna.20018

72. Bettencourt P, Pires D, Anes E. Immunomodulating microRNAs of mycobacterial infections. *Tuberculosis (Edinb)* (2016) 97:1–7. doi:10.1016/j.tube.2015.12.004

73. Wang Y, Cai Y. A survey on database resources for microRNA-disease relationships. *Brief Funct Genomics* (2017) (in press). doi:10.1093/bfgp/elw015

74. Contreras J, Rao DS. microRNAs in inflammation and immune responses. *Leukemia* (2012) 26:404–13. doi:10.1038/leu.2011.356

75. Robbins PD, Morelli AE. Regulation of immune responses by extracellular vesicles. *Nat Rev Immunol* (2014) 14:195–208. doi:10.1038/nri3622

76. Fernandez-Messina L, Gutierrez-Vazquez C, Rivas-Garcia E, Sanchez-Madrid F, de la Fuente H. Immunomodulatory role of microRNAs transferred by extracellular vesicles. *Biol Cell* (2015) 107:61–77. doi:10.1111/boc.201400081

77. de Candia P, De Rosa V, Casiraghi M, Matarese G. Extracellular RNAs: a secret arm of immune system regulation. *J Biol Chem* (2016) 291:7221–8. doi:10.1074/jbc.R115.708842

78. Weber JA, Baxter DH, Zhang S, Huang DY, Huang KH, Lee MJ, et al. The microRNA spectrum in 12 body fluids. *Clin Chem* (2010) 56:1733–41. doi:10.1373/clinchem.2010.147405

79. Arroyo JD, Chevillet JR, Kroh EM, Ruf IK, Pritchard CC, Gibson DF, et al. Argonaute2 complexes carry a population of circulating microRNAs independent of vesicles in human plasma. *Proc Natl Acad Sci U S A* (2011) 108:5003–8. doi:10.1073/pnas.1019055108

80. Turchinovich A, Weiz L, Langheinz A, Burwinkel B. Characterization of extracellular circulating microRNA. *Nucleic Acids Res* (2011) 39:7223–33. doi:10.1093/nar/gkr254

81. Zhang Y, Liu D, Chen X, Li J, Li L, Bian Z, et al. Secreted monocytic miR-150 enhances targeted endothelial cell migration. *Mol Cell* (2010) 39:133–44. doi:10.1016/j.molcel.2010.06.010

82. Wang K, Zhang S, Weber J, Baxter D, Galas DJ. Export of microRNAs and microRNA-protective protein by mammalian cells. *Nucleic Acids Res* (2010) 38:7248–59. doi:10.1093/nar/gkq601

83. Cortez MA, Bueso-Ramos C, Ferdin J, Lopez-Berestein G, Sood AK, Calin GA. microRNAs in body fluids – the mix of hormones and biomarkers. *Nat Rev Clin Oncol* (2011) 8:467–77. doi:10.1038/nrclinonc.2011.76

84. Mandel P, Metais P. Les acides nucléiques du plasma sanguin chez l'homme [The nucleic acids from blood plasma in humans]. *C R Seances Soc Biol Fil* (1948) 142:241–3.

85. Swarup V, Rajeswari MR. Circulating (cell-free) nucleic acids – a promising, non-invasive tool for early detection of several human diseases. *FEBS Lett* (2007) 581:795–9. doi:10.1016/j.febslet.2007.01.051

86. Chim SS, Shing TK, Hung EC, Leung TY, Lau TK, Chiu RW, et al. Detection and characterization of placental microRNAs in maternal plasma. *Clin Chem* (2008) 54:482–90. doi:10.1373/clinchem.2007.097972

87. Lawrie CH, Gal S, Dunlop HM, Pushkaran B, Liggins AP, Pulford K, et al. Detection of elevated levels of tumour-associated microRNAs in serum of patients with diffuse large B-cell lymphoma. *Br J Haematol* (2008) 141:672–5. doi:10.1111/j.1365-2141.2008.07077.x

88. Mitchell PS, Parkin RK, Kroh EM, Fritz BR, Wyman SK, Pogosova-Agadjanyan EL, et al. Circulating microRNAs as stable blood-based markers for cancer detection. *Proc Natl Acad Sci U S A* (2008) 105:10513–8. doi:10.1073/pnas.0804549105

89. Schwarzenbach H, Nishida N, Calin GA, Pantel K. Clinical relevance of circulating cell-free microRNAs in cancer. *Nat Rev Clin Oncol* (2014) 11:145–56. doi:10.1038/nrclinonc.2014.5

90. Ono S, Lam S, Nagahara M, Hoon DS. Circulating microRNA biomarkers as liquid biopsy for cancer patients: pros and cons of current assays. *J Clin Med* (2015) 4:1890–907. doi:10.3390/jcm4101890

91. Wang J, Chen J, Sen S. microRNA as biomarkers and diagnostics. *J Cell Physiol* (2016) 231:25–30. doi:10.1002/jcp.25056

92. Diamandis EP. The failure of protein cancer biomarkers to reach the clinic: why, and what can be done to address the problem? *BMC Med* (2012) 10:87. doi:10.1186/1741-7015-10-87

93. Koberle V, Pleli T, Schmithals C, Augusto Alonso E, Haupenthal J, Bonig H, et al. Differential stability of cell-free circulating microRNAs: implications for their utilization as biomarkers. *PLoS One* (2013) 8:e75184. doi:10.1371/journal.pone.0075184

94. Shaughnessy RG, Farrell D, Riepema K, Bakker D, Gordon SV. Analysis of biobanked serum from a *Mycobacterium avium* subsp *paratuberculosis* bovine infection model confirms the remarkable stability of circulating miRNA profiles and defines a bovine serum miRNA repertoire. *PLoS One* (2015) 10:e0145089. doi:10.1371/journal.pone.0145089

95. Unger L, Fouche N, Leeb T, Gerber V, Pacholewska A. Optimized methods for extracting circulating small RNAs from long-term stored equine samples. *Acta Vet Scand* (2016) 58:44. doi:10.1186/s13028-016-0224-5

96. McDonald JS, Milosevic D, Reddi HV, Grebe SK, Algeciras-Schimnich A. Analysis of circulating microRNA: preanalytical and analytical challenges. *Clin Chem* (2011) 57:833–40. doi:10.1373/clinchem.2010.157198

97. Wang K, Yuan Y, Cho JH, McClarty S, Baxter D, Galas DJ. Comparing the microRNA spectrum between serum and plasma. *PLoS One* (2012) 7:e41561. doi:10.1371/journal.pone.0041561

98. Blondal T, Jensby Nielsen S, Baker A, Andreasen D, Mouritzen P, Wrang Teilum M, et al. Assessing sample and miRNA profile quality in serum and plasma or other biofluids. *Methods* (2013) 59:S1–6. doi:10.1016/j.ymeth.2012.09.015

99. Binderup HG, Houlind K, Madsen JS, Brasen CL. Pre-storage centrifugation conditions have significant impact on measured microRNA levels in biobanked EDTA plasma samples. *Biochem Biophys Rep* (2016) 7:195–200. doi:10.1016/j.bbrep.2016.06.005

100. Kim DJ, Linnstaedt S, Palma J, Park JC, Ntrivalas E, Kwak-Kim JY, et al. Plasma components affect accuracy of circulating cancer-related microRNA quantitation. *J Mol Diagn* (2012) 14:71–80. doi:10.1016/j.jmoldx. 2011.09.002

101. Li S, Chen H, Song J, Lee C, Geng Q. Avoiding heparin inhibition in circulating microRNAs amplification. *Int J Cardiol* (2016) 207:92–3. doi:10.1016/j.ijcard.2016.01.129

102. Pritchard CC, Kroh E, Wood B, Arroyo JD, Dougherty KJ, Miyaji MM, et al. Blood cell origin of circulating microRNAs: a cautionary note for cancer biomarker studies. *Cancer Prev Res (Phila)* (2012) 5:492–7. doi:10.1158/1940-6207.CAPR-11-0370

103. Kim YK, Yeo J, Kim B, Ha M, Kim VN. Short structured RNAs with low GC content are selectively lost during extraction from a small number of cells. *Mol Cell* (2012) 46:893–5. doi:10.1016/j.molcel.2012.05.036

104. Pritchard CC, Cheng HH, Tewari M. microRNA profiling: approaches and considerations. *Nat Rev Genet* (2012) 13:358–69. doi:10.1038/nrg3198

105. Di Leva G, Croce CM. miRNA profiling of cancer. *Curr Opin Genet Dev* (2013) 23:3–11. doi:10.1016/j.gde.2013.01.004

106. Baker M. microRNA profiling: separating signal from noise. *Nat Methods* (2010) 7:687–92. doi:10.1038/nmeth0910-687

107. Kolbert CP, Feddersen RM, Rakhshan F, Grill DE, Simon G, Middha S, et al. Multi-platform analysis of microRNA expression measurements in RNA from fresh frozen and FFPE tissues. *PLoS One* (2013) 8:e52517. doi:10.1371/journal.pone.0052517

108. Tian T, Wang J, Zhou X. A review: microRNA detection methods. *Org Biomol Chem* (2015) 13:2226–38. doi:10.1039/c4ob02104e

109. Fiedler SD, Carletti MZ, Christenson LK. Quantitative RT-PCR methods for mature microRNA expression analysis. *Methods Mol Biol* (2010) 630:49–64. doi:10.1007/978-1-60761-629-0_4

110. Chen C, Ridzon DA, Broomer AJ, Zhou Z, Lee DH, Nguyen JT, et al. Real-time quantification of microRNAs by stem-loop RT-PCR. *Nucleic Acids Res* (2005) 33:e179. doi:10.1093/nar/gni178

111. Koshkin AA, Singh SK, Nielsen P, Rajwanshi VK, Kumar R, Meldgaard M, et al. LNA (locked nucleic acids): synthesis of the adenine, cytosine, guanine, 5-methylcytosine, thymine and uracil bicyclonucleoside monomers, oligomerisation, and unprecedented nucleic acid recognition. *Tetrahedron* (1998) 54:3607–30. doi:10.1016/S0040-4020(98)00094-5

112. Kok MG, Halliani A, Moerland PD, Meijers JC, Creemers EE, Pinto-Sietsma SJ. Normalization panels for the reliable quantification of circulating microRNAs by RT-qPCR. *FASEB J* (2015) 29:3853–62. doi:10.1096/fj.15-271312

113. Schlosser K, McIntyre LA, White RJ, Stewart DJ. Customized internal reference controls for improved assessment of circulating microRNAs in disease. *PLoS One* (2015) 10:e0127443. doi:10.1371/journal.pone.0127443

114. Marabita F, de Candia P, Torri A, Tegner J, Abrignani S, Rossi RL. Normalization of circulating microRNA expression data obtained by quantitative real-time RT-PCR. *Brief Bioinform* (2016) 17:204–12. doi:10.1093/bib/bbv056

115. Wu D, Hu Y, Tong S, Williams BR, Smyth GK, Gantier MP. The use of miRNA microarrays for the analysis of cancer samples with global miRNA decrease. *RNA* (2013) 19:876–88. doi:10.1261/rna.035055.112

116. Nalpas NC, Park SD, Magee DA, Taraktsoglou M, Browne JA, Conlon KM, et al. Whole-transcriptome, high-throughput RNA sequence analysis of the bovine macrophage response to *Mycobacterium bovis* infection in vitro. *BMC Genomics* (2013) 14:230. doi:10.1186/1471-2164-14-230

117. McLoughlin KE, Nalpas NC, Rue-Albrecht K, Browne JA, Magee DA, Killick KE, et al. RNA-seq transcriptional profiling of peripheral blood leukocytes from cattle infected with *Mycobacterium bovis*. *Front Immunol* (2014) 5:396. doi:10.3389/fimmu.2014.00396

118. Geiss GK, Bumgarner RE, Birditt B, Dahl T, Dowidar N, Dunaway DL, et al. Direct multiplexed measurement of gene expression with color-coded probe pairs. *Nat Biotechnol* (2008) 26:317–25. doi:10.1038/nbt1385

119. Wang RY, Weng KF, Huang YC, Chen CJ. Elevated expression of circulating miR876-5p is a specific response to severe EV71 infections. *Sci Rep* (2016) 6:24149. doi:10.1038/srep24149

120. Yamamoto M, Singh A, Ruan J, Gauvreau GM, O'Byrne PM, Carlsten CR, et al. Decreased miR-192 expression in peripheral blood of asthmatic individuals undergoing an allergen inhalation challenge. *BMC Genomics* (2012) 13:655. doi:10.1186/1471-2164-13-655

121. Vikram A, Kim YR, Kumar S, Li Q, Kassan M, Jacobs JS, et al. Vascular microRNA-204 is remotely governed by the microbiome and impairs endothelium-dependent vasorelaxation by downregulating Sirtuin1. *Nat Commun* (2016) 7:12565. doi:10.1038/ncomms12565

122. Keller A, Leidinger P, Bauer A, Elsharawy A, Haas J, Backes C, et al. Toward the blood-borne miRNome of human diseases. *Nat Methods* (2011) 8:841–3. doi:10.1038/nmeth.1682

123. Li N, You X, Chen T, Mackowiak SD, Friedlander MR, Weigt M, et al. Global profiling of miRNAs and the hairpin precursors: insights into miRNA processing and novel miRNA discovery. *Nucleic Acids Res* (2013) 41:3619–34. doi:10.1093/nar/gkt072

124. Siddle KJ, Tailleux L, Deschamps M, Loh YH, Deluen C, Gicquel B, et al. Bacterial infection drives the expression dynamics of microRNAs and their isomiRs. *PLoS Genet* (2015) 11:e1005064. doi:10.1371/journal.pgen.1005064

125. Huang HC, Niu Y, Qin LX. Differential expression analysis for RNA-seq: an overview of statistical methods and computational software. *Cancer Inform* (2015) 14:57–67. doi:10.4137/CIN.S21631

126. Ziemann M, Kaspi A, El-Osta A. Evaluation of microRNA alignment techniques. *RNA* (2016) 22:1120–38. doi:10.1261/rna.055509.115

127. Raabe CA, Tang TH, Brosius J, Rozhdestvensky TS. Biases in small RNA deep sequencing data. *Nucleic Acids Res* (2014) 42:1414–26. doi:10.1093/nar/gkt1021

128. Baroin-Tourancheau A, Benigni X, Doubi-Kadmiri S, Taouis M, Amar L. Lessons from microRNA sequencing using Illumina technology. *Adv Biosci Biotechnol* (2016) 07:319–28. doi:10.4236/abb.2016.77030

129. Graybill RM, Bailey RC. Emerging biosensing approaches for microRNA analysis. *Anal Chem* (2016) 88:431–50. doi:10.1021/acs.analchem.5b04679

130. Benz F, Roderburg C, Vargas Cardenas D, Vucur M, Gautheron J, Koch A, et al. U6 is unsuitable for normalization of serum miRNA levels in patients with sepsis or liver fibrosis. *Exp Mol Med* (2013) 45:e42. doi:10.1038/emm.2013.81

131. D'Haene B, Mestdagh P, Hellemans J, Vandesompele J. miRNA expression profiling: from reference genes to global mean normalization. *Methods Mol Biol* (2012) 822:261–72. doi:10.1007/978-1-61779-427-8_18

132. Reid G, Kirschner MB, van Zandwijk N. Circulating microRNAs: association with disease and potential use as biomarkers. *Crit Rev Oncol Hematol* (2011) 80:193–208. doi:10.1016/j.critrevonc.2010.11.004

133. Tang G, Shen X, Lv K, Wu Y, Bi J, Shen Q. Different normalization strategies might cause inconsistent variation in circulating microRNAs in patients with hepatocellular carcinoma. *Med Sci Monit* (2015) 21:617–24. doi:10.12659/MSM.891028

134. Mestdagh P, Van Vlierberghe P, De Weer A, Muth D, Westermann F, Speleman F, et al. A novel and universal method for microRNA RT-qPCR data normalization. *Genome Biol* (2009) 10:R64. doi:10.1186/gb-2009-10-6-r64

135. Robinson MD, Oshlack A. A scaling normalization method for differential expression analysis of RNA-seq data. *Genome Biol* (2010) 11:R25. doi:10.1186/gb-2010-11-3-r25

136. Bullard JH, Purdom E, Hansen KD, Dudoit S. Evaluation of statistical methods for normalization and differential expression in mRNA-seq experiments. *BMC Bioinformatics* (2010) 11:94. doi:10.1186/1471-2105-11-94

137. Tam S, Tsao MS, McPherson JD. Optimization of miRNA-seq data preprocessing. *Brief Bioinform* (2015) 16:950–63. doi:10.1093/bib/bbv019

138. Garmire LX, Subramaniam S. Evaluation of normalization methods in mammalian microRNA-Seq data. *RNA* (2012) 18:1279–88. doi:10.1261/rna.030916.111

139. Cleveland WS, Devlin SJ. Locally weighted regression: an approach to regression analysis by local fitting. *J Am Stat Assoc* (1988) 83:596. doi:10.1080/01621459.1988.10478639

140. Jain N, Thatte J, Braciale T, Ley K, O'Connell M, Lee JK. Local-pooled-error test for identifying differentially expressed genes with a small number of replicated microarrays. *Bioinformatics* (2003) 19:1945–51. doi:10.1093/bioinformatics/btg264

141. Bolstad BM, Irizarry RA, Astrand M, Speed TP. A comparison of normalization methods for high density oligonucleotide array data based on variance and bias. *Bioinformatics* (2003) 19:185–93. doi:10.1093/bioinformatics/19.2.185

142. Dillies MA, Rau A, Aubert J, Hennequet-Antier C, Jeanmougin M, Servant N, et al. A comprehensive evaluation of normalization methods for Illumina high-throughput RNA sequencing data analysis. *Brief Bioinform* (2013) 14:671–83. doi:10.1093/bib/bbs046

143. Zhou X, Oshlack A, Robinson MD. miRNA-Seq normalization comparisons need improvement. *RNA* (2013) 19:733–4. doi:10.1261/rna.037895.112

144. World Health Organization. *Global Tuberculosis Report 2015*. Geneva: World Health Organization (2015).

145. Leylabadlo HE, Kafil HS, Yousefi M, Aghazadeh M, Asgharzadeh M. Pulmonary tuberculosis diagnosis: where we are? *Tuberc Respir Dis (Seoul)* (2016) 79:134–42. doi:10.4046/trd.2016.79.3.134

146. Pai M, Schito M. Tuberculosis diagnostics in 2015: landscape, priorities, needs, and prospects. *J Infect Dis* (2015) 211(Suppl 2):S21–8. doi:10.1093/infdis/jiu803

147. Fu Y, Yi Z, Wu X, Li J, Xu F. Circulating microRNAs in patients with active pulmonary tuberculosis. *J Clin Microbiol* (2011) 49:4246–51. doi:10.1128/JCM.05459-11

148. Yi Z, Fu Y, Ji R, Li R, Guan Z. Altered microRNA signatures in sputum of patients with active pulmonary tuberculosis. *PLoS One* (2012) 7:e43184. doi:10.1371/journal.pone.0043184

149. Qi Y, Cui L, Ge Y, Shi Z, Zhao K, Guo X, et al. Altered serum microRNAs as biomarkers for the early diagnosis of pulmonary tuberculosis infection. *BMC Infect Dis* (2012) 12:384. doi:10.1186/1471-2334-12-384

150. Abd-El-Fattah AA, Sadik NA, Shaker OG, Aboulftouh ML. Differential microRNAs expression in serum of patients with lung cancer, pulmonary tuberculosis, and pneumonia. *Cell Biochem Biophys* (2013) 67:875–84. doi:10.1007/s12013-013-9575-y

151. Zhang X, Guo J, Fan S, Li Y, Wei L, Yang X, et al. Screening and identification of six serum microRNAs as novel potential combination biomarkers for pulmonary tuberculosis diagnosis. *PLoS One* (2013) 8:e81076. doi:10.1371/journal.pone.0081076

152. Zhang H, Sun Z, Wei W, Liu Z, Fleming J, Zhang S, et al. Identification of serum microRNA biomarkers for tuberculosis using RNA-seq. *PLoS One* (2014) 9:e88909. doi:10.1371/journal.pone.0088909

153. Vivier E, Tomasello E, Baratin M, Walzer T, Ugolini S. Functions of natural killer cells. *Nat Immunol* (2008) 9:503–10. doi:10.1038/ni1582

154. Zhang C, Xi X, Wang Q, Jiao J, Zhang L, Zhao H, et al. The association between serum miR-155 and natural killer cells from tuberculosis patients. *Int J Clin Exp Med* (2015) 8:9168–72.

155. Rittirsch D, Flierl MA, Ward PA. Harmful molecular mechanisms in sepsis. *Nat Rev Immunol* (2008) 8:776–87. doi:10.1038/nri2402

156. Parlato M, Cavaillon JM. Host response biomarkers in the diagnosis of sepsis: a general overview. *Methods Mol Biol* (2015) 1237:149–211. doi:10.1007/978-1-4939-1776-1_15

157. Benz F, Roy S, Trautwein C, Roderburg C, Luedde T. Circulating microRNAs as biomarkers for sepsis. *Int J Mol Sci* (2016) 17:E78. doi:10.3390/ijms17010078

158. Vasilescu C, Rossi S, Shimizu M, Tudor S, Veronese A, Ferracin M, et al. microRNA fingerprints identify miR-150 as a plasma prognostic marker in patients with sepsis. *PLoS One* (2009) 4:e7405. doi:10.1371/journal.pone.0007405

159. Puskarich MA, Nandi U, Shapiro NI, Trzeciak S, Kline JA, Jones AE. Detection of microRNAs in patients with sepsis. *J Acute Dis* (2015) 4:101–6. doi:10.1016/s2221-6189(15)30017-2

160. Roderburg C, Luedde M, Vargas Cardenas D, Vucur M, Scholten D, Frey N, et al. Circulating microRNA-150 serum levels predict survival in patients with critical illness and sepsis. *PLoS One* (2013) 8:e54612. doi:10.1371/journal.pone.0054612

161. Wang JF, Yu ML, Yu G, Bian JJ, Deng XM, Wan XJ, et al. Serum miR-146a and miR-223 as potential new biomarkers for sepsis. *Biochem Biophys Res Commun* (2010) 394:184–8. doi:10.1016/j.bbrc.2010.02.145

162. Wang L, Wang HC, Chen C, Zeng J, Wang Q, Zheng L, et al. Differential expression of plasma miR-146a in sepsis patients compared with non-sepsis-SIRS patients. *Exp Ther Med* (2013) 5:1101–4. doi:10.3892/etm.2013.937

163. Benz F, Tacke F, Luedde M, Trautwein C, Luedde T, Koch A, et al. Circulating microRNA-223 serum levels do not predict sepsis or survival in patients with critical illness. *Dis Markers* (2015) 2015:384208. doi:10.1155/2015/384208

164. Tacke F, Roderburg C, Benz F, Cardenas DV, Luedde M, Hippe HJ, et al. Levels of circulating miR-133a are elevated in sepsis and predict mortality in critically ill patients. *Crit Care Med* (2014) 42:1096–104. doi:10.1097/CCM.0000000000000131

165. Roderburg C, Benz F, Vargas Cardenas D, Koch A, Janssen J, Vucur M, et al. Elevated miR-122 serum levels are an independent marker of liver injury in inflammatory diseases. *Liver Int* (2015) 35:1172–84. doi:10.1111/liv.12627

166. Wang H, Meng K, Chen W, Feng D, Jia Y, Xie L. Serum miR-574-5p: a prognostic predictor of sepsis patients. *Shock* (2012) 37:263–7. doi:10.1097/SHK.0b013e318241baf8

167. Wang H, Zhang P, Chen W, Feng D, Jia Y, Xie L. Serum microRNA signatures identified by Solexa sequencing predict sepsis patients' mortality: a prospective observational study. *PLoS One* (2012) 7:e38885. doi:10.1371/journal.pone.0038885

168. World Health Organization. *Hepatitis B Fact Sheet*. (2016). Available from: http://www.who.int/mediacentre/factsheets/fs204/en

169. Ott JJ, Stevens GA, Groeger J, Wiersma ST. Global epidemiology of hepatitis B virus infection: new estimates of age-specific HBsAg seroprevalence and endemicity. *Vaccine* (2012) 30:2212–9. doi:10.1016/j.vaccine.2011.12.116

170. Ji F, Yang B, Peng X, Ding H, You H, Tien P. Circulating microRNAs in hepatitis B virus-infected patients. *J Viral Hepat* (2011) 18:e242–51. doi:10.1111/j.1365-2893.2011.01443.x

171. Song G, Jia H, Xu H, Liu W, Zhu H, Li S, et al. Studying the association of microRNA-210 level with chronic hepatitis B progression. *J Viral Hepat* (2014) 21:272–80. doi:10.1111/jvh.12138

172. Wang JY, Mao RC, Zhang YM, Zhang YJ, Liu HY, Qin YL, et al. Serum microRNA-124 is a novel biomarker for liver necroinflammation in patients with chronic hepatitis B virus infection. *J Viral Hepat* (2015) 22:128–36. doi:10.1111/jvh.12284

173. Zhang Q, Xu M, Qu Y, Li Z, Zhang Q, Cai X, et al. Analysis of the differential expression of circulating microRNAs during the progression of hepatic fibrosis in patients with chronic hepatitis B virus infection. *Mol Med Rep* (2015) 12:5647–54. doi:10.3892/mmr.2015.4221

174. Zhou J, Yu L, Gao X, Hu J, Wang J, Dai Z, et al. Plasma microRNA panel to diagnose hepatitis B virus-related hepatocellular carcinoma. *J Clin Oncol* (2011) 29:4781–8. doi:10.1200/JCO.2011.38.2697

175. Winther TN, Jacobsen KS, Mirza AH, Heiberg IL, Bang-Berthelsen CH, Pociot F, et al. Circulating microRNAs in plasma of hepatitis B e antigen positive children reveal liver-specific target genes. *Int J Hepatol* (2014) 2014:791045. doi:10.1155/2014/791045

176. Qi P, Cheng SQ, Wang H, Li N, Chen YF, Gao CF. Serum microRNAs as biomarkers for hepatocellular carcinoma in Chinese patients with chronic hepatitis B virus infection. *PLoS One* (2011) 6:e28486. doi:10.1371/journal.pone.0028486

177. Tan Y, Ge G, Pan T, Wen D, Chen L, Yu X, et al. A serum microRNA panel as potential biomarkers for hepatocellular carcinoma related with hepatitis B virus. *PLoS One* (2014) 9:e107986. doi:10.1371/journal.pone.0107986

178. Sohn W, Kim J, Kang SH, Yang SR, Cho JY, Cho HC, et al. Serum exosomal microRNAs as novel biomarkers for hepatocellular carcinoma. *Exp Mol Med* (2015) 47:e184. doi:10.1038/emm.2015.68

179. Yu F, Lu Z, Chen B, Dong P, Zheng J. microRNA-150: a promising novel biomarker for hepatitis B virus-related hepatocellular carcinoma. *Diagn Pathol* (2015) 10:129. doi:10.1186/s13000-015-0369-y

180. Li L, Guo Z, Wang J, Mao Y, Gao Q. Serum miR-18a: a potential marker for hepatitis B virus-related hepatocellular carcinoma screening. *Dig Dis Sci* (2012) 57:2910–6. doi:10.1007/s10620-012-2317-y

181. Zhang Y, Jia Y, Zheng R, Guo Y, Wang Y, Guo H, et al. Plasma microRNA-122 as a biomarker for viral-, alcohol-, and chemical-related hepatic diseases. *Clin Chem* (2010) 56:1830–8. doi:10.1373/clinchem.2010.147850

182. Zhang X, Zhang Z, Dai F, Shi B, Chen L, Zhang X, et al. Comparison of circulating, hepatocyte specific messenger RNA and microRNA as biomarkers for chronic hepatitis B and C. *PLoS One* (2014) 9:e92112. doi:10.1371/journal.pone.0092112

183. Xu J, Wu C, Che X, Wang L, Yu D, Zhang T, et al. Circulating microRNAs, miR-21, miR-122, and miR-223, in patients with hepatocellular carcinoma or chronic hepatitis. *Mol Carcinog* (2011) 50:136–42. doi:10.1002/mc.20712

184. Zhang H, Li QY, Guo ZZ, Guan Y, Du J, Lu YY, et al. Serum levels of microRNAs can specifically predict liver injury of chronic hepatitis B. *World J Gastroenterol* (2012) 18:5188–96. doi:10.3748/wjg.v18.i37.5188

185. Arataki K, Hayes CN, Akamatsu S, Akiyama R, Abe H, Tsuge M, et al. Circulating microRNA-22 correlates with microRNA-122 and represents viral replication and liver injury in patients with chronic hepatitis B. *J Med Virol* (2013) 85:789–98. doi:10.1002/jmv.23540

186. Waidmann O, Bihrer V, Pleli T, Farnik H, Berger A, Zeuzem S, et al. Serum microRNA-122 levels in different groups of patients with chronic hepatitis B virus infection. *J Viral Hepat* (2012) 19:e58–65. doi:10.1111/j.1365-2893.2011.01536.x

187. Winther TN, Bang-Berthelsen CH, Heiberg IL, Pociot F, Hogh B. Differential plasma microRNA profiles in HBeAg positive and HBeAg negative children with chronic hepatitis B. *PLoS One* (2013) 8:e58236. doi:10.1371/journal.pone.0058236

188. Winther TN, Heiberg IL, Bang-Berthelsen CH, Pociot F, Hogh B. Hepatitis B surface antigen quantity positively correlates with plasma levels of microRNAs differentially expressed in immunological phases of chronic hepatitis B in children. *PLoS One* (2013) 8:e80384. doi:10.1371/journal.pone.0080384

189. Chen Y, Li L, Zhou Z, Wang N, Zhang CY, Zen K. A pilot study of serum microRNA signatures as a novel biomarker for occult hepatitis B virus infection. *Med Microbiol Immunol* (2012) 201:389–95. doi:10.1007/s00430-011-0223-0

190. Li LM, Hu ZB, Zhou ZX, Chen X, Liu FY, Zhang JF, et al. Serum microRNA profiles serve as novel biomarkers for HBV infection and diagnosis of HBV-positive hepatocarcinoma. *Cancer Res* (2010) 70:9798–807. doi:10.1158/0008-5472.CAN-10-1001

191. Jopling CL, Yi M, Lancaster AM, Lemon SM, Sarnow P. Modulation of hepatitis C virus RNA abundance by a liver-specific microRNA. *Science* (2005) 309:1577–81. doi:10.1126/science.1113329

192. Bihrer V, Friedrich-Rust M, Kronenberger B, Forestier N, Haupenthal J, Shi Y, et al. Serum miR-122 as a biomarker of necroinflammation in patients with chronic hepatitis C virus infection. *Am J Gastroenterol* (2011) 106:1663–9. doi:10.1038/ajg.2011.161

193. Cermelli S, Ruggieri A, Marrero JA, Ioannou GN, Beretta L. Circulating microRNAs in patients with chronic hepatitis C and non-alcoholic fatty liver disease. *PLoS One* (2011) 6:e23937. doi:10.1371/journal.pone.0023937

194. van der Meer AJ, Farid WR, Sonneveld MJ, de Ruiter PE, Boonstra A, van Vuuren AJ, et al. Sensitive detection of hepatocellular injury in chronic hepatitis C patients with circulating hepatocyte-derived microRNA-122. *J Viral Hepat* (2013) 20:158–66. doi:10.1111/jvh.12001

195. Wang JH, Jiang D, Rao HY, Zhao JM, Wang Y, Wei L. Absolute quantification of serum microRNA-122 and its correlation with liver inflammation grade and serum alanine aminotransferase in chronic hepatitis C patients. *Int J Infect Dis* (2015) 30:52–6. doi:10.1016/j.ijid.2014.09.020

196. Zhang S, Ouyang X, Jiang X, Gu D, Lin Y, Kong SK, et al. Dysregulated serum microRNA expression profile and potential biomarkers in hepatitis C virus-infected patients. *Int J Med Sci* (2015) 12:590–8. doi:10.7150/ijms.11525

197. Trebicka J, Anadol E, Elfimova N, Strack I, Roggendorf M, Viazov S, et al. Hepatic and serum levels of miR-122 after chronic HCV-induced fibrosis. *J Hepatol* (2013) 58:234–9. doi:10.1016/j.jhep.2012.10.015

198. Su TH, Liu CH, Liu CJ, Chen CL, Ting TT, Tseng TC, et al. Serum microRNA-122 level correlates with virologic responses to pegylated interferon therapy in chronic hepatitis C. *Proc Natl Acad Sci U S A* (2013) 110:7844–9. doi:10.1073/pnas.1306138110

199. Estrabaud E, Lapalus M, Broet P, Appourchaux K, De Muynck S, Lada O, et al. Reduction of microRNA 122 expression in *IFNL3* CT/TT carriers and during progression of fibrosis in patients with chronic hepatitis C. *J Virol* (2014) 88:6394–402. doi:10.1128/JVI.00016-14

200. El-Diwany R, Wasilewski LN, Witwer KW, Bailey JR, Page K, Ray SC, et al. Acute hepatitis C virus infection induces consistent changes in circulating microRNAs that are associated with nonlytic hepatocyte release. *J Virol* (2015) 89:9454–64. doi:10.1128/JVI.00955-15

201. Roderburg C, Mollnow T, Bongaerts B, Elfimova N, Vargas Cardenas D, Berger K, et al. Micro-RNA profiling in human serum reveals compartment-specific roles of miR-571 and miR-652 in liver cirrhosis. *PLoS One* (2012) 7:e32999. doi:10.1371/journal.pone.0032999

202. Shrivastava S, Petrone J, Steele R, Lauer GM, Di Bisceglie AM, Ray RB. Up-regulation of circulating miR-20a is correlated with hepatitis C virus-mediated liver disease progression. *Hepatology* (2013) 58:863–71. doi:10.1002/hep.26296

203. Shwetha S, Gouthamchandra K, Chandra M, Ravishankar B, Khaja MN, Das S. Circulating miRNA profile in HCV infected serum: novel insight into pathogenesis. *Sci Rep* (2013) 3:1555. doi:10.1038/srep01555

204. El-Abd NE, Fawzy NA, El-Sheikh SM, Soliman ME. Circulating miRNA-122, miRNA-199a, and miRNA-16 as biomarkers for early detection of hepatocellular carcinoma in Egyptian patients with chronic hepatitis C virus infection. *Mol Diagn Ther* (2015) 19:213–20. doi:10.1007/s40291-015-0148-1

205. Qu KZ, Zhang K, Li H, Afdhal NH, Albitar M. Circulating microRNAs as biomarkers for hepatocellular carcinoma. *J Clin Gastroenterol* (2011) 45:355–60. doi:10.1097/MCG.0b013e3181f18ac2

206. Bihrer V, Waidmann O, Friedrich-Rust M, Forestier N, Susser S, Haupenthal J, et al. Serum microRNA-21 as marker for necroinflammation in hepatitis C patients with and without hepatocellular carcinoma. *PLoS One* (2011) 6:e26971. doi:10.1371/journal.pone.0026971

207. Tomimaru Y, Eguchi H, Nagano H, Wada H, Kobayashi S, Marubashi S, et al. Circulating microRNA-21 as a novel biomarker for hepatocellular carcinoma. *J Hepatol* (2012) 56:167–75. doi:10.1016/j.jhep.2011.04.026

208. Ge Y, Zhao K, Qi Y, Min X, Shi Z, Qi X, et al. Serum microRNA expression profile as a biomarker for the diagnosis of pertussis. *Mol Biol Rep* (2013) 40:1325–32. doi:10.1007/s11033-012-2176-9

209. Sanmarti M, Ibanez L, Huertas S, Badenes D, Dalmau D, Slevin M, et al. HIV-associated neurocognitive disorders. *J Mol Psychiatry* (2014) 2:2. doi:10.1186/2049-9256-2-2

210. Zayyad Z, Spudich S. Neuropathogenesis of HIV: from initial neuroinvasion to HIV-associated neurocognitive disorder (HAND). *Curr HIV/AIDS Rep* (2015) 12:16–24. doi:10.1007/s11904-014-0255-3

211. Kadri F, LaPlante A, De Luca M, Doyle L, Velasco-Gonzalez C, Patterson JR, et al. Defining plasma microRNAs associated with cognitive impairment in HIV-infected patients. *J Cell Physiol* (2016) 231:829–36. doi:10.1002/jcp.25131

212. Witwer KW, Sarbanes SL, Liu J, Clements JE. A plasma microRNA signature of acute lentiviral infection: biomarkers of central nervous system disease. *AIDS* (2011) 25:2057–67. doi:10.1097/QAD.0b013e32834b95bf

213. Pacifici M, Delbue S, Ferrante P, Jeansonne D, Kadri F, Nelson S, et al. Cerebrospinal fluid miRNA profile in HIV-encephalitis. *J Cell Physiol* (2013) 228:1070–5. doi:10.1002/jcp.24254

214. Ding NZ, Wang XM, Sun SW, Song Q, Li SN, He CQ. Appearance of mosaic enterovirus 71 in the 2008 outbreak of China. *Virus Res* (2009) 145:157–61. doi:10.1016/j.virusres.2009.06.006

215. Cui L, Qi Y, Li H, Ge Y, Zhao K, Qi X, et al. Serum microRNA expression profile distinguishes enterovirus 71 and coxsackievirus 16 infections in patients with hand-foot-and-mouth disease. *PLoS One* (2011) 6:e27071. doi:10.1371/journal.pone.0027071

216. Jia HL, He CH, Wang ZY, Xu YF, Yin GQ, Mao LJ, et al. microRNA expression profile in exosome discriminates extremely severe infections from mild infections for hand, foot and mouth disease. *BMC Infect Dis* (2014) 14:506. doi:10.1186/1471-2334-14-506

217. Qi Y, Zhu Z, Shi Z, Ge Y, Zhao K, Zhou M, et al. Dysregulated microRNA expression in serum of non-vaccinated children with varicella. *Viruses* (2014) 6:1823–36. doi:10.3390/v6041823

218. Tambyah PA, Sepramaniam S, Mohamed Ali J, Chai SC, Swaminathan P, Armugam A, et al. microRNAs in circulation are altered in response to influenza A virus infection in humans. *PLoS One* (2013) 8:e76811. doi:10.1371/journal.pone.0076811

219. Moran J, Ramirez-Martinez G, Jimenez-Alvarez L, Cruz A, Perez-Patrigeon S, Hidalgo A, et al. Circulating levels of miR-150 are associated with poorer outcomes of A/H1N1 infection. *Exp Mol Pathol* (2015) 99:253–61. doi:10.1016/j.yexmp.2015.07.001

220. Zhu Z, Qi Y, Ge A, Zhu Y, Xu K, Ji H, et al. Comprehensive characterization of serum microRNA profile in response to the emerging avian influenza A (H7N9) virus infection in humans. *Viruses* (2014) 6:1525–39. doi:10.3390/v6041525

221. Fabres-Klein MH, Aguilar AP, Silva MP, Silva DM, Ribon AO. Moving towards the immunodiagnosis of staphylococcal intramammary infections. *Eur J Clin Microbiol Infect Dis* (2014) 33:2095–104. doi:10.1007/s10096-014-2181-0

222. Sun J, Aswath K, Schroeder SG, Lippolis JD, Reinhardt TA, Sonstegard TS. microRNA expression profiles of bovine milk exosomes in response to *Staphylococcus aureus* infection. *BMC Genomics* (2015) 16:806. doi:10.1186/s12864-015-2044-9

223. Farrell D, Shaughnessy RG, Britton L, MacHugh DE, Markey B, Gordon SV. The identification of circulating miRNA in bovine serum and their potential as novel biomarkers of early *Mycobacterium avium* subsp *paratuberculosis* infection. *PLoS One* (2015) 10:e0134310. doi:10.1371/journal.pone.0134310

224. Casas E, Cai G, Kuehn LA, Register KB, McDaneld TG, Neill JD. Association of microRNAs with antibody response to *Mycoplasma bovis* in beef cattle. *PLoS One* (2016) 11:e0161651. doi:10.1371/journal.pone.0161651

Permissions

List of Contributors

Victoria A. McGuire and J. Simon C. Arthur
Division of Cell Signalling and Immunology, School of Life Sciences, University of Dundee, Dundee, UK

Pedro Escoll, Monica Rolando and Carmen Buchrieser
Institut Pasteur, Biologie des Bactéries Intracellulaires, Paris, France, CNRS UMR 3525, Paris, France

Virginia González-Vélez
Department Basic Sciences, Universidad Autónoma Metropolitana-Azcapotzalco, CDMX, Mèxico, Mexico

Anthony Piron
ULB Center for Diabetes Research, Faculté de Médecine, Université libre de Bruxelles (ULB), Brussels, Belgium Interuniversity Institute of Bioinformatics (IB2), Brussels, Belgium

Geneviève Dupont
Interuniversity Institute of Bioinformatics (IB2), Brussels, Belgium
Unit of Theoretical Chronobiology, Faculté des Sciences, Université libre de Bruxelles (ULB), Brussels, Belgium

K. L. Dias-Teixeira and U. G. Lopes
Institute of Biophysics Carlos Chagas Filho, Federal University of Rio de Janeiro, Rio de Janeiro, Brazil

R. M. Pereira
Institute of Microbiology Paulo de Goes, Federal University of Rio de Janeiro, Rio de Janeiro, Brazil

J. S. Silva
Department of Biochemistry and Immunology, University of São Paulo, Ribeirão Preto, Brazil

N. Fasel
Department of Biochemistry, Faculty of Biology and Medicine, Center for Immunity and Infection Lausanne, University of Lausanne, Lausanne, Switzerland

B. H. Aktas
Laboratory of Translation, Department of Hematology, Brigham and Women's Hospital, Harvard Medical School, Boston, MA, USA

Naixin Zhang and Peter E. Kima
Department of Microbiology and Cell Science, University of Florida, Gainesville, FL, USA

Yousef Nami and Babak Haghshenas
Institute of Biosciences, Universiti Putra Malaysia, Selangor, Malaysia

Minoo Haghshenas
School of Medicine, Shahid Beheshti University of Medical Sciences, Tehran, Iran

Norhafizah Abdullah
Chemical and Environmental Engineering Department, Faculty of Engineering, Universiti Putra Malaysia, Selangor, Malaysia

Ahmad Yari Khosroushahi
Drug Applied Research Center, Tabriz University of Medical Sciences, Tabriz, Iran
Department of Pharmacognosy, Faculty of Pharmacy, Tabriz University of Medical Sciences, Tabriz, Iran

Christopher N. LaRock
Department of Pediatrics, University of California San Diego, La Jolla, CA, USA

Victor Nizet
Skaggs School of Medicine and Pharmaceutical Sciences, University of California San Diego, La Jolla, CA, USA

Bruno M. Di Genova
Departamento de Microbiologia e Imunologia, Universidade Federal de São Paulo, São Paulo, Brazil

Renata R. Tonelli
Departamento de Microbiologia e Imunologia, Universidade Federal de São Paulo, São Paulo, Brazil
Instituto de Ciências Ambientais, Químicas e Farmacêuticas, Departamento de Ciências Biológicas, Universidade Federal de São Paulo, Diadema, Brazil

R. J. Allen, C. Welch and Neha Pankow
Department of Pharmacology, University of North Carolina at Chapel Hill, Chapel Hill, NC, United States

Klaus M. Hahn and Timothy C. Elston
Department of Pharmacology, University of North Carolina at Chapel Hill, Chapel Hill, NC, United States
Computational Medicine Program, University of North Carolina at Chapel Hill, Chapel Hill, NC, United States

Jian Lin, Jing Xia, Ke Y. Zhang, Yan Zeng and Qian Yang
Department of Zoology, College of Life Science, Nanjing Agricultural University, Jiangsu, China

Chong Z. Tu
Department of Histoembryology, College of Veterinary Medicine, Nanjing Agricultural University, Jiangsu, China

Julie Guignot and Guy Tran Van Nhieu
Equipe Communication Intercellulaire et Infections Microbiennes, Centre de Recherche Interdisciplinaire en Biologie (CIRB),Collège de France, Paris, France
Institut National de la Santé et de la Recherche Médicale U1050, Paris, France
Centre National de la Recherche Scientifique UMR7241, Paris, France
MEMOLIFE Laboratory of Excellence and Paris Sciences et Lettres, Paris, France

Fabio Baroni, Pamela Berni and Roberto Baricchi
Transfusion Medicine Unit, Arcispedale Santa Maria Nuova - IRCCS, Reggio Emilia, Italy

Emanuela Casali
Department of Biomedical, Biotechnological and Translational Sciences, University of Parma, Parma, Italy

Thelma A. Pertinhez
Transfusion Medicine Unit, Arcispedale Santa Maria Nuova - IRCCS, Reggio Emilia, Italy
Department of Biomedical, Biotechnological and Translational Sciences, University of Parma, Parma, Italy

Alberto Spisni
Department of Surgical Sciences, University of Parma, Parma, Italy

Boaz Job van Driel, Gongxian Liao and Cox Terhorst
Division of Immunology, Beth Israel Deaconess Medical Center, Harvard Medical School, Boston, MA, USA

Pablo Engel
Immunology Unit, Department of Cell Biology, Immunology and Neurosciences, Medical School, University of Barcelona, Barcelona, Spain

Yao Wang, Rong Li, Jiyuan Liu, Jinzhou Zhang, Yumei Cai and Sidang Liu
College of Animal Science and Veterinary Medicine, Shandong Agricultural University, Tai'an, China
Sino-German Cooperative Research Centre for Zoonosis of Animal Origin Shandong Province, Tai'an, China

Tongjie Chai, Liangmeng Wei and Ning Li
College of Animal Science and Veterinary Medicine, Shandong Agricultural University, Tai'an, China
Sino-German Cooperative Research Centre for Zoonosis of Animal Origin Shandong Province, Tai'an, China
Collaborative Innovation Centre for the Origin and Control of Emerging Infectious Diseases of Taishan Medical College, Tai'an, China

Shaolong Chen, Wenlong Xie, Kai Wu, Ping Li, Zhiqiang Ren, Lin Li, Yuan Yuan, Chunmao Zhang, Yuling Zheng, Qingyu Lv, Hua Jiang and Yongqiang Jiang
State Key Laboratory of Pathogen and Biosecurity, Beijing Institute of Microbiology and Epidemiology, Beijing, China

Carl-Fredrik Johnzon
Department of Anatomy, Physiology and Biochemistry, Swedish University of Agricultural Sciences, Uppsala, Sweden

Elin Rönnberg
Department of Medical Biochemistry and Microbiology, Uppsala University, Uppsala, Sweden

Gunnar Pejler
Department of Anatomy, Physiology and Biochemistry, Swedish University of Agricultural Sciences, Uppsala, Sweden
Department of Medical Biochemistry and Microbiology, Uppsala University, Uppsala, Sweden

Bengt Guss
Department of Biomedical Science and Veterinary Public Health, Swedish University of Agricultural Sciences, Uppsala, Sweden

Janina Mothes, Uwe Benary and Jana Wolf
Mathematical Modelling of Cellular Processes, Max Delbrück Center for Molecular Medicine, Berlin, Germany

Inbal Ipenberg, Seda Çöl Arslan and Claus Scheidereit
Signal Transduction in Tumor Cells, Max Delbrück Center for Molecular Medicine, Berlin, Germany

Anna Zawistowska-Deniziak and Katarzyna Basałaj
Witold Stefan´ ski Institute of Parasitology, Polish Academy of Sciences, Warsaw, Poland

Barbara Strojny
Division of Nanobiotechnology, Faculty of Animal Sciences, Department of Animal Feeding and Biotechnology, Warsaw University of Life Sciences, Warsaw, Poland

Daniel Młocicki
Witold Stefan´ ski Institute of Parasitology, Polish Academy of Sciences, Warsaw, Poland
Department of General Biology and Parasitology, Medical University of Warsaw, Warsaw, Poland

Sushmita Chakraborty, Bianca Kloos and Katharina F. Kubatzky
Zentrum für Infektiologie, Medizinische Mikrobiologie und Hygiene, Universitätsklinikum Heidelberg, Heidelberg, Germany

Ulrike Harre and Georg Schett
Department of Internal Medicine

Anna Pretschner, Sophie Pabel, Markus Haas, and Wolfgang Marwan
Magdeburg Centre for Systems Biology and Institute of Biology, Otto von Guericke University, Magdeburg, Germany

Monika Heiner
Computer Science Institute, Brandenburg University of Technology Cottbus-Senftenberg, Cottbus, Germany

Ronald Sluyter
School of Biological Sciences, University of Wollongong, Wollongong, NSW, Australia

Centre for Medical and Molecular Bioscience, University of Wollongong, Wollongong, NSW, Australia
Illawarra Health and Medical Research Institute, Wollongong, NSW, Australia

Carolina N. Correia, Nicolas C. Nalpas, Kirsten E. McLoughlin and John A. Browne
Animal Genomics Laboratory, UCD School of Agriculture and Food Science, University College Dublin, Dublin, Ireland

David E. MacHugh
Animal Genomics Laboratory, UCD School of Agriculture and Food Science, University College Dublin, Dublin, Ireland
University College Dublin, UCD Conway Institute of Biomolecular and Biomedical Research, Dublin, Ireland

Ronan G. Shaughnessy
UCD School of Veterinary Medicine, University College Dublin, Dublin, Ireland

Stephen V. Gordon
UCD School of Veterinary Medicine, University College Dublin, Dublin, Ireland
University College Dublin, UCD Conway Institute of Biomolecular and Biomedical Research, Dublin, Ireland

Index